572·861 2WL £60
HAR

Biotechnology &
Genetic Engineering Reviews

Volume 26

Biotechnology & Genetic Engineering Reviews

Volume 26

Editor:

STEPHEN E. HARDING

Professor of Physical Biochemistry, University of Nottingham

Associate Editor:

MICHAEL P. TOMBS

Special Professor, University of Nottingham

Nottingham
University Press

Nottingham University Press
Manor Farm, Main Street, Thrumpton
Nottingham, NG11 0AX, United Kingdom

NOTTINGHAM

First published 2010
© The several contributors named in the table of contents

British Library Cataloguing in Publication Data
Biotechnology & genetic engineering reviews - Vol. 26
I. Biotechnology – Periodicals

ISBN 978-1-904761-91-4
ISSN 0264-8725

Disclaimer

Every reasonable effort has been made to ensure that the material in this book is true, correct, complete
and appropriate at the time of writing. Nevertheless the publishers, the editors and the authors do not
accept responsibility for any omission or error, or for any injury, damage, loss or financial consequences
arising from the use of the book. Views expressed in the articles are those of the authors and not of the
Editors or Publisher.

Cover image: adapted from P Artymuik (see page 6)

Typeset by Nottingham University Press, Nottingham
Printed in Great Britain by the MPG Books Group, Bodmin and King's Lynn

CONTENTS

THE IMPACT OF STRUCTURAL PROTEOMICS ON BIOTECHNOLOGY

Babu A. Manjasetty*, Andrew P. Turnbull[1], Santosh Panjikar[2]
Proteomics & Bioinformatics Research Group, Research and Industry Incubation Center, Dayananda Sagar Institutions, Bangalore 560 078 India, [1]Cancer Research Technology Ltd, Birkbeck College, University of London, London, WC1E 7HX, UK and [2]EMBL Hamburg Outstation, c/o DESY, Notkestrasse 85, D-22603 Hamburg, Germany

BIOTECHNOLOGY AS THE ENGINE FOR THE KNOWLEDGE-BASED BIO-ECONOMY

Alfredo Aguilar[1]*, Laurent Bochereau[2] and Line Matthiessen[3]
[1]Head of Unit Biotechnologies, Directorate Biotechnologies, Agriculture and Food, Research Directorate General, European Commission, 1049 Brussels, Belgium; [2]Head of Section, Science, Technology and Education, Delegation of the European Commission to the United States, 2300 M Street NW, 20037-1434, Washington DC, USA, and [3]Head of Unit Horizontal Aspects and Coordination, Directorate Biotechnologies, Agriculture and Food, Research Directorate General, European Commission, B-1049 Brussels, Belgium

PATENT REFORM IN THE UNITED STATES

*Ann E. Mills and Patti M. Tereskerz
Center for Biomedical Ethics and Humanities, Program in Ethics and Policy in Healthcare Systems, University of Virginia, Charlottesville, VA 22901, USA*

Contributors

AGUILAR, A. *Head of Unit Biotechnologies, Directorate Biotechnologies, Agriculture and Food, Research Directorate General, European Commission, 1049 Brussels, Belgium*

BOCHEREAU, L. *Head of Section, Science, Technology and Education, Delegation of the European Commission to the United States, 2300 M Street NW, 20037-1434, Washington DC, USA*

BRAVO, A. *Instituto de Biotecnología, Universidad Nacional Autónoma de México, Ap postal 510-3 CP 62250, Cuernavaca, Morelos, México*

CANCINO-RODEZNO, A. *Instituto de Biotecnología, Universidad Nacional Autónoma de México, Ap postal 510-3 CP 62250, Cuernavaca, Morelos, México*

CHEN, J. *Key Laboratory of Industrial Biotechnology of Ministry of Education, School of Biotechnology, Jiangnan University, Wuxi, Jiangsu 214122, China*

CHISTOSERDOVA, L. *Department of Chemical Engineering, University of Washington, Seattle WA 98195, USA*

DE FELIPE, P. *Spanish Medicines Agency (AEMPS), Parque Empresarial "Las Mercedes", Campezo 1 - Edificio 8, 28022 Madrid, Spain*

DHERT, W.J.A. *Department of Orthopaedics, University Medical Center Utrecht, Utrecht, and Faculty of Veterinary Medicine, Utrecht University, The Netherlands*

DIRETTO, G. *ENEA, Casaccia Research Center, PO Box 2400, 00123 S.M. di Galeria (Roma), Italy*

ESCUIN, H. *Department of Surgery, University of California at Los Angeles (UCLA), Los Angeles, CA, USA*

GIULIANO, G. *ENEA, Casaccia Research Center, PO Box 2400, 00123 S.M. di Galeria (Roma), Italy*

HIRAMI, Y. *Laboratory for Retinal Regeneration, Center for Developmental Biology, RIKEN, 2-2-3 Minatojima-minamimachi, Chuo-ku, Kobe 650-0047, Japan and Department of Ophthalmology, Institute of Biomedical Research and Innovation, 2-2 Minatojima-minamimachi, Chuo-ku, Kobe 650-0047, Japan*

JEFFERIS, R. *School of Immunity & Infection, The College of Medical and Dental Sciences, University of Birmingham, Edgbaston, Birmingham B15 2TT UK*

KALINNA, B.H. *Centre for Animal Biotechnology, Faculty of Veterinary Science, The University of Melbourne, Parkville, 3010 VIC, Australia*

KOOPMAN, B.F.J.M. *Department of Biomechanical Engineering, Institute for Biomedical Technology, University of Twente, Enschede, The Netherlands*

LAFFERTY DOTY, S. *College of Forest Resources, University of Washington, Seattle, WA, USA*

LUKE, G.A. *Centre for Biomolecular Sciences, School of Biology, Biomolecular Sciences Building, University of St Andrews, North Haugh, St Andrews KY16 9ST, UK*

MALDA, J. *Department of Orthopaedics, University Medical Center Utrecht, Utrecht, and Faculty of Veterinary Medicine, Utrecht University, The Netherlands*

MANJASETTY, B.A. *Proteomics & Bioinformatics Research Group, Research and Industry Incubation Center, Dayananda Sagar Institutions, Bangalore 560 078 India*

MATTHIESSEN, L. *Head of Unit Horizontal Aspects and Coordination, Directorate Biotechnologies, Agriculture and Food, Research Directorate General, European Commission, B-1049 Brussels, Belgium*

MILLS, A.E. *Center for Biomedical Ethics and Humanities, Program in Ethics and Policy in Healthcare Systems, University of Virginia, Charlottesville, VA 22901, USA*

OLSON, D.M. *USDA-ARS, Crop Protection and Management Research Unit, Tifton, GA, 31794, USA*

OSAKADA, F. *Laboratory for Retinal Regeneration, Center for Developmental Biology, RIKEN, 2-2-3 Minatojima-minamimachi, Chuo-ku, Kobe 650-0047, Japan and Systems Neurobiology Laboratory, The Salk Institute for Biological Studies, 10010 North Torrey Pines Road, La Jolla, California 92037, USA*

PAMPALONI, F. *Cell Biology & Biophysics Unit, European Molecular Biology Laboratory (EMBL), Meyerhofstrasse 1, D-69117, Heidelberg, Germany*

PANJIKAR, S. *EMBL Hamburg Outstation, c/o DESY, Notkestrasse 85, D-22603 Hamburg, Germany*

PASSARINHA, L. *CICS – Centro de Investigação em Ciências da Saúde, Universidade da Beira Interior, 6201-001 Covilhã, Portugal*

PORTA, H. *Instituto de Biotecnología, Universidad Nacional Autónoma de México, Ap postal 510-3 CP 62250, Cuernavaca, Morelos, México*

QUEIROZ, J.A. *CICS – Centro de Investigação em Ciências da Saúde, Universidade da Beira Interior, 6201-001 Covilhã, Portugal*

RAINS, G.C. *University of Georgia, Tifton Campus, Biological and Agricultural Engineering Dept., Tifton, GA 31793, USA*

ROSATI, C. *Ente per le Nuove tecnologie, l'Energia e l'Ambiente (ENEA), Trisaia Research Center, S.S. 106 km 419.5, 75026 Rotondella (Matera), Italy*

ROUWKEMA, J. *Department of Biomechanical Engineering, Institute for Biomedical Technology, University of Twente, Enschede, The Netherlands*

RYAN, M.D. *Centre for Biomolecular Sciences, School of Biology, Biomolecular Sciences Building, University of St Andrews, North Haugh, St Andrews KY16 9ST, UK*

SAMARASINGHE, S. *Texas A&M University, Entomology Dept, College Station, TX, 77843, USA*

SHLOMI, T. *Department of Computer Science, Technion – Israel Institute of Technology, Haifa 32000, Israel*

SOBERÓN, M. *Instituto de Biotecnología, Universidad Nacional Autónoma de México, Ap postal 510-3 CP 62250, Cuernavaca, Morelos, México*

SOUSA, F. *CICS – Centro de Investigação em Ciências da Saúde, Universidade da Beira Interior, 6201-001 Covilhã, Portugal*

STELZER, E.H.K. *Cell Biology & Biophysics Unit, European Molecular Biology Laboratory (EMBL), Meyerhofstrasse 1, D-69117, Heidelberg, Germany*

TAKAHASHI, M. *Laboratory for Retinal Regeneration, Center for Developmental Biology, RIKEN, 2-2-3 Minatojima-minamimachi, Chuo-ku, Kobe 650-0047, Japan and Department of Ophthalmology, Institute of Biomedical Research and Innovation, 2-2 Minatojima-minamimachi, Chuo-ku, Kobe 650-0047, Japan*

TCHOUBRIEVA, E.B. *Centre for Animal Biotechnology, Faculty of Veterinary Science, The University of Melbourne, Parkville, 3010 VIC, Australia*

TERESKERZ, P.M. *Center for Biomedical Ethics and Humanities, Program in Ethics and Policy in Healthcare Systems, University of Virginia, Charlottesville, VA 22901, USA*

TOMBERLIN, J.K. *Texas A&M University, Entomology Dept, College Station, TX, 77843, USA*

TURNBULL, A.P. *Cancer Research Technology Ltd, Birkbeck College, University of London, London, WC1E 7HX, UK*

ULASIRI, D.K. *Lincoln University, Centre for Advanced Computational Solutions (C-facs), ChristChurch, New Zealand*

VAN AKEN, B. *Department of Civil and Environmental Engineering, Temple University, Philadelphia, PA*

VAN BLITTERSWIJK, C.A. *Department of Tissue Regeneration, Institute for Biomedical Technology, University of Twente, Enschede, The Netherlands*

ZHANG, D. *Key Laboratory of Industrial Biotechnology of Ministry of Education, School of Biotechnology, Jiangnan University, Wuxi, Jiangsu 214122, China*

ZHOU, Z. *Lincoln University, Centre for Advanced Computational Solutions (C-facs), ChristChurch, New Zealand*

ZHU, Y. *Department of Biosciences, TNO Quality of Life, Netherlands Organization for Applied Scientific Research, P.O. Box 360, 3700 AJ Zeist, Netherlands*

Biotechnology and Genetic Engineering Reviews - Vol. 26, 1-42 (2009)

The antibody paradigm: present and future development as a scaffold for biopharmaceutical drugs

ROY JEFFERIS

School of Immunity & Infection, The College of Medical and Dental Sciences, University of Birmingham, Edgbaston, Birmingham B15 2TT UK

Abstract

Early studies of the humoral immune response revealed an apparent paradox: an infinite diversity of antibody specificities encoded within a finite genome. In consequence antibodies became a focus of interest for biochemists and geneticists. It resulted in the elucidation of the basic structural unit, the immunoglobulin (Ig) domain, comprised of ~ 100 amino acid residues that generate the characteristic "immunoglobulin (Ig) fold". The Ig fold has an anti-parallel ß-pleated sheet (barrel) structure that affords structural stability whilst the ß-bends allow for essentially infinite structural variation and functional diversity. This versatility is reflected in the Ig domain being the most widely utilised structural unit within the proteome. Human antibodies are comprised of multiple Ig domains and their structural diversity may be enhanced through the attachment of oligosaccharides. This review summarizes our current understanding of the immunoglobulin structure/function relationships and the application of protein and oligosaccharide engineering to further develop the Ig domain as a scaffold for the generation of new and novel antibody based therapeutics.

To whom correspondence may be addressed (r.jefferis@bham.ac.uk)

Abbreviations: ADCC: Antibody dependent cellular cytotoxicity; ASGPR, Asialoglycoprotein receptor; ATA, Anti therapeutic antibody; CDC, Complement dependent cytotoxicity; CDR, Complementarity determining region; CHO, Chinese hamster ovary; CMP, Cytidine monophosphate; COG, Cost of goods; COT, Cost of treatment; DSMC, Differential scanning micro calorimetry; HACA: Human anti-chimeric antibody; HAHA: Human anti-human antibody; HAMA, Human anti-mouse antibody; ITAM, Immunoreceptor tyrosine-based activating motif; ITIM, Immunoreceptor tyrosine-based inhibition motif; MBL, Mannose binding lectin; MASP, Mannose binding lectin associated serine protease; MHC, Major histocompatability complex; NK, Natural killer cells; NSO, Mouse myeloma cell line; PCR, Polymerase chain reaction; rMAbsrecombinant antibodies; Sp2/0, Mouse myeloma cell line; Tm, Transition (melting) temperature; V_H, Heavy chain variable region; V_L, Light chain variable region;

Introduction

When I entered the field of immunology the intellectual and practical challenge was to confront the paradox of an apparent infinite diversity of antibody specificities encoded within a finite genome. This paradox has been resolved and we can readily account for the diversity of specificity in terms of genetic mechanisms and consequent protein structure. We have the tools to generate antibodies of selected specificity and engineer cells for their efficient production. The technologies for introducing random mutations, followed by selection, are so powerful that several alternative "scaffolds" have been developed for the generation of proteins with selected binding specificities. The challenge that antibodies now pose is to select and engineer them to optimise mechanisms of action *in vivo*.

We are all beneficiaries of vaccination and the generation of a protective humoral and cellular immune responses; the humoral component being comprised of antibodies that not only have specificity for the offending pathogen but also mediate its removal and destruction. It is an understanding of these mechanisms that will allow us to select and/or design antibody therapeutics that are optimised for the treatment of given disease indications. In spite of the incomplete knowledge currently available recombinant antibody therapeutics (rMAbs) are held as exemplars of translational medicine. The rMAbs currently licensed represent a significant success in terms of clinical benefit delivered and revenue (profit) generated within the biopharmaceutical industry. Additionally, it is estimated that ~ 30 % of new drugs likely to be licensed during the next decade will be based on antibody products (Carter, 2006; Moutel and Perez, 2008; Reichert and Valge-Archer, 2007). The term "antibody products" includes therapeutics composed of selected "fragments" or selected combinations of antibody domains.

High volume production with the maintenance of structural and functional fidelity of these large biological molecules results in high "cost of goods" (COG) that can limit their availability to patients, due to the strain it puts on national and private health budgets. The perceived benefits that could flow from lower COG have acted as an incentive for innovation in discovery, clone selection and rapid progress to "first in man" preclinical studies. The shortening of time-lines allows for a reduction in COG. In addition protein and glycosylation engineering can contribute to the generation of customised antibodies optimised for given disease indication and a consequent reduction in cost of treatment (COT). These developments are taking place within high labour cost economies whose products will come under increasing competition from new and bio-similar products manufactured within low labour cost economies.

The invention that led to the production of monoclonal antibodies was designed to further Caesar Milstein's research into the structural and genetic basis of antibody specificity (Kohler and Milstein, 1975). The ability to generate hybridoma cells secreting antibody specific for a pre-selected antigen allowed for dissection of polyclonal antibody responses and the generation of clones of cells, each secreting homogeneous antibodies specific for a unique structural feature (antigenic determinant or epitope) expressed by the selected antigen. Comparisons amongst such antibody populations allowed the extent and patterns of variable region sequence diversity to be determined. The potential for commercial exploitation was recognised and the landmark paper of Kohler & Milstein (1975) ended with the statement: ***Such cells can be grown* in**

vitro *in massive cultures to provide specific antibody. Such cultures could be of value for medical and industrial use.* The concluding sentence has proved to be a massive under-statement.

The first monoclonal antibody to be used as a therapeutic was a mouse IgG2a anti-human CD3 (Orthoclone-OKT3 (muromonab) (Webster *et al.*, 2006), delivered to kidney transplant patients undergoing episodes of acute rejection. Its success in these immuno-suppressed patients led to OKT-3 gaining regulatory approval and it is still in use. It was no surprise, however, that it proved to be immunogenic in humans, resulting in the development of anti-therapeutic antibody (ATA) responses; otherwise referred to as HAMA (human anti-mouse antibody). The development of HAMA precludes further exposure of patients to the therapeutic as severe adverse reactions may be precipitated. This restriction fuelled innovation and the application of genetic engineering to the progressive "humanisation" of antibodies, raised in mice, to human targets. The development of mouse/human chimeric antibodies reduced immunogenicity but a proportion of patient generated ATA; now referred to as HACA (human anti-chimeric antibody). With the advent of "fully" human antibodies, generated from phage display libraries or mice transgenic for human immunoglobulin genes, it was anticipated that immunogenicity and the development of ATA might be circumvented (Holgate and Baker, 2009; Radstake *et al.*, 2008). In practice the first licensed "fully" human antibody, Humira (*Adalimumab*), resulted in ~ 12 % incidence of ATA (HAHA – human anti-human antibody) in patients treated for rheumatoid arthritis (RA). Subsequently this was reduced to ~ 2-3 % for patients also receiving the non-steroid anti-inflammatory drug methotrexate, a mild immunosuppressant (Radstake *et al.*, 2008).

It is now appreciated that each antibody molecule, being unique in its specificity and affinity for a unique epitope, may generate immune complexes of similarly unique structure that determines the downstream effector functions activated (Woof and Burton, 2004; Nezlin and Ghetie, 2004; Jefferis, 2007; Jefferis, 2009). We are challenged, therefore, to develop protocols that reveal the full functionality of each rMAb *in vitro* and to be able to translate these finding to predict mechanisms of action *in vivo;* ultimately taking account of the multiple polymorphisms resident in an out-bred human population.

Innate and adaptive immunity

An intact immune system is essential to the integrity of the individual. We live in a hostile environment and are constantly exposed to potential pathogens; however, we are mostly unaware of these insults due to protective innate and adaptive immune responses. These responses can precipitate a cascade of inflammatory mechanisms that recruit leucocytes and cells of the reticulo-endothelial system to inactivate, ingest and destroy targeted pathogens. The response must be appropriate to the threat since uncontrolled inflammatory reactions can result in "bystander" or "collateral" damage to healthy tissue. Informed development of antibody therapeutics requires that we extrapolate from our knowledge of the action of antibodies in immune protection, in both health and disease.

Potential pathogens, bacteria, virus, fungi etc, are structurally relatively "simple" organisms whose outer surface ("membrane") is comprised of repeating structural units that form distinct architectural "patterns". Organisms from insects to human have evolved an innate immune system constituted of "pattern" recognition receptors, expressed on the surface of cells, which bind molecules presented in "patterns" and activate the cells to ingest, inactivate and destroy them (Kabelitz and Medzhitov, 2007; Gardy, *et al.*, 2009). In addition the cells may process fragments of the pathogen and "present" them to cells of the immune system to provoke a humoral and/or cellular adaptive immune response.

    The adaptive humoral immune response has two phases; a primary and a secondary response. The primary response results in the production of antibodies of the IgM class whilst the secondary response is characterized by the production of antibody of the IgG, IgA, IgE and/or IgD class; referred to as class or isotype switching. There are four subclasses of IgG (IgG1, IgG2, IgG3 & IgG4) and two subclasses of IgA (IgA1 & IgA2), giving a total of nine antibody isotypes in humans (Woof and Burton, 2004; Nezlin and Ghetie, 2004; Jefferis, 2007; Jefferis, 2009). Secondary immune responses also establish memory that provides for a rapid and amplified response to subsequent contact with the same pathogen. Protective antibody responses are extremely heterogeneous due to the presence of populations of antibodies each of unique specificity for individual epitopes expressed on the antigen. It is not possible, therefore, to unequivocally define the mechanism of action that result in immune neutralization, clearance and destruction of pathogens; however, IgG is, quantitatively, the predominant antibody isotype present in normal human serum and early studies established a simple protocol for it isolation in pure form. Consequently, IgG has been subject to intense structural and functional study.

    To date all licensed recombinant antibody therapeutics are of the IgG class and, to my knowledge, all intact antibodies in development are also of the IgG class. The predominance of the IgG1 subclass (~ 60 %) in blood allowed for detailed study of its structure and function so that initially rMAbs of the IgG1 isotype only were developed; however, the unique properties of the other IgG subclasses are also being exploited. New formats of intact rMAbs and antibody "fragments" are being developed that can deliver maximal or minimal effector activities. The biopharmaceutical industry has met the challenge to produce rMAbs, however, productivity, cost and potency remain to be optimised. All intact therapeutic antibodies currently licensed are produced by culture of transfected mammalian cells, e.g. Chinese hamster ovary (CHO), mouse NSO or mouse Sp2/0 cell lines. The quality of the antibody therapeutic product will depend on the ability of the chosen cell type and cell line to effect post-translational modifications similar or identical to those of human plasma cells. The secreted antibody may be subject to chemical changes resulting from exposure to the culture medium and enzymes released from live, senescent and dead cells (lysed) and extensive down stream processing procedures (e.g. isolation, purification, sterilisation, formulation and storage).

The polypeptide structure of human IgG

Each of the four IgG subclass proteins has the same basic four-chain structure of two light chains (~ 25 kDa) and two heavy chains (~50 kDa). At the protein sequence level the light and heavy chains are seen to be comprised of two and four repeating sequence homologous regions of ~110 amino acid residues, respectively, *Figure 1*; at the gene level the homology sequences are encoded within exons with intervening introns. Each homology region folds to give two anti-parallel β-pleated sheets bridged by an intra-chain disulphide bond and with hydrophobic side chains orientated toward the interior, *Figure 1*. This stable protein "scaffold" is referred to as the immunoglobulin fold; it is widely used throughout the proteome and allows for virtually unlimited sequence variation, within the β-bends, and the generation of unique recognition sites (Woof and Burton, 2004; Nezlin and Ghetie, 2004; Jefferis, 2007; Jefferis, 2009; Deisenhoffer, 1981; Padlan, 1996; Saphire, *et al.*, 2002).

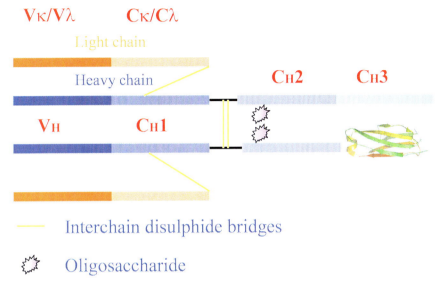

Figure 1. The four chain structure of the IgG molecule; comprised of two light chains and two heavy chains of identical sequence. The β pleated sheet structure of one CH3 domain is included.

The N-terminal homology region of the heavy and light chain of each specific antibody is of unique sequence; referred to as the variable regions of heavy (VH) and light chains (VL) respectively. Maximum sequence diversity is localised within three hypervariable or complementarity-determining regions (CDR) situated at β-bends that the immunoglobulin fold brings into spatial proximity. The VH and VL domains interact through multiple non-covalent interactions such that all six CDR are in spatial proximity and form the antigen binding site or paratope. Humans express two isotypes of light chain, kappa (κ) and lambda (λ) and four gamma (γ) IgG heavy chain isotypes (γ1, γ2, γ3, γ4). Each is characterised by one and three constant homology domains: Cκ, Cλ, CH1 or Cγ1, CH2 or Cγ2 and CH3 or Cγ3. The Cκ, Cλ domains each interact with the CH1 domain, through multiple non-covalent interactions and a single inter-chain disulphide bridge. The VH/VL-Cκ/CH1, VH/VL-Cλ/CH1 domains form a compact protein moiety that can be released from the intact IgG molecule, by proteolysis with papain, as the Fab fragment (Fragment antigen binding). The two Fab structures are linked to the Fc (Fragment crystallisable) through an open flexible "hinge" region that contains a rigid core comprised of proline and cysteine residues; the latter forming inter-heavy chain disulphide bridges, *Figure 2*. The Fc also forms an independent protein moiety stabilised by non-covalent CH3/CH3 and lateral CH2/

CH3 interactions. The CH2 domains form an open "horseshoe" structure with much of the internal space occupied by a complex diantennary oligosaccharide structure. The oligosaccharide is attached, by a covalent bond, to asparagine 297 and forms multiple non-covalent interactions with the polypeptide backbone and amino acid residue side chains of the inner surface of the CH2 domain, *Figure 2*.

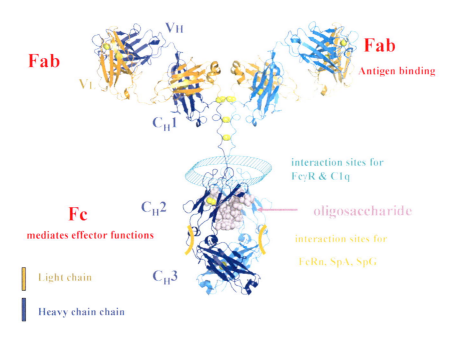

Figure 2. The α chain backbone structures the IgG molecule. The location of interaction sites for IgG-Fc ligands is indicated. Structure generated by Peter Artymiuk (University of Sheffield, UK) using PyMOL; http://pymol.sourceforge.net.

The independent mobility of each of the Fab and Fc regions allows the IgG molecule to exercise its functional divalency for antigen by engaging epitopes that may be in random spatial orientations to each other. Within an immune complex the Fab arms may be confined in spatial orientation, however, the Fc region may remain mobile and accessible for interactions with effector ligands. This model is supported by an X-ray crystallographic study of an intact IgG molecule for which coherent diffraction was obtained for the Fab regions but not for the Fc region (Marquart, *et al.*, 1980). This may be due to the Fc being in multiple orientations or retaining mobility in the crystal structure.

Overview of the quaternary structure of human IgG

The observation that antibody molecules activate effector functions only when bound to antigen, in immune complexes, led to proposals for allosteric mechanisms of action, i.e. - that antigen binding resulted in a conformation change that could be transmitted to sites outside of the antigen binding site. Possible support for this model came from early X-ray crystal structures for the human IgG1 proteins Dob and Mcg that showed contact between the Fab and Fc regions (Guddat, *et al.*, 1993). However, subsequent sequence studies revealed that these molecules were abnormal in structure and lacked the hinge sequence, encoded by the "hinge exon". Later studies of full-length IgG molecules revealed independent mobility for the Fab and Fc regions (Guddat, *et al.*, 1993; Burton, 1985). The functional hinge is considered to be comprised of three structural regions; upper and lower mobile sequences and a core comprised of cystine and proline residues that form a semi-rigid helical structure within which inter-heavy chain disulphide bridges are formed. Such a structure does not appear compatible with the transmission of conformational change from the Fab to the Fc; particularly for the extended hinge region structure of human IgG3 proteins. An associative model has gained acceptance in which multiple low affinity ligand-IgG-Fc interactions, afforded by the formation of multimeric immune complexes, results in multivalency and consequent high avidity for effector ligands. A mechanism providing accessibility to ligand binding sites in IgG-Fc has been proposed in which the Fab and Fc regions are "dislocated" with respect to each other allowing simultaneous interaction with antigen and effector molecule (Burton, 1985; Roux *et al.*, 1997). Induced conformational influences should not be entirely dismissed, however, since the mobility of the Fab/Fc regions may allow for the generation of "folded-back" orientations allowing Fab-Fc contacts and/or masking of Fc effector sites (Woof and Burton, 2004; Nezlin and Ghetie, 2004; Burton, 1985; Carrasco *et al.*, 2001; Lu *et al.*, 2007).

The inherent flexibility of the IgG molecule, together with glycoform heterogeneity, hinders crystallisation; thus whilst crystals of ten intact IgG molecules have been reported structures have been derived for only seven. Complete structures were restricted to five molecules, the abnormal IgGs Dob and Mcg (see above), the human anti-gp120 b12 antibody and two murine antibodies (Harris, *et al.*, 1997). All structures are asymmetric, illustrating independent Fab and Fc mobility, afforded by the hinge region (Guddat, *et al.*, 1993). Resolution of the Fab and Fc structure of the b12 molecule results from stabilisation provided by the close proximity (contact!) of the Fc with one Fab arm (Saphire, *et al.*, 2002).

Quaternary structure of Fab

Structures for numerous antigen specific Fab fragments have been determined by X-ray crystallography, both alone and in complex with antigen (Padlan, 1996). Early studies of Fab fragments binding small molecules (haptens) led to the lock-and-key model in which the antigen bound within a pocket; however, studies of macromolecular antigens have shown that both paratopes and epitopes may be comprised of relatively flat surfaces (Padlan, 1996). Whilst light and heavy chain CDRs are seen to contribute to epitope binding for all specificities it is not the case that all CDRs contribute equally to a given paratope. Comparison of the V_H and V_L sequences of a given Fab with that encoded within the precursor germline gene informs of the contribution of somatic hypermutation to specificity and/or affinity both within and outside of the paratope.

These studies have defined the contribution of CDRs and "framework" sequences to antigen binding and led to the development of informed protocols for humanisation of mouse antibody V regions for the generation of therapeutic antibodies (Carter, 2006; Moutel and Perez, 2008; Reichert, and Valge-Archer, 2007; Magdelaine-Beuzelin, C. *et al.*, 2009). Mapping of the surface topology of Fab regions identifies the degree of exposure of amino acid side chains and can contribute to further informed protein engineering to confer advantageous properties, e.g. solubility (Famm and Winter, 2006).

Comparison of structures of Fab and Fab-antigen complexes show that conformational change may occur within V_H and V_L domains on antigen binding (Sanguineti *et al.*, 2007; Sagawa *et al.*, 2005; Keskin, 2007). This provides further understanding of epitope recognition and binding and may reveal residues underlying the paratope, the vernier zone, that are essential to its architecture, mobility and hence specificity and/ or affinity (Makabe *et al.*, 2008). The junction of the V_H/C_H1 and V_L/C_L, referred to as the "switch" residues, is characterised by a change in direction of the polypeptide chains, referred to as the "elbow angle". The elbow angle is characteristic for a given Fab but can vary widely between Fabs; lambda light chains appear to be compatible with larger elbow angles than kappa light chains (Stanfield *et al.*, 2006). There is evidence that conformational changes resulting from antigen binding can be transmitted to the C_H1/C_L domain; also, the reciprocal finding that differences in C_H1 structure can influence antigen-binding affinity (Pritsch *et al.*, 1996).

An understanding of the structure and function of the Fab region of antibodies facilitated genetic engineering to progressively "humanise" antibodies raised in mice (Carter, 2006; Moutel and Perez, 2008; Reichert, and Valge-Archer, 2007; Pritsch *et al.*, 1996; Dubel, 2007; Presta, 2008; Magdelaine-Beuzelin, C. *et al.*, 2009). The engineered V_H and V_L gene sequences being ligated to the constant region gene sequences of heavy and light chains, respectively, to generate humanised antibodies. In practice such "humanized" V_H and V_L domains usually resulted in reduced affinity and/or specificity such that some mouse residues had to be re-introduced, with consequent potential immunogenicity. Libraries of human V_H and V_L gene sequences have been generated from human peripheral blood lymphocytes, using PCR protocols, and their protein products expressed in phage display allowing for selection of combinations of human V_H and V_L sequences with specificity for a selected target; the V_H and V_L sequences can subsequently be expressed with selected human constant regions (Jostok, *et al.* 2004). Possibly the ultimate source of human antibodies specific for human targets can be realised from immunisation of mice that are transgenic for the expression of human variable and constant region genes; the endogenous Ig genes having been inactivated (Lonberg, 2008). It may not be possible to dictate the human constant region expressed but subsequent engineering can rectify this.

Current experience suggests that even "fully" human antibodies can be immunogenic, at least in a proportion of patients. Initially, one may wonder why this should be so; however, on reflection it may seem inevitable. The hallmark of an antibody is its specificity, not just for a particular target but also for a unique structural feature on that target – the epitope. This is achieved within a secondary immune response that is characterised by somatic hypermutation and selection. Thus each antibody is a structurally unique molecule with a unique epitope-binding site (the paratope). Antibodies generated from phage display libraries or transgenic mice are unique to an individual, human or mouse, and may be perceived as "foreign" to a unique recipient, i.e. the patient. Antibody therapeutics are manufactured in xenogeneic tissue (Chinese hamster ovary CHO cells; mouse NS0 cells etc) that may yield product not having the required human type co- and post-translational modifications and/or add non-human co- and post-translational modifications (Walsh and Jefferis, 2006).

Functional and genetic (core) hinge structures

The core hinge sequence is encoded by a single exon for IgG1, IgG2 and IgG4 but by several exons for IgG3. The functional hinge is comprised of mobile C-terminal residues of the CH1 domain (upper hinge), a rigid core and mobile N-terminal residues of the CH2 domain (lower hinge) (Burton, 1985). This flexible "spacer" sequence allows for independent mobility of Fab and Fc protein moieties and binding to their respective ligands. The IgG4 molecule presents a unique case in which disulphide exchange within the hinge region may occur and allow for dissociation/re-association between heavy-light chain pairs; referred to as "half-molecules" (Aalberse and Schuurman, 2002; van der Neut Kolfschoten *et al.*, 2007). Random re-association of heavy-light chain pairs can result in the generation of antibodies of dual specificity, both *in vitro* and *in vivo*. This characteristic can be overcome by replacing the -C-P-S-C- sequence present within the hinge of the IgG4 molecule by the IgG1 sequence –C-P-P-C-. The length of the hinge region and the number of inter-heavy chain disulphide bonds differs significantly between the human IgG subclasses and influences mobility and average solution conformation of the Fab and Fc moieties, with respect to each other (Burton, 1985; Burton and Woof, 1992). This may include the ability to assume a "dislocated" form that provides access to effector ligand binding sites localised to the hinge proximal region of the CH2 domain. The extended hinge region of IgG3 molecules has been the subject of genetic/protein engineering to generate hinge regions of different length; no direct relationship between hinge length and an ability to bind and activate the C1 component of complement was demonstrated, however, at least one inter-heavy chain disulphide bridge is required (Michaelsen *et al.*, 1994; Brekke, *et al.*, 1995; Redpath *et al.*, 1998).

Quaternary structure of IgG-Fc: the protein moiety

An early insight into the structure of IgG-Fc was obtained from X-ray crystallographic studies of human IgG1-Fc (Deisenhofer, 1981), generated by cleavage of polyclonal IgG at the Lys_{222}-Thr_{223} peptide bond, within the hinge region, and extending to the C-terminus residue 446. It was reported that interpretable electron density was not obtained for residues 223-237, that comprise most of the core and lower hinge region, or the C-terminal residues 444-446; it was not known at this time that the CH3 exon coded for a C-terminal 447 lysine residue that is removed by endogenous carboxypeptidase B. The α carbon IgG structure shown in *Figure 2* is a composite of the Deisenhofer (1991) IgG-Fc structure for residues 238-443 with the addition of a computer generated structure for the hinge sequence (Lund, *et al.*, 1991) and an Fab structure (Marquart, *et al.*, 1980). Salient structural features are:

i.   CH2 domains were less well ordered than CH3 domains. Defined structure was obtained for CH2 structure at the CH2-CH3 interface with increasing disorder being observed in the vicinity of the hinge region, indicated by increased temperature factors or missing electron density. One CH2 domain was more ordered than the other, due to a crystal contact with a neighbouring CH2 domain.

ii. The area of contact between the CH2 and CH3 domains comprises about two-thirds the lateral area of contact between the two CH3 domains. This suggests that the CH2-CH3 contact contributes to the relative stability observed for the C-terminal proximal region of CH2 domains; as opposed to the "softness" of those parts of the CH2 domain remote from the CH2-CH3 interface.

iii. Hydrophobic and hydrophilic interactions between the carbohydrate and inner protein surface of the CH2 domain "substitutes" for the domain pairing observed for other domains

iv. Higher structural resolution was obtained for β-sheets regions than for than for β-bends.

These interpretations and conclusions are confirmed in other X-ray structures obtained for human IgG-Fc alone (Harris, *et al.*, 1997; Saphire, *et al.*, 2002; Matsumiya *et al.*, 2007) and in complex with Staphylococcal protein A (SpA) (Deisenhofer, 1981), Streptococcal protein G (SpG) (Sauer-Eriksson *et al.*, 1995), rheumatoid factor (RF) (Duquerroy *et al.*, 2007; Sohi *et al.*, 1996; Corper, *et al.*, 1997) and human FcγRIIIb (Sondermann *et al.*, 2000; Radaev *et al.*, 2001). The internal mobility of the lower hinge and hinge proximal regions of the CH2 domains ("softness of structure) may result in the generation of an equilibrium of high order conformers that may differentially bind unique ligands, e.g. homologous Fcγ receptors. Thus previous proposals that different ligands may bind through "overlapping non-identical sites" (Lund *et al.*, 1991) may be modified to suggest that they each bind to a unique conformer with contacts being made through some common amino acid residue side chains and/or main chain atoms (see below).

The IgG-Fc has been shown to be functionally divalent for ligands binding at the CH2-CH3 interface e.g. the neonatal Fc receptor (FcRn), RF, SpA, SpG (Woof and Burton, 2004; Nezlin and Ghetie, 2004; Jefferis, 2009; Deisenhofer, 1981; Sauer-Eriksson *et al.*, 1995; West, *et al.*, 2000; Matsumiya *et al.*, 2007; Duquerroy *et al.*, 2007; Sohi *et al.*, 1996; Corper, *et al.*, 1997), *Figure 2*. Due to the symmetry of the IgG-Fc these interaction sites are at ~180° to each other and each is accessible to bind macromolecular ligands to form multimeric complexes. By contrast the FcγRIII binding site is asymmetric and both heavy chains are engaged such that monomeric IgG is univalent for Fcγ receptors and the C1 component of complement; this is essential in order to obviate continuous activation of inflammatory cascades by circulating IgG *in vivo*. Residues of the lower hinge region that are disordered in the Fc crystals are ordered in the Fc-FcγRIII complex and directly involved in binding the receptor (Sondermann *et al.*, 2000; Radaev *et al.*, 2001).

The IgG-Fc oligosaccharide moiety

The presence of a core diantennary heptasaccharide has been shown to be essential for optimal activation of Fcγ receptors and the C1 component of complement. The oligosaccharide of normal polyclonal IgG, attached at asparagine 297, is highly heterogeneous and comprised of the core heptasaccharide with variable addition of fucose, galactose, bisecting N-acetylglucosamine and sialic acid. *Figure 3*; sialylation is modest with < 10 % of structures being monosialylated or disialylated (Jefferis *et al.*, 1990; Rudd *et al.*, 1997; Routier *et al.*, 1998; Jefferis 2009). From Fig 3a it can be calculated that a total of 32 unique oligosaccharide structures are possible, on each heavy chain; one must not be dismissive of minor oligosaccharide structures since each could be the predominant form present on an antibody secreted from a single plasma cell. Considering neutral oligosaccharides only and assuming random pairing of heavy chain glycoforms, a total of 128 IgG glycoforms ((16 x 16)/2) bearing neutral oligosaccharides may be generated; given the symmetry of the IgG molecule the two heavy chains are equivalent.

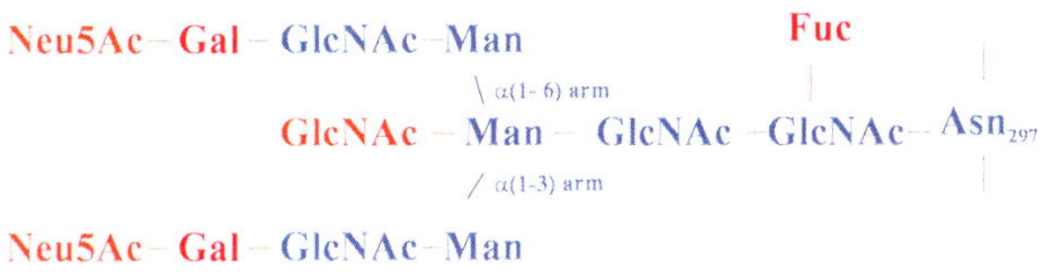

Figure 3a. The complex diantennary oligosaccharides that may be present in the IgG-Fc of normal polyclonal IgG: ■■■ core heptasaccharide; ■■■ outer arm sugar residues that may be present. GlcNAc, N-acetylglucosamine; Man, mannose; Fuc, fucose; Gal, galactose; Neu5Ac, N-acetylneuraminic acid

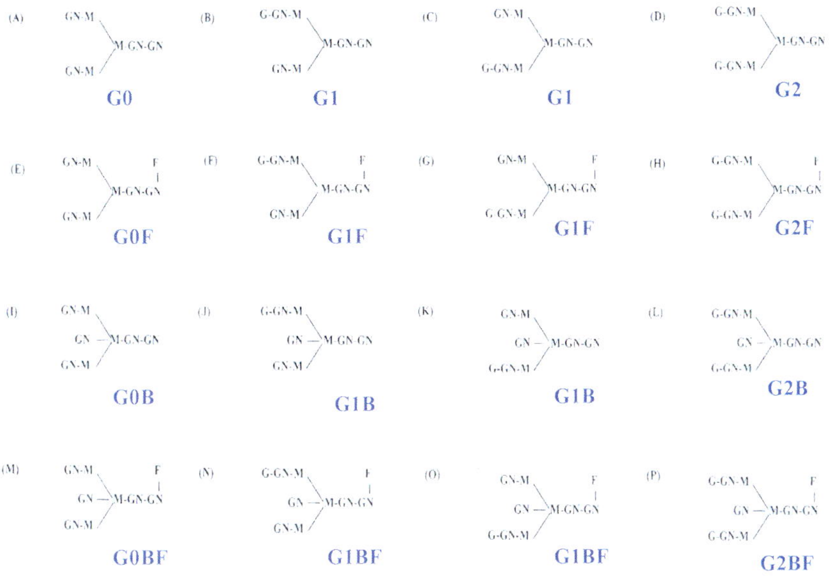

Figure 3b. The potential "library" of neutral complex diantennary oligosaccharides released from normal human IgG-Fc. GN, N-acetylglucosamine; M, mannose; F, fucose; G, galactose.

Unfortunately, there are several systems of nomenclature currently in use to represent oligosaccharide structures and, consequently, antibody glycoforms (http://glycomics. scripps.edu/CFGnomenclature.pdf; Crispin *et al.*, 2007; Campbell *et al.*, 2008; http:// glycobase.ucd.ie). Carbohydrate chemists and specialist mass spectrometry scientists, amongst others, have developed different nomenclatures. A shorthand system has evolved, for the non-specialist glycobiologists, particularly with respect the presence or absence of galactose, fucose and bisecting N-acetylglucosamine residues. Thus, the core heptasaccharide, devoid of galactose, fucose and N-acetylglucosamine is designated G0 (zero galactose) and when bearing one or two galactose residues are designated G1 and G2, respectively. The shorthand nomenclature for the heptasaccharide + fucose/galactose are: G0F, G1F and G2F; when bisecting N-acetylglucosamine is present a B is added, e.g. G0B, G0BF, G1BF etc. This shorthand nomenclature is illustrated in *Figure 3b*. The approximate composition of oligosaccharides released from normal polyclonal human IgG-Fc is G0, 3 %; (G1), 3 %; G2, 6 %; G0F, 23 %; G1F, 30 %; G2F, 24 %; G0BF, 3 %; G1BF, 4 %; G2BF, 7 % (Farooq, *et al.*, 1997; Routier, *et al.*, 1998). It is similarly important to define the glycoform of the whole IgG molecule since it has been shown that symmetrical and asymmetrical pairing of heavy chain glycoforms occurs (Masuda, *et al.*, 2000; Mimura, *et al.*, 2007) it has been hypothesised that enhanced ADCC may be observed for IgG in which only one heavy chain bears oligosaccharide devoid of fucose (Ferrara *et al.*, 2006); thus, IgG-Fc bearing (G0/G0F) oligosaccharides should be as potent as one bearing (G0)2 oligosaccharides. The shorthand nomenclature is adopted within this review.

Methods able to determine the glycoform profile of specific IgG antibody populations produced within a normal polyclonal response have only recently been reported; significant variations in the glycoform profiles for IgG antibodies having specificity for platelet antigens being observed (Wuhrer *et al.*, 2009). Analysis of monoclonal human IgG proteins, isolated form the sera of patients with multiple myeloma, suggest that each plasma cell clone secretes IgG with an individual (unique!) glycoform profile; including extremes for the presence or absence of fucose, galactose, bisecting N-acetylglucosamine and sialic acid. (Jefferis *et al.*, 1990; Farooq, *et al.*, 1997). Interestingly, for individual patients the oligosaccharide profile of the IgG paraprotein was shown to vary at different stages of the disease whilst that of the normal polyclonal component remained stable (Farooq *et al.*, 1997). This may provide an index of intraclonal variation that has not been evident from studies of protein or DNA sequences. The oligosaccharide profiles of hybridoma and recombinant IgG proteins produced in mammalian cells are significantly influenced by the cell type, the culture method and conditions employed, however, the (G0F)2 glycoform predominates (see below). Under conditions of stress, e.g. nutrient depletion, acid pH etc., deviant glycosylation may be observed e.g. the presence of high mannose forms and/or incomplete occupancy (Pascoe *et al.*, 2007).

IgG-Fc protein/oligosaccharide interactions

The oligosaccharide has yielded electron density in all except one of the IgG-Fc molecules and complexes subjected to X-ray crystallographic analysis (Deisenhofer, 1981; Woof and Burton, 2004; Nezlin and Ghetie, 2004; Jefferis, 2009; Deisenhofer, 1981;

Matsumiya *et al.*, 2007; Sauer-Eriksson *et al.*, 1995; Duquerroy *et al.*, 2007; Sohi *et al.*, 1996; Corper, *et al.*, 1997; Sondermann *et al.*, 2000; Radaev *et al.*, 2001; Krapp *et al.*, 2003); the exception being a complex formed between an IgG4-Fc and the Fab fragment derived from an IgM rheumatoid factor (Sohi *et al.*, 1996; Corper, *et al.*, 1997). In each case the core heptasaccharide has been resolved together with fucose linked $\alpha(1-6)$ to the primary N-acetylglucosamine residue and galactose present, on the $\alpha(1-6)$ arm; it is not known whether additional sugar residues were present, but mobile. Sugar residues of the core and $\alpha(1-6)$ arm make multiple hydrophobic and hydrophilic non-covalent interactions with inner face of the CH2 domain. A total of 72 interactions are possible, including potential for six CH2 protein/oligosaccharide hydrogen bonds and six hydrogen bonds between sugar residues (Deisenhofer, 1981; Padlan, 1990). The $\alpha(1-3)$-Man-GlcNAc arms are orientated towards the internal space between the CH2 domains and weak lateral interactions between sugar residues have been suggested for some structures but not others (Deisenhofer, 1981, Girardi, *et al.*, 2009). Although direct interactions between FcγRIII and the oligosaccharide are not observed (Sondermann *et al.*, 2000; Radaev *et al.*, 2001) aglycosylated IgG-Fc does not activate this receptor. It is concluded, therefore, that the protein and oligosaccharide each exert mutual influence on the conformation of the other.

Differential scanning micro-calorimetry and isothermal microcalorimetry has been employed to compare the stability and FcγR binding activity of a series of homogeneous, normal, truncated and aglycosylated forms of IgG-Fc (Mimura *et al.*, 2000; Mimura *et al.*, 2001). Fully galactosylated, (G2)2 and agalactosylated (G0)2 glycoforms of IgG-Fc exhibited two transition temperatures, Tm1 and Tm2 of 71.4^0 and 82.2^0, representing the unfolding of the CH2 and CH3 domains, respectively. These data suggest that whilst the galactose residue on the $\alpha(1-6)$ arm has substantial contacts with the protein structure it does not impact CH2 domain stability. Sequential removal of the terminal GlcNAc and the two arm mannose residues, generating a (GlcNAc-2Man)2 glycoform resulted in destabilisation of the CH2 domain and a lowering of Tm1 to 67.7^0, Tm2 was unchanged. The thermodynamic parameters describing CH2 thermal denaturation of all IgG-Fc glycoforms was consistent with a cooperative unfolding. By comparison the unfolding of the CH2 domain of aglycosylated IgG1-Fc was non-cooperative, involving at least one intermediate (Mimura *et al.*, 2000). It was proposed that this intermediate is a partially unfolded CH2 domain pair possessing hinge proximal disordered/unfolded loops that may account for the compromised functional activities of deglycosylated IgG and IgG-Fc (Mimura *et al.*, 2000; Mimura *et al.*, 2001; Lund *et al.*, 1990; Krapp *et al.*, 2003).

These structural and binding studies show that the IgG-Fc (GlcNAc2Man)2 glycoform is sufficient for generation of lower hinge conformers compatible with FcγRI and C1 binding and activation. It is proposed that, in addition to the covalent bond, each GlcNAc2Man trisaccharide has the potential to form 31 non-covalent contacts with the protein, at least three (3) being hydrogen bonds (Padlan, 1990). These data are consistent with X-ray crystal data obtained for the same truncated IgG-Fc glycoforms that showed progressive increases in the temperature factors for the CH2 domain, as evidence of progressive structural disorder (destabilisation) (Mimura *et al.*, 2000; Krapp *et al.*, 2003). Interestingly, in this study all attempt to obtain a crystal form of deglycosylated IgG-Fc failed. This may be indicative of the structural destabilisation that results from complete removal of oligosaccharides. Truncation of the sugar residues results in the mutual approach of CH2 domains with the generation of a "closed" conformation; in contrast to the "open" conformation observed for the fully galactosylated IgG-Fc (Krapp *et al.*, 2003). These interpretations of the data should be tempered by consideration of the temperature dependence of partial specific volume of protein and carbohydrate. It is asserted, ".... the coefficient of thermal expansion for the carbohydrate chain of the glycoprotein (IgG-Fc) is much larger than that for the protein" (Mimura *et al.*, 2000); all crystal structures of IgG-Fc have been generated from data collection at temperatures well below physiological. Given the unique positioning of the oligosaccharide, between the two CH2 domains, it is possible that the protein/oligosaccharide interactions could be significantly different at 37°C; one might anticipate that the expansion of the oligosaccharide would enhance contact with the protein moiety and lateral interactions between the two oligosaccharide structures, reducing internal mobility for each.

A structural change localised to the hinge proximal region of the CH2 domain was observed on comparison of the NMR spectra of glycosylated and aglycosylated IgG3-Fc (Lund *et al.*, 1990); assigned chemical shifts for histidine residues were employed as reporter groups for local structural change. Two histidine residues in IgG3-Fc are within bends of the CH2 domain, at residues 268 and 285, respectively, and three are within the CH2/CH3 interface region. A change in local environment of His 268 only was observed for the aglycosylated IgG3-Fc; this residue is not a contact residue for the oligosaccharide. This evidence for a very restricted structural difference is in agreement with the lack of detectable change in epitope expression between glycosylated and aglycosylated Fc, when probed with a panel of monoclonal antibodies (Walker *et al.*, 1989). Similarly, an NMR study of a truncated IgG-Fc glycoform bearing only a fucosylated primary GlcNAc revealed a "subtle" conformational change in the lower hinge region, in comparison with the wild type glycosylated protein. This glycoform did not bind soluble recombinant FcγRIIIa (Yamaguchi *et al.*, 2006).

Cellular IgG-Fc receptors

Two distinct functions for Ig-Fc receptors may be distinguished: i) to bind antibody/antigen complexes and initiate effector functions leading to their removal and destruction, ii) to mediate transport across epithelial membranes, i.e. transcytosis.

I) IgG-Fc RECEPTORS (FcγR) MEDIATING ANTIGEN CLEARANCE

Three types/classes of membrane bound human FcγR (FcγRI (CD64), FcγRII (CD32), FcγRIII (CD16)) and six sub-types (FcγRI, FcγRIIA, FcγRIIB1, FcγRIIB2, FcγRIIC, FcγRIIIA, FcγRIIIB) have been defined by immunochemical, biochemical and gene sequencing studies (Radaev and Sun, 2002; Sondermann *et al.*, 2001; Woof and Burton, 2004; Nimmerjahn and Ravetch, 2007; Nimmerjahn and Ravetch, 2008). These receptors may be constitutively expressed, up regulated and/or induced by cytokines generated and released within an inflammatory response. It is virtually impossible, therefore, to predict the FcγR mediated sequalae of immune complex formation within a polyclonal protective antibody response. Effector functions triggered are diverse and include inactivation and removal (e.g. phagocytosis, respiratory burst and cytolysis), accessory functions such as the enhancement of antigen presentation by dendritic cells, and the down-regulation of growth and differentiation of lymphocytes. It is evident, therefore, that FcγRs play an important role in the induction, establishment and resolution of protective immune responses.

Multiple parameters determine the structure and biological activities of immune complexes: i) valency of the antibody, ii) average affinity/avidity of the antibody population, iii) isotype of the antibody, iv) valency or epitope density of the antigen, v) valency of individual effector ligands, v) cumulative valency when multiple ligands are engaged, e.g. FcγR and complement receptors etc, vii) proportions of each antibody isotype within a polyclonal response (Lucisano Valim and Lachmann, 1991; Michaelsen *et al.*, 1991; Voice and Lachmann, 1997). The availability of panels of recombinant antibody molecules of each human Ig isotype of identical epitope specificity has allowed the influence of some of these parameters to be determined and summarizing *in vitro* studies of neutrophil activation Voice and Lachmann (1997) concluded "...*in vivo* these results may translate into differential activation of neutrophils by soluble immune complexes, dependent on their characteristics, leading to subtle nuances in the aetiology, pathology and control of the immune response..". These findings caution reserve when attempting to extrapolate from *in vitro* studies to *in vivo* responses.

FcγRIA is frequently referred to as the high affinity receptor for IgG since, in early studies, the binding of monomeric IgG1 and IgG3 is readily demonstrated; the affinity of IgG4 was reported to be an order of magnitude lower whilst binding of IgG2 could not be detected. The FcγRIA receptor is constitutive expressed on mononuclear phagocytes and dendritic cells; however, expression can be up-regulated and/or induced on leucocytes by the action of cytokines. FcγRIIA is the most widely expressed FcγR and is found on most hemopoietic cells. Polymorphic variants of FcγIIA are identified by the presence of histidine (FcγRIIA-131H) or arginine (FcγRIIA-131R) at amino acid residue 131. The higher affinity of the FcγRIIA-131H form for IgG2 results in differing ability to respond to engagement by IgG2 immune complexes (Warmerdam *et al.*, 1991). Comparative studies with PMNL from donors either homozygous for FcγRIIA-131R or FcγIIA-131H showed the latter had higher phagocytic capacity for *Streptococcus pneumoniae* opsonised with IgG2 antibody. A polymorphism within the membrane distal domain of FcγRIIA does not appear to affect activity (Rodriguez *et al.*, 1999). The FcγRIIB receptor is expressed on B-lymphocytes and monocytes and ligation of this receptor results in growth and differentiation inhibition.

Initially the FcγRIIIA receptor was reported to bind and be activated by IgG1 and IgG3 only; however, recognition of a polymorphism in the receptor and the differential influence of IgG glycoforms has radically changed our understanding, with important practical consequences. The avidity of binding of IgG differs between the FcγRIIIA—158V and FcγRIIIA-158F polymorphic variants (Koene *et al.*, 1997). It was demonstrated, *in vitro*, that IgG1 antibody is more efficient at mediating ADCC through homozygous FcγRIIIA-158V bearing cells than homozygous FcγRIIIA-158F or heterozygous FcγRIIIA-158V/FcγRIIIA-158F cells (Koene *et al.*, 1997). It was anticipated, therefore, that similar differences in ADCC efficacy might pertain *in vivo*, depending on the polymorphic form of FcγRIIIA expressed. This prediction was verified in a study of the clinical response of patients receiving Rituxan who had relapsed after conventional treatment for non-Hodgkin's lymphoma (Cartron *et al.*, 2002); B cell depletion in patients with systemic lupus erythematosis was also enhanced for those homozygous for FcγRIIIA-158V (Anolik *et al.*, 2003). Similarly, FcγRIIIA polymorphisms were shown to influence the response of Crohn's disease patients to infliximab (Louis *et al.*, 2004) and red blood cell clearance by anti-D antibody (Miescher *et al.*, 2004). By contrast FcγRIIIA polymorphism was shown not to influence the response to humira in patients with chronic lymphocytic leukaemia (Lin *et al.*, 2005). Three parameters were thus defined that influence B cell clearance; intrinsic properties of the target cell, the polymorphic form of the FcγRIIIA expressed on effector cells and the glycoform of the rMAb (see below).

All FcγR are trans-membrane molecules, except FcγRIIIB that is glycosylphosphatidylinositol (GPI)-anchored within the membrane of neutrophils. FcγRI and FcγRIIIA are members of the multi-chain immune recognition receptor (MIRR) family and are present in the membrane as hetero-oligomeric complexes: an IgG/antigen complex binding α chain and a signalling γ chain; the FcγRIIIA α chain of NK cells is associated with a signalling ζ chain. FcγRIIA and FcγRIIB molecules are comprised of an α chain only (Nimmerjahn and Ravetch, 2007; Nimmerjahn and Ravetch, 2008). The FcγR alpha chains show a high degree of sequence homology in their extra cellular domains (70-98%) but differ significantly in their cytoplasmic domains. The cytoplasmic domains of γ chains and the FcγRIIA α chain express the immunoreceptor tyrosine-based activation motif (ITAM) that is involved in the early stages of intracellular signal generation. By contrast the FcγRIIB receptor α chain expresses an immunoreceptor tyrosine-based inhibition motif (ITIM) (Woof and Burton, 2004; Nimmerjahn and Ravetch, 2007; Nimmerjahn and Ravetch, 2008). Cellular activation may be dependent on the balance between the relative levels of expression of these two isoforms and hence the balance of signals generated through the ITAM and ITIM motifs (Nimmerjahn and Ravetch, 2007; Nimmerjahn and Ravetch, 2008). Since the FcγRIIIB is linked to the membrane by a GPI moiety it cannot signal directly but there is evidence that on ligation it can associate with trans-membrane molecules on the surface of neutrophils to deliver activation signals (Edwards *et al.*, 1997; Coxon *et al.*, 2000; Shibata-Koyama, *et al.*, 2009). Monomeric IgG may bind and occupy FcγRs but is functionally monovalent and cannot cross-link the receptors; activation or inhibition signals only result when FcγRs are cross-linked by multivalent ligand forms i.e. IgG immune complexes. There is considerable interest in the design of IgG molecules that differentially bind and activate FcγR (Shields *et al.*, 2001; Siberil *et al.*, 2007; Desjarlais *et al.*, 2007).

FcγR BINDING SITES ON IgG

Structural definition of ligand interaction sites of IgG-Fc allows for informed design and engineering of new therapeutics with properties that may be optimal for a given disease indication. The first ligand investigated was the high affinity FcγRI receptor, since the binding of radio-labelled IgG could be quantitated. Initially, candidate residues were selected on the basis of sequence comparison of solvent accessible residues in IgG isotypes of human and rodent origin that differed in their ability to bind FcγRI (Woof *et al.*, 1986) and the lower hinge sequence Leu234-Leu-Gly-Gly237 was proposed as critical. Subsequently alanine scanning confirmed this proposal and extended it to include FcγRII and FcγRIII (Lund, *et al.*, 1991; Shields *et al.*, 2001). X-ray crystallographic analysis of IgG-Fc in complex with a soluble recombinant form of FcγRIII, provided proof of the direct involvement of the lower hinge regions and hinge proximal CH2 domain residues (Sondermann *et al.*, 2000; Radaev *et al.*, 2001). The interaction site on the IgG-Fc is asymmetric and binds discrete conformations of the lower hinge residues of each heavy chain. Other residues of the hinge proximal region of the CH2 domain also form contacts with FcγRIII. Interestingly, one structure reveals a possible contribution of the primary N-acetylglucosamine residue to binding (Sondermann *et al.*, 2000) whilst the other primary publication held that there is no direct contact (Radaev *et al.*, 2001); in a subsequent review the authors of the latter publication stated that the oligosaccharides contributes ~ 100 $Å^2$ to the contact interface (Radaev and Sun, 2002; Radaev and Sun, 2001); this conclusion resulted from refinement of the crystal data (P. Sun – personal communication). These investigators also demonstrated that whilst the binding of deglycosylated IgG-Fc to the *E. coli* derived aglycosylated Fc RIIIa was undetectable the binding of deglycosylated whole IgG was only decreased, 10–15-fold (Radaev and Sun 2001). This serves to remind us to exercise caution when tempted to extrapolate from *in vitro* experimental data to *in vivo* biological outcomes. Taken together these data suggest that any direct contribution of the oligosaccharide to binding is minimal and that the oligosaccharide contributes indirectly through its influence on protein conformation. The involvement of both heavy chains in the formation of an asymmetric binding site provides a structural explanation for an essential requirement - that the IgG should be univalent for the FcγR; if monomeric IgG were divalent it could cross link cellular receptors and hence constantly activate inflammatory reactions.

ROLE OF IgG GLYCOFORMS IN RECOGNITION BY CELLULAR FcγRs

Since glycosylation of IgG-Fc is essential to recognition and activation of effector ligands quantitative and/or qualitative functional differences between glycoforms might be anticipated. Analysis of the oligosaccharides released from IgG isolated from the sera of patient with a number of inflammatory diseases, including rheumatoid arthritis (Parek *et al.*, 1985), inflammatory bowel disease (Go *et al.*, 1994), vasculitis (Holland *et al.*, 2002), coeliac disease (Cremata *et al.*, 2003), periodontal disease (Novak *et al.*, 2005) etc has revealed increases in the levels of non-galactosylated glycoforms. In addition differentiation between rheumatic disease(s) is claimed for "sugar mapping" (Alavi and Axford ,2006; Alavi and Axford, 2008). Studies employing recombinant monoclonal antibodies have established that individual glycoforms can have a very significant impact on functional activity (Jefferis, *et al.*, 1998; Umana *et al*, 1999; Davies *et al.*, 2001; Shields *et al.*, 2002; Shinkawa *et al.*, 2003; Niwa *et al.*, 2005; Satoh *et al.*, 2006; Anthony *et al.*, 2008; Shibata-Koyama *et al.*, 2009). This leads one to speculate whether the immune system responds not only by production of an optimal isotype but also an optimal glycoform. Analytical protocols employing mass spectrometry are now providing the tools to determine the glycoform profile of specific antibody populations and differences in the glycoform profiles of anti-platelet antibodies, between individuals, has been reported (Wuhrer *et al.*, 2009); it remains to be seen whether these differences are related to disease activity.

A minority of oligosaccharides released from polyclonal IgG-Fc are sialylated whilst ~ 70 % bear one or two galactose residues (Jefferis *et al.*, 1990; Rudd *et al.*, 1997; Routier *et al.*, 1998). The paucity of sialylation is presumed to reflect the intimate integration of the oligosaccharides within the IgG-Fc structure such that the steric/spatial requirements of the $\alpha(2,6)$ sialyltransferase cannot be met, rather than being due to any deficit in the sialylation machinery. This conclusion is supported by the finding that when both IgG-Fc and IgG-Fab are glycosylated the latter bears highly galactosylated and sialylated structures, demonstrating that the glycosylation machinery is fully functional. The presence or absence of terminal galactose and/or sialic acid residues does not influence catabolism since IgG is not catabolised in the liver, via the asialoglycoprotein receptor (ASGPR), but by multiple cell types expressing FcRn. The balance between structure and accessibility is well illustrated for a panel of IgG-Fcs in which individual amino acid residues making contacts with the oligosaccharide were replaced by alanine. In each case hypergalactosylated and highly sialylated glycoforms resulted, suggesting some relaxation of structure allowing access to glycosyl transferases (Lund *et al.*, 1996).

Recent studies suggest that sialylated human IgG-Fc may be anti-inflammatory, relative to asialylated IgG-Fc. In two reports it was claimed that the anti-inflammatory activity resulted from reduced affinity of binding to IgG-Fc receptors (Kaneko *et al.*, 2006; Scallon *et al.*, 2007). However, it was subsequently shown that the anti-inflammatory activity is mediated by sialylated glycoforms of IgG-Fc binding and activating the C-type lectin SIGN-R1, expressed on mouse macrophages. The human analogue of SIGN-R1 is DC-SIGN (Anthony *et al.*, 2008).

Differences in galactosylation is a major source of IgG-Fc glycoform heterogeneity. The distribution for normal adult serum derived IgG-Fc is agalactosyl (G0 + G0F + G0BF), 20 – 25 %, mono-galactosyl (G1 + G1F + G1bF), 35 – 45 % and di-galactosyl (G2 + G2F + G2BF), 10 – 20 % (Jefferis *et al.*, 1990; Rudd *et al.*, 1997; Routier *et al.*, 1998). Lower levels of IgG-Fc galactosylation are observed for older adults and there is also a small but significant gender difference (Yamada *et al.*, 1997). An increase in IgG-Fc galactosylation occurs over the course of a normal pregnancy with levels returning to the adult norm following parturition (Williams *et al.*, 1995; Kibe *et al.*, 1996). Hypogalactosylation of IgG-Fc is reported for a number of inflammatory states associated with autoimmune diseases (Parek *et al.*, 1985; Go *et al.*, 1994; Holland *et al.*, 2002; Cremata *et al.*, 2003; Novak *et al.*, 2005; Alavi and Axford ,2006; Alavi and Axford, 2008). The extent of IgG-Fc galactosylation observed for monoclonal human myeloma IgG proteins is highly variable, indicating that the level of IgG-Fc galactosylation is a clonal property (Farooq *et al.*, 1997. The antibody products of CHO, Sp2/0 and NS0 cell lines used in commercial production of recombinant antibody are generally highly fucosylated but hypogalactosylated, relative to normal human IgG (Umana *et al,* 1999; Davies *et al.*, 2001; Shields *et al.*, 2002; Shinkawa *et al.*, 2003; Niwa *et al.*, 2005; Satoh *et al.*, 2006; Shibata-Koyama *et al.*, 2009); it is necessary therefore, to consider the possible impact of differential IgG-Fc galactosylation on functional activity.

Initially, the coincidence of IgG-Fc hypogalactosylation and the presence of autoantibodies, rheumatoid factors (RF), with specificity for IgG-Fc in the serum of patients with rheumatoid arthritis (RA) fuelled speculation of cause and effect, i.e. that the lack of galactose on the $\alpha(1,6)$ arm generates a denatured and immunogenic form of IgG. The finding of hypogalactosylation in a number of other autoimmune/inflammatory diseases, in the absence of RF and RA, and its presence in normal serum derived IgG negates this proposition, however; levels of galactosylation can be an index of inflammation that has diagnostic, prognostic and management potential (Alavi and Axford, 2006; Alavi and Axford, 2008). Several studies have probed the influence of the presence or absence of galactose residues on IgG-Fc structure and function, with conflicting reports. A NMR study of galactosylated (G1F, G2F) and agalactosylated (G0F) glycoforms of IgG-Fc reported the mobility of the glycan to be comparable to that of the backbone polypeptide chain, with the exception of the galactose residue on the $\alpha(1,3)$ arm, which was highly mobile; it was concluded that agalactosylation does not induce any significant change in glycan mobility and that it remains "buried" within the protein structure (Yamaguchi *et al.*, 1998). This report is consistent with binding and stability studies showing minimal differences between G0 and G2 glycoforms (Mimura *et al.*, 2000; Mimura *et al.*, 2001; Maenaka *et al.*, 2001) and crystal structures (Krapp *et al.*, 2003) of a series of truncated glycoforms of IgG-Fc. A sophisticated NMR study probed changes in local environments on the binding of soluble recombinant FcγRIII to G2F and G0F glycoforms of IgG1-Fc and reported chemical shift differences > 0.2 ppm for Lys248 and Val308 residues (Yamaguchi, *et al.*, 2006); this is a very localised change distant from the interaction site for the FcγRIIIa moiety. The finding of a changed environment for these residues is interesting since they were not predicted, from the crystal structure, to make contacts with the $\alpha(1-6)$ arm galactose residue; small perturbations for the oligosaccharide contact residues Lys246, Asp249, Thr256 were also observed. Conflicting data was reported from an earlier NMR study which concluded that glycans with $\alpha(1,6)$ galactose residues had the same relaxation time as the protein backbone, whilst in the absence of $\alpha(1-6)$ arm galactose the glycan had relaxation rates 30 times slower, indicating high mobility and freedom from interactions with the protein structure (Wormold *et al.*, 1997).

Studies of the influence of terminal galactose residues on IgG-Fc effector functions have also resulted in conflicting reports; probably because the impact of core fucosylation was not appreciated in earlier studies (Mimura *et al.*, 2000; Mimura *et al.*, 2001; Lund *et al.*, 1990; Krapp *et al.*, 2003; Walker *et al.*, 1989; Yamaguchi *et al.*, 2006; Kumpel *et al*, 1995; Groenink *et al.*, 1996; Kumpel *et al.*, 2007). Only a "slight" differences in binding/adherence to FcγRI, FcγRII or FcγRIII bearing cells was reported for highly galactosylated anti-D antibody, compared to pauci-galactosylated anti-D, however, a consistent reduction in lysis of erythrocytes mediated through FcγRIIIA on NK cells was reported (Kumpel *et al*, 1995; Groenink *et al.*, 1996; Kumpel *et al.*, 2007). No difference in binding or receptor mediated signalling though FcγRII was reported for G0 and G2 glycoforms of the anti-CD52 humanised IgG1 Campath-1H antibody (Boyd *et al.*, 1995).

As previously stated recombinant IgG antibody therapeutics secreted by mammalian cell lines adapted for commercial production are hypogalactosylated, relative to normal IgG-Fc. The possible consequences for *in vivo* activity are extrapolated from *in vitro* assays and animal experiments. Removal of terminal galactose residues from Campath-1H was shown to reduce CDC but to be without effect on FcγR mediated functions (Boyd *et al.*, 1995). Similarly, the ability of rituximab to kill tumour cells by CDC was shown to be increased by a factor of two (2) for the (G2F)2 glycoform, in comparison to the (G0F)2 glycoform (http://www.fda.gov/cder/biologics/review/ritugen112697-r2.pdf). The product that gained licensing approval was comprised of ~ 25 % of the G1F oligosaccharide, therefore, regulatory authorities required that galactosylation of the manufactured product be controlled to within a few % of this value.

THE INFLUENCE OF FUCOSE AND BISECTING N-ACETYLGLUCOSAMINE ON IgG-Fc ACTIVITIES

A comparison of the ability of antibody produced in different cell lines to mediate ADCC showed the product of rat YB2/0 cells to be more active than the product of CHO or Sp2/0 cells (Lifely *et al.*, 1995). Analysis of the oligosaccharide profiles of these antibodies showed a possible correlation of ADCC activity with the ability of YB2/0 cells to produce IgG glycoforms bearing a bisecting N-acetylglucosamine residue. This rationale appeared to be vindicated by the demonstration of increased FcγRIII mediated ADCC for antibody produced in CHO cells that had been transfected with the human β 1,4-N-acetylglucosaminyltransferase III (GnTIII) gene and shown to be producing antibody bearing bisecting N-acetylglucosamine residues (Umana *et al.*, 1999; Davies *et al*, 2001). A profound increase in FcγRIII mediated ADCC was also reported for antibody produced in a mutant CHO cell line that failed to add either bisecting GlcNAc or fucose (Shields *et al.*, 2002). Comparison of the ability of fucosylated and non-fucosylated glycoforms bearing bisecting GlcNAc of IgG antibody to mediate ADCC led to the conclusion that it is the absence of fucose rather than the presence of bisecting GlcNAc that accounts for increased FcγRIIIA mediated ADCC (Shinkawa *et al.*, 2003); the rationale being that expression of GnTIII in CHO cells with the addition of bisecting N-acetylglucosamine inhibits the endogenous α(1,6) – fucosyltransferase and the addition of fucose (Ferrara *et al.*, 2006). Transfection of CHO cells with genes for chimeric transferases that localize the GnTIII transferase to an earlier Golgi compartments resulted in increased addition of bisecting GlcNAc with increased inhibition of the addition of fucose and consequent increased ADCC (Ferrara *et al.*, 2006). It is difficult to reconcile these data with the conclusion of Davies *et al.* (2001) who credited the improved ADCC to an IgG glycoform bearing bisecting GlcNAc and fucose. A possible explanation for this discrepancy could be that only the major glycoforms were positively characterized in this study and that the increased ADCC activity observed was due to the presence of a minor but increased population of non-fucosylated glycoforms. It should be noted that the glycoform bearing bisecting GlcNAc but the absence of fucose is a minor component (< 3 %) of oligosaccharides released from normal polyclonal human IgG-Fc but may be a predominant glycoform for monoclonal myeloma IgG proteins.

Surface plasmon resonance studies show that non-fucosylated IgG-Fc binds soluble recombinant FcγRIII with higher affinity than does the fucosylated form whilst aglycosylated IgG-Fc shows no evidence of binding (Okazaki *et al.*, 2004; Niwa, *et al.*, 2005). A similar study of the binding of different glycoforms of FcγRIII showed that glycosylation at asparagine 162 of the receptor influenced IgG-Fc binding, with aglycosylation at this site resulting in increased affinity for the normal fucosylated glycoform of IgG-Fc (Drescher *et al.*, 2003; Ferrara *et al.*, 2006b). This residue is at the interface between the FcγRIIIA receptor and IgG-Fc and it was suggested that the presence of fucose on IgG-Fc might result in steric inhibition of glycosylated FcγRIIIA binding (Ferrara *et al.*, 2006b). It was concluded that high affinity IgG-Fc/FcγRIIIA binding requires an interaction of sugar residues attached at Asn 162 with surface structures of the non-fucosylated IgG-Fc glycoform and that due to the asymmetry of the IgG-Fc/FcγRIIIA interaction one non-fucosylation heavy chain, within an IgG molecule, would be sufficient for tight binding. Interestingly, it was shown that the increased affinity for the non-fucosylated glycoform of IgG-Fc was negated when FcγRIIIA was not glycosylated at Asn 162 (Ferrara *et al.*, 2006b). The presence or absence of fucose was shown not to influence the binding affinity of IgG-Fc for the inhibitory FcγRIIB receptor and it was suggested that IgG antibody glycoform might be a sensitive modulator of FcγRIIIA mediated ADCC; through tissue specific production of different Asn 162 glycoforms of FcγRIIIA (Ferrara *et al.*, 2006b; Edberg and Kimberley, 1997). Although most studies were conducted with IgG1 subclass proteins increased ADCC was also demonstrated for non-fucosylated IgG3 and IgG4 antibodies, with some activity also being observed for IgG2 (Niwa *et al.*, 2005). Increased ADCC activity was also reported for non-fucosylated glycoforms of CH1/CL deleted fusion proteins and could, presumably, be extended to ligand-IgG-Fc fusion proteins in general (Bitonti *et al*, 2006; Shoji-Hosaka *et al.*, 2006).

Non-fucosylated oligosaccharides account for ~ 10 % of those released from normal polyclonal IgG-Fc. Given random pairing between different heavy chain glycoforms a maximum of ~ 10 % of assembled IgG molecules may be anticipated to be comprised of one non-fucosylated heavy chain, with variable galactosylation. Given the asymmetry of the interaction of IgG-Fc with FcγRIII the presence of one non-fucosylated heavy chain should result in increased affinity and ADCC function (Ferrara *et al.*, 2006b). Studies of human IgG myeloma proteins, however, show that antibody producing plasma cell clones can secret predominantly non-fucosylated IgG glycoforms (Jefferis *et al.*, 1990; Farooq *et al.*, 1997). Thus, polyclonal IgG may similarly contain populations of IgG comprised of two non-fucosylated heavy chain glycoforms. The significance of fucosylation has been established from studies of monoclonal antibody therapeutics, *in vitro* (Niwa *et al.*, 2005; Natsume *et al.*, 2005; Natsume *et al.*, 2006; Susuki *et al.*, 2007; Okazaki *et al.*, 2004); however, improved functional efficacy for non-fucosylated antibody therapeutics has been reported for *ex vivo* studies (Susuki *et al.*, 2007). It has been established that IgG at normal serum concentrations can inhibit NK cell mediated ADCC; however, the increased affinity of afucosylated IgG for FcγRIII overcomes this inhibitory effect and accounts for the improvement in activity (Ferrara *et al.*, 2006, Niwa *et al.*, 2005; Natsume *et al.*, 2005; Natsume *et al.*, 2006; Susuki *et al.*, 2007; Okazaki *et al.*, 2004; Masuda *et al.*, 2007). A stable fucosyl transferase "knockout" CHO-K1 cell line has been generated and developed as a commercial vehicle for the production of non-fucosylated antibody products:

(http://www.lonza.com/group/en/company/news/archive/news_2007/strategic_collaboration.html; http://www.kyowa-kirin.co.jp/english/news/kyowa/er080715_01.html).

The foregoing discussion was developed around functionality with respect to mononuclear peripheral blood leucocytes; however, a different claim has been made for polymorphonuclear cells, namely neutrophils (Peipp *et al.*, 2008). A study employed two batches of a monoclonal antibody with specificity for epidermal growth factor receptor; one with a high fucose content and the other lower fucose content. It was reported that the batch with the higher fucose content was more active in neutrophil mediated ADCC than that with low fucose content. A contrary result is reported for neutrophil mediated phagocytosis of killed CD20⁺ lymphoma B cells with exposure to non-fucosylated rituximab resulting in enhanced phagocytosis, in comparison with exposure to the fucosylated form (Shibata-Koyama *et al.*, 2009).

FcRn

I) TRANSCYTOSIS

The FcRn receptor was first identified from studies of the transport of IgG across the gut of new-born rats and designated the neonatal Fc receptor; subsequently the human homologues was identified in human placenta and was presumed to effect transport IgG from mother to foetus (Simister, 1989; Ghetie and Ward, 2002; Nezlin and Ghetie, 2004; Roopenian and Akilesh, 2007). Interestingly, this Fc receptor is a structural homologue of MHC Class I molecules, rather than the three classes of cellular FcγR previously defined (West *et al.*, 2000). The interaction site on IgG-Fc is at the CH2/CH3 interface and the CH3 sequence -H-N-H-Y-H- (433-436) is of functional significance since titration of these histidine residues accounts for the observed binding of IgG to FcγRn at pH 6.0-6.5 and its release at pH 7.0-7.5 (Simister, 1989; Ghetie and Ward, 2002; Nezlin and Ghetie, 2004; Roopenian and Akilesh, 2007; Vaccaro, *et al.*, 2005). The interaction of IgG-Fc with FcRn appears not to be influenced by the natural glycoform profiles, or indeed the presence or absence of oligosaccharides

Each of the four human IgG subclasses are transferred across the placenta; however, with differing facility. Cord blood levels of IgG1 may be higher than in matched maternal blood whilst IgG3 and IgG4 levels are equivalent. The level of IgG2 is ~ 60 % of the concentration in maternal blood (Simister, 2003). It is of interest to note that IgG3 is transferred with equal efficacy to IgG1 although it has a shorter half-life; suggesting that its interaction with FcRn in the environment of the placenta may be different from that in endosomes in the catabolic pathway. It is also observed that during pregnancy the level of galactosylation of maternal IgG increases and that there is preferential transport of galactosylated IgG across the placenta (Williams *et al.*, 1995; Kibe *et al.*, 1996). This provides circumstantial evidence to suggest that the affinity of IgG for FcγRn may differ between glycoforms, under conditions operative at the interface between the mother and the placenta.

The long half-life of IgG antibodies is being exploited through the generation of fusion proteins, e.g. single chain Fv-Fc (scFv-Fc)$_2$, cytokine-IgG-Fc therapeutics. The presence of the IgG-Fc region contributes to improved stability, pharmacokinetics and pharmacodynamics. A further development opens a new route for administration. It has been shown that FcRn is expressed in the central and upper airways of the lung and that drug-IgG-Fc fusion proteins delivered to these sites can be transported, by transcytosis, to the systemic circulation. This is an exciting development with considerable promise and significance (Shoji-Hosaka *et al.*, 2006; Bitonti and Dumont, 2006; http://www.syntnx.com/home.php; Woodnutt *et al.*, 2008).

II) CATABOLISM

The catabolic pathway of human IgG antibodies is mediated by FcRn, which is expressed on many tissues (Simister, 1989; Ghetie and Ward, 2002; Roopenian and Akilesh, 2007; West, *et al.*, 2000; Nezlin and Ghetie, 2004; Vaccaro, *et al.*, 2005; Simister 2003). The mechanism of action is essentially that proposed, as a theoretical model, by Brambell and Hemmings (1964). It was proposed that normal cellular activity of uptake of fluid with the formation of pinocytotic vacuoles includes uptake of IgG by cells expressing FcRn. Subsequent lowering of pH results in saturation binding of IgG to FcRn and protection from degradation by enzymes present in the vacuole, unbound IgG being degraded. When the membrane of the vacuole is re-expressed as the cellular membrane the IgG bound to FcRn is exposed to pH 7.2 and dissociates into the tissue fluid. The catabolic half-life of human IgG3 is reported as ~7 days, in contrast to 21 days for IgG1, 2 & 4; however, these data have been obtained for IgG3 molecules of G3m(b) and G3m(g) allotype in which arginine replaces histidine at residue 435 (Jefferis and Lefranc, 2009) which could result in a lowered affinity for FcRn. Since IgG3 is efficiently transported across the placenta it would appear that IgG3-FcRn interactions differ when effecting placental transport or catabolism; it would be of interest to determine the catabolic half-life of IgG3 molecules of the G3m(s,t) allotype which has a histidine residue at 435 (Jefferis *et al.*, 1984; Jefferis and Lefranc, 2009). The catabolic half-life of IgG mediated through FcRn does not appear to be influenced by IgG-Fc glycoform (Nezlin and Ghetie, 2004; Roopenian and Akilesh, 2007); however, it should be emphasised that only glycoforms of IgG bearing neutral oligosaccharides have been evaluated. It may be anticipated that sialylation could influence IgG-FcRn interactions since they would introduce a negative charge in the vicinity of the histidine residues involved in FcRn binding. Protein engineering has been successfully applied to increase the affinity of IgG-Fc binding to FcRn and hence to increase the catabolic half-life (Hinton *et al.*, 2006). Prolongation of the half-life of an IgG therapeutic could translate into reduced frequency of dosing and attendance at clinic; thus reducing the cost of treatment (CoT).

There is evidence that glycoproteins expressing terminal N-acetylglucosamine residues may be cleared through the mannose receptor, although it is hypothesised that GlcNAc residues expressed on G0 forms of IgG-Fc may not be accessible (Dong *et al.*, 1999) to the receptor. Enhanced clearance of glycoforms of the IgG-Fc/TNF receptor fusion protein therapeutic (Lenercept) having exposed terminal N-acetylglucosamine residues has been reported and it was concluded, or surmised, that clearance was mediated through the mannose receptor (MR) (Jones *et al.*, 2007; Keck *et al.*, 2008). By contrast (G0F)2 glycoforms of IgG have been shown to be less susceptible to enzymatic degradation, by papain, than the (G2F)2 glycoform (Raju and Scallon, 2007); however, aglycosylated IgG is more susceptible to degradation by this enzyme (Raju and Scallon, 2006). The latter studies may not be immediately relevant to normal *in vivo* activities; however, it can serve to alert us to monitoring susceptibility to enzymatic cleavage when generating engineered forms of IgG-Fc.

Complement activation

The C1 component of complement is comprised of the three proteins C1q, C1r and C1s. Binding of immune complexes to the C1q component activates C1 to generate an enzyme that initiates the classical pathway. It is established that C1q/C1 activation is dependent on IgG-Fc glycosylation and that glyco-engineering can be employed to modulate activity (Burton, 1985; Boyd *et al.*, 1995; Woof and Burton, 2004; Nezlin and Ghetie, 2004; Jefferis, 2009; http://www.fda.gov/cder/biologics/review/ritugen112697-r2.pdf). An early study reported that complement activation by rituximab varies with the level of galactosylation (http://www.fda.gov/cder/biologics/review/ritugen112697-r2.pdf); similar observations have not been reported for other recombinant antibodies produced by Chinese hamster ovary cells, but they have been rumoured. Activation is dependent on multiple parameters that determine the nature of the immune complexes formed. Immune complexes incorporating IgG1 and/or IgG3 antibody are highly active whilst for IgG2 only complexes formed in antigen excess may be active; there is a consensus that IgG4 does not activate the classical pathway (Lucisano Valim and Lachmann, 1991; Michaelsen *et al.*, 1991; Voice and Lachmann, 1997; Aase and Michaelsen, 1994; Aalberse and Schuurman, 2001; Campbell, 2004; Bajtay *et al.*, 2006; Macor and Tedesco, 2007; van der Neut Kolfschoten *et al.*, 2007). A hybrid IgG1/IgG3 molecule has been shown to exhibit enhanced activity, relative to either IgG1 or IgG3 alone (Natsume *et al.*, 2008). Activation of C1q/C1 initiates a cascade of enzymatic reactions with cleavage of complement components to generate fragments that bind to the immune complex whilst liberating other fragments that that recruit leucocytes to augment the inflammatory response. In addition to expressing FcγR the leucocytes may also express receptors for complement fragments bound to the complex, thus further opsonising it (Bajtay *et al.*, 2006; Macor and Tedesco, 2007). The complement cascade also results in the formation of a multimeric "membrane attack complex" that inserts into cellular and bacteria walls to generate a "pore" that allows the ingress of water and consequent lysis (Podack and Deyev, 2007).

Protein engineering studies suggest that the human IgG1 interaction site for C1q/C1 is localised to the hinge proximal region of the CH2 domain (Duncan and Winter, 1988; Morgan *et al.*, 1995; Lund *et al.*, 1996; Shields *et al.*, 2001). This proposal is supported by the demonstration that replacement of the Pro 331 residue of IgG4 by serine converts it to a molecule that can activate C1; proline 331 is localised at the junction of the b6 bend and the fγ3 β-strand and is topographically proximal to histidine 268 (Aalberse and Schuurman, 2001; van der Neut Kolfschoten *et al.*, 2007). It is likely that replacement of proline by serine has a significant effect on local secondary/tertiary in the hinge proximal region. Extensive studies of mutant chimeric human IgG3 proteins have established that the efficiency of C1 activation is not directly determined by the length of the hinge region but that at least one inter-heavy chain disulphide bridge is required (Michaelsen *et al.*, 1994; van der Neut Kolfschoten *et al.*, 2007; Wang and Weiner, 2008).

The serum protein mannan binding lectin (MBL) and the cellular MR each recognises and bind arrays of GlcNAc, in addition to mannose (Nezlin and Ghetie,

2004; Malhotra *et al.*, 1995; Garred *et al.*, 2000; Arnold *et al.*, 2005; Saevarsdotti *et al.*, 2004; Arnold *et al.*, 2006). The MBL molecule is a structural homologue of C1q that forms complexes with MASP-1, MASP-2 and MASP-3 (MBL associated serine proteases) molecules that are the homologues of C1s and C1r. The MBL/MASP complexes circulates in the blood and when activated triggers the complement cascade through the initial binding and cleavage of C4, as for the C1 complex. A degalactosylated form of IgG1 was shown to bind and activate the MBL/MASP complex and initiate a consequent cascade (Malhotra *et al.*, 1995). It is possible that immune complexes comprised of G0/G0F IgG-Fc may engage and activate MBL *in vivo* such that in inflammatory diseases, characterised by increased levels of G0F IgG-Fc glycoforms, activation of the lectin pathway may contribute to and perpetuate inflammation (Garred *et al.*, 2000; Arnold *et al.*, 2005; Saevarsdotti *et al.*, 2004; Arnold *et al.*, 2006).

The mannose receptor

The mannose receptor (MR) is a C-type lectin, expressed at the surface of macrophages, endothelial and dendritic cells that recognises arrays of mannose and N-acetylglucosamine residues (Taylor *et al.*, 2003; Gazi, and Martinez-Pomares, 2009). Of particular interest is uptake by dendritic cells since they are "professional" antigen presenting cells that initiate protective immune responses, but may also be implicated in the generation of autoimmunity (Cambi and Figdor, 2003; Gazi, and Martinez-Pomares, 2009). It may also be relevant to the generation of immune responses to monoclonal antibody therapeutics; thus, the predominance of G0F glycoforms, with exposed terminal N-acetylglucosamine residues, may similarly result in immune complex uptake by dendritic cells and the presentation of non-self peptides generated from the antibody, e.g. mouse V regions, idiotypic determinants etc. The TNF receptor component of Lenercept has been shown to bear oligosaccharides having terminal N-acetylglucosamine residues and to be predisposed to clearance through the mannose receptor (Jones *et al.*, 2007; Keck *et al.*, 2008). This role for MR has been disputed and there is evidence that enhanced antigen presentation is through the processing of glycoproteins taken up by DC-SIGN, another C-type lectin molecules (Cambi and Figdor, 2003; Anthony *et al.*, 2008).

IgG-Fab glycosylation

It is established that ~ 30 % of polyclonal human IgG molecules bear *N*-linked oligosaccharides within the IgG-Fab region, in addition to the conserved glycosylation site at Asn 297 in the IgG-Fc (Youings *et al.*, 1996; Holland *et al.*, 2006; Jefferis, 2007; Jefferis, 2009). When present they are attached within the variable regions of the kappa (Vκ), lambda (Vλ) or heavy (VH) chains; sometimes both. In the immunoglobulin sequence database ~ 20 % of expressed IgG variable regions have *N*-linked glycosylation consensus sequences (Asn-X-Thr/Ser; where X can be any amino acid except proline). Interestingly, these consensus sequences are mostly not germline encoded but result from somatic hypermutation – suggestive of positive selection for improved antigen binding. Analysis of polyclonal human IgG-Fab reveals the presence of diantennary oligosaccharides that are extensively galactosylated and substantially sialylated, in contrast to the oligosaccharides released from IgG-Fc (Youings *et al.*, 1996; Holland *et al.*, 2006; Jefferis, 2007; Jefferis, 2009). This pattern was maintained for IgG-Fab prepared from hypogalactosylated IgG isolated from the sera of patients with Wegner's granulomatosis or microscopic polyangiitis (Holland *et al.*, 2006). Thus, factors within the local environment of IgG producing plasma cells, *in vivo*, influence the efficacy of glyco-processing of IgG-Fc but not IgG-Fab during passage through the Golgi apparatus. The functional significance for IgG-Fab glycosylation of polyclonal IgG has not been fully determined but data emerging for monoclonal antibodies suggests that Vκ, Vλ or VH glycosylation can have a neutral, positive or negative influence on antigen binding (Jefferis, 2007; Jefferis, 2009). The differences observed for polyclonal IgG-Fc and IgG-Fab glycoforms has been maintained for recombinant antibodies produced in CHO cells and myeloma IgG proteins (Mimura *et al.*, 2000; Mimura *et al.*, 2001; Yamaguchi *et al.*, 2006; Mimura *et al.*, 2007)

The influence of glycosylation on the thermal stability of human IgG1-Fc has been demonstrated in DSMC and X-ray crystallographic studies (Mimura *et al.*, 2000; Mimura *et al.*, 2001; Krapp *et al.*, 2003). Since it is generally observed that the oligosaccharide present in glycoproteins contributes to solubility and stability it is possible that IgG-Fab glycosylation may similarly be beneficial, particularly when formulating IgG therapeutics at concentrations of 100 – 150 mg/ml. Such high concentration formulations allow the development of self-administration protocols and can reduce dosing intervals, resulting in reduced COT. Controlling glycoform fidelity at two sites offers a further challenge to the biopharmaceutical industry.

The licensed antibody therapeutic Erbitux (cetuximab), bears an *N*-linked oligosaccharide at Asn 88 of the VH region; interestingly there is also a glycosylation consensus sequence at Asn 41 of the VL but it is not occupied (Qian *et al.*, 2007). Analysis of the IgG-Fc and IgG-Fab oligosaccharides of Erbitux, produced from Sp2/0 cells, reveal highly significant differences in composition. Whilst the IgG-Fc oligosaccharides are typical, i.e. comprised predominantly of diantennary G0F oligosaccharides, the IgG-Fab oligosaccharides are extremely heterogeneous and include complex diantennary, triantennary and hybrid oligosaccharides; non-human oligosaccharides were also present, e.g. galactose in α(1,3) linkage to galactose and *N*-glycylneuraminic acid residues.

A recent study reported that of 76 patients treated with Erbitux 25 had hypersensitivity reactions to the drug and this was shown to be due to the presence of IgE anti-gal α(1,3) gal antibodies. Interestingly, environmental factors appeared to influence the development of IgE anti-gal α(1,3) gal responses and IgE antibodies were detected in pre-treatment samples from 17 of the patients (Chung *et al.*, 2008). The incidence varied significantly between treatment centres and may be linked to differences in predominant infectious agents present in local environments.

A detailed analysis of the glycoforms of a humanised IgG rMAb bearing oligosaccharides at Asn 56 of the VH and Asn 297, also produced in Sp2/0 cells, reveals the expected IgG-Fc glycoform profile of predominantly G0F oligosaccharides; however, eleven oligosaccharides were released from the IgG-Fab, including diantennary and triantennary oligosaccharides bearing gal α(1,3) gal, *N*-glycylneuraminic acid and *N*-acetyl galactosamine residues (Huang *et al.*, 2006). The consistent observation of higher levels of galactosylation and sialylation for IgG-Fab *N*-linked oligosaccharides, in comparison to IgG-Fc, is thought to reflect increased exposure and/or accessibility.

In view of these experiences it would seem that the perceived virtues of the NS0 and Sp2/0 cells might best be pursued by engineering to inactivate the gal $\alpha(1-3)$ and *N*-glycylneuraminic acid transferases.

The double challenge to produce of rMAbs having appropriately glycosylated IgG-Fc and IgG-Fab has led to some companies engineering out VH or VL glycosylation motifs when present in candidate rMAbs (Carter *et al.*, 1992); however, present reports suggest that CHO cells can glycosylate VH and/or VL motifs in a similar manner to that observed for normal polyclonal IgG (Lim *et al.*, 2008).

IgG binding proteins produced by pathogens.

Unique host/parasite relationships have evolved that include the production, by bacteria and virus, of molecules that specifically recognise and bind host immunoglobulins (Woof and Burton, 2004; Nezlin and Ghetie, 2004; Jefferis, 2009; Deisenhofer, 1981; Padlan, 1996; Shields *et al.*, 2002; Burton, 1985; Nitsche-Schmitz *et al.*, 2007; Olsen *et al.*, 1998; Lubinski *et al.*, 1998; Atalay *et al.*, 2002; Sprague *et al.*, 2004; Rhodes and Trowsdale, 2007). These phenomena are of profound significance for an understanding of host/parasite interactions that may compromise the immune system and include:

i.   The ability of a pathogen to coat its surface with host protein in order to evade immune recognition.

ii.   The complexed surface immunoglobulin may facilitate entry into cells through Ig-Fc and/or complement receptors.

iii. The induction of virus encoded IgG-Fc receptor proteins on virus-infected cells may facilitate the uptake of virus/IgG immune complexes, thus amplifying infection.

Elucidation of the structural basis for Ig recognition could contribute to an under-standing of common and/or unique elements in these mutual recognition phenomena. Detailed structural data is available only for SpA and SpG binding to IgG-Fc (Deisenhofer, 1981; Sauer-Eriksson *et al.*, 1995) and SpA and SpG binding to a Fab fragments (Derrick *et al.*, 1994; Graille *et al.*, 2000). Inferred structural specificity is available for herpes simplex virus Fc receptor and hepatitis C virus core protein binding to IgG-Fc (Armour *et al.*, 2002; Namboodiri *et al.*, 2007) Interestingly, the bacterial and viral IgG binding protein share overlapping interaction sites on IgG-Fc, namely the inter-CH2/CH3 domain region. This is also the region of the molecule that expresses the interaction site for FcRn; therefore, when engineering IgG-Fc to modulate interactions with FcRn, and influence catabolic half-life, the consequent impact on binding other ligands should be taken into account.

Antibody production in non-mammalian systems

It will be evident that an ability to produce selected homogenous glycoforms of recombinant antibody molecules could be advantageous. Manipulation of culture conditions for mammalian cells can have a limited, but significant, influence on the glycoform profile of product (Andersen and Reilly, 2004; Pascoe, *et al.*, 2007) and may allow for manipulation of the glycoform profile over the time of a production run. Cell engineering is being undertaken in order to "knock-out" and/or "knock-in" genes encoding for selected glycosyltransferases; as illustrated above. The development of alternative production vehicles to mammalian cells has been hindered by their inability to effect human type PTMs (Walsh and Jefferis, 2006); however, systematic cell or whole organism engineering is overcoming these limitations.

PLANTS

Human protein sequences can be produced with fidelity in plants (Liénard *et al.*, 2007) but post-translational modifications present a problem, particularly for glycosylation. Plants can add human type "core" diantennary oligosaccharides but they also add $\alpha(1,3)$ fucose and $\beta(1,2)$ xylose sugar residues that are not expressed in humans and are reported to be immunogenic and, possibly, allergenic. Several strategies have been employed to overcome these disadvantages (Fitchette *et al.*, 2007; Liénard *et al.*, 2007).

Homogenous human rMAb glycoforms have been produced in the plant *Lemna minor* (duckweed) employing RNA inhibition (RNAi) of fucose and xylose transferases (Cox *et al.*, 2006). Successful double "knockout" of the $\alpha(1,3)$ fucose and $\beta(1,2)$ xylose transferases has been reported, with the production of human vascular endothelial growth factor, in *Physcomitrella patens* (moss) (Kaprivova *et al.*, 2004) and endogenous glycoproteins in *Arabidopsis thaliana* (Van Droogenbroeck *et al.*, 2007). Each of these systems may yield rMAb with enhanced ADCC due to the non-addition of $\alpha(1,6)$ fucose residues.

INSECT CELLS

The utility of the Sf2 insect cell line has been extended with the generation of the SfSWT-1 line that is transgenic for genes encoding five mammalian glycosyltransferases, including sialyl-transferases. In a further development a transgenic line was generated that can also synthesise CMP sialic acid and hence sialylate glycoproteins in the absence of an exogenous source, e.g. bovine serum (Hill *et al.*, 2006).

YEAST

Yeasts synthesise glycoproteins bearing high mannose oligosaccharides, the number of mannose residues varying between individual strains. *Pichia pastoris* has been successfully engineered to produce selected homogeneous glycoforms of human IgG-Fc (Potgieter *et al.*, 2009).

PROKARYOTES

Small protein therapeutics may be economically produced in E. coli, e.g. insulin. The production of full-length IgG antibodies has also been reported, however, they are not glycosylated and lack the potential to activate effector functions (Hamilton *et al.*, 2006). It has recently been reported that *Campylobacter jejuni* has the capacity to add oligosaccharides to proteins and that this capacity can be transferred to E. coli (Georgiou and Segatori, 2005; Langdon *et al.*, 2009. The resulting glycosylation patterns are radically different to those of mammalian cells; however, it suggests that it may be possible to engineer *E. coli* to express mammalian type oligosaccharides.

Concluding remarks

It is salutary to contemplate the influence that a conservative amino acid replacement (glycine > alanine, 235) and/or the presence or absence of a fucose sugar residue can have on the functional activity of the human IgG-Fc. It is also salutary to contemplate the multiple ligands that IgG-Fc may engage and activate – to the benefit or determent of the individual. The search for an understanding of the interactions between structure and function has called for innovative and original thinking. We now have the tools to evaluate many activities *in vitro* but the challenge to establish models and/ or modalities for determining mechanisms activated *in vivo* remains. This chapter has dealt in some detail on the impact of glycosylation on antibody function; however, our understanding is far from complete. It is interesting to note that disparate ligands may bind to the Fc through common amino acid residue contacts; within the hinge proximal region for FcγR and C1q and the CH2/CH3 interface for FcRn, SpA, SpG, RFs and FcγR encoded with the genomes of some viruses; the addition of sialic acid might further influence Fc/ligand interactions at this interface. A rationalisation for the topography of ligand binding sites may be the functional necessity for circulating IgG to be monovalent for FcγRs and C1q, to prevent continuous cellular activation whilst providing opportunity for divalency at the inter-CH2/CH3 region. Given the influence of the oligosaccharide content on functional activity it may be exploited as a "rheostat" with homogeneous glycoforms being selected for a predetermined functional profile optimal for a given disease indication. These insights into IgG therapeutic antibody interactions with IgG-Fc-ligands will allow for the development of improved protocols for their *in vivo* applications.

References

AALBERSE, R.C AND SCHUURMAN, J. (2002). IgG4 breaking the rules. *Immunology* **105**, 9-19,

AASE, A. AND MICHAELSEN, T.E. (1994). Opsonophagocytic activity induced by chimeric antibodies of the four human IgG subclasses with or without help from complement. *Scandanavian Journal of Immunology* **39**, 581.

ALAVI, A. AND AXFORD, J.S. (2006). The pivotal nature of sugars in normal physiology and disease. *Wiener Medizinische Wochenschrift* **156**, 19-33.

ALAVI, A, AND AXFORD, J.S. (2008). Glyco-biomarkers: potential determinants of cellular physiology and pathology. *Disease Markers* **25**, 193-205

ANDERSEN, D.C. AND REILLY, D.E. (2004). Production technologies for monoclonal antibodies and their fragments. *Current Opinion in Biotechnology* **15**, 456-62

ANOLIK, J.H., CAMPBELL, D., FELGAR, R.E. *ET AL*. (2003). The relationship of Fc RIIIa genotype to degree of B cell depletion by Rituximab in the treatment of systemic lupus erythematosus. *Arthritis and Rheumatism* **48**, 455-59.

ANTHONY, R.M., WERMELING, F., KARLSSON, M.C. *ET AL*. (2008). Identification of a receptor required for the anti-inflammatory activity of IVIG. *Proceedings of the National Academy of Science USA* **105**, 19571-8.

ARMOUR, K.L., ATHERTON, A., WILLIAMSON, L.M. *ET AL*. (2002). The contrasting IgG-binding interactions of human and herpes simplex virus Fc receptors. *Biochemical Society Transactions* **30**, 495-500.

ARNOLD, J.N., WORMALD, M.R., SUTER, D.M. *ET AL*. (2005a). Human serum IgM glycosylation: identification of glycoforms that can bind to mannan-binding lectin. *Journal of Biological Chemistry* **280**, 29080-87.

ARNOLD, J.N., DWEK, R.A., RUDD, P.M. *ET AL*. (2005b). Mannan binding lectin and its interaction with immunoglobulins in health and in disease. *Immunology Letters* **106**, 103-10.

ATALAY, R., ZIMMERMANN, A., WAGNER, M. *ET AL*. (2002). Identification and expression of human cytomegalovirus transcription units coding for two distinct Fcgamma receptor homologs. *Journal of Virology* **76**, 8596-08.

BAJTAY, Z., CSOMOR, E., SANDOR, N. *ET AL*. (2006). Expression and role of Fc- and complement-receptors on human dendritic cells. *Immunology Letters* **104**, 46-52

BITONTI, A.J. AND DUMONT, J.A. (2006). Pulmonary administration of therapeutic proteins using an immunoglobulin transport pathway. *Advances in Drug Delivery Reviews* **58**, 1106-18.

BOYD, P.N., LINES, A.C. AND PATEL, A.K. (1995). The effect of the removal of sialic acid, galactose and total carbohydrate on the functional activity of Campath-1H. *Molecular Immunology* **32**, 1311–1318.

BRAMBELL, F.W.R., HEMMINGS, W.A. AND MORRIS, I.G. (1964). A theoretical model of gammaglobulin catabolism. *Nature* **203**, 1352-1355.

BREKKE, O.H. MICHAELSEN, T.E. AND SANDLIE, I. (1995). The structural requirements for complement activation by IgG: does it hinge on the hinge? *Immunology Today* **16**, 85-90.

BURTON DR. (1985) Immunoglobulin G - functional sites. *Molecular Immunology* **22**, 161-206.

BURTON, D.R. AND WOOF, J.M. (1992). Human antibody effector function. *Advances in Immunology* **51**, 1-84.

CAMBI, A. AND FIGDOR, C.G. (2003). Levels of complexity in pathogen recognition by C-type lectins *Curent. Opinion in Cell Biology* **15**, 539-46.

CAMPBELL, M.P., ROYLE, L., RADCLIFFE, C.M. *et al*. (2008). Glycobase and autoGU: Tools for HPLC based glycan analysis. *Bioinformatics* **24**, 1214-6.

CARRASCO, B., GARCIA DE LA TORRE, J., DAVIS, K.G. *ET AL*. (2001). Crystallohydrodynamics for solving the hydration problem for multi-domain proteins: open physiological conformations for human IgG. *Biophysical Chemistry* **93**, 181-96

CARROLL, M.C. (2004) The complement system in regulation of adaptive immunity. *Nature Immunology* **5**, 981-6.

CARTER, P., PRESTA, L., GORMAN, C.M. *ET AL*. (1992). Humanization of an anti-p185HER2 antibody for human cancer therapy. *Proceedings of the National Academy of Science USA*. **89**, 4285-89.

CARTER, P.J. (2006). Potent antibody therapeutics by design. *Nature Reviews: Immunology* **6**, 343-357.

CARTRON, G., DACHEUX, L., SALLES, G. *ET AL.* (2002). Therapeutic activity of humanized anti-CD20 monoclonal antibody and polymorphism in IgG Fc receptor FcγRIIIa gene. *Blood* **99**, 754-758.

CHUNG, C.H., MIRAKHUR, B., CHAN, E. *ET AL.* (2008). Cetuximab-induced anaphylaxis and IgE specific for galactose-alpha-1,3-galactose. *New England Journal of Medicine* **58**, 1109-17.

CORPER, A.L., SOHI, M.K., BONAGURA, V.R. *ET AL.* (1997). Structure of human IgM rheumatoid factor Fab bound to its autoantigen IgG Fc reveals a novel topology of antibody-antigen interaction. *Nature Structural Biology* **4**, 374-81.

COX, K.M., STERLING, J.D., REGAN, J.T. *ET AL.* (2006). Glycan optimization of a human monoclonal antibody in the aquatic plant *Lemna minor*. *Nature Biotechnology* **24**, 1591-97.

COXON, P.Y., RANE, M.J., POWELL, D.W. *ET AL.* (2000). Differential mitogen-activated protein kinase stimulation by Fc gamma receptor IIa and Fc gamma receptor IIIb determines the activation phenotype of human neutrophils. *Journal of Immunology* **164**, 6530-37.

CREMATA, J. A, SORELL, L. AND MONTESINO, R. (2003). Hypogalactosylation of serum IgG in patients with coeliac disease. *Clinical and Experimental Immunology* **133**, 422-29.

CRISPIN, M., STUART, D.I. AND JONES, E.Y. (2007). Building meaningful models of glycoproteins. *Nature Structural & Molecular Biology* **14**, 354.

DASGUPTA, S., NAVARRETE, A.M., BAYRY, J. *ET AL.* (2007). A role for exposed mannosylations in presentation of human therapeutic self-proteins to CD4+ T lymphocytes. *Proceedings of the National Academy of Sciences USA* **104**, 8965-70.

DAVIES, J., JIANG, L., LABARRE, M.J. *ET AL.* (2001). Expression of GTIII in a recombinant anti-CD20 CHO production cell line: Expression of antibodies of altered glycoforms leads to an increase in ADCC thro' higher affinity for FcRIII. *Biotechnology and Bioengineering* **74**, 288-94.

DEISENHOFFER, J. (1981). Crystallographic refinement and atomic models of a human Fc fragment and its complex with fragment B of protein A from Staphylococcus aureus at 2.9- and 2.8-Å resolution. *Biochemistry* **20**, 2361-70.

DERRICK, J.P. AND WIGLEY, D.B. (1994). The third IgG-binding domain from streptococcal protein G. An analysis by X-ray crystallography of the structure alone and in a complex with Fab. *Journal of Molecular Biology* **243**, 906-18.

DONG, X., STORKUS, W.J. AND SALTER, R.D. (1999). Binding and uptake of agalactosyl IgG by mannose receptor on macrophages and dendritic cells. *Journal of Immunology* **163**, 5427-34.

DRESCHER, B., WITTE, T. AND SCHMITT, R.E. (2003). Glycosylation of FcγRIII in N163 as mechanism of regulating receptor affinity. *Immunology* **110**, 335-40.

DUBEL, S. (2007). Recombinant therapeutic antibodies. *Applied Microbiology and Biotechnology* **74**, 723-9.

DUNCAN, A.R. AND WINTER, G. (1988). The binding site for C1q on IgG. *Nature* **332**, 738-40

DUQUERROY, S., STURA, E.A., BRESSANELLI, S. *ET AL.* (2007). Crystal structure of a human autoimmune complex between IgM rheumatoid factor RF61 and IgG1 Fc reveals a novel epitope and evidence for affinity maturation. *Journal of Molecular Biology* **368**, 1321-31.

Desjarlais, J.R., Lazar, G., Zhukovsky, E.A. *et al.* (2007). Optimizing engagement of the immune system by anti-tumor antibodies: an engineer's perspective. *Drug Discovery Today* **12**, 898-910

Edberg, J.C. and Kimberly, R.P. (1997). Cell type-specific glycoforms of Fc gamma RIIIa (CD16): differential ligand binding. *Journal of Immunology* **158**, 3849-57.

Edwards, S.W., Watson, F. and Gasmi, L. (1997). Activation of human neutrophils by soluble immune complexes: role of Fc gamma RII and Fc gamma RIIIb in stimulation of the respiratory burst and elevation of intracellular Ca2+. *Annals of the New York Academy of Sciences* **832**, 341-57.

Famm, K. and Winter, G. (2006). Engineering aggregation-resistant proteins by directed evolution. *Protein Engineering Design and Selection* **19**, 479-81.

Farooq. M., Takahashi, N., Arrol, H., *et al.* (1997). Glycosylation of polyclonal and paraprotein IgG in multiple myeloma. *Glycoconjugate Journal* **14**, 489-92.

Ferrara, C., Brunker, P., Moser, S., *et al.* (2006a). Modulation of therapeutic antibody effector functions by glycosylation engineering: influence of Golgi enzyme localization domain and co-expression of heterologous beta1, 4-N-acetylglucosaminyltransferase III and Golgi alpha-mannosidase II. *Biotechnology Bioengineering* **93**, 851-61.

Ferrara, C., Stuart, F., Sondermann, P. *et al.* (2006b) The carbohydrate at FcgammaRIIIa Asn-162. An element required for high affinity binding to non-fucosylated IgG glycoforms. *Journal of Biological Chemistry* **281**, 5032-36.

Fitchette, A.C., Dinh, O.T., Faye, L., *et al.* (2007). Plant proteomics and glycosylation. *Methods in Molecular Biology* **355**, 317-42.

Gardy, J.L., Lynn, D.J., Brinkman, F.S. *et al.* (2009). Hancock RE. Enabling a systems biology approach to immunology: focus on innate immunity. *Trends in Immunology* May 8. (Epub ahead of print)

Garred, P., Madsen, H.O., Marquart, H., *et al.* (2000). Two edged role of mannose binding lectin in rheumatoid arthritis: a cross sectional study. *Journal of Rheumatology* **27**, 26–34.

Gazi, U. and Martinez-Pomares, L. (2009) Influence of the mannose receptor in host immune responses**.** *Immunobiology* Jan 20. (Epub ahead of print)

Georgiou, G. and Segatori, L. (2005). Preparative expression of secreted proteins in bacteria: status report and future prospects. *Current Opinion in Biotechnology* **16**, 538-45.

Ghetie, V. and Ward, E.S. (2002). Transcytosis and catabolism of antibody. *Immunology Research.* **25**, 97-113.

Girardi, E., Holdom, M.D., Davies, A.M., *et al.,* (2009) The crystal structure of rabbit IgG-Fc. *Biochemical Journal* **417**, 77-83.

Go, M.F., Schrohenloher, R.E. and Tomana, M. (1994). Deficient galactosylation of serum IgG in inflammatory bowel disease: correlation with disease activity. *Journal of Clinical Gastroenterology* **18**, 86-87.

Graille, M., Stura, E.A., Corper, A.L. *et al.* (2000). Crystal structure of a Staphylococcus aureus protein A domain complexed with the Fab fragment of a human IgM antibody: structural basis for recognition of B-cell receptors and superantigen activity. *Proceedings of the National Academy of Science USA* **97**, 5399-404.

GROENINK, J., SPIJKER, J., VAN DEN HERIK-OUDIJK, I.E. *ET AL*. (1996). On the interaction between agalactosyl IgG and Fc gamma receptors. *European Journal of Immunology* **26**, 1404-7.

GUDDAT, L.W., HERRON, J.N. AND EDMUNDSON, A.B. (1993). Three-dimensional structure of a human immunoglobulin with a hinge deletion. *Proceedings of the National Academy of Science USA* **90**, 4271-75.

HAMILTON, S.R., DAVIDSON, R.C., SETHURAMANN, N. *ET AL*. (2006). Humanization of yeast to produce complex terminally sialylated glycoproteins. *Science* **313**, 1441-3.

HARRIS, L.J. *ET AL*., (1997). Refined structure of an intact IgG2a monoclonal antibody. *Biochemistry* **36**, 1581-97.

HILL, D.R., AUMILLER, J.J., SHI, X. *ET AL*., (2006). Isolation and analysis of a baculovirus vector that supports recombinant glycoprotein sialylation by SfSWT-1 cells cultured in serum-free medium. *Biotechnology and Bioengineering* **95**, 37-47.

HINTON, P.R. XIONG, J.M., JOHLFS, M.G. *ET AL*. (2006). An engineered human IgG1 antibody with longer serum half-life. *Journal of Immunology* **176,** 346-56.

HOLGATE, R.G. AND BAKER, M.P. (2009). Circumventing immunogenicity in the development of therapeutic antibodies. *IDrugs* **12**, 233-7

HOLLAND, M., TAKADA, K., OKUMOTO, T. *ET AL*. (2002). Hypogalactosylation of serum IgG in patients with ANCA-associated systemic vasculitis. *Clinical and Experimental Immunology* **29**, 183-190.

HOLLAND, M., YAGI, H., TAKAHASHI, N. *ET AL*. (2006). Differential glycosylation of polyclonal IgG, IgG-Fc and IgG-Fab isolated from the sera of patients with ANCA associated systemic vasculitis. *Biochimica et Biophysica Acta* **1760**, 669-77.

HUANG, L., BIOLSI, S., BALES, K.R. *ET AL*. (2006). Impact of variable domain glycosylation on antibody clearance: An LC/MS characterization. *Analytical Biochemistry* **349**, 197-207

JEFFERIS, R., STEINITZ, M. AND NIK JAFFAR, M.I.B. (1984). VIII. A monoclonal rheumatoid factor having specificity for a discontinuous epitope determined by Histidine/ Arginine interchange at residue 435 of immunoglobulin G. *Immunology Letters* **7**, 191-94.

JEFFERIS, R., LUND, J., MIZUTANI, H. *ET AL*. (1990). Comparative study of the N-linked oligosaccharide structures of human IgG subclass proteins. *Biochemical Journal* **268**, 529-37.

JEFFERIS, R., LUND, J. AND POUND, J.D. (1998). IgG-Fc mediated effector functions: molecular definition of interaction sites for effector ligands and the role of glycosylation. *Immunological Reviews* **163**, 59-76.

JEFFERIS R. (2007). Antibody therapeutics: isotype and glycoform selection. *Expert Opinion on Biological Therapeutics* **7**:1401-13.

JEFFERIS, R. (2009a). Glycosylation as a strategy to improve antibody-based therapeutics. *Nature Reviews. Drug Discovery* **8**, 226-34

JEFFERIS, R. AND LEFRANC, M-P. (2009b) Human immunoglobulin allotypes: possible implications for immunogenicity. mAb In press.

JONES, A.J., PAPAC, D.I., CHIN, E.H., *ET AL*. (2007). Selective clearance of glycoforms of a complex glycoprotein pharmaceutical caused by terminal N-acetylglucosamine is similar in humans and cynomolgus monkeys. *Glycobiology* **17**, 529-40.

JOSTOK, T. *ET AL*., (2004). Rapid generation of functional human IgG antibodies derived from Fab-on-phage display libraries. *Journal of Immunological Methods* **289**, 65-80.

Kabelitz, D. and Medzhitov, R. (2007). Innate immunity--cross-talk with adaptive immunity through pattern recognition receptors and cytokines. *Current Opinion in Immunology* **19**, 1-3.

Kaneko, Y., Nimmerjahn, F. and Ravetch, J. (2006). Anti-inflammatory activity of immunoglobulin G resulting from Fc sialylation. *Science* **313**, 670-3.

Kaprivova, A., Stemmer, C., Altmann, F., *et al.* (2004). Targeted knockouts of *Physcomitrella* lacking plant-specific immunogenic N-glycans. *Plant Biotechnology Journal* **2**, 517-23.

Keskin, O. (2007). Binding induced conformational changes of proteins correlate with their intrinsic fluctuations: a case study of antibodies. *BMC Structural Biology* **7**, 31.

Koene, H.R., Kleijer, M., Algra, J., *et al.* (1997). FcγRIIIa-158V/F polymorphism influences the binding of IgG by natural killer cell FcγRIIIa, independently of the Fcγ RIIIa-48L/R/H phenotype. *Blood* **90**, 1109-14.

Keck, R., Nayak, N., Lerner, L., *et al.* (2008). Characterization of a complex glycoprotein whose variable metabolic clearance in humans is dependent on terminal N-acetylglucosamine content. *Biologicals* **36**, 49-60.

Kibe, T., Fujimoto, S. and Ishida C. (1996). Glycosylation and placental transport of immunoglobulin G. *Journal of Clinical Biochemistry and Nutrition* **21**, 57-63.

Kohler, G. and Milstein, C. (1975). Continuous cultures of fused cells secreting antibody of predefined specificity. *Nature* **256**, 495-97.

Koki Makabe K., Nakanishi1, T., Tsumoto, K. *et al.* (2008). Thermodynamic consequences of Mutations in Vernier Zone Residues of a Humanized Anti-human Epidermal Growth Factor Receptor Murine Antibody, 528. *Journal of Biological Chemistry* **283**, 1156-66.

Krapp, S., Mimura, Y., Jefferis, R., *et al.* (2003). Structural analysis of human IgG glycoforms reveals a correlation between oligosaccharide content, structural integrity and Fcγ-receptor affinity. *Journal of Molecular Biology* **325**, 979-89.

Kumpel, B.M., Wang, Y., Griffiths, H.L., *et al.,* (1995). The biological activity of human monoclonal IgG anti-D is reduced by beta-galactosidase treatment. *Human Antibodies and Hybridomas* **6**, 82-8

Kumpel, B.M. (2007). Efficacy of RhD monoclonal antibodies in clinical trials as replacement therapy for prophylactic anti-D immunoglobulin: more questions than answers. *Vox Sang* **93**, 99-111.

Langdon, R.H., Cuccui, J. and Wren, B.W. (2009) N-linked glycosylation in bacteria: an unexpected application. *Future Microbiology* **4**, 401-12

Liénard. D., Sourrouille, C., Gomord, V. *et al.*, (2007) Pharming and transgenic plants. *Biotechnology Annual Reviews* **13**, 115-47.

Lifely, M.R., Hale, G. and Boyse, S. (1995). Glycosylation and biological activity of CAMPATH-1H expressed in different cell lines and grown under different culture conditions. *Glycobiology* **5**, 813-22.

Lim, A., Reed-Bogan, A. and Harmon, B.J. (2008). Glycosylation profiling of a therapeutic recombinant monoclonal antibody with two N-linked glycosylation sites using liquid chromatography coupled to a hybrid quadrupole time-of-flight mass spectrometer. *Analytical Biochemistry* **375**, 163-72.

Lin, T.S., Flinn, I.W., Modali, R., *et al.* (2005). *FCGR3A* and *FCGR2A* polymorphisms may not correlate with response to Alemtuzumab in chronic lymphocytic leukemia.

Blood **105**, 289-91.

LONBERG N. (2008) Fully human antibodies from transgenic mouse and phage display platforms. *Current Opinion in Immunology* **20**, 450-9.

LU, Y., HARDING, S.E., MICHAELSEN, T.E. *ET AL*. (2007). Solution conformation of wild-type and mutant IgG3 and IgG4 immunoglobulins using crystallo-hydrodynamics: possible implications for complement activation. *Biophysical Journal* **93**, 3733-44.

LUCISANO VALIM Y.M. AND LACHMANN, P.J. (1991) The effect of antibody isotype and antigenic epitope density on the complement-fixing activity of immune complexes: a systematic study using chimaeric anti-NIP antibodies with human Fc regions. *Clinical and Experimental Immunology* **84**, 1-8.

LUBINSKI, J., NAGASHUNUMUGAN, T., FRIEDMAN, H.M. (1998). Viral interference with antibody and complement. *Seminars in Cellular Development and Biology* **9**, 329-337.

LUND, J., WINTER, G., JONES, P.T., *ET AL*. (1991). Human FcγRI and FcγRII interact with distinct but overlapping sites on human IgG. *Journal of Immunology* **147**, 2657-62.

LUND, J., TANAKA, T., TAKAHASHI, N. *ET AL*. (1990). A protein structural change in aglycosylated IgG3 correlates with loss of huFc gamma R1 and huFc gamma R111 binding and/or activation. *Molecular Immunology*. **27**, 1145-53.

LUND. J., TAKAHASHI, N., POUND, J., *ET AL*. (1996). Multiple interactions of IgG with its core oligosaccharide can modulate recognition by complement and human FcγRI and influence the synthesis of its oligosaccharide chains. *Journal of Immunology* **157**, 4963-69.

LOUIS, E.L., GHOUL, Z., VERMEIRE, S., *ET AL*. (2004). Association between polymorphism in IgG FcγIIIa coding gene and biological response to Infliximab in Crohn's disease. *Alimentary Pharmacology and Therapeutics* **19**, 511-19.

MACOR, P. AND TEDESCO, F. (2007). Complement as effector system in cancer immunotherapy. *Immunology Letters* **111**, 6-13

MAENAKA, K., VAN DER MERWE, P.A., STUART, D.I. *ET AL*. (2001). The human low affinity Fcgamma receptors IIa, IIb, and III bind IgG with fast kinetics and distinct thermodynamic properties. *Journal of Biological Chemistry* **276**, 44898-904.

MAGDELAINE-BEUZELIN, C., GOODALL, M., JEFFERIS, R. *ET AL*. (2009). IgG1 heavy chain-coding gene polymorphism, marker of G1m allotypes, and development of antibodies-to-Infliximab. *Pharmacogenetics and Genomics* **19**, 383-7.

MALHOTRA, R., WORMALD, M.R., RUDD, P.M. *ET AL*. (1995). Glycosylation changes of IgG associated with rheumatoid arthritis can activate complement via the mannose-binding protein. *Nature Medicine* **1**, 237-43.

MAKADE, K., *(AU: 3 AUTHORS BEFORE ET AL)* *ET AL*. (2008). Thermodynamic consequences of mutations in vernier zone residues of a humanized anti-human epidermal growth factor receptor murine antibody, 528. *Journal of Biological Chemistry* 283, 1156-66.

MARQUART, M., DEISENHOFER, J., HUBER, R. *ET AL*. (1980). Crystallographic refinement and atomic models of the intact immunoglobulin molecule Kol and its antigen-binding fragment at 3.0 Å and 1.0 Å resolution. *Journal of Molecular Biology* **141**, 369-91.

MASUDA, K., YAMAGUCHI, Y., KATO, K. *ET AL*., (2000). Pairing of oligosaccharides in the Fc region of immunoglobulin G. *FEBS Letters* **473**, 349-57.

MASUDA, K., KUBATO, T., KANEKO, E. *ET AL.* (2007). Enhanced binding affinity for FcgammaRIIIa of fucose-negative antibody is sufficient to induce maximal antibody-dependent cellular cytotoxicity. *Molecular Immunology* **44**, 3122-31.

MATSUMIYA, S., YAMAGUCHI, Y., SAITO, J. *ET AL.* (2007). Structural Comparison of Fucosylated and Nonfucosylated Fc Fragments of Human Immunoglobulin G1. *Journal of Molecular Biology* **368**, 767-79

MICHAELSEN TE, BREKKE OH, AASE A. *ET AL.* (1994). One disulfide bond in front of the second heavy chain constant region is necessary and sufficient for effector functions of human IgG3 without a genetic hinge. *Proceedings of the National Academy of Science USA*. **91**, 9243-47.

MICHAELSEN, T.E., GARRED, P. AND AASE, A. (1991). Human IgG subclass pattern of inducing complement-mediated cytolysis depends on antigen concentration and to a lesser extent on epitope patchiness, antibody affinity and complement concentration. *European Journal of Immunology* **21**, 11-6.

MIESCHER, S., SPYCHER, M.O., AMSTUTZ, H. *ET AL.* (2004). A single recombinant anti-RhD IgG prevents RhD immunization: association of RhD-positive red blood cell clearance rate with polymorphisms in the FcγRIIA and FcγIIIA genes. *Blood* **103**, 4028-35.

MIMURA, Y., ASHTON, P.R., TAKAHASHI, N. *ET AL.* (2007). Contrasting glycosylation profiles between Fab and Fc of a human IgG protein studied by electrospray ionization mass spectrometry. *Journal of Immunological Methods* **326**, 116-26.

MIMURA Y., CHURCH S., GHIRLANDO R. *ET AL.* (2000). The influence of glycosylation on the thermal stability and effector function expression of human IgG1-Fc: properties of a series of truncated glycoforms. *Molecular Immunology* **37**, 697-706.

MIMURA, Y., ASHTON, P.R., TAKAHASHI, N., *ET AL.* (2007). Contrasting glycosylation profiles between Fab and Fc of a human IgG protein studied by electrospray ionization mass spectrometry. *J Immunol Methods* **326**, 116-26.

MIMURA, Y., SONDERMANN, P., GHIRLANDO, R. *ET AL.* (2001). The role of oligosaccharide residues of IgG1-Fc in FcγIIb binding. *Journal of Biological Chemistry* **276**, 45539-47.

MORGAN, A., JONES, N.D., NESBITT, A.M. *ET AL.* (1995). The N-terminal end of the C_H2 domain of chimeric human IgG1 anti-HLA-DR is necessary for C1q, FcγRI and FcγRII binding. *Immunology* **86**, 319-24.

MOUTEL, S. AND PEREZ, F. (2008). Antibodies--Europe. Engineering the next generation of antibodies. *Biotechnology Journal* **3**, 298-300.

NAMBOODIRI, A.M., BUDKOWSKA, A., NIERTERT, P.J. *ET AL.* (2007). Fc gamma receptor-like hepatitis C virus core protein binds differentially to IgG of discordant Fc (GM) genotypes. *Molecular Immunology* **44**, 3805-08.

NATSUME, A., WAKITANI, M., YAMANE-OHNUKI, N. *ET AL.* (2005). Fucose removal from complex-type oligosaccharide enhances the antibody-dependent cellular cytotoxicity of single-gene-encoded antibody comprising a single-chain antibody linked the antibody constant region. *Journal of Immunological Methods* **306**, 93-103.

NATSUME, A., WAKITANI, M., YAMANE-OHNUKI, N. *ET AL.* (2006). Fucose removal from complex-type oligosaccharide enhances the antibody-dependent cellular cytotoxicity of single-gene-encoded bispecific antibody comprising of two single-chain antibodies linked to the antibody constant region. *Journal of Biochemistry* **140**, 359-368.

NATSUME, A., IN, M., TAKAMURA, H. *ET AL.* (2008). Engineered antibodies of IgG1/IgG3

The antibody paradigm 37

mixed isotype with enhanced cytotoxic activities. *Cancer Research* **68**, 3863-72.

NEZLIN, R. AND GHETIE, V. (2004). Interactions of immunoglobulins outside the antigen-combining site. *Advances in Immunology* **82**, 155-215.

NIMMERJAHN, F. AND RAVETCH, J. (2007). Fc-receptors as regulators of immunity. *Advances in Immunology* **96**,179-204.

NIMMERJAHN, F. AND RAVETCH, J. (2008). Fcgamma receptors as regulators of immune responses. *Nature Reviews in Immunology* **8**, 34-47.

NITSCHE-SCHMITZ, D.P., JOHANSSON, H.M., SASTALLA, I. *ET AL.* (2007). Group G streptococcal IgG binding molecules FOG and protein G have different impacts on opsonization by C1q. *Journal of Biological Chemistry* **282**, 17530-36.

NIWA, R., NATSUME, A., UEHARA, A. *ET AL.* (2005). IgG subclass-independent improvement of antibody-dependent cellular cytotoxicity by fucose removal from Asn297-linked oligosaccharides. *Journal of Immunological Methods* **306**, 151-60.

NOVAK, J., TOMANA, M., SHAH, G.R. *ET AL.* (2005). Heterogeneity of IgG glycosylation in adult periodontal disease. *Journal of Dental Research* **84**, 897-901.

OKAZAKI, A., SHOJI-HOSAKA, E., NAKAMURA, K. (2004). Fucose depletion from human IgG1 oligosaccharide enhances binding enthalpy and association rate between IgG1 and FcgammaRIIIa. *Journal of Molecular Biology* **336**, 1239-49.

OLSEN, JK., SANTOS, R.A., GROSE, C. (1998). Varicella-zoster virus glycoprotein gE: endocytosis and trafficking of the Fc receptor. *Journal of Infectious Disease* **178**, Suppl 1:S2-6.

PADLAN, E. A. (1990). Fc receptors and the action of antibodies. Ed. H. Metzger H, pp 12-30. Washington D.C.: *American Society for Microbiology.*

PADLAN, E.A. (1996). X-ray crystallography of antibodies. *Advances in Protein Chemistry* **49**, 57-133.

PAREKH, R.B., DWEK, R.A. AND SUTTON, B.J. (1985). Association of rheumatoid arthritis and primary osteoarthritis with changes in the glycosylation pattern of total serum IgG. *Nature* **316**, 452-57.

PASCOE, D.E., ARNOTT, D., PAPOUTSAKIS, E.T. *ET AL.* (2007). Proteome analysis of antibody-producing CHO cell lines with different metabolic profiles. *Biotechnology and Bioengineering* **98**, 391-410

PEIPP, M., LAMMERTS VAN BUEREN, J.J., SCHNEIDER-MERCK, T. (2008). Antibody fucosylation differentially impacts cytotoxicity mediated by NK and PMN effector cells. *Blood* 112, 2390-9.

PODACK, E.R., DEYEV, V. AND SHIRATSUCHI, M. (2007) Pore formers of the immune system. Advances in Experimental Medicine and Biology 598, 325-41.

POTGIETER, T.I., CUKAN, M., DRUMMOND, J.E. (2009). Production of monoclonal antibodies by glycoengineered Pichia pastoris. *Journal of Biotechnology* **139**, 318-25.

PRESTA, L. (2008). Molecular engineering and design of therapeutic antibodies. *Current Opinion in Immunology* **20**, 460-70.

PRITSCH, O., HUDRY-CLERGEON, H., BUCKLE, H. *ET AL.* (1996). Can Immunoglobulin C_H1 Constant Region Domain Modulate Antigen Binding Affinity of Antibodies? *Journal of Clinical Investigation* **98**, 2235-43

QIAN, J., LIU, T., YANG, L. *ET AL.* (2007). Structural characterization of N-linked oligosaccharides on monoclonal antibody cetuximab by the combination of orthogonal matrix-assisted laser desorption/ionization hybrid quadrupole-quadrupole time-of-flight tandem mass spectrometry and sequential enzymatic digestion.

Analytical Biochemistry **364,** 8-18

Radaev, S., Motyka, S., Fridman, W.H. *et al.* (2001). The structure of human type FcγIII receptor in complex with Fc. *Journal of Biological Chemistry* **276,** 16469-77.

Radaev, S. and Sun, P. (2002). Recognition of immunoglobulins by Fc gamma receptors. *Molecular Immunology* **38,** 1073-83.

Radaev, S. and Sun, P.D. (2001). Recognition of IgG by Fcgamma receptor. The role of Fc glycosylation and the binding of peptide inhibitors. *Journal of Biological Chemistry* **276,** 16478-83.

Radstake, T.R., Svenson, M., Eijsbouts, A.M. (2008). Formation of antibodies against infliximab and adalimumab strongly correlates with functional drug levels and clinical responses in rheumatoid arthritis. *Annals of Rheumatic Diseases* 2008 Nov 19. (Epub ahead of print)

Raju, T.S. and Scallon, B. (2007). Fc glycans terminated with N-acetylglucosamine residues increase antibody resistance to papain. *Biotechnology Progress* **23,** 964-71.

Raju, T.S. and Scallon, B. (2006). Glycosylation in the Fc domain of IgG increases resistance to proteolytic cleavage by papain. *Biochemical Biophysical Research Communications* **341,** 797-803.

Redpath, S., Michaelsen, T.E., Sandlie, I. *et al.* (1998). The influence of the hinge region length in binding of human IgG to human Fcgamma receptors. *Human Immunology* **59,** 720-7

Reichert, J.M. and Valge-Archer, V.E. (2007). Development trends for monoclonal antibody cancer therapeutics. *Nature Reviews: Drug Discovery* **6,** 349-56.

Rhodes, D.A. and Trowsdale, J. (2007). TRIM21 is a trimeric protein that binds IgG Fc via the B30.2 domain. *Molecular Immunology* **44,** 2406-14

Rodriguez, M.E., van der Pol, W-L., Sanders, L.A.M. (1999). Crucial role of FcγRIIa (CD32) in assessment of functional anti-*Streptococcus pneumoniae* antibody activity in human sera. *Journal of Infectious Diseases* **179,** 423-33.

Roopenian, D.C. and Akilesh, S. (2007). FcRn: the neonatal Fc receptor comes of age. *Nature Reviews: Immunology* **7,** 715-25.

Roux, K.H., Strelets, L. and Michaelsen, T.E. (1997). Flexibility of human IgG. subclasses. *Journal of Immunology* **159,** 3372-82.

Routier, F.H., Hounsell, E.F., Rudd, P.M. *et al.* (1998). Quantitation of human IgG glycoforms isolated from rheumatoid sera: a critical evaluation of chromatographic methods. *Journal of Immunological Methods* **213,** 113-30.

Rudd, P.M. and Dwek, R.A. (1997). Glycosylation: heterogeneity and the 3D structure of proteins. *Critical Reviews in Biochemistry and Molecular Biology* **32,** 1-100.

Saevarsdottir, S., Vikingsdottir, T. and Valdimarsson, H. (2004). The potential role of mannan-binding lectin in the clearance of self-components including immune complexes. *Scandanavian Journal of Immunology* **60,** 23-9.

Sanguineti, S., Centeno, Crowley, J.M. *et al.* (2007). Specific recognition of a DNA immunogen by its elicited antibody. *Journal of Molecular Biology* **370,** 183-95.

Sagawa, T., Oda, M., Morii, H. *et al.* (2005). Conformational changes in the antibody constant domains upon hapten-binding. *Molecular Immunology* **42,** 9-18.

Saphire, E.O., Stanfield, R.L., Crispin, M.D. *et al.* (2002). Contrasting IgG structures reveal extreme asymmetry and flexibility. *Journal of Molecular Biology* **319,** 9-18.

Satoh, M., Iida, S. and Shitara, K. (2006). Non-fucosylated therapeutic antibodies as next-generation therapeutic antibodies. *Expert Opinion on Biological Therapy* **6**, 1161-73.

SAUER-ERIKSSON, A.E., KLEYWEGT, G.J., UHL, M. *ET AL.* (1995). Crystal structure of the C2 fragment of streptococcal protein G in complex with the Fc domain of human IgG. *Structure* **3**, 265-278.

SCALLON, B., TAM, S.H., McCARTHY, S.G. *ET AL.* (2007). Higher levels of sialylated Fc glycans in immunoglobulin G molecules can adversely impact functionality. *Molecular Immunology* **44**, 1524-34

SHIBATA-KOYAMA, M., IIDA, S., MISAKA, H. *ET AL.* (2009). Nonfucosylated rituximab potentiates human neutrophil phagocytosis through its high binding for FcgammaRIIIb and MHC class II expression on the phagocytotic neutrophils. *Experimental Hematology* **37**, 309-21

SHIELDS, R.L., NAMENUK, A.K., HONG, K. *ET AL.* (2001). High resolution mapping of the binding site on human IgG1 for FcγRI, FcγRII, FcγRIII, and FcRn and design of IgG1 variants with improved binding to the FcγR. *Journal of Biological Chemistry* **276**, 6591-604.

SHIELDS, R.L., LAI, J., KECK, R. *ET AL.* (2002). Lack of Fucose on Human IgG1 N-Linked oligosaccharide improves binding to human FcγRIII and Antibody-dependent cellular toxicity. *Journal of Biological Chemistry* **277**, 26733-40.

SHINKAWA, T, NAKAMURA, K, YAMANE, N, *ET AL.*(2003) The absence of fucose but not the presence of galactose or bisecting N-acetylglucosamine of human IgG1 complex-type oligosaccharides shows the critical role of enhancing antibody-dependent cellular cytotoxicity. *Journal of Biological Chemistry* **278**, 3466-73.

SHOJI-HOSAKA, E., KOBAYASHI, Y., WAKITANI, Y. *ET AL.* (2006). Enhanced Fc-dependent cellular cytotoxicity of Fc fusion proteins derived from TNF receptor II and LFA-3 by fucose removal from Asn-linked oligosaccharides. *Journal of Biochemistry (Tokyo)* **140**, 777-83.

SIBERIL, S., DUTERTRE, C.A. AND FRIDMAN, W.H. (2007). The key to optimize therapeutic antibodies? *Critical Reviews in Oncology and Hematology* **62**, 26-33.

SIMISTER, N.E. AND MOSTOV, K.E. (1989). An Fc receptor structurally related to MHC class I antigens. *Nature* **337**, 184-7

SIMISTER, N.E. (2003). Placental transport of immunoglobulin G. *Vaccine* **21**, 3365-69.

SOHI, M.K., CORPER, A.L., WAN, T. *ET AL.* (1996). Crystallization of a complex between the Fab fragment of a human immunoglobulin M (IgM) rheumatoid factor (RF-AN) and the Fc fragment of human IgG4. *Immunology* **88**, 636-41.

SONDERMANN, P., HUBER, R., OOSTHUIZEN, V. *ET AL.* (2000). The 3.2-A crystal structure of the human IgG1 Fc -FcγRIII complex. *Nature* **406**, 267-273.

SONDERMANN, P., KAISER, J. AND JACOB, U. (2001). Molecular basis for immune complex recognition: a comparison of Fc-receptor structures. *Journal of Molecular Biology* **309**, 737-49

SPRAGUE, E.R., MARTIN, W.L. AND BJORKMAN, P.J. (2004). pH dependence and stoichiometry of binding to the Fc region of IgG by the herpes simplex virus Fc receptor gE-gI. *Journal of Biological Chemistry* **279**, 14184-93.

STANFIELD, R.L., ZEMLA, A., WILSON, I.A. *ET AL.* (2006). Antibody elbow angles are influenced by their light chain class. *Journal of Molecular Biology* **357**, 1566-74.

SUZUKI, E., NIWA, R., SAJI, S. *ET AL.* (2007). A nonfucosylated anti-HER2 antibody augments antibody-dependent cellular cytotoxicity in breast cancer patients. *Clinical Cancer Research* **13**, 1875-82.

Taylor, M.E. and Drickamer, K. (2003). Structure-function analysis of C-type animal lectins. *Methods in Enzymology* **363**, 3-16.

Umana, P., Jean-Mairet, J., Moudry, R. *et al.* (1999). Engineered glycoforms of an anti-neuroblastoma. IgG1 with optimized antibody-dependent cellular cytotoxic activity. *Nature Biotechnology* **17**, 176-80.

Vaccaro, C., Zhou, J., Ober, R.J. *et al.* (2005). Engineering the Fc region of immunoglobulin G to modulate in vivo antibody levels. *Nature Biotechnology* **23**, 1283-88.

van der Neut Kolfschoten, M., Schuurman, J., Losen, M. *et al.* (2007). Anti-Inflammatory Activity of Human IgG4 Antibodies by Dynamic Fab Arm Exchange. *Science* **317**, 1554 –7

Van Droogenbroeck, B., Cao, J., Stadlmann, J. *et al.* (2007). Aberrant localization and underglycosylation of highly accumulating single-chain Fv-Fc antibodies in transgenic Arabidopsis seeds. *Proceedings of the National Academy of Science USA* **104**, 1430.

Voice, J.K. and Lachmann, P.J. (1997). Neutrophil Fc gamma and complement receptors involved in binding soluble IgG immune complexes and in specific granule release induced by soluble IgG immune complexes. ***European Journal of Immunology*** **27**, 2514-23

Walker, M.R., Lund, J., Thompson, K.M. *et al.* (1989). Aglycosylation of human IgG1 and Igg3 monoclonal antibodies can eliminate recognition by human cells expressing Fc gamma RI and/or Fc gamma RII receptors. *Biochemical Journal* **259**, 347-53.

Walsh, G. and Jefferis, R. (2006). Post-translational modifications in the context of therapeutic proteins. *Nature Biotechnology* 24, 1241-52.

Wang, S.Y. and Weiner, G. (2008). Complement and cellular cytotoxicity in antibody therapy of cancer. *Expert Opinion on Biological Therapy* **8**, 759-68.

Warmerdam, P.A.M., van de Winkel, J.G.J., Vlug, A. *et al.* (1991). A single amino acid in the second IgG like domain of the human Fcγ receptor II is critical for human IgG2 binding. *Journal of Immunology* **147**, 1338–43.

Webster, A.C., Pankhurst, T., Rinaldi, F. *et al.* (2006). Monoclonal and polyclonal antibody therapy for treating acute rejection in kidney transplant recipients: a systematic review of randomized trial data. *Transplantation* **81**, 953-65.

West, A..P. and Bjorkman, P.J. (2000). Crystal structure and immunoglobulin G binding properties of the human major histocompatibility complex-related Fc receptor. *Biochemistry* **39**, 9698-708.

Williams, P.J., Arkwright, P.D. and Rudd, P.M. (1995). Short communication: Selective transport of maternal IgG to the foetus. *Placenta* **16**, 749-56.

Woof, J.M., Partridge, L.J., Jefferis, R. *et al.* (1986). Localisation of the monocyte binding region on immunoglobulin G. *Molecular Immunology* **23**, 319-30.

Woof, J.M. and Burton, D.R. (2004). Human antibody-Fc receptor interactions illuminated by crystal structures. *Nature Reviews: Immunology* **4**, 89-99.

Wormald, M.R., Rudd, P.M., Harvey, D.J. *et al.* (1997). Variations in oligosaccharide-protein interactions in immunoglobulin G determine the site-specific glycosylation profiles and modulate the dynamic motion of the Fc oligosaccharides. *Biochemistry* **36**, 1370-80.

WOODNUTT, G., VIOLAND, B. AND NORTH, M. (2008) Advances in protein therapeutics. *Current Opinion in Drug Discovery and Development* **11(6)**, 754-61

WUHRER, M., PORCELIJN, L., KAPUR, R. *ET AL.* (2009). Regulated glycosylation patterns of IgG during alloimmune responses against human platelet antigens. Journal of Proteome Research 8, 450-6

YAMADA, E., TSUKAMOTO, Y., SASAKI, R. *ET AL.* (1997). Structural changes of immunoglobulin G oligosaccharides with age in healthy human serum. *Glycoconjugate Journal* **14**, 401–5.

YAMAGUCHI, Y., KATO, K., SHINDO, M. *ET AL.* (1990). Dynamics of the carbohydrate chains attached to the Fc portion of immunoglobulin G as studied by NMR spectroscopy assisted by selective 13C labeling of the glycans. *Journal of Biomolecular NMR* **12**, 385-94.

YAMAGUCHI, Y., NISHIMURA, M., NAGANO, M. *ET AL.* (2006). Glycoform-dependent conformational alteration of the Fc region of human immunoglobulin G1 as revealed by NMR spectroscopy. *Biochimica et Biophysica Acta* **1760**, 693-700.

YOUINGS, A., CHANG, S.C., DWEK, R.A. *ET AL.* (1996). Site-specific glycosylation of human immunoglobulin G is altered in four rheumatoid arthritis patients. *Biochemical Journal* **314**, 621-30.

Biotechnology and Genetic Engineering Reviews - Vol. 26, 43-64 (2009)

Transgenic plants and associated bacteria for phytoremediation of chlorinated compounds

BENOIT VAN AKEN [1,*] AND SHARON LAFFERTY DOTY [2]

[1] *Department of Civil and Environmental Engineering, Temple University, Philadelphia, PA.* [2] *College of Forest Resources, University of Washington, Seattle, WA*

Abstract

Phytoremediation is the use of plants for the treatment of environmental pollution, including chlorinated organics. Although conceptually very attractive, removal and biodegradation of chlorinated pollutants by plants is a rather slow and inefficient process resulting in incomplete treatment and potential release of toxic metabolites into the environment. In order to overcome inherent limitations of plant metabolic capabilities, plants have been genetically modified, following a strategy similar to the development of transgenic crops: genes from bacteria, fungi, and mammals involved in the metabolism of organic contaminants, such as cytochrome P-450 and glutathione

---

To whom correspondence may be addressed (bvanaken@temple.edu)

Abbreviations: ABTS, 2,2-azino-*bis*(3-ethylbenzothiazoline-6-sulphonic acid); BPA, bisphenol A; BTEX, benzene, toluene, ethylbenzene, xylene; *c*DCE, *cis*-dichloroethylene; cDNA, complementary DNA; CDNB, 1-chloro-2,4-dinitrobenzene; CERCLA, Comprehensive Environmental Response, Compensation, and Liability Act; 2-CP, 2-chlorophenol; 4-CP, 4-chlorophenol; CT, carbon tetrachloride; 2,4-DCP, 2,4-dichlorophenol; 2,6-DCP, 2,6-dichlorophenol; DDT, 1,1,1-trichloro-2,2-*bis*-(4'-chlorophenyl) ethane; DNAPL, dense non-aqueous phase liquid; ECS, γ-glutamylcysteine synthetase; EDB, ethylene dibromide; 2,4-D, 2,4-dichlorophenoxyacetic acid; EPA, Environmental Protection Agency; GMO, genetically modified organism; GS, glutathione synthetase; GSH, reduced glutathione; GST, glutathione *S*-transferase; HPLC, high pressure liquid chromatography; ISR, induced systemic resistance; log K_{ow}, octanol-water partition coefficient; MnP, manganese-peroxidase; PAH, polyaromatic hydrocarbon; PCB, polychlorinated biphenyl; PCE, tetrachloroethene; PCP, pentachlorophenol; *pcp*, pentachlorophenol monooxygenase; ppb, part per billion; PETN, pentaerythritol tetranitrate; TCAA, trichloroacetic acid; TCC, tetrachlorocatechol; TCE, trichloroethylene; TeCA, 1,1,2,2-tetrachloroethane; TCEOH, trichloroethanol; 2,4,5-TCP, 2,4,5-trichlorophenol; 2,4,6-TCP, 2,4,6-trichlorophenol; TNT, 2,4,6-trinitrotoluene; TOM, toluene *o*-monooxygenase; VC, vinyl chloride.

S-transferase, have been introduced into higher plants, resulting in significant improvement of tolerance, removal, and degradation of pollutants. Recently, plant-associated bacteria have been recognized playing a significant role in phytoremediation, leading to the development of genetically modified rhizospheric and endophytic bacteria with improved biodegradation capabilities. Transgenic plants and associated bacteria constitute a new generation of genetically modified organisms for efficient and environmental-friendly treatment of polluted soil and water. This review focuses on recent advances in the development of transgenic plants and bacteria for the treatment of chlorinated pollutants, including chlorinated solvents, polychlorinated phenols, and chlorinated herbicides.

Introduction

Phytoremediation is an emerging technology that uses plants and associated bacteria for the treatment of soil and groundwater contaminated by toxic pollutants. Phytoremediation was first developed for the removal of toxic metals. Although metals are not susceptible to biodegradation, they can be efficiently absorbed by the roots through natural mineral uptake mechanisms (Salt *et al.*, 1998). The concept of using plants for remediation of organic pollutants emerged a few decades ago with the recognition that plants were capable of metabolizing toxic pesticides (Cole, 1983). Since then, phytoremediation has been shown to efficiently reduce chemical hazards associated with various classes of organic pollutants, including chlorinated compounds, pesticides, explosives, and polycyclic aromatic hydrocarbons (Schnoor *et al.*, 1995; Salt *et al.*, 1998; Macek *et al.*, 2000; Meagher, 2000; Pilon-Smits, 2005).

Even though phytoremediation has today acquired the status of a proven technology for the treatment of heavy metal and organic-contaminated soil, it also suffers serious limitations that impair further development of field applications (Schnoor *et al.*, 1995; Salt *et al.*, 1998). As autotrophic organisms, plants usually lack the enzymatic machinery necessary for efficient metabolism of organic compounds, often resulting in slow removal and incomplete degradation of contaminants (Eapen *et al.*, 2007). Inherent plant limitations for the metabolism of xenobiotic compounds led to the idea of modifying plants genetically by the introduction of bacterial or mammalian genes involved in the degradation of toxic chemicals, following a strategy that has been applied for decades with transgenic crops expressing resistance against insects or pesticides (Dietz and Schnoor, 2001). Even though transgenic plants have not been used for field applications of phytoremediation, successful plant transformation protocols and increasing need for affordable remediation strategies will likely help overcome legal and technical barriers currently limiting the technology. Transgenic plants and associated bacteria for phytoremediation constitute a new generation of genetically modified organisms (GMO) for efficient and environmental friendly decontamination of soil and water (Macek *et al.*, 2008).

The use of transgenic plants for phytoremediation has been covered recently in several reviews (Cherian and Oliveira, 2005; Pilon-Smits, 2005; Eapen *et al.*, 2007; Doty, 2008; Macek *et al.*, 2008; James and Strand, 2009; Kawahigashi, 2009; Van Aken, 2009). The present article summarizes new developments of plants engineered for enhancing remediation of chlorinated compounds, which constitute a major class of soil and groundwater contaminants in the U.S. (Ajo-Franklin *et al.*, 2006; Cooper and Jones, 2008). As a new trend of phytoremediation, the use of transgenic bacteria associated with plants will also be presented (Cherian and Oliveira, 2005; Eapen *et al.*, 2007; Doty, 2008; Weyens *et al.*, 2009b).

Phytoremediation: cleaning up pollution with plants

Living organisms are commonly exposed to natural or xenobiotic toxic chemicals. As a consequence, they have developed multiple detoxification mechanisms to prevent harmful effects from exposure to these compounds. Bacteria, more than higher life forms, are extremely versatile organisms, which allow them to constantly develop new metabolic pathways for the degradation of a large range of xenobiotic pollutants (Limbert and Betts, 1996). While provided with lower adaptation capabilities, higher organisms, such as plants and mammals, also possess detoxification mechanisms to counteract the harmful effects of toxic contaminants (Sandermann, 1994).

Bioremediation exploits the natural capability of living organisms to degrade toxic chemicals. Traditional remediation technologies of polluted sites requires soil excavation and transport, prior to treatment by incineration, landfilling or compositing, which is costly, damaging for the environment, and, in many cases, practically infeasible due to the range of the contamination (Gerhardt *et al.*, 2009). There is therefore a considerable interest in developing cost-effective alternatives based on microorganisms or plants. Because of its potential for the sustainable mitigation of environmental pollution, bioremediation has been recently listed among the 'top ten technologies for improving human health' (Daar *et al.*, 2002).

Plant-mediated bioremediation of metal-contaminated soil was proposed first in the 1970s (Brooks *et al.*, 1977). Because of their natural capability to absorb minerals, plants efficiently remove heavy metals from polluted soils. However, unlike organic pollutants, elemental metals cannot be 'degraded'. They typically accumulate in plant tissues, which have to be harvested and safely disposed, a process termed *phytoextraction* (Salt *et al.*, 1998).

Although the metabolism of xenobiotics by plants has been known for a long time (Castelfranco *et al.*, 1961), the idea that plants can be used to detoxify organic compounds emerged in the 1980s with the metabolism of 1,1,1-trichloro-2,2-*bis*-(4'-chlorophenyl) ethane (DDT) and benzo(*a*)pyrene (Cole, 1983). Since then, phytoremediation acquired the status of a proven technology for the remediation of soil and groundwater contaminated by a variety of organic compounds, including pesticides, chlorinated solvents, polyaromatic hydrocarbons (PAHs), and explosives (Schnoor *et al.*, 1995; Salt *et al.*, 1998). Phytoremediation has been extensively reviewed in the literature (Schnoor *et al.*, 1995; Salt *et al.*, 1998; Macek *et al.*, 2000; Meagher, 2000; Pilon-Smits, 2005).

Phytoremediation encompasses a range of processes beyond direct plant uptake and metabolism, and it is best described as plant-mediated remediation (Schnoor *et al.*, 1995; Salt *et al.*, 1998; Macek *et al.*, 2000; McCutcheon and Schnoor, 2003). While definitions and terminology vary, the different phytoremediation processes can be summarized as in *Figure 1*: pollutants in soil and groundwater can be taken up inside plant tissues (*phytoextraction*) or adsorbed to the roots (*rhizofiltration*); pollutants inside plant tissues can be transformed by plant enzymes (*phytotransformation*) or can volatilize into the atmosphere (*phytovolatilization*); pollutants in soil can be degraded by microbes in the root zone (*rhizosphere bioremediation*) or incorporated to soil material (*phytostabilization*) (Salt *et al.*, 1998; Dietz and Schnoor, 2001; Mc-Cutcheon and Schnoor, 2003; Pilon-Smits, 2005).

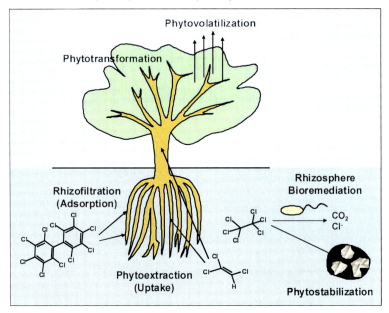

Figure 1. Phytoremediation involves several processes: Pollutants in soil and groundwater can be taken up inside plant tissues (*phytoextraction*) or adsorbed to the roots (*rhizofiltration*); pollutants inside plant tissues can be transformed by plant enzymes (*phytotransformation*) or can volatilize into the atmosphere (*phytovolatilization*); pollutants in soil can be degraded by microbes in the root zone (*rhizosphere bioremediation*) or incorporated to soil material (*phytostabilization*) (Salt *et al.*, 1998; Dietz and Schnoor, 2001; Pilon-Smits, 2005). Source: Van Aken (2008).

Being exposed to a variety of natural allelochemicals and xenobiotics, plants have developed diverse detoxification mechanisms of organic compounds (Singer, 2006). Based on the observation that plants can metabolize pesticides, Sandermann (1994) introduced the *green liver* concept suggesting a detoxification sequence similar to what occurs in the mammalian liver (*Figure 2*): Initial *activation* of the toxic chemical by oxidation, reduction or hydrolysis (*Phase I*), transferase-catalyzed *conjugation* of the activated compound with a molecule of plant origin forming an adduct less toxic and more soluble than the parent chemical (*Phase II*), and *sequestration* of the conjugate in the vacuole or incorporation into the cell wall (*Phase III*) (Cole, 1983; Sandermann, 1994; Coleman *et al.*, 1997).

Phytoremediation offers several advantages over other remediation strategies: low cost because of the absence of energy-consuming equipment and limited maintenance, no damaging impact on the environment because of the *in situ* nature of the process, and large public acceptance as an attractive *green technology* (Gerhardt *et al.*, 2009). In addition, phytoremediation offers potential beneficial side-effects, such as erosion control, site restoration, carbon sequestration, and feedstock for biofuel production (Dietz and Schnoor, 2001; Doty *et al.*, 2007). As autotrophic organisms, plants use sunlight and carbon dioxide as energy and carbon sources. From an environmental standpoint, plants can be seen as 'natural, solar-powered, pump-and-treat systems' for cleaning up contaminated soils (Eapen *et al.*, 2007).

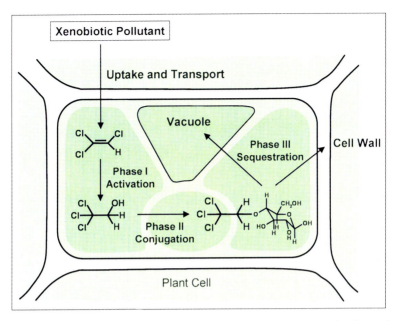

Figure 2. The three phases of the *green liver* model. Hypothetical pathway representing the metabolism of trichoroethylene (TCE) in plant tissues: Phase I: *Activation* of TCE by oxidation to trichloroethanol; Phase II: *Conjugation* with a plant molecule (sugar); Phase III: *Sequestration* of the conjugate into the cell wall or the vacuole. Source: Van Aken (2008).

However, the technology also suffers drawbacks: phytoremediation is limited to shallow contamination and 'moderately hydrophobic' compounds susceptible to be absorbed by the roots (Schnoor *et al.*, 1995; Burken and Schnoor, 1998b). More importantly, remediation by plants is often slow and incomplete: as a corollary to their autotrophic metabolism, plants usually lack the biochemical machinery to achieve *full* mineralization of most organic pollutants, especially the most recalcitrant, such as polychlorinated biphenyls (PCBs) and PAHs (Schnoor *et al.*, 1995). Phytoremediation can therefore lead to non desirable effects, such as the accumulation of toxic pollutants and metabolites that may be released to the soil, enter the food chain, or volatilize into in the atmosphere (Newman *et al.*, 1997; Pilon-Smits, 2005; Yoon *et al.*, 2006; Eapen *et al.*, 2007). Engineering plants expressing microbial or mammalian genes involved in xenobiotic metabolism would therefore improve the efficiency and environmental safety of phytoremediation, potentially leading to wider application in the field (Dietz and Schnoor, 2001). Finally, additional constraints to phytoremediation are not of technical order, but are the current regulations, competition with other methods, and proprietary rights (Marmiroli and McCutcheon, 2003).

Chlorinated compounds: sources, toxicity, biodegradation

CHLORINATED SOLVENTS

Chlorinated solvents are the most common contaminants at U.S. hazardous sites (Ajo-Franklin *et al.*, 2006). Their water solubility, although low, is 4 - 5 orders of magnitude greater than the U.S. Environmental Protection Agency's (EPA) maximum allowable contamination levels. As a consequence, chlorinated solvents constitute a serious threat to human health and the environment (Ajo-Franklin *et al.*, 2006). While chlorinated solvents are theoretically susceptible to biodegradation, they can persist in the environment for decades. Most chlorinated solvent contaminations occur in the form of dense non-aqueous phase liquid (DNAPL) (Ajo-Franklin *et al.*, 2006). DNAPLs contain many different compounds, of which the most important ones are trichloroethene (TCE), tetrachloroethene (PCE), and carbon tetrachloride (CT) (*Figure 3*). TCE - the most common groundwater contaminants in the U.S. - is toxic to mammals (hepatotoxicity) and it is suspected to be carcinogenic. The EPA has established a 5 ppb limit for TCE in drinking water. Degradation products of chlorinated solvents are sometimes more toxic than the parent compound, as it is the case for vinyl chloride (VC) produced by dechlorination of TCE. Another important chlorinated compound is chloroform, a toxic by-product formed during water chlorination. Most chlorinated solvents are listed as EPA Priority Pollutants (http://oaspub.epa.gov/) and/or figure in the top-50 of the 2005 Comprehensive Environmental Response, Compensation, and Liability Act (CERCLA) Priority List of Hazardous Substances (http://www.atsdr.cdc.gov/).

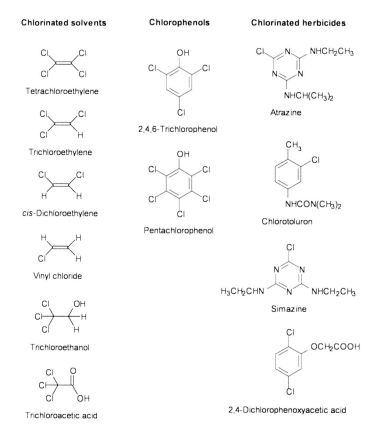

Figure 3. Chemical formula of selected chlorinated compounds.

POLYCHLORINATED PHENOLS

Polychlorophenols have been used largely as bactericides, algicides, molluscides, fungicides, and disinfectants. As a consequence of their broad outside application, they are widespread pollutants of soil and groundwater (Cooper and Jones, 2008; Rubilar *et al.*, 2008). Another major source of chlorinated phenols is the bleaching of pulp in paper mills. As highly toxic compounds, they have potential adverse effects on human and environmental health: embryo-larvae lethal doses determined for chlorinated phenols range from 17 µg L^{-1} for pentachlorophenol (PCP) to 800 µg L^{-1} for less toxic congeners (Cooper and Jones, 2008). PCP has been used extensively as a fungicide. PCP was first registered in the U.S. as a wood preservative in 1936 and, since then, it has been used in the manufacture of ropes, paints, insulation materials, and brick walls. PCP is classified by the EPA as a 'possible human carcinogen'. In addition, technical grade PCP typically contains about 10% impurities consisting of other chlorophenol congeners, as well as highly toxic dibenzo-*p*-dioxins and dibenzofurans (Cooper and Jones, 2008).

CHLORINATED PESTICIDES

Pesticides, including herbicides, insecticides, and fungicides, are an integral part of modern agriculture as they are labor-saving and increase the yield and quality of crops (Kawahigashi *et al.*, 2007). However, chlorinated pesticides are notoriously toxic for higher life forms and persistent in the environment. Agricultural use of herbicides affects large areas resulting in non-point source pollution. It is estimated that between 1 and 5% of field-applied herbicides escape the site of application by runoff, potentially contaminating ground and surface water. Long-term exposure to low levels of herbicide has unknown effects, but is suspected to negatively impact phytoplankton and disturb the composition of aquatic and terrestrial ecosystems (Kawahigashi, 2009). The most notable chlorinated pesticide is the insecticide DDT that was used in massive quantities after World War II. Because of its toxicity and persistence in the environment, DDT was banned in the U.S. since 1972. Beside DDT, many classes of pesticides have been manufactured: Triazine herbicides, such as atrazine and simazine, act on plants by inhibiting photosynthesis. Chloroacetanilide herbicides, such as alachlor and metolachlor, block the synthesis of chain fatty acids. Another class of chlorinated herbicides consists of chlorophenoxy compounds, such as 2,4-dichlorophenoxy acetic acid (2,4-D), that were used as military defoliants (Relyea, 2005).

Phytoremediation of chlorinated compounds

Phytoremediation is particularly well adapted for the treatment of most chlorinated compounds. Contamination plumes of chlorinated solvents (DNAPLs) are typically shallow, moving slowly, and at dose that does not generally induce phytotoxicity (Ajo-Franklin *et al.*, 2006). Briggs *et al.* (1982) and Burken and Schnoor (1998a) developed experimental relationships based on the octanol-water partition coefficient (log K_{ow}) to predict the uptake and translocation of organic chemicals by plants. Based on their relationships, 'moderately hydrophobic' compounds ($0.5 < \log K_{ow} < 3$), which include most chlorinated compounds, are efficiently taken up and translocated into plant tissues (Ajo-Franklin *et al.*, 2006; Kawahigashi, 2009).

PHYTOREMEDIATION OF CHLORINATED SOLVENTS

Many reports have shown the capability of plants to metabolize chlorinated solvents. Using hybrid poplar plants and root cell cultures, Newman *et al.* (1997) first demonstrated the oxidative metabolism of TCE, similar to what is observed in the mammalian liver. TCE oxidation products were detected, including trichloroethanol (TCEOH) and trichloroacetic acid (TCAA). Similar metabolites were found in subsequent experiments using a variety of plant organisms and cell cultures (Nzengung and Jeffries, 2001; Shang *et al.*, 2001; Doty *et al.*, 2003). The detection of oxidized metabolites resulting from the addition of hydroxyl (-OH) groups suggests the intervention of cytochrome P-450 monooxygenases (*Phase I* of the *green liver* model) (Newman *et al.*, 1997; Nzengung and Jeffries, 2001; Shang *et al.*, 2001; Doty *et al.*, 2003). Besides oxidative metabolism, plant reductive dehalogenation of chlorinated solvents has been reported, following a mechanism similar to microbial anaerobic respiration (Nzengung and Jeffries, 2001; Wolfe and Hoehamer, 2003). While the active enzyme has not been identified, the authors suggested the intervention of a plant dehalogenase. Shang *et al.* (2001) reported glycosylation of TCEOH (generated from TCE oxidation) in both TCE-exposed tobacco and poplar tissues. Glycosyltransferase and glutathione *S*-transferase can potentially catalyze the conjugation of activated (hydroxylated) chlorinated solvents with sugar molecules or reduced glutathione (*Phase II* of the *green liver* model) (Sandermann, 1994). Plant glycosyltransferase-mediated conjugations have also been reported in the detoxification of chlorophenols and DDT metabolites. Sequestration of chlorinated compounds and their metabolites (*Phase III* of the *green liver* model) was suggested based on the observation that a fraction of [^{14}C]-TCE was incorporated into the cell wall and large biomolecules (Newman *et al.*, 1997; Nzengung and Jeffries, 2001). Recently, James *et al.* (2009) reported dehalogenation of tetrachloroethylene (PCE) by hybrid poplar trees under controlled field conditions. The authors observed a 99% reduction of chlorinated ethenes and detected trichloroethene (TCE) and *cis*-dichloroethylene (*c*DCE) in the treated effluent. Based on the recovery of free chloride, they concluded to an almost complete dehalogenation of initial PCE, whereas 98% of PCE applied to a lab control chamber of unplanted soil was recovered in the effluent water or volatilized into the atmosphere.

Although limited information is currently available regarding the enzymes involved in the metabolism of chlorinated solvents in plants, preliminary microarray results indicate that exposure of poplar trees to TCE induces genes encoding enzymes potentially involved in the three phases of the *green liver* model (Kang *et al.*, 2009).

Several publications have described higher degradation of chlorinated solvents in vegetated soil (rhizosphere) as compared with non-vegetated soils (Anderson and Walton, 1995; Godsy *et al.*, 2003). For instance, Godzy *et al.* (2003) analyzed TCE degradation by microbes from the rhizosphere of cottonwood trees located at a chlorinated ethene-contaminated site. The results showed that only microbes from the rhizosphere of mature trees (~22-years old) were capable of dechlorinating TCE to *c*DCE and vinyl chloride (VC), whereas microbes associated with newly planted trees were not. In another study, soil of unplanted transects through a TCE-contaminated site revealed much higher levels of remaining TCE compared to soil of transects planted with native English Oak and Common Ash trees (Weyens *et al.*, 2009a).

Although a major critique about phytoremediation is its low efficiency, there have been several promising field trials for the treatment of chlorinated solvent-contaminated sites. For instance, in a five-year study, Eastern cottonwood trees were used for the treatment of low-level TCE-contamination in groundwater at the Naval Air Station Joint Reserve Base (Fort Worth, TX). The study concluded that the primary process responsible for TCE mitigation was *in situ* hydrological control of the contaminant plume, although reductive dechlorination in the root zone and enzymatic transformation within leaf tissue were also reported (Eberts *et al.*, 2003; Eberts *et al.*, 2005). Another field experiment was conducted using hybrid poplar trees for the treatment of groundwater contaminated with a mixture of halogenated volatile compounds, including TCE, *c*DCE, and 1,1,2,2-tetrachloroethane (TeCA), at the J-Field Superfund site, Aberdeen Proving Ground (Edgewood, MD). This five-year study showed that vegetation was able to control plume migration, mainly through the transpiration of groundwater. In addition, contaminants were significantly degraded in the rhizosphere and inside plant tissues, while detectable concentrations of volatile halogenated compounds were transpired into the atmosphere (Hirsh *et al.*, 2003).

PHYTOREMEDIATION OF POLYCHLORINATED PHENOLS

Several publications have reported the potential of phytoremediation for the treatment of soil and water contaminated by polychlorophenols (Newman and Reynolds, 2004). Different plant-mediated remediation mechanisms have been described, including rhizosphere biodegradation, uptake, and metabolism of chlorophenols inside plant tissues. Miller *et al.* (2002) compared the PCP toxicity and rates of mineralization in crested wheatgrass (*Agropyron cristatum* × *desertorum*) grown in sterile soil and soil inoculated with bacteria from a PCP-contaminated soil. Inoculated seedlings exhibited a higher tolerance to PCP and mineralized three-fold more [^{14}C]-PCP than sterile seedlings. On the other hand, only 10% of [^{14}C]-PCP were mineralized by sterile seedlings, suggesting that rhizosphere biodegradation was the main mechanism of PCP metabolism. Jansen *et al.* (2004) showed that exposure of duckweed (*Spirodela punctata*) to 2,4,6-trichlorophenol (2,4,6-TCP) induced the secretion of extracellular peroxidases, resulting in a higher dechlorination rate. The authors concluded that the ability of *S. punctata* to respond to toxic chemicals by secretion of extracellular peroxidase may constitute a protection strategy against xenobiotic stress.

Also, different studies have demonstrated that plants can metabolize chlorophenols. Ucisik and Trapp (2008) exposed willow trees (*Salix viminalis*) to increasing concentrations of 4-chlorophenol (4-CP) - a precursor of the synthesis of the herbicide, 2,4-D, resulting in an almost complete disappearance of the chemical, whereas treeless experiments resulted in only 6 to 10% reduction. In other experiments, different chlorophenyl glycosides were extracted from tissues of common duckweed (*Lemna minor*) exposed to 2,4-dichlorophenol (2,4-DCP) and 2,4,5-trichlorophenol (2,4,5-TCP), showing that plants were capable of uptake and glycosyltransferase-mediated transformation of chlorophenols. Identification of these products suggests that chlorophenols were further sequestrated into vacuoles or cell wall (Day and Saunders, 2004). Using hairy root cultures of wild carrot (*Daucus carota*), de Araujo *et al.* (2002) showed that *Agrobacterium rhizogenes*-transformed roots were able to remove more than 90% of exogenous phenolic compounds, including phenol, 2-chlorophenol (2-CP), 2,6-dichlorophenol (2,6-DCP), and 2,4,6-TCP, from the culture medium within 120 h after treatment. The authors also observed that the metabolism of these compounds was accompanied by an increase in peroxidase activity.

PHYTOREMEDIATION OF CHLORINATED PESTICIDES

Pesticides are removed from the environment by degradation by bacteria and plants. Phytoremediation of pesticides has been extensively studied and reviewed in the literature (Cole, 1983; Sandermann, 1994; Coleman *et al.*, 1997; Schroder *et al.*, 2001; Karthikeyan *et al.*, 2004). Chlorinated pesticides can be degraded by a variety of plants, including trees, aquatic plants, and crops/grasses. Mechanisms of removal involve mainly rhizosphere degradation, uptake and translocation, and sequestration or metabolism inside plant tissues (Salt *et al.*, 1998; Macek *et al.*, 2000; Karthikeyan *et al.*, 2004; Pilon-Smits, 2005).

The metabolism of pesticides in higher plants is known for decades and typically follows the three phases of the *green liver* model (Sandermann, 1994; Coleman *et al.*, 1997; Schroder *et al.*, 2001; Kawahigashi *et al.*, 2007). Sandermann *et al.* (1994) reported the plant metabolism of the herbicide, 2,4-D, including hydroxylation of the aromatic ring (*Phase I*), conjugation with *O*-manolyl-glucoside (*Phase II*), and deposition into the vacuole (*Phase III*). Burken and Schnoor (1998b) showed that [^{14}C]-atrazine was taken up, hydrolyzed, and dealkylated into less toxic metabolites inside hybrid poplar trees: only 21 and 10% of the parent atrazine was found in leaves after 50 and 80 days of exposure, respectively, which provides evidence of the metabolism of atrazine in poplar tissues.

The major enzymes involved in the degradation of chlorinated pesticides are cytochrome P-450 monooxygenases and glutathione *S*-transferases involved in *Phases I* and *II* of the *green liver* model, respectively. Cytochrome P-450 is one the largest families of plant enzymes that, among other functions, catalyzes the detoxification of a variety of xenobiotic compounds through hydroxylation and oxidative dealkylation (Sandermann, 1994; Coleman *et al.*, 1997; Schroder *et al.*, 2001; Kawahigashi *et al.*, 2007). Cytochrome P-450s have been reported to oxidize many chlorinated pesticides, including chlorotoluron, linuron, atrazine, and isoproturon (Kawahigashi *et al.*, 2007).

Transgenic plants for phytoremediation of chlorinated compounds

Although chlorinated solvents and herbicides, such as atrazine and metolachlor, have been shown to be removed by plants, only rather slow biodegradation rates have been achieved in field trials, potentially leading to accumulation and volatilization of toxic compounds (Doty *et al.*, 2007; Kawahigashi, 2009). Genetic transformation of plants for enhanced phytoremediation capabilities is typically achieved by the introduction of new external genes whose products are involved in various detoxification processes (Cherian and Oliveira, 2005; Eapen and D'Souza, 2005). Microbes and mammals are heterotrophic organisms that possess the enzymatic machinery necessary to achieve a near-complete mineralization of organic molecules. Microbial and mammalian catabolic genes can be used to complement the metabolic capabilities of plants (Eapen *et al.*, 2007). Transgenic plants have been engineered for phytoremediation of heavy metals and organic pollutants (Eapen and D'Souza, 2005; Eapen *et al.*, 2007). Early examples include tobacco plants expressing a yeast metallothionein gene and showing a higher tolerance to cadmium (Misra and Gedamu, 1989), *Arabidopsis thaliana* overexpressing a zinc transporter protein and showing a two-fold higher accumulation of zinc in roots (van der Zaal *et al.*, 1999), and tobacco plants expressing bacterial reductases (PETN reductase and nitroreductase) and showing a higher tolerance to the explosive TNT (Hannink *et al.*, 2001). The use of transgenic plants for phytoremediation applications has been reviewed recently in several articles (Cherian and Oliveira, 2005; Eapen *et al.*, 2007; Doty, 2008; Macek *et al.*, 2008; James and Strand, 2009; Kawahigashi, 2009; Van Aken, 2009). Table 1 presents a non-exhaustive list of transgenic plants and bacteria used for phytoremediation of chlorinated compounds.

Table 1. Summary of the publications on transgenic plants and bacteria for the phytoremediation of chlorinated compounds

Compound	Gene	Source	Host Organism	Reference
Various herbicides	Cytochrome P-450 (CYP)	Mammal	Rice (*Oryza sativa*) Potato (*Solanum tuberosum*)	Ohkawa *et al.*, (1999)
Trichloroethene (TCE) & ethylene dibromide (EDB)	Cytochrome P-450 2E1	Human	Tobacco (*Nicotiana tobacco*)	Doty *et al.* (2000)
Trichloroethene (TCE)	Toluene *o*-monooxygenase (TOM)	*Burkholderia cepacia*	Bacteria from the rhizosphere of poplar trees (*Populus Canadensis*) and Southern California shrub *Rhizobium* sp.	Shim *et al.* (2000)
Trichloroethene (TCE)	Cytochrome P-450 2E1	Rabbit	*Atropa belladonna*	Banerjee *et al.* (2002)
Pentachlorophenol (PCP)	Manganese-peroxidase (MnP)	wood-decaying fungus, *Coriolus versicolor*	Tobacco (*Nicotiana tobacco*)	Iimura *et al.* (2002)
Toluene	Toluene-degradation plasmid (*p*TOM)	*Burkholderia cepacia*	Natural endophyte of yellow lupine	Barac *et al.* (2004)
Various herbicides: • Atrazine • Metolachlor • 1-Chloro-2,4-dinitrobenzene (CDNB)	γ-Glutamylcysteine synthetase (ECS) Glutathione synthetase (GS)		Indian mustard (*Brassica juncea*)	Flocco *et al.* (2004)
Pentachlorophenol (PCP)	Pentachlorophenol monooxygenases	*Sphingobium chlorophenolicum*	*Pseudomonas gladioli* colonizing rhizosphere of Chinese chive (*Allium tuberosum*)	Nakamura *et al.* (2004)
Trichlorophenol (TCP) & other phenolic chemicals	Laccase	Roots of cotton (*Gossypium arboreum*)	*Arabidopsis thaliana*	Wang *et al.* (2004)
Acetochlor & Metolachlor	γ-Glutamylcysteine synthetase (ECS)	Bacteria	Hybrid poplar (*Populus tremula* × *Populus alba*)	Gullner *et al.* (2005)
Alachlor	Glutathione S-transferase (GST I)	Maize (*Zea mays*)	Tobacco (*Nicotiana tabacum*)	Karavangeli *et al.* (2005)

Table 1. Contd.

Compound	Gene	Source	Host Organism	Reference
Various herbicides: • Atrazine • Metalochlor • Norflurazon	Cytochrome P-450 (CYP1A1)	Human	Rice (*Oryza sativa*)	Kawahigashi *et al.* (2005)
Pentachlorophenol (PCP) & bisphenol A (BPA)	Laccase	*Coriolus versicolor*	Tobacco (*Nicotiana tobacco*)	Sonoki *et al.* (2005)
Various herbicides: • Atrazine • Metalochlor	Cytochrome P-450 (CYP1A1)	Human	Rice (*Oryza sativa*)	Kawahigashi *et al.* (2006)
Various halogenated compounds: • Trichloroethene (TCE) • Vinyl chloride (VC) • Carbon tetrachloride (CT) • Chlorofrom • Benzene	Cytochrome P-450 2E1	Rabbit	Hybrid poplar (*Populus deltoides* × *Populus alba*)	Doty *et al.* (2007)
Various herbicides: • Atrazine • Chlorotoluron • Norflurazon	Cytochrome P-450 (CYP1A1)	Human	Rice (*Oryza sativa*)	Kawahigashi *et al.* (2007)

Doty *et al.* (2000) first developed tobacco plants (*Nicotiana tabacum*) expressing a human cytochrome P-450 2E1. Mammalian cytochromes P-450 2E1 are known to be involved in the oxidation of a wide range of toxic halogenated compounds, including TCE, CT, and chloroform. Tobacco leaf tissues were transformed with a nopaline-type *Agrobacterium tumefaciens* strain containing a binary vector with a P-450 2E1 cDNA. Whole transgenic plants regenerated from transformed tissues were shown to metabolize TCE 640-fold faster than wild-type plants. Transgenic plants also exhibited an increased uptake and debromination of ethylene dibromide (EDB). Banerjee *et al.* (2002) developed transgenic hairy root cultures of *Atropa belladonna* by the introduction of a rabbit cytochrome P-450 2E1. As measured by the production of the metabolites, chloral and TCEOH, transgenic cultures were shown to metabolize TCE more rapidly than wild-type plants.

However, although *A. thaliana* and tobacco are well characterized, laboratory model plants, they are not well adapted for phytoremediation applications, given their small proportions and shallow root system. Poplar and aspen (*Populus* sp.), on the contrary, are widely distributed, fast-growing, and high biomass plants ideal for phytoremediation applications (Schnoor, 2000). To overcome the limitation inherent to low biomass plants, Doty *et al.* (2007) successfully performed the genetic transformation of poplar plants (*Populus deltoides* × *Populus alba*) overexpressing mammalian cytochrome P450 2E1 (CYP2E1) for the enhanced metabolism of different volatile toxic compounds, including TCE, VC, carbon tetrachloride (CT), chloroform, and benzene. Among the different clones tested, a transgenic line expressed CYP2E1 at a 3.7 to 4.6-fold higher level and exhibited the highest level of TCE metabolism (> 100 higher than non-transgenic controls). When cultivated in hydroponic solution spiked with toxic compounds, this line was capable of extracting about 90% of TCE (as compared to < 3% by non-transgenic controls), 99% of chloroform (as compared to 20% by controls), and 92 to 94% of carbon tetrachloride (as compared to 20% by controls). Transgenic plants were also shown to remove volatile compounds from contaminated air at a higher rate than non-transgenic controls (Doty *et al.*, 2007).

Transgenic crop plants have been developed both for enhanced cross-resistance to pesticides and removal of residual toxic agrochemicals (i.e., phytoremediation) (Kawahigashi *et al.*, 2007). Mechanisms of herbicide resistance in plants include primarily modification of the target site and enhanced detoxification (Ohkawa *et al.*, 1999). Although resistance in weed is based mainly on target modification, resistance in crops is based on detoxification involving cytochrome P-450 or glutathione *S*-transferase. For instance, weed resistance to atrazine is associated to a point mutation of the chloroplastic *psb*A gene encoding part of the photosystem II, therefore reducing its affinity for atrazine. Unfortunately, an attempt to transform canola (*Brassica napus*) by the insertion of atrazine-resistant *psb*A resulted in a significant drop in yield (Ohkawa *et al.*, 1999). On the other hand, crop tolerance is often related to the capability of detoxifying pesticides, such as cytochrome P-450-mediated hydroxylation of chlorotoluron (*Phase I*) and glutathione *S*-transferase-mediated conjugation of atrazine (*Phase II*) (Ohkawa *et al.*, 1999).

Kawahigashi *et al.* (2007) introduced a human cytochrome P-450 (CYP1A1) involved in the detoxification of various xenobiotic compounds into rice plants

(*Oryza sativa*), resulting in cross-resistance towards different classes of chlorinated herbicides, including atrazine and chlorotoluron. Transgenic plants exhibited enhanced metabolism of the herbicides, chlorotoluron and norflurazon, as compared with wild-type plants. The same group showed that the simultaneous introduction of three human cytochromes P-450s with overlapping substrate specificity resulted in cross-tolerance of transgenic rice against 12 herbicides with different mechanisms of action and structures, including atrazine, metolachlor, and norflurazon (Kawahigashi *et al.*, 2005). Transgenic plants were also capable of germination in medium containing chloroacetanilide herbicides, whereas non-transgenic rice did not grow (Kawahigashi *et al.*, 2005). Beside higher tolerance, transgenic rice also showed enhanced metabolism and higher removal rate of herbicides in hydroponic and greenhouse experiments. Transgenic rice expressing mammalian cytochrome P-450s (pIKBACH) has been used in large-scale phytoremediation trials, showing a significant reduction of atrazine and metolachlor in contaminated soil as compared with wild-type planted soil or non-planted soil (Kawahigashi *et al.*, 2006).

TRANSGENIC PLANTS EXPRESSING *PHASE II* ENZYMES OF THE *GREEN LIVER* MODEL

As an illustration of transgenic plants expressing enzymes involved in *Phase II* of the *green liver* model, γ-glutamylcysteine synthetase (ECS) and glutathione synthetase (GS) were overexpressed in Indian mustard (*Brassica juncea*), showing higher levels of glutathione and total nonprotein thiols (Flocco *et al.*, 2004). ECS and GS are involved in the synthesis of the tripeptide glutathione, known to play a key role in detoxification of various toxic organic compounds. Both ECS and GS transgenic plants showed enhanced tolerance to chlorinated herbicides, including atrazine, metolachlor, and 1-chloro-2,4-dinitrobenzene (CDNB). Introduction of a bacterial ECS into hybrid poplar (*P. tremula* × *P. alba*) also resulted in higher tolerance when growing on acetochlor and metolachlor-contaminated soil. Transgenic poplars contained higher γ-glutamylcysteine and glutathione levels than wild-type plants (Gullner *et al.*, 2005). Using a similar approach, a maize (*Zea mays*) glutathione *S*-transferase (GST I) subunit was expressed in transgenic tobacco plants, resulting in the expression of the active enzyme capable of catalyzing glutathione conjugation with the herbicide alachlor. Transgenic tobacco also showed an increased tolerance to alachlor (Karavangeli *et al.*, 2005).

TRANSGENIC PLANTS EXPRESSING SECRETORY ENZYMES

Plant roots naturally release a range of enzymes capable of degrading chemicals in the rhizosphere (Schnoor, 1997; Gianfreda and Rao, 2004; Wang *et al.*, 2004). An alternative approach for improving phytoremediation performances consists of engineering plants to express and secrete microbial enzymes that will be released into the environment to achieve *ex-planta* bioremediation. Fungi naturally perform extracellular digestion of macromolecules by releasing catabolic enzymes, such as peroxidases, laccases, and cellulases, that remain active outside living cells. Based on this observation, Iimura *et al.* (2002) transferred a manganese-peroxidase gene (MnP) from the wood-decaying fungus, *Coriolus versicolor*, into tobacco plants, resulting in the successful expression of the MnP transgene. The authors reported that a transgenic line was expressing MnP activity at a level 54-fold higher than control lines. Transgenic lines were capable of removing pentachlorophenol (PCP) (250 μM) from the hydroponic solution (86% reduction) at rates about 2-fold higher than control lines (38% reduction). Similarly, a secretory laccase isolated from the roots of cotton plants (*Gossypium arboreum*) was overexpressed in transgenic *A. thaliana* plants that showed enhanced tolerance and biodegradation of trichlorophenol (TCP) (Wang *et al.*, 2004). Transgenic lines exhibited higher laccase activity as determined by the oxidation of 2,2-azino-*bis*(3-ethylbenzothiazoline-6-sulphonic acid) (ABTS). Additionally, secretory laccase activity in transgenic plants was responsible for enhanced resistance and biotransformation of TCP, along with other phenolic chemicals. Following a similar approach, Sonoki *et al.* (2005) expressed an extracellular laccase of *Coriolus versicolor* into tobacco plants, resulting in the expression and secretion of the active enzyme into hydroponic medium. HPLC analyses showed that the transgenic lines were capable of enhanced removal of PCP and bisphenol A (BPA) (20 μmol g^{-1} plant dry weight).

Transgenic plant-associated bacteria for phytoremediation

TRANSGENIC RHIZOSPHERE BACTERIA

Plants are known to increase both microbial numbers and activity in soil, which can result in an increase of biodegradation activity (Anderson *et al.*, 1993; Limbert and Betts, 1996). Root exudates and root turnover increase soil organic carbon, which is beneficial both for microbial growth and co-metabolic biodegradation. Root exudates also contain organic acids, alcohols, and phenolic compounds that increase bioavailability and biodegradation, by solubilization of hydrophobic pollutants (Chaudhry *et al.*, 2005). Specific compounds in root exudates have also been suggested to induce microbial enzymes and stimulate biodegradation capabilities (Fletcher and Hedge, 1995). Finally, roots introduce oxygen in the rhizosphere, which is necessary for oxidative biodegradation by oxygenases and laccases (Anderson *et al.*, 1993; Shimp *et al.*, 1993; Chaudhry *et al.*, 2005).

Shim *et al.* (2000) modified different *Rhizobium* strains and plant-associated bacteria from the rhizosphere of poplar trees (*P. canadensis*) and Southern California shrub by the introduction of a toluene *o*-monooxygenase gene (TOM) from *Burkholderia cepacia*. All modified bacteria exhibited the capability to degrade TCE and were able to colonize and compete with poplar, wheat, and barley rhizophere bacteria. In addition, *Pseudomonas* recombinant bacteria isolated from the poplar rhizosphere stably expressed TOM after 29 days.

Following a similar approach, Nakamura *et al.* (2004) introduced *pcp* genes encoding pentachlorophenol monooxygenases from a known PCP degrader, *Sphingobium chlorophenolicum*, into *Pseudomonas gladioli*, a bacterium naturally colonizing the rhizosphere of Chinese chive (*Allium tuberosum*). The transformant bacterium was able to colonize Chinese chive and extensively degrade PCP in the liquid medium. Degradation experiments showed that PCP in soil planted with Chinese chive colonized by *P. gladioli* transformants was removed to a larger extent, as compared with soil planted with Chinese chive only. Tetrachlorocatechol (TCC) was detected in plant tissues, suggesting active metabolism of PCP.

Unlike rhizopheric (i.e., associated with the rhizosphere) and phyllophilic (i.e., living on the surface of the leafs) bacteria, endophytic bacteria or endophytes colonize the internal tissues of plants, but showing no sign of infection or negative effect (Ryan *et al.*, 2008). Nearly all plant species have been shown to harbor endophytic bacteria in their internal tissues establishing symbiotic or commensal relationships: endophytes provide plants with growth-promoting compounds, including phosphates, plant hormones, and siderophores; diazotrophic endophytes fix atmospheric nitrogen; and endophytes prevent or mitigate the effect of plant pathogens, possibly by the activation of plant induced systemic resistance (ISR) (Van Aken *et al.*, 2004; Doty *et al.*, 2007; Ryan *et al.*, 2008). In addition, many endophytes have been shown to play a role in the metabolism of toxic xenobiotic pollutants, therefore potentially enhancing phytoremediation. For instance, an endophytic *Methylobacterium* isolated from poplar trees was shown to actively metabolize nitrosubstituted explosives, raising the question of the role of bacteria in plant biodegradation capabilities (Van Aken *et al.*, 2004).

A study by Moore *et al.* (2006) describes numerous hybrid cottonwood endophytes isolated from a BTEX-contaminated site, some of them exhibiting tolerance to BTEX and TCE. Recently, Weyens *et al.* (2009a) reported a higher reduction of TCE in a groundwater plume across areas vegetated with English Oak (*Quercus robur*) and Common Ash (*Fraxinus excelsior*) as compared with non-vegetated soil. Analysis of bacteria isolated from the rhizosphere and plant tissues showed an enrichment of TCE-degraders, suggesting their implication in TCE reduction in vegetated soil.

While no report has been published to date on the transformation of plant endophytes for phytoremediation of chlorinated compounds, Barac *et al.* (2004) described the conjugative transformation of natural endophytes for improved *in planta* degradation of toluene, which could potentially be useful for the remediation of chlorinated solvents. A toluene-degradation plasmid (*p*TOM) of *Burkholderia cepacia* was transferred by conjugation into a natural endophyte of yellow lupine. Results showed that engineered endophytic bacteria strongly degraded toluene, which was associated with a significant decrease of toluene phytotoxicity for the host plant. In addition, the authors reported a 50 to 70% reduction of toluene transpiration, suggesting higher metabolism inside plant tissues and potentially resulting in mitigation of environmental risks.

Conclusions

With an ever growing world population, human needs have required enhanced performances of plants and microbes using the resources of genetic engineering. Transgenic bacteria have been used for the industrial production of pharmaceutical molecules and human proteins, and transgenic plants have been used for the expression of beneficial traits, such as Bt-maize expressing a toxin-producing gene from *Bacillus thuringiensis* and conferring resistance to insects. From an environmental standpoint, agricultural plants expressing genes involved in the biodegradation of pesticides are the first transgenic organisms that have been used for phytoremediation applications. Recently, non-agricultural plants have been developed to mitigate pollution of soil and groundwater by toxic agrochemicals and other xenobiotic pollutants. Further developments of transgenic organisms for phytoremediation may involve the introduction of broad-substrate catabolic genes, natural or engineered, for the simultaneous remediation of a range of pollutants, such as usually found in contaminated sites, e.g., chlorinated solvent, metals, and nitroaromatics. In addition, biodegradation of many xenobiotics are catalyzed by similar, broad-substrate enzymes, such as cytochrome P-450 monooxygenases, glutathione *S*-transferases, and fungal peroxidases, that can potentially be used for the treatment of multiple pollutants. Moreover, the introduction of multiple transgenes involved in different phases of the metabolism of xenobiotics in plants, i.e., uptake by roots and the different phases of the *green liver* model, would allow enhancing both the removal *and* metabolism of several toxic compounds and could therefore help overcome a major limitation inherent to phytoremediation, i.e., the threat that accumulated toxic compounds would volatilize or otherwise contaminate the food chain. An important barrier to the application of transgenic plants for bioremediation in the field is associated with the true or perceived risk of horizontal gene transfer to related wild or cultivated plants. Therefore, it is likely that the next generation of transgenic plants will involve systems preventing such a transfer, for instance by the introduction of transgenes into chloroplastic DNA or the use of conditional lethality genes (Davison, 2005). Since bacteria naturally exchange plasmids via conjugation, endophytes that gain genes involved in pollutant degradation might not be considered 'genetically modified' and may be subject to fewer restrictions in usage.

Acknowledgements

The authors are grateful to the Strategic Environmental Research and Development Program (SERDP; award numbers 02 CU13-17 and ER-1499), the University of Iowa Superfund Basic Research Program (NIEHS; award number P42ES05605), West Virginia University Research Corporation (PSCoR Award), and the University of Washington Superfund Basic Research Program (NIEHS; award number P42ES04696).

References

Ajo-Franklin, J.B., Geller, J.T., and Harris, J.M. (2006). A survey of the geophysical properties of chlorinated DNAPLs. *Journal of Applied Geophysics* **59**, 177-189

Anderson, T.A., Guthrie, E.A., and Walton, B.T. (1993). Bioremediation in the Rhizosphere. *Environmental Science & Technology* **27**, 2630-2636

Anderson, T.A., and Walton, B.T. (1995). Comparative Fate of C-14 Trichloroethylene in the Root-Zone of Plants from a Former Solvent Disposal Site. *Environmental Toxicology and Chemistry* **14**, 2041-2047

Banerjee, S., Shang, T.Q., Wilson, A.M., Moore, A.L., Strand, S.E., Gordon, M.P., and Doty, S.L. (2002). Expression of functional mammalian P450 2E1 in hairy root cultures. *Biotechnology and Bioengineering* **77**, 462-466

Barac, T., Taghavi, S., Borremans, B., Provoost, A., Oeyen, L., Colpaert, J.V., Vangronsveld, J., and van der Lelie, D. (2004). Engineered endophytic bacteria improve phytoremediation of water-soluble, volatile, organic pollutants. **22**, 583-588

Briggs, G., Bromilow, R., and Evans, A. (1982). Relationship between lipophilicity and root uptake and translocation of non-ionised chemicals by barley. *Pestic. Sci.* **13**, 495-504

Brooks, R., Lee, J., Reeves, R., and Jaffre, T. (1977). Detection of nickelferous rocks by analysis of herbarium specimens of indicators plants. *J. Geochem. Explor.* **7**, 49-57

Burken, J.G., and Schnoor, J.L. (1998a). Predictive relationships for uptake of organic contaminants by hybrid poplar trees. *Environmental Science & Technology* **32**, 3379-3385

Burken, J.G., and Schnoor, J.L. (1998b). Uptake and fate of organic contaminants by hybrid poplar trees. *Abstracts of Papers of the American Chemical Society* **213**, 106-ENVR

Castelfranco, P., Foy, C., and Deutsch, D. (1961). Non-enzymatic detoxification of 2-chloro-4,6-bis(ehtylamino)-S-triazine (simazine) by extract of Zea maize. *Weeds* **9**, 580-591

Chaudhry, Q., Blom-Zandstra, M., Gupta, S., and Joner, E.J. (2005). Utilising the synergy between plants and rhizosphere microorganisms to enhance breakdown of organic pollutants in the environment. *Environmental Science and Pollution Research* **12**, 34-48

Cherian, S., and Oliveira, M.M. (2005). Transgenic plants in phytoremediation: Recent advances and new possibilities. *Environmental Science & Technology* **39**, 9377-9390

Cole, D.J. (1983). Oxidation of xenobiotics in plants. *Prog Pest Biochem Toxicol* **3**, 199-253

Coleman, J., Blake-Kalff, M., and Davies, T. (1997). Detoxification of xenobiotics by plants: Chemical modification and vacuolar compartimentation. *Trends Plant Sci* **2**, 144-151

Cooper, G.S., and Jones, S. (2008). Pentachlorophenol and cancer risk: Focusing the lens on specific chlorophenols and contaminants. **116**, 1001-1008

Daar, A.S., Thorsteinsdottir, H., Martin, D.K., Smith, A.C., Nast, S., and Singer, P.A. (2002). Top ten biotechnologies for improving health in developing countries. *Nat Gen* **32**, 229-232

Davison, J. (2005). Risk mitigation of genetically modified bacteria and plants designed for bioremediation. *J Ind Microbiol Biotechnol* **32**, 639-650

Day, J.A., and Saunders, F.M. (2004). Glycosidation of chlorophenols by Lemna minor. **23**, 613-620

de Araujo, B.S., Charlwood, B.V., and Pletsch, M. (2002). Tolerance and metabolism of phenol and chloroderivatives by hairy root cultures of Daucus carota L. **117**, 329-335

Dietz, A., and Schnoor, J. (2001). Advances in phytoremediation. *Environmental Health Perspectives* **109**, 163-168

Doty, S.L. (2008). Enhancing phytoremediation through the use of transgenics and endophytes. *New Phytol* **179**, 318-333

Doty, S.L., James, C.A., Moore, A.L., Vajzovic, A., Singleton, G.L., Ma, C., Khan, Z., Xin, G., Kang, J.W., Park, A.Y., Meilan, R., Strauss, S.H., Wilkerson, J., Farin, F., and Strand, S.E. (2007). Enhanced phytoremediation of volatile environmental pollutants with transgenic trees. *Proc Natl Acad Sci USA* **104**, 16816-16821

Doty, S.L., Shang, T.Q., Wilson, A.M., Moore, A.L., Newman, L.A., Strand, S.E., and Gordon, M.P. (2003). Metabolism of the soil and groundwater contaminants, ethylene dibromide and trichloroethylene, by the tropical leguminous tree, Leuceana leucocephala. *Water Research* **37**, 441-449

DOTY, S.L., SHANG, T.Q., WILSON, A.M., TANGEN, J., WESTERGREEN, A.D., NEWMAN, L.A., STRAND, S.E., and GORDON, M.P. (2000). Enhanced metabolism of halogenated hydrocarbons in transgenic plants containing mammalian cytochrome P450 2E1. *Proc Natl Acad Sci USA* **97**, 6287-6291

EAPEN, S., and D'SOUZA, S.F. (2005). Prospects of genetic engineering of plants for phytoremediation of toxic metals. *Biotechnology Advances* **23**, 97-114

EAPEN, S., SINGH, S., and D'SOUZA, S.F. (2007). Advances in development of transgenic plants for remediation of xenobiotic pollutants. *Biotechnol Adv* **25**, 442-451

EBERTS, S.M., HARVEY, G.J., JONES, S.A., and W., B.S. (2003) Multiple-Process Assessment for a Chlorinated-Solvent Plume. *In* S McCutcheon, J Schnoor, eds, Phytoremediation. Transformation and Control of Contaminants. John Wiley, Hoboken, pp 589-633

EBERTS, S.M., JONES, S.A., BRAUN, C.L., and HARVEY, G.J. (2005). Long-term changes in ground water chemistry at a phytoremediation demonstration site. *Ground Water* **43**, 178-186

FLETCHER, J., and HEDGE, R. (1995). Release of phenols from perennial plant roots and their potential importance in bioremediation. *Chemosphere* **31**, 3009-3016

FLOCCO, C.G., LINDBLOM, S.D., and SMITS, E. (2004). Overexpression of enzymes involved in glutathione synthesis enhances tolerance to organic pollutants in Brassica juncea. **6**, 289-304

GERHARDT, K.E., HUANG, X.D., GLICK, B.R., and GREENBERG, B.M. (2009). Phytoremediation and rhizoremediation of organic soil contaminants: Potential and challenges. **176**, 20-30

GIANFREDA, L., and RAO, M.A. (2004). Potential of extracellular enzymes in remediation of polluted soils: A review. *Enzyme and Microbial Technology* **35**, 339-354

GODSY, E.M., WARREN, E., and PAGANELLI, V.V. (2003). The role of microbial reductive dechlorination of TCE at a phytoremediation site. *International Journal of Phytoremediation* **5**, 73-87

GULLNER, G., GYULAI, G., BITTSANSZKY, A., KISS, J., HESZKY, L., and KOMIVES, T. (2005). Enhanced inducibility of glutathione S-transferase activity by paraquat in poplar leaf discs in the presence of sucrose. *Phyton-Annales Rei Botanicae* **45**, 39-44

HANNINK, N., ROSSER, S.J., FRENCH, C.E., BASRAN, A., MURRAY, J.A.H., NICKLIN, S., and BRUCE, N.C. (2001). Phytodetoxification of TNT by transgenic plants expressing a bacterial nitroreductase. *Nature Biotechnology* **19**, 1168-1172

HIRSH, S.R., COMPTON, H.R., MATEY, D.H., WROBEL, J.G., and H., S.W. (2003) Five-Year Pilot Study: Aberdeen Proving Ground, Maryland. *In* S McCutcheon, J Schnoor, eds, Phytoremediation. Transformation and Control of Contaminants. John Wiley, Hoboken, pp 635-659

IIMURA, Y., IKEDA, S., SONOKI, T., HAYAKAWA, T., KAJITA, S., KIMBARA, K., TATSUMI, K., and KATAYAMA, Y. (2002). Expression of a gene for Mn-peroxidase from Coriolus versicolor in transgenic tobacco generates potential tools for phytoremediation. **59**, 246-251

JAMES, C.A., and STRAND, S.E. (2009). Phytoremediation of small organic contaminants using transgenic plants. *Current Opinion in Biotechnology* **20**, 237-241

JAMES, C.A., XIN, G., DOTY, S.L., MUIZNIEKS, I., NEWMAN, L., and STRAND, S.E. (2009). A mass balance study of the phytoremediation of perchloroethylene-contaminated groundwater. *Environmental Pollution*

JANSEN, M.A.K., HILL, L.M., and THORNELEY, R.N.F. (2004). A novel stress-acclimation response in Spirodela punctata (Lemnaceae): 2,4,6-trichlorophenol triggers an increase in the level of an extracellular peroxidase, capable of the oxidative dechlorination of this xenobiotic pollutant. **27**, 603-613

KANG, J.W., DOTY, S.L., WILKERSON, H.-W., FARIN, F.M., BAMMLER, T., BEYER, R.P., and STRAND, S.E. (2009). Differential Gene Regulation in Response to Trichloroethylene (TCE) in Hybrid Poplar and Transgenic Poplar. Manuscript in preparation. *Manuscript in preparation*

KARAVANGELI, M., LABROU, N.E., CLONIS, Y.D., and TSAFTARIS, A. (2005). Development of transgenic tobacco plants overexpressing maize glutathione S-transferase I for chloroacetanilide herbicides phytoremediation. **22**, 121-128

KARTHIKEYAN, R., DAVIS, L.C., ERICKSON, L.E., AL-KHATIB, K., KULAKOW, P.A., BARNES, P.L., HUTCHINSON, S.L., and NURZHANOVA, A.A. (2004). Potential for plant-based remediation of pesticide-contaminated soil and water using nontarget plants such as trees, shrubs, and grasses. **23**, 91-101

KAWAHIGASHI, H. (2009). Transgenic plants for phytoremediation of herbicides. *Current Opinion in Biotechnology* **20**, 225-230

KAWAHIGASHI, H., HIROSE, S., INUI, H., OHKAWA, H., and OHKAWA, Y. (2005). Enhanced herbicide cross-tolerance in transgenic rice plants co-expressing human CYP1A1, CYP2B6, and CYP2C19. **168**, 773-781

KAWAHIGASHI, H., HIROSE, S., OHKAWA, H., and OHKAWA, Y. (2006). Phytoremediation of the herbicides atrazine and metolachlor by transgenic rice plants expressing human CYP1A1, CYP2B6, and CYP2C19. **54**, 2985-2991

KAWAHIGASHI, H., HIROSE, S., OHKAWA, H., and OHKAWA, Y. (2007). Herbicide resistance of transgenic rice plants expressing human CYP1A1. **25**, 75-84

LIMBERT, E., and BETTS, W. (1996). Influence of substrate chemistry and microbial metabolic diversity on the bioremediation of xenobiotic contamination. *The Genetic Engineer and Biotechnologist* **16**, 159-180

MACEK, T., KOTRBA, P., SVATOS, A., NOVAKOVA, M., DEMNEROVA, K., and MACKOVA, M. (2008). Novel roles for genetically modified plants in environmental protection. *Trends in Biotechnology* **26**, 146-152

MACEK, T., MACKOVA, M., and KAS, J. (2000). Exploitation of plants for the removal of organics in environmental remediation. *Biotechnology Advances* **18**, 23-34

MARMIROLI, N., and MCCUTCHEON, S. (2003) Making phytoremediation a successful technology. *In* S McCutcheon, J Schnoor, eds, Phytoremediation. Transformation and Control of Contaminants. John Wiley, Hoboken, pp 85-119

MCCUTCHEON, S., and SCHNOOR, J. (2003) Overview of phytotransformation and control of wastes. *In* S McCutcheon, J Schnoor, eds, Phytoremediation. Transformation and Control of Contaminants. John Wiley, Hoboken, pp 3-58

MEAGHER, R.B. (2000). Phytoremediation of toxic elemental and organic pollutants. *Current Opinion in Plant Biology* **3**, 153-162

MILLER, E.K., and DYER, W.E. (2002). Phytoremediation of pentachlorophenol in the crested wheatgrass (Agropyron cristatum x desertorum) rhizosphere. **4**, 223-238

MISRA, S., and GEDAMU, L. (1989). Heavy-metal tolerant transgenic *Brassica napus* L. and *Nicotiana tabacum* L. plants. *Theor Appl Genet* **78**, 161-168

MOORE, F.P., BARAC, T., BORRERNANS, B., OEYEN, L., VANGRONSVELD, J., VAN DER LELIE,

D., Campbell, C.D., and Moore, E.R.B. (2006). Endophytic bacterial diversity in poplar trees growing on a BTEX-contaminated site: The characterisation of isolates with potential to enhance phytoreniediation. **29**, 539-556

Nakamura, T., Motoyama, T., Suzuki, Y., and Yamaguchi, I. (2004). Biotransformation of pentachlorophenol by Chinese chive and a recombinant derivative of its rhizosphere-competent microorganism, Pseudomonas gladioli M-2196. **36**, 787-795

Newman, L.A., and Reynolds, C.M. (2004). Phytodegradation of organic compounds. *Current Opinion in Biotechnology* **15**, 225-230

Newman, L.A., Strand, S.E., Choe, N., Duffy, J., Ekuan, G., Ruszaj, M., Shurtleff, B.B., Wilmoth, J., Heilman, P., and Gordon, M.P. (1997). Uptake and biotransformation of trichloroethylene by hybrid poplars. *Environmental Science & Technology* **31**, 1062-1067

Nzengung, V.A., and Jeffries, P.M. (2001). Sequestration, phytoreduction, and phytooxidation of halogenated organic chemicals by aquatic and terrestrial plants. *International Journal of Phytoremediation* **3**, 13-40

Ohkawa, H., Tsujii, H., and Ohkawa, Y. (1999). The use of cytochrome P450 genes to introduce herbicide tolerance in crops: a review. **55**, 867-874

Pilon-Smits, E. (2005). Phytoremediation. *Annual Review of Plant Biology* **56**, 15-39

Relyea, R.A. (2005). The impact of insecticides and herbicides on the biodiversity and productivity of aquatic communities. **15**, 618-627

Rubilar, O., Diez, M.C., and Gianfreda, L. (2008). Transformation of chlorinated phenolic compounds by white rot fungi. **38**, 227-268

Ryan, R.P., Germaine, K., Franks, A., Ryan, D.J., and Dowling, D.N. (2008). Bacterial endophytes: recent developments and applications. **278**, 1-9

Salt, D., Smith, R., and Raskin, I. (1998). Phytoremediation. *Annu Rev Plant Physiol Plant Mol Biol* **49**, 643-668

Sandermann, H. (1994). Higher plant metabolism of xenobiotics: The 'green liver' concept. *Pharmacogenetics* **4**, 225-241

Schnoor, J. (1997) Phytoremediation. *In* Technology Evaluation Report, TE-98-01. GWRTAC Ground-Water Remediation Technologies Analysis Center

Schnoor, J. (2000) Degradation by plants - Phytoremediation. *In* H-J Rehm, G Reed, eds, Biotechnology, Ed 2nd Vol 11b. Wiley-VCH, Weinheim, pp 371-384

Schnoor, J., Licht, L., McCutcheton, S., Wolfe, N., and Carreira, L. (1995). Phytoremediation of organic and nutrient contaminants. *Environ Sci Technol* **29**, 318A-323A

Schroder, P., Scheer, C., and Belford, B.J.D. (2001). Metabolism of organic xenobiotics in plants: conjugating enzymes and metabolic end points. *Minerva Biotecnologica* **13**, 85-91

Shang, T.Q., Doty, S.L., Wilson, A.M., Howald, W.N., and Gordon, M.P. (2001). Trichloroethylene oxidative metabolism in plants: the trichloroethanol pathway. *Phytochemistry* **58**, 1055-1065

Shim, H., Chauhan, S., Ryoo, D., Bowers, K., Thomas, S.M., Canada, K.A., Burken, J.G., and Wood, T.K. (2000). Rhizosphere competitiveness of trichloroethylene-degrading, poplar-colonizing recombinant bacteria. *Applied and Environmental Microbiology* **66**, 4673-4678

Shimp, J., Tracy, J., Davis, L., Lee, E., Huang, W., Erickson, L., and Schnoor, J. (1993). Beneficial effects of plants in the remediation of soil and groundwater contaminated with organic materials. *Critical Reviews in Environmental Science and Technology* **23**, 41-77

SINGER, A. (2006) The chemical ecology of pollutants biodegradation. *In* M Mackova, D Dowling, T Macek, eds, Phytoremediation and Rhizoremediation: Theoretical Background. Springer, Dordrecht, pp 5-21

SONOKI, T., KAJITA, S., IKEDA, S., UESUGI, M., TATSUMI, K., KATAYAMA, Y., and IIMURA, Y. (2005). Transgenic tobacco expressing fungal laccase promotes the detoxification of environmental pollutants. **67**, 138-142

UCISIK, A.S., and TRAPP, S. (2008). Uptake, removal, accumulation, and phytotoxicity of 4-chlorophenol in willow trees. **54**, 619-627

VAN AKEN, B. (2008). Transgenic plants for phytoremediation: helping nature to clean up environmental pollution. *Trends in Biotechnology* **26**, 225-227

VAN AKEN, B. (2009). Transgenic plants for enhanced phytoremediation of toxic explosives. *Current Opinion in Biotechnology* **20**, 231-236

VAN AKEN, B., YOON, J.M., JUST, C.L., and SCHNOOR, J.L. (2004). Metabolism and mineralization of hexahydro-1,3,5-trinitro-1,3,5-triazine inside poplar tissues (Populus deltoides x nigra DN-34). *Environmental Science & Technology* **38**, 4572-4579

VAN DER ZAAL, B.J., NEUTEBOOM, L.W., PINAS, J.E., CHARDONNENS, A.N., SCHAT, H., VERKLEIJ, J.A.C., and HOOYKAAS, P.J.J. (1999). Overexpression of a novel Arabidopsis gene related to putative zinc-transporter genes from animals can lead to enhanced zinc resistance and accumulation. *Plant Physiology* **119**, 1047-1055

WANG, G.D., LI, Q.J., LUO, B., and CHEN, X.Y. (2004). Ex planta phytoremediation of trichlorophenol and phenolic allelochemicals via an engineered secretory laccase. **22**, 893-897

WEYENS, N., TAGHAVI, S., BARAC, T., LELIE, D.v.d., BOULET, J., ARTOIS, T., CARLEER, R., and VANGRONSVELD, J. (2009a). Bacteria associated with oak and ash on a TCE-contaminated site: characterization of isolates with potential to avoid evapotranspiration of TCE. *Environmental Science and Pollution Research*

WEYENS, N., VAN DER LELIE, D., TAGHAVI, S., and VANGRONSVELD, J. (2009b). Phytoremediation: plant–endophyte partnerships take the challenge. *Current Opinion in Biotechnology* **20**, 237-254

WOLFE, N.L., and HOEHAMER, C.F. (2003) Enzymes used by plants and microorganisms to detoxify organic compounds. *In* S McCutcheon, J Schnoor, eds, Phytoremediation: Control and Transport of Contaminants. John Wiley, Hoboken, pp 499-528

YOON, J.M., VAN AKEN, B., and SCHNOOR, J.L. (2006). Leaching of contaminated leaves following uptake and phytoremediation of RDX, HMX, and TNT by poplar. *International Journal of Phytoremediation* **8**, 81-94

Biotechnology and Genetic Engineering Reviews - Vol. 26, 65-82 (2009)

Defense and death responses to pore forming toxins

ANGELES CANCINO-RODEZNO, HELENA PORTA, MARIO SOBERÓN AND ALEJANDRA BRAVO*

Instituto de Biotecnología, Universidad Nacional Autónoma de México, Ap postal 510-3 CP 62250, Cuernavaca, Morelos, México.

Abstract

Pore forming toxins (PFT) are important virulence factors produced by bacteria to kill eukaryotic cells by forming holes in the cellular membrane. They represent a diverse group of proteins with a wide range of target cells. Although the amino acid sequence is not conserved among the different PFT, many of them share some aspects of their mechanism of action. In general, the mode of action of PFT involves receptor recognition, activation by proteases, and aggregation into oligomeric-structures that insert into the membrane to form ionic pores. Beside the pore formation activity, PFT may have other effects during its interaction with their target cells such as intra-cellular signaling or transport of other enzymatic components, as in the case of anthrax or diphtheria toxins produced by *Bacillus anthracis* and *Corynebacterium diphtheria*, respectively (Parker and Feil, 2005).

Although PFT have evolved as a pathogenic mechanism, some of them have great impact in society since they have different applications in biotechnology or are used as therapeutic agents, or as tools in the study of cell biology (Schiavo and van der Goot, 2001). On the other side, their target organisms have evolved different mechanisms

* To whom correspondence may be addressed (bravo@ibt.unam.mx)

Abbreviations: PFT, Pore forming toxins; CDC, cholesterol dependent cytolysins; PA, anthrax protective antigen; MACPF, membrane-attack complex of complement and perforin; NLR, Nod-like receptors; MAPK, mitogen activated protein kinase; TNF-α, tumor necrosis factor; PLY, pneumolysin; ILY, intermedilysin; LLO, listeriolysin; SLO, streptolysin O; HlyA, α-hemolysin; PVL, Panton-Valentine leukocidin; Hla, α-hemolysin; CPE, *Clostridium prefringens* enterotoxin; Bt-R$_1$, *Manduca sexta* cadherin receptor for Cry1Ab; GPI, glycosyl phosphatidyl-inositol; RNAi, RNA interference; Akt, protein kinase B; UPR, endoplasmic reticulum stress response to unfolded proteins; CHO, Chinese hamster ovary cells; SREBP, sterol responsive element binding proteins; VCC, *Vibrio cholerae* cytolysin; PRR, pattern recognition receptors.

to counter toxin action. Understanding the mechanism of action of PFT as well as the host responses to toxin action would provide ways to deal with these pathogens or with emerging pathogens and more importantly to improve the action of toxins that have biotechnological applications. In this review we will describe the intracellular effects induced by some PFT and the cellular responses evolved by eukaryotic cell to overcome PFT action.

Introduction

An essential feature of PFT is their capacity to interact with the target membrane and the conformational changes that lead to pore formation. In some sense PFT are unique proteins that are synthesize as soluble proteins that are capable of breaking the hydrophobic barrier of the cellular membrane becoming integral membrane-proteins. PFT are usually conformed by multiple domains. Some domains may have different and complementary functions as receptor binding, translocation, pore formation and other enzymatic activities such as ADP-ribosylating, methaloprotease or adenylate cyclase activities among others (Parker and Feil, 2005).

An initial step in pore formation includes interaction with the target-membrane, although the mechanism is unclear some generalities can be drawn, such as the presence of exposed aromatic residues that bind to the membrane bilayer as the cholesterol dependent cytolysins (CDC) produced by different genera of Gram-positive bacteria, including *Clostridium, Bacillus, Streptococcus, Listeria* and *Arcanobacterium* (Palmer, 2001), or the presence of hydrophobic helical hairpins as the case of diphtheria toxin from *C. diphtheria* or colicins produced by some *Escherichia coli* strains (Parker and Feil, 2005).

Regarding membrane insertion, there are several examples of PFT that span the membrane bilayer by forming β-barrel structures, named β-PFT. Other PFT involve clusters of α-helices, named α-PFT. In most of the β-PFT, the insertion into the membrane to form the final pore involves only a small part of the protein and the rest of the protein, including the receptor and membrane binding domains, remain outside of the membrane. In some of the β-PFT, the trasmembrane region represent a small-unstructured loop motive in the soluble protein that refolds to form the transmembrane β-strand as the case of anthrax protective antigen (PA) toxin or aerolysin from *Aeromonas hydrophila* (Parker and Feil, 2005). In other β-PFT the transmembrane β-strands already exist in the pre-pore, such as the α-toxin and leukocidins from *Staphylococcus aureus* (Parker and Feil, 2005). Finally, in the case of the CDC toxin family the transmembrane β-strands are formed from existing strands connected by short α-helices involving an important conformational change to β-strand (see insert of *Figure 1*)(Parker and Feil, 2005). Some toxins require six or seven β-hairpins, each from a different monomer, to form a 12 or 14 β-strand-barrel as leukotoxins, α-toxin, or anthrax PA. Other PFT requires a much higher number of β-strands to form the β-barrel, as is the case of the CDC-toxin family that require more than 35 monomers, each one inserting a pair of β–hairpins in the final pore. In contrast, helical α-PFT like colicins, have a dynamic structure that involves insertion of preformed α-helices into the membrane (Parker and Feil, 2005).

Recently, the structure of some PFT have been compared with cellular proteins that are also capable of breaking the hydrophobic barrier of the membrane as the case of the membrane-attack complex of complement and perforin (MACPF) that shows some similarities at the 3-D structural level with CDC-toxins (*Figure 1*), or the apoptosis inhibitor Bcl-XL protein that shows some similarities with colicin (*Figure 2*) (Anderluh and Lakey, 2008). The analysis of the superposition of the 3-D structures of human-MACPF and CDC toxin intermedilysin (ILY) from *Streptococcus intermedius*, shows that these proteins share 149 residues, that in the CDC toxin correspond to the trans-membrane regions 1 and 2 (*Figure 1*) (Slade *et al.*, 2008).

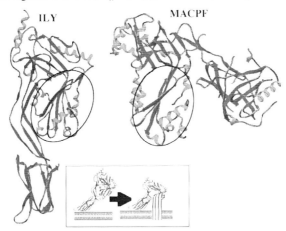

Figure 1. Ribbon representation of the intermedilysin (ILY) from *Streptococcus intermedius* (PDB number 1S3R) and human membrane-attack complex of complement and perforin (MACPF) (PDB Number 2RD7). This figure was produced by the computer program Swiss Pdb-Viewer. Both proteins show a small region, encircled that shows some similarities at the 3-D structural level. In the CDC toxin, ILY, this region corresponds to the transmembrane region involved in pore formation. Insert shows a schematic representation of the large conformational change that CDC may have during pore formation.

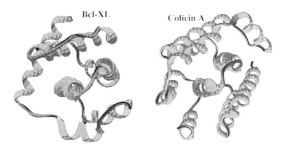

Figure 2. Ribbon representation of the apoptosis inhibitor Bcl-xl (PDB number 1r2d) that shows some similarities with colicin A (PDB number 1col). This figure was produced by the computer program Swiss Pdb-Viewer.

Defense mechanisms evolved by eukaryotic cells to bacterial infections

Eukaryotic cells have several resources to respond to different biotic and abiotic stresses. In the case of pathogen invasion the immune system is constantly challenged to destroy the pathogen and infected cells avoiding the destruction of non-infected tissue (Pedersen, 2007). Two general defense responses have been described, the innate and the adaptive immune responses.

INNATE IMMUNE RESPONSE

The innate immune system allows the eukaryotic cells to have a quick and broad spectrum of responses. The innate immune cells such as macrophages and dendritic cells directly suppress the pathogenic microorganisms through phagocytosis or by induction of the expression of cytokines, chemokines and costimulatory molecules that participate in pathogen elimination (Akira *et al.*, 2006). In the innate immune system, the major humoral components are the complement enzymes of the alternative, lectin-binding and classical pathways, which involve more than 30 plasma and cell-surface proteins (Honeyman and Harrison, 2004). The complement activation leads to opsonization and killing of pathogens by forming a membrane pore named the MACPF, which results in the lysis of pathogens (Huw, 2008). It is intriguing that eukaryotes evolved also pore-forming proteins as an strategy to defend themselves from pathogen attacks.

The innate response represents the early events that involve pattern recognition receptors (PRR) that recognize general features that microorganisms share named pathogen-associated molecular patterns also known as PAMP in different cellular compartments, such as the plasma membrane, endosomes or cytoplasm. These receptors are evolutionarily conserved and include: Toll-like receptors, retinoic acid-inducible gene-I-like receptors and Nod-like receptors (NLR). These receptors have different cytoplasmic domains, responsible for transducing signals inside of the cell (Honeyman and Harrison, 2004; Huw, 2008). For instance, the effector-domain of NLR bind to the corresponding domains of their downstream signaling cascade inducing the formation of a large multiprotein complex called the inflammasome that recruited caspases resulting in activation of the transcription factor NF-κB and the mitogen activated protein kinase (MAPK) pathways. The free NF-κB binds then to its target genes, facilitating their transcription, leading to the production of proinflammatory cytokines such as tumor necrosis factor (TNF-α), and interleukins IL-1β and IL-6. The MAPK pathways include ERK, JNK and p38 subfamilies that also play a pivotal role in the biosynthesis of cytokines (Dong *et al.,* 2002). The pyroptosis is the programmed cell death process that involves the activation of caspase-1 and the inflammasome complex formation.

An alternative cellular mechanism to control bacterial infections is autophagy, which is considered an innate immune mechanism that acts in the intracellular space. Autophagy restricts the infection by sequestering intracellular pathogens in a double membrane structure, named autophagosoma that will be delivered to lysosomes for degradation (Deretic, 2006).

ADAPTIVE IMMUNE RESPONSE

Only when innate host defense is overwhelmed, the induction of an adaptive immune response is required (Pedersen, 2007). In this condition, the responses of the innate immune system instruct the development of long-lasting pathogen-specific adaptive immune responses. The adaptive system is mediated by antigen receptors on lymphocytes, which produces a more sustained and comprehensive response. This response takes several days to mount, but take advantage of a very large repertoire of antigen receptors. One important characteristic of the adaptive system is that it retains a memory of exposure to each microbe and ensures the system is mobilized more rapidly upon a subsequent infection by the same pathogen (Huw, 2008).

The adaptive immune system consists of B- and T-cells, which provide pathogen specific immunity to the host through somatic rearrangement of antigen receptor genes. B-cells produce specific antibodies to neutralize pathogens, whereas T-cells provide the cytokine milieu to clear pathogen-infected cells through their cytotoxic effects or via signals to B-cells (Hoebe *et al.,* 2004).

Intracellular effects induced by PFT

The pore formation activity of PFT directly permeabilize the plasma membrane, leading to changes in cytoplasmic ions composition, such as Ca^{2+} or K^+ ions and also induce changes in osmotic pressure. Recent studies have shown that high concentrations of PFT destroy target cells. However, before the irreversible damage occurs, the cell may trigger sophisticated mechanisms and display signal-transduction pathways as responses to subcytolytic concentrations of PFT. The host responses may include adaptive or innate immunity responses as well as cellular non-immune defenses (Aroian and van der Goot, 2007).

Cells of the immune system such as lymphocytes, macrophages or dendritic cells produce numerous inflammatory molecules and also may undergo apoptosis after PFT action. For instance, interleukins IL-8 and IL-1β are induced after treatment with α-toxin, aerolysin, pneumolysin (PLY) from *S. pneumoniae*, and Listeriolysin (LLO) from *Listeria monocytogenes* (Shoma et al 2008, Chopra et al 2000, Bhadki et al 1989; Gekara et al., 2007) TNF-α is induced after treatment with aerolysin and Streptolysin O (SLO) from *S. pneumoniae (*Stassen et al 2003: Chopra et al 2000).

Non-immune cells, such as epithelial cell, vacuolated in response to PFT, and induce MAPK p38 activation (Ratner et al 2006; Husmann et al 2006; Huffman et al 2004; Bischof et al 2008)

Toxin-induced membrane permeabilization may activate pathways involved in cell survival or related to cell death. We will review both of these intracellular effect induced by PFT.

Mechanisms of cell death induced by PTF

APOPTOTIC DEATH INDUCED BY PFT AT SUBLYTIC CONCENTRATIONS

Apoptosis is described as a programmed death process that avoids eliciting inflammation. It is considered to be a regulated and controlled process. The morphology associated with apoptosis is characterized by nuclear and cytoplasmic condensation and cellular fragmentation. The apoptosis pathway implies a metabolic cascade started by initiator caspases (caspase-2, -8, -9 and -10) followed by activation of the effector caspases (caspase-3, -6 and -7) that will cleave different cellular substrates to produce the morphological and biochemical features associated with apoptosis. For example, a DNA ladder is produced by cleavage of genomic DNA between nucleosomes, as a product of a caspase-activated DNase; detachment of apoptotic cells resulted from cleavage of the components of the adhesion complex; and plasma membrane blebbing resulted from caspase mediated activation of gelsolin an actin depolymerizing-enzyme (Fink and Cookson, 2005).

Several PFT induce cell death by two mechanisms, oncosis and apoptosis, depending on the cell type and on the dose of toxin. When used at low concentrations some PFT induce apoptosis, in contrast to high toxin dose, where cells died quickly by oncosis or necrosis. The participation of the pore formation activity of PFT in the induction of these responses was confirmed by using point mutants affected in pore formation, or by knockout mutants that do not produce the PFT or by reducing toxin expression by antisense RNA. Also, heat inactivated toxins were unable to induce apoptosis or cell death (Genestier *et al.,* 2005; McClane and Chakrabarti, 2004; Katayama *et al.,* 2007; Wiles *et al.*, 2008; Menzies and Kourteva, 2000).

The Panton-Valentine Leukocidis (PVL) from *S. aureus*, is a β-PFT composed by two components, the LukS-PV and the LukF-PV, forming an octameric β-barrel pore. Purified PVL induced apoptosis in neutrophils, monocytes and macrophages when used at sublytic concentrations. This process involves activation of caspases-9 and -3. In contrast, high concentrations of this toxin induced necrotic alterations forming large-pores (*Table 1*) (Genestier *et al.,* 2005). The *Clostridium prefringens* enterotoxin (CPE) at low dose induces typical apoptotic cell death in mammalian Caco2 cells. The observed effects include DNA fragmentation with a ladder-pattern, chromatin condensation, mitochondrial membrane depolarization and activation of caspase-3 and -7. Similarly to PVL, a high dose of CPE induces cell death by oncosis showing random DNA shearing. Studies performed with CPE knockout mutant confirmed the importance of this toxin in inducing plasma membrane permeability alterations and in promoting cell death (*Table 1*) (McClane and Chakrabarti, 2004). The parasporin-1 produced by *Bacillus thuringiensis*, the α-hemolysin (HlyA) produced by *Escherichia coli* (UPEC) and α-toxin also induces apoptosis cell death when used at low dose (Katayama *et al.,* 2007; Wiles *et al.,* 2008; Menzies and Kourteva, 2000). Parasporin-1 is toxic to certain mammalian cancer cell lines such as HeLa, and induces apoptotic cell death by a mechanism that involves a large influx of extracellular Ca^{+2} and activation of caspase-3. Synthetic caspase-inhibitors are able to block parasporin-1 cytotoxic activity (Katayama *et al.,* 2007). The HlyA toxin induces apoptosis at sub-lytic concentrations on different target host cells such as neutrophils, T-lymphocytes, renal and epithelial cells (Wiles *et al.,* 2008) and the α-toxin triggers apoptosis in endothelial and lymphocytes cells. TUNEL and DNA ladders were observed when cells were treated with at 5 to 20 nM of α-toxin. High concentrations of α-toxin in-duced cell lysis by necrosis. Toxin negative mutants or down regulation experiments of expression of α-toxin using an anti-sense fragment of α-toxin gene, show no DNA ladder formation when compared with a clinical *S. aereus* isolate suggesting that low concentration of α-toxin is responsible of triggering apoptosis in these cells (*Table 1*) (Menzies and Kourteva, 2000).

The aerolysin triggers apoptosis in T-lymphomas at sub-nM concentrations and this effect was overcome at high toxin concentrations, where cells died quickly and apoptotic pathway was not triggered (Nelson et al 1999). The authors used different variants of aerolysin to demonstrate that binding interaction with toxin receptors is not enough to trigger the apoptosis response, nor the clustering of receptors that may be induced by toxin oligomerization induced the intracellular signal for apoptosis activation. It was shown that aerolysin cause apoptosis of T-cells by the production

Table 1. Summary of the different responses triggered by PFT that cause cell death.

Toxin	Cell Response	Concentration	Type of cell	Characteristics	Involvement of pore formation	Reference
PVL	Apoptosis	5 nM	Neutrophils, Monocytes	Activation of caspase-9, -3, chromatin condensation	Recombinant PVL, Δluk-PV deleted mutant	Genestier et al., 2005
CPE	Necrosis Apoptosis	200 nM 1 µg/ml	Macrophages Caco2	DNA fragmentation, chromatin condensation, activation caspase-3, -7. Random DNA shearing	knockout mutant	McClane and Chakrabarti, 2004
SLO	Oncosis	10 µg/ml				
PLY	Pyroptosis	1-16 µg/ml	Macrophages	Activation caspase-1	Pure toxin, Δslo	Timmer et al., 2008
	Apoptosis	Low dose 0.5 µg/ml	Endothelial Neuroblastoma	MAPK p38 activates caspase-6, -9 and -3, DNA fragmentation, chromatin condensation, TUNEL	Ca^{2+} influx, Δply deleted mutant.	Guessan et al., 2005 Stringaris et al., 2002
	Pyroptosis	Low dose 0.1 µg/ml	Macrophages	Activation of caspase-1, IL-1α, IL-1β, IL-18	knockout mutant	Shoma et al., 2008
Leukotoxin	Apoptosis	0.2 U[a]	Leukocytes	Inhibition of Akt activation	lktC deficient mutant	Atapattu and Czuprynski, 2005
HlyA	Apoptosis	Sublytic	Neutrophils, T-lymphocyte Renal. 5637 Bladder-epithelial	Inactivate Akt signaling by stimulation of host phosphatases	ΔhlyA mutant HlyA not-HlyC-activated. Point-mutants	Wiles et al., 2008
α-toxin	Apoptosis	Sublytic	5637bladder-Epithelial	Inactivate Akt signaling	Influx of Ca^{2+}	Wiles et al., 2008
	Apoptosis	1 µg/ml 5 to 20 nM	Epithelial Lymphocytes	TUNEL, DNA ladders	Toxin negative mutants. Heat-inactivated toxin	Menzies and Kourteva, 2000
Aerolysin	Necrosis Apoptosis Necrosis	>50 nM Sub-nM high dose	Endothelial T-lymphomas	Cell detachment but no-TUNEL DNA fragmentation	Increase in Ca^{2+}, Point-mutants	Nelson et al., 1999
Anthrax LT	Apoptosis	sublytic	Leukocytes, Macrophages	Inactivate Akt signaling		Wiles et al., 2008
	Pyroptosis	1 µg/ml		Caspase-1, cytokine activation, DNA cleavage	K^+ efflux	Fink et al., 2008
Cry1Ab	Oncosis	180 nM	H5 ovarian insect cells	Membrane blebbing, activation of adenylyl cyclase. G-protein and protein kinase A. increase in AMPc.	Not-determined	Zhang et al., 2006
Parasporin1	Apoptosis	1-10 mg/ml	HeLa	Activation of caspase-3	Influx of Ca^{+2}	Katayama et al., 2007

[a] One U of Leokotoxin was defined as the leukotoxin dilution causing 50% killing of 106 leukocytes BL-3.

of small number of channels in the cell membrane. Channel formation resulted in a rapid increase in intracellular Ca^{2+}, which was speculated to be the signal triggering apoptosis (*Table 1*) (Nelson *et al.*, 1999).

Finally, in the case of PLY toxin it was demonstrated that it induces apoptosis in human endothelial and neuroblastoma cells. The apoptosis pathway was activated through MAPK p38 and JNK pathways. PLY apoptotic cells shows activation of caspases -6, -9 and a late activation of caspase-3. Inhibition of p38 and JNK phosphorylation strongly reduced caspases activation and apoptosis (Guessan *et al.*, 2005). DNA fragmentation and chromatin condensation were observed, and caspase inhibition exert significant protection from PLY induced toxicity. It was reported that extracellular Ca^{2+} influx mediated by PLY pores correlates with activation of MAPK p38 since inhibition of Ca^{2+} influx in a Ca^{2+} free bathing solution and inhibition of MAPK p38 activity with selective inhibitors such as SB203580, rescued neuronal cells from death, suggesting that Ca^{2+} pores induced by PLY, activate MAPK p38 response and correlated with apoptosis death (*Table 1*) (Stringaris *et al.*, 2002).

ONCOSIS DEATH INDUCED BY PFT

Cell death by oncosis is characterized by cellular and organelle swelling, blebbing and increase in membrane permeability. Oncosis may result from toxic agents that interfere with ATP generation, or cause uncontrolled cellular energy consumption. Also, altered intracellular Ca^{+2} levels can induce oncotic cell death. An essential element of oncosis is its inflammatory nature.

As stated above several PFT (PVL, CPE, α-toxin, aerolysin) induce oncotic cell death when assayed in high concentrations, although the mechanism that is responsible for oncosis induction remains unknown (*Table 1*) (Genestier *et al.*, 2005; McClane and Chakrabarti, 2004; Menzies and Kourteva, 2000; Nelson *et al.*, 1999).

The only example that explains the mechanism of oncosis death has been reported for the insecticidal Cry1Ab toxin produced by *Bacillus thuringiensis* (Zhang *et al.*, 2006). These studies were performed in H5 insect cell line expressing one receptor of Cry1Ab toxin, the *Manduca sexta* cadherin receptor ($Bt-R_1$). The authors demonstrated that cell death was triggered by the interaction of the monomeric Cry1Ab toxin with $Bt-R_1$. This binding event activates a Mg^{+2} dependent signal cascade pathway that is responsible for oncotic cell death. The cytological changes include membrane blebbing, appearance of ghost nuclei, cell swelling and lysis. It was shown that binding of Cry1Ab to $Bt-R_1$ activates a G-protein, which in turn activates adenylyl cyclase promoting the production of intracellular cAMP. The increased levels of cAMP activated protein kinase A that is ultimately responsible for causing cell dead (Zhang *et al.*, 2006). The authors proposed that cell death resulted from a complex cellular response not by osmotic lysis produced by toxin pore formation activity. However, this oncotic process has not been studied in insect-midgut cells and may be particular for this specific cell line, which was originated from ovarian cells of *Trichoplusia ni*. An opposite model explains that the mechanism of Cry1Ab toxin action in insect larvae is due to pore formation activity of the toxin (Bravo *et al.*, 2004, Bravo *et al.*, 2007). In this model, it was proposed that binding with of Cry1Ab toxin to $Bt-R_1$ facilitates an additional protease cleavage at the N-terminal end of the toxins eliminating helix α-1 of domain I (Gomez *et al.*, 2002). This cleavage induces assembly of an oligomeric form of the toxin. The oligomers bind to secondary receptors, which are glycosylphosphatidyl-inositol (GPI)-anchored proteins, such as aminopeptidase N in *M. sexta* and alkaline phosphatase in the lepidopteran *Heliothis virescens* or the dipteran *Aedes aegypti* (Bravo *et al.*, 2007). After the oligomers bind to secondary receptors, they insert into membrane microdomains, where GPI anchored receptors are localized, and create pores in the apical membrane of midgut cells causing osmotic shock, bursting of the midgut cells that result in insect death (Bravo *et al.*, 2004). Recently it was shown that genetic engineered modified Cry1Ab and Cry1Ac toxins that were deleted of amino-terminal end including helix α-1 (*i.e.* Cry1AbMod and Cry1AcMod) formed oligomers in the absence of Bt-R$_1$ receptor (Soberón *et al.*, 2007). Interestingly, these modified toxins kill *M. sexta* insects, in which the Bt-R$_1$ protein was silenced by RNA interference (RNAi) and which resulted in resistance to Cry1Ab. The Cry1AMod toxins also kill *Pectinophora gossypiella* resistant insects in which the Cry1A-resistance had been found to be due to deletions in the cadherin Bt-R$_1$ gene (Morin *et al.*, 2003, Soberon *et al.*, 2007). In addition, Cry1Ab mutants affected in oligomerization or in pore formation are non-toxic despite the fact that they bind to Bt-R$_1$ receptor with similar affinity as the wild type toxin and recently it was demonstrated that certain Cry1Ab point mutants that affect pore formation exert a dominant negative phenotype over the native Cry1Ab toxin (Jimenez-Juarez *et al.*, 2007; Rodriguez-Almazan *et al.*, 2009). Overall these results indicate that binding to Bt-R$_1$ is not enough to trigger cell death in the insect larvae and that oligomerization and pore formation activity are essential steps in the mode of action of Cry toxins. Nevertheless it is still possible that intracellular signals may be triggered by pore-formation induced by Cry toxin in the insect larvae, as has been shown for other PFT (Arioan and van der Goot 2007; Soberón, et al., 2009).

PYROPTOSIS INDUCED CELL DEATH BY PFT

Pyroptosis is a caspase-1-dependent inflammatory form of cell death that involves the secretion of the pro-inflammatory-cytokines such as IL-1β and IL-18 after formation of inflammasome complexes. (Fernandes-Alnemri *et al.*, 2007).

Different PFT activates pyroptosis in target cells. PLY, SLO and lethal toxin (LT) from *B. anthracis* activate caspase-1 in macrophages (*Table 1*) (Shoma *et al.*, 2008; Timmer *et al.*, 2008; Fink *et al.*, 2008). It was shown that treatment with PLY induces the production of IL-1α, IL-1β and IL-18 and a *S. pneumoniae* mutant lacking the *ply* gene was unable to induce this response, suggesting that PLY plays an important role in promoting cell death though pyroptosis (Shoma *et al.*, 2008). In agreement with this observation it was reported that caspase-1 deficient macrophages were more resistant to cell-death induced by another CDC toxin, the SLO toxin (Timmer *et al.*, 2008). The response of macrophages to LT include the stimulation of the inflamma-some, cytokine activation, DNA cleavage and lysis mediated by the formation of membrane pores between 1.1 and 2.4 nm that could be blocked by osmoprotectants and glycine (Fink *et al.*, 2008). Interestingly, it was shown that LT proteolytic activity is necessary but not sufficient to trigger this caspase-1 dependent response and that this response depends on pore formation activity of the anthrax PA toxin and K$^+$ efflux (Fink *et al.*, 2008).

PFT INACTIVATE THE HOST CELL SURVIVAL AKT PATHWAY PROMOTING CELL DEATH BY APOPTOSIS

Akt, also known as protein kinase B or PKB, is a key regulator of host cell survival pathway. Akt plays an important role in controlling cell cycle, endocytosis and vesicular trafficking, apopotosis inhibition and inflammatory responses to bacterial infections. Akt is activated by phosphorylation downstream PI3-kinase and G-protein coupled-receptors. This cell-survival pathway is inactivated by sublytic concentrations of different PFT such as HlyA, aerolysin, α-toxin and leukotoxin in leukocytes and epithelial cells (*Table 1*) (Wiles *et al.*, 2008, Atapattu and Czuprynski, 2005). The PFT stimulates specific host phosphatases that inactivate Akt as a result of pore formation and influx of Ca^{+2} ions. The pore forming activity of HlyA depends on the presence of an activator protein named HlyC, that is an acyltransferase that acylates HlyA resulting in pore-formation activation. HlyA produced in absence of HlyC insert into the host membrane but is unable to form pores. Under these conditions it is unable to inactivate Akt pathway indicating that pore formation is required to stimulate Akt dephosphorylation. In addition, inactive HlyA mutants unable to make pores or the addition of dextran to external medium, which was shown to block HlyA pores, prevented HlyA induced dephosphorylation of Akt. Akt has several downstream targets such as glycogen synthase kinase involved in cell proliferation, metabolism pathways and inflammatory responses. Then the inactivation of Akt by PFT may adversely affect host inflammatory responses and induce apoptotic response at sublytic concentrations (Wiles *et al.*, 2008)

Defense mechanisms evolved by eukaryotic cells to PFT

ACTIVATION OF MAPK P38 PATHWAY

Epithelial cells start an early immune response, involving activation MAPK p38 pathway after treatment with low concentrations of several PFT. It was demonstrated that the osmotic stress produced after formation of few pores by different PFT in the target cells, induce a MAPK p38-phosphorylation-response that is crucial to prevent bacterial infection (Ratner *et al.*, 2006).

The activation of MAPK p38 after treatment with PFT was first described in *Caenorhabditis elegans* treated with the *B. thuringiensis* Cry5B toxin (*Table 2*) (Huffman *et al* 2004). Later it was shown that other PFT such as aerolysin, PLY, SLO, α-hemolysin (HlyA) from *S. aureus*, and anthrolysin O from *B. anthracis*, also induce a MAPK p38 response in epithelial cells of baby hamster or human embryonic kidney cells when assayed a low doses (*Table 2*) (Huffman *et al.*, 2004, Ratner *et al.*, 2006). It was shown that nM concentration of these toxins are not sufficient to cause cytolysis, but are able to initiate a proinflamatory response at the beginning of the infection involved in defense to PFT and cell survival. It is important to mention that the observed phosphorylation of MAPK p38 protein was correlated with the pore formation activity of these PFT, since toxin deficient mutants or point mutations in toxin regions essential for pore formation activity were unable to induce this p38-response. In addition, the presence of high molecular weight osmolytes, as dextran, in the extracellular medium or calcium chelating compounds, as EGTA, inhibit the

Table 2. Summary of the different responses triggered by PFT that cause cell survival.

Toxin	Cell Response	Concentration	Type of cell	Characteristics	Involvement of pore formation	Reference
SLO	Stress	100 ng/ml	Epithelial	Activation of MAPK p38		Ratner et al., 2006
	Stress	250 ng/ml	Mast cells	MAPK p38 induce TNF-α synthesis	Slo- deficient mutant	Stassen et al., 2003
	Stress	100 ng/ml	HaCaT	Activation of MAPK p38		Husmann et al., 2006
	Endocytosis	200 ng/ml	Kidney, HeLa	Dynamin-independent	Ca^{2+} influx	Idone et al., 2008
PLY	Stress	50 ng/ml	Epithelial	Activation of MAPK p38	Ply-deleted mutant	Ratner et al., 2006
LLO	Stress	Low dose	Mast cells	TNF-α synthesis	Ca^{2+} influx	Gekara et al., 2007
α-toxin	Stress	10 ng/m	HaCaT	Activation of MAPK p38	Point-mutant	Husmann et al., 2006
	Biosynthesis of lipids	30 nM	HeLa	Activation of caspase-1 that activates SREBP	K$^+$ efflux	Gurcel et al., 2006
	Endocytosis-exocytosis	500 ng/ml	Cos7, HaCaT	Endocytosis dynamin-dependent	Point-mutants	Husmann et al., 2009
Aerolysin	Stress	25-50 nM	Epithelial	Activation of MAPK p38		Huffman et al., 2004
	Biosynthesis of lipids	0.2 nM	HeLa	MAPK p38 activates UPR system		Bischof et al., 2008
VCC	Biosynthesis of lipids	0.2 nM	CHO, HeLa	Activation of caspase-1 that activates SREBP	Point-mutants	Gurcel et al., 2006
	Autophagy	60 pM	CHO, Caco2	Extensive vacuolation	VCC-Null mutant	Gutierrez et al., 2007
Hla	Stress	100 pg/ml	Ephitelial	Activation of MAPK p38	Hla deficient mutant	Ratner et al., 2006
Cry5B	Stress	low dose	C. elegans	Activation of MAPK p38		Huffman et al., 2004
	Biosynthesis of lipids	low dose		MAPK p38 activates UPR system		Bischof et al., 2008

stress response, suggesting that osmotic stress probably involves Ca^{2+} influx (Ratner *et al.*, 2006).

The activation of MAPK p38 in *C. elegans* or in HeLa cells, after Cry5 or aerolysin treatment respectively, induces the endoplasmic reticulum stress response to unfolded proteins (UPR). The UPR system is a complex response to different stress situations in the cell. The UPR pathway protects cells for accumulation of unfolded proteins during toxin intoxication and increases phospholipids biogenesis to defend against the toxin. One of the transducers-branch that activates UPR response, named *ire-1* arm, is specifically activated by MAPK p38 after treatment of nematode *C. elegans* with the Cry5 PFT. It was proposed that cells have adapted the UPR pathway to promote cellular defense to PFT (Bischof *et al.*, 2008).

The protective role of MAPK p38 pathway in switching on a survival response was also observed in human epithelial cells (HaCaT) and in marrow-derived mast cells after treatment with α-toxin or SLO, respectively (*Table 2*) (Husmann *et al.*, 2006; Stassen *et al.*, 2003). Treatment HaCaT cells with α-toxin at concentrations lower than 10 ng/ml, activates MAPK p38 phosphorilation. This activation correlates with pore formation activity since a pore formation mutant that is still able to bind the cells did not produce that effect. The inhibition of MAPK p38 with the specific SB203580-inhibitor, inhibits the recovery process of cells to membrane damage (Husmann *et al.*, 2006). In mast cells, low doses of SLO activate MAPK p38 pathway and also degranulation. The MAPK p38 up regulates cytokines mRNA expression such as tumor necrosis factor alpha (TNF-α). The production of TNF-α plays an important role in host defense in the murine model since it recruits inflammatory cells that have critical roles in innate and adaptive immunity (Stassen *et al.*, 2003). The TNF-α synthesis in mast cells is also induced by another CDC toxin, the LLO (*Table 2*). This response depends in LLO induced Ca^{2+} influx from extracellular milieu and release from intracellular stores (Gekara *et al.*, 2007).

ACTIVATION OF LIPID BIOSYNTHESIS

Epithelial cells respond to PFT treatment by the activation of central regulators involved in membrane biogenesis (Gurcel *et al.*, 2006). It was reported that in Chinese hamster ovary cells (CHO) and HeLa cells the treatment with aerolysin or α-toxin triggers K^+ efflux. The decrease in intracellular K^+ is linked to the assembly of two types of multiprotein inflammasome complexes (named IPAF and NALP3) that activate caspase-1. The caspase-1 is involved the activation of the central regulator of lipid metabolism, the sterol responsive element binding proteins (SREBP). Caspase-1 triggers the export of SREBP from endoplasmic reticulum and its activation by Golgi proteases. The activated SREBP function as transcription factors that migrate to the nucleus promoting the synthesis of genes involved in membrane biogenesis. The activation of this lipid pathway is directly correlated with cell survival (*Table 2*) (Gurcel *et al.*, 2006). Pore formation is necessary for activation of SREBP transcription factor since toxin mutants affected in insertion into the membrane or in oligomer formation that lack pore formation activity did not activate this lipid pathway (Gurcel *et al.*, 2006).

In another report it was demonstrated that pretreatment with interferon IFN-α before treatment of lung epithelial cells with α-toxin prevents cell-death. The IFN-α-induced protection involves an up-regulation of lipid metabolism that depends on protein synthesis and on the activity of fatty acid synthase (Yarovinsky *et al.*, 2008). This response is independent of MAPK p38 since SB203580 inhibitor has no effect on IFN-α-induced protection (Yarovinsky *et al.*, 2008).

ELIMINATION OF PFT BY ENDOCYTIC MECHANISMS

When the plasma membrane of eukaryotic cells is mechanically injured a process that involves exocytosis is triggered and this process is totally dependent in a transient Ca^{2+} influx. It is proposed that cells are rapidly resealed by the delivery of intracellular membrane to the damaged-cell surface, thus membrane fusion provide a patching for cell repair (McNeil and Kirchhause, 2005). However this mechanism would not fix a stable transmembrane lesion caused by a PFT and it was demonstrated that repair of transmembrane pores produced by several PFT requires other processes different from the exocytosis-resealing mechanism of mechanical wounds. There are two different mechanisms involved in repairing transmembrane pores induced by PFT in plasma membrane, both involved endocytosis but their main difference resides in their dependence of Ca^{2+} influx. In the case of PFT that induce Ca^{2+} influx through pore formation, such as SLO that produce pores permeable to Ca^{2+} ions, the lesion could be repaired in kidney and HeLa cells by a process that involves endocytosis to remove the SLO containing pores from the plasma membrane (*Table 2*) (Idone *et al.*, 2008). The endocytosis of early endosomes involved in PFT pore repair is a dynamin-independent mechanism (Idone *et al.*, 2008). By contrast, in the case of PFT that make small pores in the plasma membrane, which are non permeable to Ca^{2+} ions such as the α-toxin, a different endocytic process is involved in the repair of this pore lesion. In this process the eukaryotic cells internalize the pore-complex and then returned it to the extracellular milieu as an exosome structure (*Table 2*) (Husmann *et al.*, 2009). It has been reported that endocytosis and the later exocytosis as exosomes, rescue HaCaT, and Cos7 cells from α-toxin attack. Cells (Huh7 or HEK293) that have low efficiency at internalizing the PFT did not survive after toxin treatment. This endocytic mechanism is dependent of dynamin and could be specifically inhibited with the inhibitor dynasore, impairing the recovery of ATP-levels after permeabilization with α-toxin and enhancing cell death (Husmann *et al.*, 2009).

PROTECTIVE ROLE OF AUTOPHAGY

As stated above autophagy is a survival mechanism since it may restrict the infection by sequestering the pathogens in the autophagosoma and further degradation in lysosomes (Deretic, 2006). *Vibrio cholerae* cytolysin (VCC) is a PFT that generates extensive vacuolation in epithelial cells (Löner *et al.*, 2009). Recently it was shown that VCC modulate autophagy in CHO and Caco2 cells as a cellular defense pathway against this toxin and autophagy inhibition resulted in decreased survival of Caco2 cells upon VCC intoxication (Gutierrez *et al.*, 2007). The autophagosome formation is induced at low concentrations the toxin as an attempt of the cells to survive, whereas high concentrations of VCC lead to cell lysis (*Table 2*) (Gutierrez *et al.*, 2007).

Concluding remarks

When PFT are present at high concentrations, cells cannot deal with high pore formation activity in the plasma membrane and die quickly by oncosis or necrosis since they could not control or compensate the permeability changes. By contrast when the PFT are present in low sublytic concentrations the cells have different strategies to respond against them and promote survival as autophagy, endocytosis, up-regulation of lipid metabolism or activation of stress responses directed by MAPK p38. However low concentrations of PFT can also induce cell death by apoptosis, or pyroptosis depending in the cell type. The main difference between these two processes is the triggering of an inflammatory response that is present only in pyroptosis but not in apoptosis programmed cell death. *Table 1* shows that induction of apoptosis by low doses of PFT is not restricted to immune cells such as neutrophils, monocytes, macrophages, T- lymphocytes, leukocytes, since it has been also observed in other cell types such as epithelial cells as Caco2, HeLa, renal and endothelial cells. In contrast pyroptosis has only been observed in macrophages. It was proposed that induction of cell death by apoptosis could be an escaping mechanism that bacteria may have selected to avoid the inflammasome response that could control the infection.

Among the different signal transduction pathways that are activated by PFT, MAPK p38 activation plays an important role. It is clear that MAPK p38 may trigger survival responses to PFT action in several cell types. However, in some others, this pathway activates apoptosis as shown in neuronal cells after treatment with PLY. *Table 2* shows that the cell survival-responses induced by MAPK p38 activation are frequently observed in different cells after treatment with low doses of PFT. In marrow-derived mast cells, activation of MAPK p38 induces TNF-α synthesis. TNF-α recruits inflammatory cells that have critical roles in innate and adaptive immunity. In *C. elegans* and epithelial cells MAPK p38 induces the UPR stress response which protects the cells for accumulation of unfolded proteins during toxin intoxication and increases biogenesis of phospholipids to repair the lesion.

It is interesting that in all cases the signal that triggers the survival or death responses at low dose of PFT is related to the changes in the membrane permeability observed after pore formation activity of PFT. These data are supported by different point mutations in PFT affecting their pore formation activity or knockout mutants where PFT are missing. In addition, the influx of Ca^{2+} or the efflux of K^+ seems to be important signals to trigger these responses (*Table 2*). High doses of PFT also kill susceptible cells by producing severe membrane damage due to high pore formation activity (*Table 1*). It is likely that *in vivo* conditions will resemble those of low dose of PFT, triggering cells defense trough -live and -death pathways.

Although the PFT are neither similar at their amino acid sequence nor their structural levels, their mechanism of action is similar since this relies in changing the membrane permeability of their target cells and as a consequence cells have evolved similar mechanisms to avoid or repair membrane damage that would ensure cell survival upon pathogen attack.

Acknowledgements

This work was supported by CONACyT (U48631-Q); DGAPA-UNAM (IN218608-3, IN210208); NIH 1R01 AI066014 and USDA. 2207-35607-17780.

References

Akira, S., Uematsu, S. and Takeuchi, O. (2006). Pathogen recognition and innate immunity. *Cell* **124**, 783-801.

Anderluh, G. and Lakey J.H. (2008). Disparate proteins use similar architectures to damage membranes. *Trends in Biochemical Sciences* **33**, 482-90.

Aroian, R. and van der Goot, F.G. (2007). Pore-forming toxins and cellular non-immune defences (CNIDs). *Current Opinion in Microbiology* **10**, 57-61.

Atapattu, D.N. and Czuprynski, C.J. (2005). *Mannehemia haemolytica* leokotoxin induces apoptosis of bovine lymphoblastoid cells BL-3 via caspase-9 dependent mitochondrial pathway. *Infection and Immunity* **73**, 5504-13.

Bhakdi, S., Muhly, M., Korom, S. and Hugo, F. (1989). Release of interleukin-1beta associated with potent cytocidal action of *Streptococcal* alpha-toxin on human monocytes. *Infection and Immunity* **57**, 3512-9.

Bischof, L.J., Kao, Ch-Y., Ferdinand, C.O.L. *ET AL*. (2008). Activation of the unfolded protein response is required for defenses against bacterial pore forming toxin in vivo. *PLoS Pathogens* **4**, e1000176.

Bravo, A., Gómez, I., Conde, J. *ET AL*. (2004). Oligomerization triggers binding of a *Bacillus thuringienis* Cry1Ab pore-forming toxin to aminopeptidase N receptor leading to insertion into membrane microdomains. *Biochimica et Biophysica Acta* **1667**, 38-46.

Bravo, A., Gill, S. S. and Soberón, M. (2007). Mode of action of *Bacillus thuringiensis* toxins and their potential for insect control. *Toxicon* **49**, 423-35.

Chopra, A.K., Xu, X., Ribaldo. D. *ET AL*. (2000). The cytotoxic enterotoxin of *Aeromonas hydrophila* induces proinflammatory cytokine production and activates arachidonic acid metabolism in macrohages. *Infection and Immunity* **68**, 2808-18.

Deretic, V. (2006). Autophagy as immune defense mechanism. *Current Opinion in Immunology* **18**, 375-82.

Dong, C., Davis, R.J., and Flavel, R.A. (2002). MAP kinases in the immune response. *Annual Review in Immunology* **20**, 55-72.

Fernandes-Alnemri, T., Wu, J., Yi, J.W. *ET AL*. (2007). The pyroptosome: a supramolecular assembly of ASC dimmers mediating inflammatory cell death via caspase-1 activation. *Cell death and differentiation* **14**, 1590-604.

Fink, S.L. and Cookson, B.T. (2005). Apoptosis, pyroptosis and necrosis: mechanistic description of dead and dying eukaryotic cells. *Infection and Immunity* **73**, 1907-16.

Fink, S.L., Bergsbaken, T., and Cookson, B.T. (2008). Anthrax lethal toxin and *Salmonella* elicit the common cell death pathway of caspose-1-dependent pyroptosis via distinct mechanisms. *Proceedings of the National Academy of Sciences of the United States of America* **105**, 4312-17.

Gekara, N.O., Westphal, K., Ma, B. *ET AL*. (2007). The multiple mechanisms of Ca^{2+} signalling by listeriolysin O, the cholesterol-dependent cytolysin of *Listeria monocytogenes*. *Cellular Microbiology* **9**, 2008-21.

Genestier, A.L., Michallet, M.C., Prevost, G. *ET AL*. (2005). *Staphylococcus aureus* Panton-Valentine leukocidin directly targets mitochondria and induces BAX-independent apoptosis of human neutrophils. *Journal Clinical Investigation* **115**, 3117-27.

Gómez. I., Sánchez. J., Miranda. R. *ET AL.* (2002). Cadherin-like receptor binding facilitates proteolytic cleavage of helix α-1 in domain I and oligomer pre-pore formation of *Bacillus thuringiensis* Cry1Ab toxin. *FEBS letters* **513**, 242-46.

Guessan P.D.N., Schmeck, B., Ayim, A. *ET AL.* (2005). *Streptococcus pneumoniae* R6x induced p38 and JNK-mediated caspase-ependent apoptosis in human endothelial cells. *Thrombosis and Haemosttasis* **94**, 295-303

Gurcel. L., Abrami. L., Girardin, S. *ET AL.* (2006). Caspase-1 Activation of lipid metabolic pathways in response to bacterial pore-forming toxins promotes cell survival. *Cell* **126**: 1135–45.

Gutierrez, M.G., Saka, H.A., Chinen, I. *ET AL.* (2007). Protective role of autophagy against *Vibrio cholerae* cytolysin, a pore-forming toxin from *V. cholerae*. *Proceedings of the National Academy of Sciences of the United States of America* **104**, 1829–34.

Honeyman, M.C. and Harrison, L.C. (2004). *Immunity: Humoral and Cellular* In *Encyclopedia of the life Sciences* pp 1-10. New York: John Wiley & Sons. Doi: 10.1038/npg.els.0001235

Hoebe, K., Janssen, E., and Beutler, B. (2004). The interface between innate and adaptive immunity. *Nature Immnonology* **5**, 971-4.

Huffman, D.L., Abrami, L., Sasik, *ET AL.* (2004). Mitogen-activated protein kinase pathways defend against bacterial pore-forming toxins. *Proceedings of the National Academy of Sciences of the United States of America* **101**, 10995-1000.

Husmann, M., Beckmann, E., Boller, K. *ET AL.* (2009). Elimination of bacterial pore-forming toxin by sequential endocytosis and exocytosis. *FEBS Letters* **583**, 337-44.

Husmann, M., Dersch, K., Bobkiewicz, W. *ET AL.* (2006). Differential role of p38 mitogen activated protein kinase for cellular recovery from attack by pore-forming *S. aureus* α-toxin or streptolysin O. *Biochemical and Biophysical Research Communications* **344**, 1128–34.

Huw, D.D. (2008). Immune System. In: *Encyclopedia of Life Sciences* eds. John Wiley and Sons, pp1-10. Chichester: DOI: 10.1002/9780470015902.a0000898.pub2

Idone, V., Tam, Ch., Goss, J. *ET AL.* (2008). Repair of injured plasma membrane by rapid Ca^{2+}-dependent endocytosis. *Journal of Cell Biology* **180**, 905-14.

Jiménez-Juárez, N., Muñoz-Garay, C., Gómez, I. *ET AL.* (2007). *Bacillus thuringiensis* Cry1Ab mutants affecting oligomer formation are non toxic to *Manduca sexta* larvae. *Journal of Biological Chemistry* **282**, 21222-9.

Katayama, H., Kusaka, Y., Yokota H. *ET AL.* (2007). Parasporin-1 a novel cytotoxic protein from *Bacillus thuringiensis* induces Ca^{2+} influx and sustained elevation of the cytoplasmic Ca^{2+} concentration in toxin sensitive cells. *Journal of Biological Chemistry* **282**, 7742-52.

Löner, S., Walev, I., Boukhallouk, F. *ET AL.* (2009). Pore formation by *Vibrio cholera* cytolysin follows the same archetypical mode as β-barrel toxins from Gram positive organism. *FASEB Journal* fj.08-12688 published online March 10, 2009

McClane, B. A. and Chakrabarti, G. (2004). New insights into the cytotoxic mechanisms of *Clostridium perfringens* enterotoxin. *Anaerobe* **10**, 107-14.

McNeil, P.L. and Kirchhause, T. (2005). An emergency response team for membrane repair. *Nature Reviews.Molecular Cell Biology* **6**, 499-505.

Menzies, B. E. and Kourteva, I. (2000). *Staphylococcus aureus* alpha-toxin induces apoptosis in endothelial cells. *FEMS Immunology and Medical Microbiology* **29**, 39-45.

MORIN, S., BIGGS, R. W., SHRIVER, L. *ET AL.* (2003). Three cadherin alleles associated with resistance to *Bacillus thuringiensis* in pink bollworm. *Proceedings of the National Academy of Sciences of the United States of America* **100**, 5004-9.

NELSON, K.L., BRODSKY, R.A. and BUCKLEY, J.T. (1999). Channels formed by subnanomolar concentrations of the toxin aerolysin trigger apoptosis of T lymphomas. *Cellular Microbiology* **1**, 69-74.

PALMER, M. (2001). The family of thiol-activated, cholesterol-binding cytolysins. *Toxicon* **39**, 1681-9.

PARKER, M.W. and FEIL, S.C. (2005). Pore forming proteins toxins: from structure to function. *Progress in Biophysics and Molecular Biology* **88**, 91-142.

PEDERSEN, A. E. (2007). Immunity to Infection. In: *Encyclopedia of Life Sciences*, pp 1-8. New York: John Wiley & Sons. Doi: 10.1002/9780470015902.a0000478.pub2

RATNER, A.J., HIPPE, K.R., AGUILAR, J.L. *ET AL.* (2006). Epithelial cells are sensitive detectors of bacterial pore-forming toxins. *Journal of Biological Chemistry* **281**, 12994–8.

RODRÍGUEZ-ALMAZÁN, C., ZAVALA, L.E., MUÑOZ-GARAY, C. *ET AL.* (2009). Dominant negative mutants of *Bacillus thuringiensis* Cry1Ab toxin function as anti-toxins: Demonstration of the role of oligomerization in toxicity. *PloS ONE* **4**, e5545.

SCHIAVO, G. and VAN DER GOOT, F.G. (2001). The bacterial toxin toolkit. *Nature Reviews Molecular Cell Biology* **2**, 530-7.

SOBERÓN, M., GILL, S.S., and BRAVO, A. (2009). Signaling versus punching hole: How do *Bacillus thuringiensis* toxins kill insect midgut cells? *Cellular and Molecular Life Sciences* **66**, 1337-49.

SOBERÓN, M., PARDO-LÓPEZ, L., LÓPEZ, I. *ET AL.* (2007). Engineering modified Bt toxins to counter insect resistance. *Science* **318**, 1640-2.

SHOMA, S., TSUCHIYA, K., KAWAMURA, I., *ET AL.* (2008). Critical involvement of pneumolysin in production of interleukin-1alpha and caspase-1-dependent cytokines in infection with *Streptococcus pneumoniae in vitro*: a novel function of pneumolysin in caspase-1 activation. *Infection and Immunity* **76**, 1547-57.

SLADE, D.J., LOVELACE, L.L., CHRUSZCZ M., *ET AL.* (2008). Crystal structure of the MACPF domain of human complement protein C8α in complex with C8γ subunit. *Journal of Molecular Biology* **379**, 331-342.

STASSEN, M., MULLER, C., RICHTER, C. *ET AL.* (2003). The streptococcal exotoxin streptolysin O activates mast cells to produce tumor necrosis factor alpha by p38 mitogen-activated protein kinase- and protein kinase C-dependent pathways. *Infection and Immunity* **71**, 6171-7.

STRINGARIS, A. K., GEISENHAINER, J., BERGMANN, F. *ET AL.* (2002). Neurotoxicity of pneumolysin, a major pneumococcal virulence factor, involves calcium influx and depends on activation of p38 mitogen-activated protein kinase. *Neurobiology of Disease* **11**, 355-68.

TIMMER, A.M., TIMMER, J.C., PENCE, M.A. *ET AL.* (2009). Streptopysin O promotes group a *Streptococcus* immune evasion by accelerated macrophage apoptosis. *Journal of Biological Chemistry* **284**, 862-71.

WILES, T.J., DHAKAL, B.K., ETO, D.S. and MULVEY, M.A. (2008). Inactivation of host Akt/protein kinase B signaling by bacterial pore-forming toxins. *Molecular Biology of the Cell* **19**, 1427-38

Yarovinsky, T.O., Monick, M.M., Husmann, M. and Hunninghake, G.W. (2008). Interferons increase cell resistance to *Staphylococcal* alpha-toxin. *Infection and Immunity* **76**, 571-7.

Zhang, X., Candas, M., Griko N.B. *ET AL.* (2006). A mechanism of cell death involving an adenylyl cyclase/PKA signaling pathway is induced by the Cry1Ab toxin from *Bacillus truringiensis. Proceedings of the National Academy of Sciences of the United States of America* **103**, 9897-902.

Biotechnology and Genetic Engineering Reviews - Vol. 26, 83-116 (2009)

Biomedical application of plasmid DNA in gene therapy: A new challenge for chromatography

F. SOUSA, L. PASSARINHA, J.A. QUEIROZ*

CICS – Centro de Investigação em Ciências da Saúde, Universidade da Beira Interior, 6201-001 Covilhã, Portugal

Abstract

Gene therapy and DNA vaccination are clinical fields gradually emerging in the last few decades, in particular after the discovery of some gene-related diseases. The increased relevance of biomedical applications of plasmid DNA (pDNA) to induce therapeutic effects has had a great impact on biopharmaceutical research and industry. Although there are several steps involved in the pDNA manufacturing process, the several unit operations must be designed and integrated into a global process. After the plasmid has been designed according to the requirements for clinical administeration to humans, it is biosynthesised mainly by an *E. coli* host. The overriding priority of the production process is to improve plasmid quantity - the production conditions need to be optimised to guarantee pDNA stability and biological activity.

The complexity and diversity of biomolecules present on the pDNA-containing extracts represent the main concern and limitation to achieve pure and biologically active pDNA. There has been a recent intenstification of the improvement of existing purification procedures or the establishment of novel schemes for plasmid purification.

---

* To whom correspondence may be addressed (jqueiroz@ubi.pt)

Abbreviations: AC, affinity chromatography; AEC, anion-exchange chromatography; cccDNA, covalently closed circular DNA; *E. coli, Escherichia coli*; DCW, dry cell weight; DO, dissolved oxygen; EMEA, European Agency for the Evaluation of Medical Products; EU, endotoxin units; FDA, Food and Drug Administration; gDNA, genomic DNA; GMP, Good Manufacturing Practices; HIC, hydrophobic interaction chromatography; HPLC, high performance liquid chromatography; IMAC, immobilized metal ion-affinity chromatography; LB, Luria-Bertani; oc, open circular; OD, optical density; pDNA, plasmid DNA; RNAi, interference RNA; sc, supercoiled; SDCAS, semi-defined medium containing casamino acids; SDS, sodium dodecyl sulfate; SDSOY, semi-defined medium containing soya amino acids; SEC, size-exclusion chromatography; TFF, tangential flow filtration; THAC, triple-helix affinity chromatography.

This review focuses on the progress and relevance of chromatographic methodologies in the purification of pDNA-based therapeutic products. The review will attempt to assemble their different contributions of the different chromatographic procedures that are being used in the pDNA purification area. The advantages and disadvantages of the different chromatographic techniques, as well as the most significant improvements in response to the challenge of purifying pDNA will be discussed, emphasizing the future directions in this field.

Introduction

Novel technologies using genes themselves as drugs to treat infections, hereditary genetic disorders, cancers, heart diseases and many other devastating illnesses, have brought together the efforts of biotechnology, genetic, pharmaceutical and medical researchers to develop efficient therapeutic products and new production platforms. With the aid of these advances, new promising and potentially revolutionary human DNA therapeutics have emerged in a way to therapeutically manipulate the gene set-up in many organisms, to alter the DNA content and introduce new useful characteristics.

The increased interest in gene therapy or DNA vaccination as alternative therapies has been accompanied by the expansion of new methodologies. Although, the first gene therapy reports were constructed by using DNA as therapeutic products, more recently the applicability of RNA molecules coding for tumor associated antigens is now also being exploited for specific vaccination approaches (Weide *et al.* 2008). A novel mechanism for regulation of gene expression has, for example, recently been discovered, involving the ribonucleic acid interference (RNAi). The RNAi mechanism uses two forms of small RNA molecules, micro RNA which prevents protein synthesis, and small interfering RNA which degrades messenger RNA (McCarthy *et al.* 2009).

Although many clinical trials of gene therapy and DNA vaccination have displayed exciting efficacies, some safety problems have hindered DNA-based products from reaching the market and their clinical application (Hu *et al.* 2006). On the course from early pre-clinical research to final commercial products, gene therapy tools and production methods have undergone tremendous changes to improve safety and efficacy. Hence, to reach a successful biomedical application, the need to improve or optimise the global biosynthesis and manufacturing process has arisen. In fact, the potency of these therapeutic products based on DNA insertion has been significantly enhanced by recent advances in the genetic, immunological and biotechnological research fields. Some of the more noteworthy approaches are based on the implementation of plasmid constructs, achieved through genetic engineering, on the improvement of the plasmid manufacturing process and on the discovery of novel and more suitable delivery systems. All the operations involved in plasmid design, production and purification must be planned and adjusted to control critical conditions. The purification degree of the therapeutic plasmid product is one of the most significant requisites, placing a high degree of relevance on the selection of the purification strategy to guarantee plasmid DNA (pDNA) stability and biological activity.

THE RELEVANCE OF pDNA ON EMERGING GENE THERAPIES

In recent years, society in general, medical, industrial and research groups have assisted in an increased interest in the development of DNA vaccination and gene therapy strategies, as well as in the recombinant production of therapeutic agents. Gene therapy is expected to have a major impact on human healthcare in the future and become a promising therapeutic option for treating and improving control of cancer and genetic diseases (Guo *et al.* 2008). Although some conventional treatments are effective, gene therapy could be more economical and more convenient because it provides higher targeting and prolonged duration of action, which allows biological effects to be more subtle and better localized to the most appropriate cells (Manthorpe *et al.* 2005; Mountain 2000; Patil *et al.* 2005; Sebestyen *et al.* 2007). However, the development of gene-based products as an alternative for the conventional drugs presented new challenges for researchers, clinicians and regulatory authorities. While there is a long history with adverse effects of pharmaceutical drugs, the successful commercial implementation of gene therapy drugs has only recently have been reported. In 2003, the Chinese drug regulatory agency approved the first gene therapy product for head and neck squamous cell carcinoma, which subsequently became available at 2004 (Patil *et al.* 2005). In this particular gene-based treatment, the delivery system is capable of inserting the p53 gene into tumour cells, thereby inducing cell death. The approval of other gene-based therapeutic products has subsequently occurred, and since gene therapy begins to produce its first clinical successes the interest for achieving suitable vectors is also rapidly expanding.

Recombinant viral vectors are the natural preferred vehicles for heterologous gene delivery for immune responses and have been extensively studied and developed, mimicking real-life infection (Anderson and Schneider 2007; Brave *et al.* 2007). However, several drawbacks are associated with the application of viruses, since their intrinsic immunogenicity restricts their repeated usage, their delivery capability is restricted to relatively small amounts of DNA (Braun 2008) and they have a small risk of random integration with oncogene activation and consequent leukemia (Brannon-Peppas *et al.* 2007; Patil *et al.* 2005; Tangney *et al.* 2006). Largely due to the setbacks encountered with viral vectors, emphasis has been placed on improving the efficacy of non-viral systems because of their favourable safety profile. In particular, naked-DNA vaccination has arisen as an effective strategy in the preventive medicine field with promising future prospects. The ability of pDNA to activate both the humoral and cellular immune systems against the encoded antigen (Liu 2003; Manthorpe *et al.* 2005; McDonnell and Askari 1996; Weide *et al.* 2008) have resulted in an intensive study of new strategies aimed at increasing the DNA vaccine immunogenicity.

DNA-based therapies present better potential for multivalent treatments, are comparatively less expensive and easy to produce using a generic manufacturing process based on bacterial culture and DNA preparation protocols (Manthorpe *et al.* 2005; Zanin *et al.* 2007). In addition, the quality control is standardized, storage conditions are independent of the encoding genes (Manthorpe *et al.* 2005) and these therapeutic products are also relatively safe (Zanin *et al.* 2007). Hence, as non-viral vectors are safer, less immunogenic and have greater potential for delivery of larger genetic units (Tangney *et al.* 2006), they can also be used in various gene therapy approaches.

However, given the initial lack of efficiency of non-viral vectors in experimental studies and in clinical settings, the overall outcome has clearly indicated that improved synthetic vectors and delivery techniques are required for successful clinical gene therapy (Braun 2008). Recently, the development of a gene delivery methodology based on a hydrodynamics procedure has resulted in high level of gene expression and suppression of liver metastasis after intravenous injection of naked pDNA (Yonenaga *et al.* 2007). Another example with clinical significance was the expression of vascular endothelial growth factor using naked plasmid DNA that enhanced the recovery of local blood flow and stimulated the proliferation of vascular endothelial cells (Chang *et al.* 2008). In 2007, it was reported the first clinical success using a pDNA vector to treat critical limb ischemia by inducing the expression of angiogenic growth factors (Morishita *et al.* 2004).

As summarized in *Table 1*, the promising preclinical results encouraged non viral-based therapies into clinical trials for a number of diseases with complex etiologies, such as cystic fibrosis (Lee *et al.* 2005), anemia (Sebestyen *et al.* 2007), cancer (Yonenaga *et al.* 2007), and neurodegenerative diseases such as Alzheimer's disease and Parkinson's disease. The application of non-viral vectors in DNA vaccination strategies has also resulted in progress, focusing on several target diseases such as influenza (Drape *et al.* 2006; Hoare *et al.* 2005; Kodihalli *et al.* 1999), malaria (Wang *et al.* 1998) Ebola (Martin *et al.* 2006), HIV (Mwau *et al.* 2004), or human papillomavirus (Sheets *et al.* 2003).

Table 1. Examples of plasmid DNA clinical trials currently in progress

Plasmid / Drug	Target Disease	Trial Phase	Purpose	Related Reference
pGA2/JS2 Plasmid DNA Vaccine	HIV infection	Phase I	Induce an immune response against HIV proteins	Gurunathan *et al.* 2000
VRC-AVIDNA036-00-VP	H5N1 Influenza infection	Phase I	Induce immunity against H5N1	Subbarao *et al.* 2006
HGF plasmid	Critical limb ischemia	Phase II	Evaluate the improvement of blood perfusion	Powell *et al.* 2008
pDNA encoding gp100	Cancer – Melanoma	-	Increasing T-cell responses against cancer antigens	Rosenberg *et al.* 2003
pGM169/GL67A	Cystic Fibrosis	Phase I Phase II	Evaluate gene transfer to the lungs of patients with cystic fibrosis	Alton *et al.* 1999

RELATION OF PDNA TOPOLOGY WITH ITS BIOLOGICAL FUNCTION

Although the development of the DNA assisted clinical strategies may be performed by using different DNA vectors, the application of pDNA is being more accepted because of the advantages associated with its non-viral origin. Plasmids are high molecular weight (M $>10^6$ Da), double-stranded DNA constructs containing transgenes, which encode specific proteins. In nature, plasmids are found in bacteria where pDNA is located separately from the chromosomal DNA. All essential functions of the bacterial cell are described by the genetic code of the chromosomal DNA, and only certain specific functions are expressed by genes of the plasmid.

An advantage of using pDNA in genetic therapies is that no major formulation or alteration of plasmid molecules is required. Approximately half of the plasmid encodes the two major elements used by bacteria during production, namely a bacterial origin of replication and a selectable marker, usually an antibiotic resistance factor (Kaslow 2004; Liu 2003). The other half of the plasmid ring comprises a complete eukaryotic expression element, consisting of a promoter, a region encoding the gene of interest and a termination sequence (Kaslow 2004; Liu 2003).

The mechanism of action of pDNA in gene therapy or DNA vaccination requires that plasmid molecules gain access into the nucleus after entering the cytoplasm (Patil *et al.* 2005). This is one of the steps that can be limiting in efficient gene expression. Therefore, the design and engineering of plasmids to obtain maximum transfection efficiency is being extensively investigated. Several parameters may influence the transfection efficiency and gene expression being the compaction degree one of the most stated.

Inside the eukaryotic cells the DNA structure must be compact, for this reason the original right-handed twist DNA molecule is reorganized into a more complex structure which results from the additional twist in the molecule in the opposite sense resulting in negatively supercoiled (sc) DNA. Biologically, negative supercoiling is advantageous because it promotes the unwinding and strand separation necessary during some biological phenomena (Clark 2005). It is currently accepted that DNA topology is of fundamental importance for a wide range of biological processes including DNA transcription, replication, recombination, control of gene expression and genome organization (Palecek *et al.* 2004). In fact, DNA-protein binding is often supercoiling dependent, and the excess energy contained in scDNA is relieved by protein binding. A number of DNA-protein complexes only occur because the binding is promoted by negative supercoiling due to stabilization of writhing (Palecek *et al.* 2004). Recent studies suggest a possible relation between the p53 function and the DNA topology. Palecek and co-workers have hypothesized that the DNA supercoiling degree may play a significant role in the complex p53-regulatory network (Palecek *et al.* 2004).

Similarly, plasmids present the active supercoiled structure also known as covalently closed circular DNA (cccDNA). If one strand of this form is nicked, the supercoiling can unravel, resulting other plasmid conformation described as open circular (oc). Linear forms can also be generated if both strands are cleaved once at approximately the same position (Schleef and Schmidt 2004). Both oc and linear forms may be damaged at different gene locations randomly, which make these forms less efficient in transferring gene expression, as exemplified for instance in a recent study on the impact of supercoiling on the efficacy of a rabies DNA vaccine (Cupillard *et al.* 2005). This is especially true if promoter or gene coding regions have been destroyed (Cherng et al. 1999; Schleef and Schmidt 2004). These and other pDNA structures depend on certain special characteristics within the DNA sequence, as well as supercoiling stress or unfavourable environment conditions, like extreme pH or high temperature. Changing the temperature has an effect on the helical repeat of the DNA, because a temperature increase results in more thermal motion promoting a gradual unwinding of the DNA helix (Clark 2005). As demonstrated by recent circular dichroism analysis

(Sousa *et al.* 2007), supercoiled pDNA will become absolutely less supercoiled with the temperature increased. As these conditions can contribute to an alteration in DNA conformation, an accurate control is extremely important to guarantee the DNA stability and consequently its biological activity. Therefore, the only intact and undamaged form is the supercoiled pDNA, and the prediction of the potential applications of this product has motivated the development of new technologies to efficiently respond to the increased demand for the production of large amounts of highly pure pDNA (Manthorpe *et al.* 2005). For this reason, pDNA has become a very promising gene delivery vector because it can be genetically manipulated, produced by cultivation of plasmid harbouring *Escherichia coli* (*E. coli*) and purified in a subsequent downstream processing event (Liu 2003; Rolland 2005).

Plasmid DNA biosynthesis

Nowadays it is well established that non-viral gene therapy requires considerable amounts (gram scale) of pharmaceutical-grade pDNA per patient since the efficacy and duration of gene expression is presently relatively low. Additionally, it has been suggested that relaxed forms are less desirable for transfection than supercoiled structure and should therefore be minimized in a gene therapy product, especially if their potencies cannot be determined. In order to fulfil these points, the reliance on biosynthesis for producing the desirable quantities of supercoiled pDNA isoform, for *in vitro* testing and proof-of-concept preclinical trials is becoming more widespread (Patil *et al.* 2005.

The manufacturing process for clinical-grade plasmid DNA vaccines encompasses a number of key steps (*Figure 1*). Initially, a vector consisting of a plasmid backbone, including typical elements, such as the origin of replication, an antibiotic resistance gene, a stronger eukaryotic promoter and a polyadenylation signal sequence, and encoding the therapeutic gene(s) of interest, is generated. Subsequently, a bacterial cell bank containing the plasmid is established to create a uniform inoculum for further process development and large-scale fermentation. Specifically, any new bioreactor system must be able to support the cultivation of cells at high density to achieve economical plasmid production. Through optimization of the biological system, growth environment and the growth mode, improvements can be achieved in biomass productivity, plasmid yield, plasmid quality and manufacturing costs. Subsequently, the cells are lysed using one of a plurality of disruption methods. As reviewed below, these two initial unit operations are critical for the improvement plasmid quantity and for the adjustment of the production and lysis conditions in order to guarantee pDNA stability and biological activity.

OVERVIEW OF PRODUCTION FLOWSHEET STRATEGIES

As gene therapy and DNA vaccines advance towards approval by the Food and Drug Administration (FDA) it is essential to devise industrial processes whereby DNA can be economically manufactured at the gram scale or beyond (Carnes *et al.* 2006). Nowadays, regulatory requirements stipulate that the production of any agent intended for use in human clinical trials, such as pDNA, must be performed under current "Good Manufacturing Practices" (cGMP) This implies that the manufacturing process is fully in-control, performs as intended and includes the use of methodologies that ensure the identity, safety, purity and potency of the manufactured product. In general, the biotechnological manufacture of pDNA is separated into two stages: upstream processing during which pDNA is produced by cells genetically engineered to contain the gene of interest and downstream processing during which the pDNA is isolated and purified. The flow diagram presented in *Figure 1* outlines the main steps involved in the development of a pDNA manufacturing process. To date, most efforts toward process development have focused on downstream processing. However, the production and quality of a final product are also determined by upstream processing. The features which are important for an efficient expression of a plasmid vector are high segregational stability in order to reduce contamination and high copy number that favours the structural form and DNA sequence (Tejeda-Mansir and Montesinos 2008). Plasmid fermentation processes ideally maximize both the volumetric yield (mg/L) and specific yield (mg/g DCW) of high quality supercoiled plasmid since other isoforms are difficult to remove during purification and are considered undesirable by regulatory agencies. Specifically, high volumetric yields (parameter: biomass concentration) facilitate smaller and more economical fermentations, while high specific yield (parameter: pDNA copy number) drastically improves plasmid purity and yield in downstream processing.

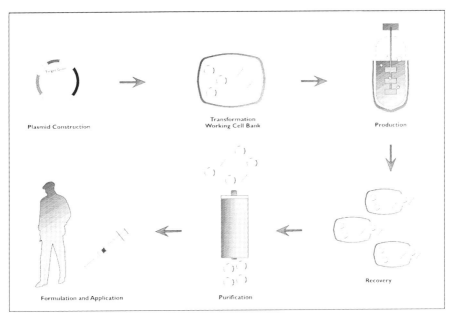

Figure 1. Flow diagram of the processes involved in the upstream and downstream manufacture of plasmid DNA vectors at laboratorial and industrial scales.

At the laboratory scale, production and purification of plasmids are generally viewed as relatively easy and simple procedures. However, plasmid production under non-optimized laboratory conditions generally leads to volumetric titers ranging from 5 to 40 mg/L (Prather *et al.* 2003). In the context of milligram range doses and large patient populations, these processes and their associated productivity levels are inadequate for economically viable plasmid production. Although a few publications mention large scale plasmid production processes achieving productivity levels of several hundreds of milligrams per liter (Listner *et al.* 2006; Rozkov *et al.* 2004), which represents a significant improvement over standard laboratory processes.

In general, plasmids are maintained in cells by exerting a selective pressure upon an organism. For instance, a plasmid may be engineered to carry an antibiotic resistance gene or, another common method, use a mutant host that is unable to synthesize an essential amino acid (an auxotroph) (Tejeda-Mansir and Montesinos 2008). Also, therapeutic plasmids should be stable in their monomeric form ($\geq 90\%$) and resistant to multimerization. It has been suggested that the formation of multimers in a cell reduces the apparent copy number, leading to higher instability. To circumvent this, multimerization can be controlled by inclusion of a multimer resolution system that utilizes a site specific recombinase, or in alternative preparing the seed stocks at low temperature in order to decrease metabolic burden in bioreactor production (Przybylowski *et al.* 2007).

The choice of the bacterial host strain is another important factor to be considered for fermentation of pDNA. Typically, DNA vaccine plasmids are manufactured using NIH automatic exempt attenuated *E. coli* K12 strains such as DH5α (Carnes *et al.* 2006), DH5 (Listner *et al.* 2006), DH1 (Cooke *et al.* 2004), JM108, SCS1-L (Singer *et al.* 2009) or DH10B (Lahijani *et al.* 1996). Also, BL21 (an *E. coli* B strain) has recently been reported as a high yielding plasmid production host (Phue *et al.* 2008). In addition it has been extensively described that for pDNA production auxotrophic strains are suitable for growth in semi-defined media but not for production in minimal media since they require expensive supplementation. And further, it is essential that strains do not possess mucoid properties, since they become difficult to harvest and foul downstream processing.

The control of plasmid copy number is essential to tune the metabolic load and to avoid irreversibly overstraining the cell factory, which would compromise production. Therefore, some authors have described the reduction of growth rate as a relevant tool, in order to increase pDNA copy number and also the percentage of supercoiling (Lin-Chao and Bremer 1986; Seo and Bailey 1985).

A common misconception is that high yields obtained with shake flask strategies also lead to high fermentation yields. In fact, some studies demonstrated that strain specific plasmid productivity in a shake flask is poorly predictive of fermentation efficiency, and host performance varies between different fermentation processes. The biological basis for this phenomenon is unknown, but it is critical that vector changes to increase yield be driven by fermentation rather than shake flask evaluation (Carnes and Williams 2007). For instance, DH5α has been described by Huber and co-workers (2005) as a poorer strain in a panel of 7 hosts for production of 3 different plasmids, using LB shake flask media. The authors mentioned that DH5α lead to 1.3 mg/OD550 /L; 65% sc, DH1 3.0 mg/OD550 /L; 84% sc; while JM108 present the highest performance (8.2 mg/OD550/L; 91% sc). By contrast, in an inducible 30-42°C semi-defined batch fermentation process, DH1 and DH5α were both registered as high producing hosts (Carnes *et al.* 2006). These observations are also found for BL21 host strains. In an extensive shake flask evaluation of plasmid specific yield and % supercoiling in 17 hosts with three different pUC origins, BL21 (DE3) was the poor producer due to the RecA+ gene product (Yau *et al.* 2008). In contrast, Phue and co-workers reported high yields (1.9 g/L) of predominantly supercoiled plasmid in a 30°C to 42°C inducible process (Phue *et al.* 2008). Although these batch fermentations are usually simple and short, they have fundamental disadvantages that result in limited pDNA yields. This is due to substrate inhibition and salt precipitation at high nutrient concentrations. With the demands for pDNA in numerous clinical trials approaching the 100-gram scale or greater, a shake flask fermentation, with typical yields of 10-20 mg/L, becomes unfeasible.

A comparison of several fermentation strategies and the respective plasmid yields are presented in *Table 2*. Some processes such as those described by (Huber *et al.* 2005) are not comparable, since maximum specific plasmid yield occur relatively early in the process and is lower during later biomass accumulation. As we have indicated in *Table 2*, lower temperatures (such as 30°C) may be used in the batch stage of fed-batch fermentation to reduce maximum specific growth rate and decrease plasmid copy number (Carnes *et al.* 2006). This strategy combined with induction at 42°C reduces metabolic burden and plasmid loss during the batch phase by minimizing growth difference between plasmid-containing and plasmid-free cells. Typically, a higher temperature (42°C) is employed to induce selective plasmid amplification with some replication origins such as pUC, and pMM1 (Hamann *et al.* 2000). Also, alternative strategies have been developed to reduce plasmid mediated metabolic stress, to alleviate growth rate inhibition at 37°C. For example, engineering a strain to overproduce intermediaries of the pentose-phosphate pathway (required for plasmid replication) by overexpression of glucose-6-phosphate dehydrogenase has partially alleviated plasmid mediated growth rate reduction (Flores *et al.* 2004). As we can see in *Table 2*, three fed-batch processes have recently been described providing more than 500 mg/L plasmid yields. Specifically, patent work from Merck Ltd. has described highly productive clonal subtypes cultivated by a fed-batch fermentation process in chemically defined medium that lead to maximum specific yields of 30-32 mg pDNA/g DCW achieved with feed rates from 8-12 g/L/h (Chartrain *et al.* 2005). In addition other patents have disclosed fed-batch fermentation, in which plasmid-containing *E. coli* cells are grown at a reduced temperature during part of the fed-batch phase, followed by a temperature up-shift in order to accumulate pDNA (Carnes and Williams 2006).

Table 2. Plasmid specific yields from several fermentation strategies

Fermentation Process	Volumetric plasmid yield (mg/L)	Specific plasmid yield (mg/gDCW)	Reference
SD - 30°C growth 42°C induction	2220	51	Carnes *et al.* 2006
SD – 30°C growth 42°C induction	1923	10	Phue *et al.* 2008
D – 37°C throughout	1600	39	Listner *et al.* 2006
D - 37°C growth mode FB 60% glycerol constant 8.0g/L/h	200-1300	30-32	Chartrain *et al.* 2005
SD - 37°C growth mode FB exponential C limiting	1500-2100	43	Carnes and Williams 2006
SD - 30°C to 42°C, two stages C	1020	22	Carnes 2007

SD- Semidefined medium; D- Defined medium; FB- Fed-Batch; C-Continuous; DCW- Dry Cell Weight

A number of recent reports have discussed the fermentation strategies focused on production of plasmid, but have not addressed the effect of fermentation conditions on the quality of the resulting pDNA. The fermentation processes require a balanced medium that supplies adequate amounts of nutrients needed for energy, biomass and cell maintenance. In particular, media for plasmid production should support high nucleotide pools in cells and supply energy for replication while minimizing other cell activity (Wolff *et al.* 1990). Typically, the nutritional requirements are satisfied by a minimal (defined) media, a semi-defined media or a complex media. Specifically, fermentation processes using minimal media are highly reproducible and plasmid copy number may even be higher when using these conditions. In semi-defined media a decrease in reproducibility is not necessarily a problem for plasmid production; however, reliable sources should be used to prevent too much variability over time. Otherwise, culture media containing complex components underlies normal deviations that make the cultivation process less reproducible and problems with contaminant removal in downstream processing. In accordance with a recent study, it has been surprisingly found that rather high magnesium salt concentrations such as 80 mM - preferably of magnesium sulphate - yield excellent results with regard to the homogeneity of the sc plasmid monomers (Schmidt *et al.* 2002). Previous studies have also investigated the effect of the rheology of medium fermentation on the characteristics of sc pDNA obtained during the lysis step. A semi-defined medium containing casamino acids (SDCAS) for eaxmple was found to support higher plasmid stability in contrast to a similar medium containing soya amino acids (SDSOY) or LB. Specifically, differences were observed in the cell harvest characteristics, pDNA primary recovery, pDNA yield and quality between cells grown on LB and on SDCAS medium (O'Kennedy *et al.* 2000).

Normally, the effects of the components in a medium on cell growth and plasmid production should be evaluated to solve the trade-off between higher copy number and reduced growth rate. Acetate accumulation is also a major concern for medium design and high density fermentation of recombinant *E.coli* because it inhibits cell growth and protein expression (Xu *et al.* 2005). Nevertheless, sometimes acetate addition improves pDNA yield in the experimental system. This suggests that low rates lead to high specific pDNA yield provided the cell growth is not seriously inhibited by acetate (Xu *et al.* 2005). The use of reduced growth rate is the unifying principle in high quality, high yield plasmid fermentations. Indeed, high growth rates have been associated with acetate production, plasmid instability, and lower percentages of supercoiled plasmid. A reduced growth rate alleviates growth rate-dependent plasmid instability by providing time for plasmid replication to be synchronized properly with cell division (Tejeda-Mansir and Montesinos 2008). Specifically, in the production of larger plasmids, fed-batch cultures are grown to lower cell densities (40-80) because slower growth rates are often required for optimal production. Yields can be much lower (75-100 mg/L), but are 50-80 times higher than what is achieved in batch fermentation.

Without question, fed-batch fermentation is especially useful for plasmid production since higher biomass yields are obtained since substrate can be supplied at a rate such that it is nearly completely consumed, never reaching inhibitory concentrations (Carnes and Williams 2007). A feedback control nutrient feeding strategy based on pH and dissolved oxygen (DO) has also been used to regulate the cell growth rate by controlling the interactivity of the nutrient feed rate, pH and DO. This process increased the total yield of pDNA by approximately 4 fold as compared to batch fermentation (Wolff *et al.* 1990).

One of the simplest and most effective feeding strategies is exponential feeding. The fermentation begins with a batch mode where the cells are grown at maximum specific growth (μ_{max}) until the substrate is exhausted, at which point the nutrient feeding begins. As an alternative at the industrial level, feeding strategies are either feedback controlled (e.g. DO-stat, pH stat, metabolic activity, biomass and substrate concentration), or predetermined (e.g. constant, linear, stepwise, or exponential feeding). Recently a growth-controlled fed-batch process was proposed for plasmid production and yielded up to 250 mg/L of pDNA (O'Mahony *et al.* 2007).

In general, pDNA supercoiling is known to be affected by oxygen (Passarinha *et al.* 2006) and temperature. Indeed, a single drop in dissolved oxygen concentration down to 5% of air saturation has been found to lead to a rapid loss in plasmid stability (Carnes 2007). The optimal temperature for *E. coli* growth is 37°C, although lower temperatures (30-37°C) may be used in batch fermentation to cause a reduced maximum specific growth rate. Wherein, plasmid production is maintained at a low level to avoid retardation of growth. Batch fermentation at 30°C using glycerol will typically result in a μ_{max} ≤0.3 h^{-1}, which is sufficient to prevent deleterious acetate accumulation and growth rate associated plasmid instability (Thatcher *et al.* 1997).

PRIMARY ISOLATION

Currently, the impact of the engineering environment on the bacterial cell suspension prior to the lysis step must be considered as a pertinent aspect in the implementation of a suitable pDNA bioprocess. Although bacterial lysis is commonly considered a disconnected step in plasmid biosynthesis, process intensification would benefit from an integration of production with the capture of the bacteria. The main options for harvesting large volumes of culture are continuous centrifugation or microfiltration (Kong *et al.* 2008). Previous studies have shown that *E. coli* cells harvested with an intermittent-discharge, continuous flow disc-stack centrifuge are subjected to stresses that are sufficiently high to cause lysis, especially if cells grown in a defined medium (Chan *et al.* 2006). Typically, the centrifugation method used had a strong influence on sc pDNA yield and enhances in productivity could be obtained by an integrated approach between fermentation and centrifugation conditions. In fact, Kong and co-workers (2008) have shown that *E. coli* cells harvested using a solid-bowl centrifuge followed by ressuspension in TE buffer for not more than 2 h and at temperatures lower than 13°C gave the highest sc pDNA yields (Kong *et al.* 2008).

Over the last three decades or so, a concerted effort has been devoted to lysis and further recovery of pDNA in order to maximize the purity at harvest. Typically, most current plasmid disruption schemes adopt a variant of the alkaline lysis protocol originally proposed by Birnboim and Doly (1979) and adapted by Lahijani and co-workers (1996) to release the pDNA from the bacteria. In this type of protocol, disruption is achieved by high pH in the presence of a detergent (e.g., 0.2 M NaOH / 1% SDS). Cellular debris, gDNA and proteins are precipitated by addition of potassium acetate at pH 5.0, and finally the pDNA is recovered by precipitation with alcohol (isopropanol or ethanol). This method has received great attention in that it offers concurrent removal of host cell genomic DNA and proteins by denaturation and selective precipitation of their denatured forms, although the final lysate tends to contain significant amounts of proteins, RNA and endotoxins, which has subsequently to be removed in the subsequent downstream process. Also, this approach requires mixing optimization because the lysates are viscous and shearing of gDNA must be avoided.

Heat lysis has been suggested as an alternative for plasmid release, starting with the boiling method originally proposed by Holmes and Quigley (1981). Specifically, this method avoids the denaturing/renaturing cycle of the pDNA inherent to alkaline lysis. Despite this, heating however may serve to deactivate potentially harmful impurities such as DNases – on the other hand the process may also damage the target product. Although the boiling lysis can release more plasmids than that from the alkaline lysis, the yield and purity of the plasmid is inconsistent and the difficult operation makes this method undesirable even for research scale preparation. As in alkaline lysis, the pDNA must be further purified, concentrated, and subjected to polishing steps based on extraction, chromatography or isopycnic centrifugation (Prazeres and Ferreira 2004). Also, high-pressure homogenizers are used to continuously disrupt cells. However, this strategy can result in the degradation of pDNA caused by severe shear stress, leading to gDNA fragments with similar sizes as the target pDNA (Clemson and Kelly 2003). Despite these effects, glucose and sucrose are often included in the relevant buffers in order to protect plasmids against shearing.

As described below, the lysis step is seen as one of the most critical operations in large-scale pDNA processing, as it becomes more complex to engineer with increasing scale where some inherent problems remain to be solved (Hoare *et al.* 2005). Firstly, for an efficient alkaline lysis, it is necessary to achieve rapid mixing (local pH extremes above 12.5 will irreversibly damage most plasmids) of several buffers as well as rapid changes of the temperature. Secondly, the salt and organic solvents used in the alkaline method have to be removed and discarded afterwards, a problem that will increase with the scale of operation. Finally, the gelatinous precipitate containing cell debris and impurities which is moderately shear-sensitive is typically removed by high-speed centrifugation (Feliciello and Chinali 1993), but a unit operation extremely difficult to scale up. The application of filtration may be used as alternative but at large scale has been described to entail considerable loss in pDNA even when fine filtering agents or additional flotation steps have been used, especially in the presence of divalent cations. For these reasons, alkaline lysis becomes increasingly difficult to control, suffers from lack of reproducibility and may involve significant loss of plasmid (Prazeres *et al.* 1999).

In addition, the large-scale potential of heat lysis has been investigated by Lee and Sagar (1999). In fact, the method proposed by those authors is extremely attractive since it avoids the accompanying denaturation and renaturation of the pDNA. Nevertheless, the requirement to remove the fragile precipitate remains a challenge and an additional microfiltration step is required to obtain a clear lysate. As an alternative, Hilbrig and co-workers (2004) have proposed an integrated method for cell capture and heat lysis where the biomass is harvested by filtration in the presence of a filter aid. The great advantage with this is that no detergent and no further clarification is deemed necessary. This procedure was subsequently improved by O'Mahony and co-workers (2005), where the lysis takes place in situ on the filter cake, which allowed an easy integration and heat disruption at acid pH. Under these circumstances, it becomes possible not only to obtain directly a clear lysate, but also to pre-purify the pDNA to a significant degree.

Nowadays the development of continuous lysis procedures (Zhu *et al.* 2005) have been emerging, which includes the use of a flow-through cell harvest procedure employing a continuous thermal treatment, minimizing the number of centrifugation steps, and providing a scalable, efficient and cost-effective plasmid-preparation system. Due to its flow-through design and consistency from batch to batch, this protocol can be easily scaled-up and can lead to higher yields of pDNA due to three main factors: (i) lysis buffer, (ii) temperature and (iii) the cycle time. Indeed, the characteristics of pDNA obtained from the continuous thermal lysis were shown to be nearly identical to those from the conventional alkaline lysis, a finding confirmed not only by enzyme digestions, cloning and transfection, but also by their expression in eukaryotic cells.

After lysis, the clarification and concentration steps are designed to remove host proteins and nucleic acids, and reduce further the volume of the process stream prior to chromatography. Capture has been accomplished by alcohol precipitation followed by chloroform or phenol purification. Neither of these procedures is desirable for the treatment of large commercial-scale volumes of lysate while that are not recommended for the manufacture of pharmaceutical-grade pDNA. Normally, after isopropyl alcohol precipitation and centrifugation the DNA is contaminated with RNA, however by using hollow fiber membranes, the pDNA is concentrated while RNA is filtered through. Additionally with several washes of 70% of ethanol, the plasmid products contain significantly less contaminated RNA. As a result, more efficient RNA removal steps upstream of the chromatography stage are required.

In general, filtration technology is not normally regarded as a separation tool in plasmid purification and is often considered only for clarification and sterilization. However, some studies (Eon-Duval *et al.* 2003) have shown that by combining the precipitating effect of calcium chloride salt on high-molecular weight RNA with the clearance of low-molecular-weight RNA by tangential flow filtration (TFF), it is possible to reduce RNA to undetectable levels. Despite these promising results, plasmid size and degree of supercoiling may affect the performance of TFF and therefore conditions may have to be reassessed individually for each plasmid.

Preparative-scale purification of plasmid DNA has been attempted by diverse methods, including precipitation with solvents, salts and detergents. Despite these improvements, chromatography is seen as the most suitable downstream processing method for plasmid DNA. Indeed, during the last years this technique is described as an established, scalable technology with a proven track record in pharmaceutical production (Przybylowski *et al.* 2007).

Chromatographic perspectives enhancing pDNA purification

Recent advances in biotechnological areas have led to the investigation and implementation of new technologies for the production of complex biomolecules, which have the potential to assist human health care. The evaluation of the global process attributes a significant relevance to the downstream operations, being considered one of the most important and expensive steps. Although precipitation and ultrafiltration techniques may be involved in the isolation process and aqueous two-phase systems

(Barbosa *et al.* 2008) have been applied in the purification step, liquid chromatography is undoubtedly the most widely used technique.

The development of any plasmid purification strategy must first consider the complex origin of the plasmid-containing extract, since the cell components (gDNA, RNA, proteins and endotoxins) also released during the lysis are regarded as contaminants. A prerequisite for the therapeutic application of pDNA is the final recovery of a highly purified, homogeneous preparation of sc pDNA (>97%), to conform with the strict guidelines established by regulatory agencies (Stadler *et al.* 2004). Generally, the Food and Drug Administration (FDA) and the European Agency for the Evaluation of Medical Products (EMEA) recommend that the final pDNA therapeutic product should be free from host genomic DNA (<2 µg/mg pDNA), host proteins (<3 µg/mg pDNA), RNA (<0.2 µg/mg pDNA) and endotoxins (<10 EU/mg pDNA) (Stadler *et al.* 2004). The relevant approaches used therefore require a clearly adjusted quality assurance in pDNA manufacturing (Schleef and Schmidt 2004). Therefore, several chromatographic methodologies, such as size-exclusion, ion-exchange, hydrophobic interaction, and affinity have already been applied, either as an isolated step or integrated into an overall purification process (*Table 3*). Some difficulties are commonly encountered during pDNA purification, mainly due to the physicochemical similarity found between pDNA and the contaminants that are negatively charged and present similar size and hydrophobicity (Stadler *et al.* 2004). These issues represent a significant constraint for the purification of pDNA and explains the requirement for more than one chromatographic step to purify this product to the desired level.

Hence, the purification of pDNA by chromatography using conventional chromatographic media faces some limitations, not only related to the characteristics of the molecules involved, pDNA and impurities, but also related with the limitations of available stationary phases, mainly because of their low capacity to bind large biomolecules, such as pDNA (Diogo *et al.* 2005; Jungbauer and Hahn 2008). For instance, in the purification of pDNA it has to be considered that the hydrodynamic radius of a pDNA is larger than that of an average protein, and has very low diffusivity (Jungbauer and Hahn 2008). Furthermore, for packed columns with soft chromatographic beads, scale-up is limited by mechanical factors, such as bed instability and mass transfer problems. These shortcomings lead to the recent development of monoliths intended for industrial application (Jungbauer and Hahn 2008). This support consists of a single piece of a porous material, and the main reasons for the increased interest in such supports are the low back pressures associated that permit their use at high flow rates and the rapid mass transfer properties (Jungbauer and Hahn 2004). These characteristics can facilitate fast separations, short analysis times and a decreasing band broadening making efficient chromatographic separations (Mallik and Hage 2006). In fact, the application of monolithic columns to purify pDNA is an emerging area and can represent some technical advantages, because this large macromolecule presents very different biophysical properties than proteins. Together with monoliths (Bencina *et al.* 2004; Branovic *et al.* 2004; Jungbauer and Hahn 2004; Urthaler *et al.* 2005), other new supports have been developed to overcome the diffusion limitation as well as to improve the binding capacity of the support for the target molecules. Such superporous supports (Gustavsson *et al.* 1999; Tiainen *et al.* 2007b), and adsorptive membranes (Giovannini *et al.* 1998; Teeters *et al.* 2003) are advanced approaches that have been shown to be able to partially solve the problems associated with low capacity.

The challenges associated with efficient plasmid purification are based on an ongoing desire for the improvement or establishment of novel purification schemes, and with regards to the recovery of the DNA product to suitable degree purity acceptable for therapeutic application. The decision to use one or more chromatographic steps with contrasting selectivity is determined by the nature and distribution of the residual impurities and contaminants (*Table 3*), as well as by the anticipated plasmid dosage.

Table 3. Comparison of main applications of different published chromatographic techniques to purify pDNA

	Chromatographic support	Application	Reference
Gel Filtration	Superose 6 prep grade	Resolution of high molecular weight nucleic acids (gDNA and pDNA) from smaller molecules	Ferreira *et al.* 1997
	Sephacryl S1000 SF	Reduction of the contamination of pDNA with gDNA and RNA, using high salt concentration	Li *et al.* 2007
Ion Exchange	Q-Sepharose Big Beads	Separation of pDNA from gDNA and RNA, with a prior clarification of the sample	Ferreira *et al.* 1999
	Q-Sepharose	Isolation of sc plasmid from low molecular weight RNA, gDNA and other plasmid forms	Prazeres *et al.* 1998
	Fractogel DEAE	pDNA purification without RNAse application, using a prior precipitation and TFF	Eon-Duval and Burke 2004
HIC	Non-porous packing (TSKgel Butyl-NPR)	Separation of oc and sc pDNA isoforms	Iuliano *et al.* 2002
	Sepharose-gel derivatized with 1,4-butanediol-diglycidylether	Separation of pDNA from gDNA, RNA, proteins	Diogo *et al.* 2000; 2001a; 2001b
	Biporous support Phenyl-based matrix	Separation of pDNA from RNA and proteins	Li *et al.* 200)
	Superporous cross-linked cellulose – HIC CELBEADS	Separation of pDNA from gDNA, RNA, proteins	Deshmukh and Lali 2005
Affinity	IMAC matrices	Isolation of pDNA and gDNA from RNA, endotoxins and plasmid denatured forms	Murphy *et al.* 2003b ; Cano *et al.* 2005 ; Tan *et al.* 2007
	THAC matrices	Isolation of pDNA; Reduction of RNA and gDNA contamination	Wils *et al.* 1997 Schluep and Cooney 1998
	Protein-based matrices	Isolation of pDNA; Elimination of RNA and proteins	Woodgate *et al.* 2002 ; Darby *et al.* 2007
	Amino acids-based matrices	Purification of sc pDNA from clarified *E. coli* lysates	Sousa *et al.* 2005; 2006; 2008a; 2009c

SIZE-EXCLUSION CHROMATOGRAPHY

The application of size-exclusion chromatography (SEC) for nucleic acids purification is facilitated by the fact that the high molecular weight gDNA or pDNA macromolecules elute close to the exclusion limit of the matrix, whilst the low molecular weight impurities (RNA, proteins, endotoxins) show much greater retention by the column matrices. Supercoiling also reduces the hydrodynamic radius of the pDNA. By exploiting this concept and by selecting a SEC support with appropriate selectivity, it is possible to fractionate the different DNA molecules, with baseline separations of RNA, endotoxins and proteins. SEC has the additional advantage of enabling the process buffer to be exchanged for a formulation or storage buffer (Horn *et al.* 1995), ensuring rigorous process control over small amounts of contaminants introduced during the process.

It is worth noting that up to the 1990's, SEC experiments had not been able to yield significant results in pDNA purification trials, due to the lack of stationary phases adequate for the separation of nucleic acids with high molecular mass and complex conformation (Bywater *et al.* 1983; Micard et al. 1985). This situation was however changed with the introduction of rigid and highly porous composite polyacrylamide/dextran stationary phases.

Sephacryl S-1000 (Amersham Biosciences) is one of the stationary phases most widely used in the SEC of pDNA (Bywater *et al.* 1983). This stationary phase presents an exclusion limit of 20 kbp for linear DNA and is reported to be simple, inexpensive and reproducible, yielding milligram amounts of highly pure sc pDNA from partially purified *E. coli* lysate (Diogo *et al.* 2005). In Sephacryl S-1000, the fractionation of the isoforms was found to be dependent on the molecular weight of the pDNA. Specifically, with small pDNA (4.4 Kbp) the isoforms partially overlap, whereas for sizes above 10 Kbp the separation is complete (Vo-Quang *et al.* 1985). Alternatively in the case of lysates extensively contaminated with RNA, some overlapping can occur between the DNA and RNA peaks with a consequent decrease in yield (Raymond *et al.* 1988). Typically, chromatographic trials in these supports are lengthy and up to 200-300 min may be necessary to complete a run. Although a pDNA trial was complete in 45 min, isoform separation was found to be poor.

Another widely used SEC media for pDNA purification is Superose 6B. Typically, this support has a much lower size-exclusion limit for DNA (450 bp), more resistance to pressure and pDNA elution can be achieved in only 20 min (Vo-Quang *et al.* 1985). Occasionally, the performance of Superose 6 can be superior to Sephacryl S-1000 while better process yields were obtained with higher flow rates. In spite of these advantages, preliminary treatments of the lysate with RNase are essential and the separation of gDNA and non-sc isoforms is still difficult to accomplish with this support. Indeed, the discrepancy in pore diameter between these support media would also appear to make Sephacryl S-1000 a better choice for resolving intact gDNA from pDNA.

Recently, the selectivity for pDNA versus RNA has been drastically increased by combining the use of Sepharose 6 Fast Flow with an elution buffer using a high ammonium sulphate concentrations ($\geq$ 1.5 M). This results in more than one-third of the column volume being able to be loaded per run, resulting in an unusually high productivity for a typical SEC step (Stadler *et al.* 2004). As for the case of Sephacryl S-1000, the Fractogel TSK from Merck has been claimed to separate sc pDNA from impurities and other pDNA isoforms: nevertheless the method is highly time-consuming and some overlapping (sc and oc forms) has been found to be higher for a larger pDNA such as 26.5 Kbp (Diogo *et al.* 2005).

In an SEC step, a judicious choice of the target fractions has to be performed in order to enable the recovery of almost-pure supercoiled plasmid. Indeed, inherent limitations in this chromatographic step (e.g. lower capacities and higher dilution factors) results in loss of pDNA owing to detection limitations. In light of this brief description, SEC is viewed essentially as an adjuvant step in the downstream processing of pDNA.

ANION-EXCHANGE CHROMATOGRAPHY

Anion-exchange remains the most popular chromatography technique as it offers the advantages of rapid separation, no solvent requirement, sanitisation with sodium hydroxide and wide selection of industrial media. In particular, plasmid purification by anion-exchange chromatography (AEC) takes advantage of the polyanionic nature of DNA, due to the presence of phosphate groups on the nucleic acid backbone and is therefore, conveniently captured on a resin derivatised with positively charged functional groups. Over the last few years, a considerable number of anion-exchange adsorbents have been constructed, employing non-porous silica fibers (e.g. polymeric amines poly-ethyleneimine or chitosan) (Tiainen *et al.* 2007c), and have been examined with the aim of evaluating the factors that influence desorption and recovery of pDNA. Classically, the ideal AEC support should not be based on primary amines but on tertiary or quaternary amines (Tiainen *et al.* 2007a). For instance, although the poly(ethyleneimine) matrix has been shown to promote pDNA retention very efficiently, the interaction is unfortunately irreversible (Tiainen *et al.* 2007a). An explanation is the existence of hard-binding primary amine ligands or the presence of the highly charged plasmids that will dislocate the pKa of the ligand amines, making the ligand retain its charge also at pH values high above the pKa for the free ligand.

Overall, AEC as a tool for the preparation of pharmaceutical-grade pDNA unfortunately suffers from several drawbacks. Specifically, the loading of crude lysate containing large amounts of impurities directly onto an anion-exchanger is not recommended and primary purification involving precipitation or filtration steps is really essential (Eon-Duval and Burke 2004). Also, the low capacity for plasmid of most commercial chromatography media is also a concern for large-scale production. Even so, this setback could be tackled by adding surface-extenders or using large-pore supports based on polystyrene (Levy *et al.* 2000) and agarose (i.e., superporous agarose beads) (Tiainen *et al.* 2007b).

In general, this stage can be ideal for removing RNA, oligoribonucleotides and some proteins, but other polyanionic molecules with a similar chemical composition and structure (gDNA and RNA) (Sofer and Hager), or charge, such as endotoxins (Wicks *et al.* 1995), may co-purify with pDNA because of their similar binding affinities. Therefore, during the last two decades, several protocols for both packed (Chandra *et al.* 1992) and expanded bed operations (Theodossiou *et al.* 2001; Thwaites *et al.* 2002) have been developed. Despite these efforts, the incorporation of AEC in a global purification flow sheet will be more difficult to accomplish with plasmids (usually large ones) that have sizes closer to those of gDNA fragments. Indeed, some authors have reported recoveries below 5% when exploring anion-exchange poly(p-chloromethylstyrene) beads (Unsal *et al.* 2000) or acute difficulties in removing adsorbed DNA from anion-exchange expanded-bed matrices (Theodossiou and Thomas 2002). Moreover, the resolution of pDNA from high molecular weight RNA molecules by AEC is insufficient (Chandra *et al.* 1992). In fact, the supercoiled conformation of plasmids can reduce the overall charge density and may be responsible for early co-elution (Eon-Duval and Burke 2004). This problem can be solved by the addition of RNase. However, the incorporation of extraneous enzymes from e.g. bovine sources imposes serious validation problems in the large-scale purification of gene vectors.

Several authors have claimed that AEC may be considered for large-scale processing only if the capacity for plasmid is sufficient. For instance, a typical anion-exchanger capacity of 40-200 µg/mL (Prazeres *et al.* 1999) is well below that for globular proteins (up to ≥ 100 mg/mL) and is a challenge for scale-up. Curiously, Eon-Duval and Burke (2004) have found that the dynamic capacity for plasmid can be increased significantly by using a suitable salt concentration in the load, while preventing the binding of RNA to the adsorbent (Eon-Duval and Burke 2004). The breakthrough experiments showed that Q Sepharose FF had a low capacity (≤ 1 mg/ml) when compared with Q ceramic HyperD F (≥ 5.3 mg/ml) (Eon-Duval and Burke 2004). In the case of Q Sepharose FF, binding is achieved mostly through the interaction with functional groups on the surface on the beads resulting in low capacity, whereas with Q Ceramic HyperD F the plasmids not only bind to the hydrogel coating the surface of the particle, but also to the hydrogel filling the pores. Alternatively, with the Frac-togel DEAE support (Eon-Duval and Burke 2004), high selectivity and capacity has been achieved and within a wider range of loading salt concentration, high recovery, robustness and reproducibility.

It has been convincingly shown by confocal microscopy (Ljunglof *et al.* 1999) that separation of pDNA by porous chromatography media is not efficient, i.e. the macromolecules tend to be too large to diffuse into the pores and hence bind as an outer layer only. Gustavsson and co-workers (2004) have taken advantage of this drawback and introduced a flow-through purification protocol for pDNA on AEC with a non-charged outer surface. This new type of chromatography matrix combines size exclusion and anion-exchange principle. In comparison to SEC this method has a much higher volumetric capacity and delivers the plasmids in a concentrated form (Gustavsson *et al.* 2004). Indeed, this strategy is referred as a great promise as single-unit primary recovery step for the purification and concentration of pDNA from clarified alkaline lysates.

Recently, some supports have been developed to possess a very wide macropore structure that provides convective flow inside the pores. Therefore, diffusion by mass transfer is controlled and plasmids are transported into the interior of the beads by convective flow. Up until now, convection-aided monolithic supports, such as polymethacrylate-based media, have been claimed to be very effective and efficient for pDNA separations (Urthaler *et al.* 2005). The great advantage is the high recovery of pDNA, which is usually above 95% and a satisfactory high dynamic binding capacity. Indeed, DEAE-charged monoliths with 1.5 µm pores have given a plasmid binding capacity of 8 mg/mL (Urthaler *et al.* 2005). Other efficient matrices such as flow-pore-equipped beads, membranes (Teeters *et al.* 2003) and small-size particles (Jungbauer and Hahn 2004) have structures that ensure an excellent availability for plasmid isolation trials. In the case of small-size particles, a column packed with 10 µm beads will provide a ten times greater surface area than a column with 100 µm beads. Indeed, 15 µm anion-exchange beads gave capacities in the range of 2 mg/mL (Levy *et al.* 2000). Alternatively, membrane adsorbents (Teeters *et al.* 2003) and polystyrene-based Poros 50 HQ show an excellent plasmid binding capacity, in the region of 10 mg/mL. Also, the smaller diameter superporous agarose beads have been found to have four to five times higher plasmid binding capacity than the corresponding homogeneous agarose beads (Tiainen *et al.* 2007b). Some studies have demonstrated that, in spite of the lower capacity of superporous quaternary amine-substituted beads in comparison with superporous poly(ethyleneimine)-substituted beads, the plasmid recovery can be highly satisfactory and in the range of 70-100% (Tiainen *et al.* 2007b).

As reviewed above, large-scale purification of pDNA by monolithic and superporous AEC takes advantages of good flow properties, plasmid binding capacity and plasmid recovery. Nevertheless, the mechanism of pDNA separation using these new supports is still not totally understood because of contradictory experimental data. This knowledge can nonetheless bring new future inghights for the identification of anion-exchange adsorbents with suitable characteristics for the large-scale downstream of pharmaceutical-grade pDNA.

HYDROPHOBIC INTERACTION CHROMATOGRAPHY

Hydrophobic interaction chromatography (HIC) is a well established bioseparation technique at laboratory, preparative and industrial scales (Gustavsson *et al.* 1999; Xiao *et al.* 2007). HIC takes advantage of biomolecular hydrophobicity and promoting separation on the basis of hydrophobic interactions between immobilized hydrophobic ligands and non-polar regions on biomolecules (Queiroz *et al.* 2001). Retention occurs at high salt concentrations, driven mainly by a displacement of ordered water molecules around the biomolecules and the ligands which leads to an increase in entropy (Gustavsson *et al.* 1999) and in attraction to non-polar groups on a stationary phase. The requirement for high salt concentration is often viewed as a disadvantage, especially regarding the industrial application of this method, because the use of salt may not be convenient due to the associated costs and the environmental impact. Elution is achieved by decreasing the salt concentration of the mobile phase, which weakens the hydrophobic interactions (Jungbauer *et al.* 2005; Queiroz *et al.* 2001; Xiao *et al.* 2007). This technique is often used because, despite the complex mechanism involved in the interaction, the structural damage to the biomolecules is minimal and their biological activity is maintained using HIC, due to weaker interaction than ion-exchange or reversed-phase chromatography (Queiroz *et al.* 2001). Thus, HIC can provide purification based on hydrophobicity with minimal solvent requirements and with the potential of minimal product degradation (Xiao *et al.* 2007). Temperature, pH, and salts can have a significant impact on HIC retention and selectivity (Jungbauer *et al.* 2005; Xiao *et al.* 2007), and hence, by selecting and optimizing such conditions, biomolecules can be efficiently separated.

In the particular case of nucleic acids, HIC has become an important new separation modality for the isolation of pDNA. In this way, by exploiting the differences in hydrophobicity of pDNA, single-stranded nucleic acid species and endotoxins, HIC can be applied in pDNA purification (Diogo *et al.* 2001a; Iuliano *et al.* 2002).

The major factors affecting the retention of nucleic acids on HIC are their size, base composition and structure (Ferreira *et al.* 2000). Experiments carried out with an agarose-based support derivatized with a mildly hydrophobic ligand have proven that is possible to separate native pDNA from the more hydrophobic nucleic acid impurities (RNA, gDNA, oligonucleotides, denatured pDNA) (Diogo *et al.* 1999; Diogo *et al.* 2001a). By using a high ammonium sulphate concentration, it was found that pDNA elutes immediately, whilst impurities remained more retained, eluting later by decreasing the salt concentration. This behaviour was explained by the fact that pDNA molecules have their hydrophobic bases packed inside the double helix, and thus the hydrophobic interaction with the support is minimal. On the other hand, single stranded nucleic acid impurities show a higher exposure of the hydrophobic bases, and interact with the hydrophobic ligands (Diogo *et al.* 2001a). In accordance with the relation to base composition, partial denaturation of the double-helix at AT-rich locations, forming single-stranded regions within the molecule, also leads to the exposure of the bases to the ligands and thus to an increase in hydrophobic interaction strength (Ferreira *et al.* 2000).

Hydrophobic interaction chromatography has already been applied to the purification of a cystic fibrosis gene therapy vector (pCF1-CFTR) of clinical grade (Diogo *et al.* 2000) and a pDNA vaccine against rabies (Diogo *et al.* 2001b). Biological activity studies and protection experiments performed in mice with the vaccine have shown that the experimental vaccine displays immunogenic activity and potency similar to, or higher than that of the vaccine prepared with a commercial kit (Diogo *et al.* 2001b). Based on the same mechanism, hydrophobic interaction HPLC methods have also been developed for the quantification of pDNA and assessment of its purity in process streams (Diogo *et al.* 2003; Iuliano *et al.* 2002). As an example, the development of an HPLC technique based on HIC with a non-porous packing (Iuliano *et al.* 2002) has allowed the quantification of open circular and supercoiled plasmid isoforms for plasmids with different sizes.

The immobilization of hydrophobic ligands to a biporous medium has recently been shown to purify pDNA by HIC (Li *et al.* 2005). In this study, the authors referred to the structure of the biporous matrix, which allows convective flow of mobile phase through the pores, leading to an enhanced intraparticle mass transport (Li *et al.* 2005). The performance of superporous agarose beads has also been evaluated in hydrophobic interaction chromatography applications (Gustavsson *et al.* 1999) with regard to improvement of resolution and capacity of supports for application on either a preparative or analytical scale. Deshmukh and Lali (2005) have also described the development of a scalable adsorptive separation technology, using rigid cross-linked cellulose beads for single step purification of pDNA from cell lysates, exploiting a combination of factors like hydrophobicity and macroporosity of adsorbent matrix CELBEADS (Deshmukh and Lali 2005). All the improvements of this chromatographic technique have made possible not only the implementation of several efficient purification processes, but also the establishment of crucial analytical methods to monitor and control pDNA quality.

AFFINITY CHROMATOGRAPHY

Affinity chromatography (AC) is one of the most versatile and adaptable types of liquid chromatography, since it is the only technique that uses a specific binding agent to purify a biomolecule on the basis of its biological function or individual chemical structure (Schiel *et al.* 2006). Although the early affinity definition was related to the interactions similar to that occurring in many biological systems, such as the binding of an enzyme with a substrate or of an antibody with an antigen (Schiel *et al.* 2006), the meaning of the affinity concept, in the biomolecules separation context, has undergone evolutionary changes over the years, especially as a way to answer to the challenges of purifying new biomolecules with clinical and therapeutic interest.

Undoubtedly, the properties of pDNA molecules, their particularities, and the sensitivity of their structures to changes in physical and chemical environmental conditions will determine the interaction with a particular chromatographic support. One of the most important concerns when working on purification of pDNA is the adjustment of all working conditions to improve the performance of the biotechnological strategy, whilst at the same time always guaranteeing the structural and functional stability of pDNA. Furthermore, a key factor for achieving successful pDNA purification by affinity chromatography is the selection of the ligand used within the column. Since its discovery, affinity chromatography has grown to include a wide variety of ligands for analytical and preparative applications (Schiel *et al.* 2006). However, the biological origin of some affinity ligands is viewed as a limitation (Lowe *et al.* 2001), since these ligands tend to be fragile and associated with low binding capacity. For this reason, the design of synthetic ligands, which would combine the selectivity of natural ligands with the high capacity and durability of synthetic systems, is an emerging area (Lowe *et al.* 2001) for improving AC.

The specific interactions occurring between ligand and target molecules can be the result of either electrostatic and/or hydrophobic interactions, van der Waals' forces and hydrogen bonding. As a consequence of this diversity of possible interactions, the elution step can be performed specifically, using a competitive ligand, or nonspecifically, by changing the pH, ionic strength or polarity, depending on the matrix and the chemical characteristics of biomolecules. The specific nature of the underlying interactions is a major advantage of AC, since it results in a high selectivity and high resolution (Platonova and Tennikova 2005). Overall, in a single step, affinity purification can offer immense advantages over other less selective and time-consuming multi-step procedures. Different types of AC, such as immobilised metal ion-affinity chromatography (IMAC), triple-helix affinity chromatography (THAC), protein-DNA affinity chromatography and amino acid-DNA affinity chromatography have been employed for the purification of pDNA.

Several research groups have recently attempted to isolate nucleic acids molecules and to purify pDNA exploiting IMAC technology. It has been reported that IMAC matrices were able to selectively adsorb single-stranded nucleic acids through metal ion interactions with aromatic base nitrogens (Murphy *et al.* 2003a), whereas oligonucleotide duplexes, pDNA and gDNA showed low IMAC binding affinity (Murphy *et al.* 2003a). The feasibility of using IMAC for the purification of pDNA directly from an alkaline cell lysate was also reported (Tan *et al.* 2007), but its recovery was only possible by introducing a pre-treatment step. The main disadvantage of IMAC is the impossibility to isolate pDNA from a gDNA-containing extract, due to the similar double stranded structure (Cano *et al.* 2005).

Triple helix affinity chromatography (THAC) is also an alternative affinity technique to purify pDNA, based on the sequence-specific interaction of a triple-helix forming oligonucleotide with pDNA (Schluep and Cooney 1998; Wils *et al.* 1997).

A pyrimidine oligonucleotide is covalently linked to the chromatographic matrix and binds to duplex DNA via the major groove and through the formation of Hoogsteen hydrogen bonds. However, the triple-helix interaction is only possible if a suitable target homopurine sequence has been previously inserted in the pDNA (Schluep and Cooney 1998; Wils *et al.* 1997). Some reports provide evidence that it is possible to purify sc pDNA with THAC and to significantly reduce the level of contaminating RNA, endotoxins and gDNA in a single step (Wils *et al.* 1997). However, the relatively low recovery and the slow kinetics of the triple-helix formation are the main disadvantages of the strategy.

In protein-DNA chromatography, two major strategies have been described. The first uses a bifunctional protein-based affinity linker consisting of a zinc finger (ZF) DNA-binding protein that was fused to glutathione S-transferase (GST-ZF) (Woodgate *et al.* 2002). The ZF domain of the protein binds to a specific sequence, while the GST domain binds to a glutathione Sepharose affinity matrix. The second strategy was developed by exploiting the natural interaction between the lac operon sequence contained in the pDNA and its repressor, the lacI protein (Forde *et al.* 2006; Hasche and Voss 2005). In these cases, the elution is mainly performed by competition and the limitations are associated to low yields and remaining gDNA contaminations. Han and Forde (2008) have developed an affinity approach by immobilizing a peptide to a monolithic support to investigate the purification of pDNA in an attempt to improve the capacity. Their study resulted in the recovery of pDNA with a high purity level, however the sc isoform was not totally isolated from oc pDNA and a significant amount of pDNA was lost in the flowthrough (Han and Forde 2008).

Recently, a new affinity chromatography approach was implemented, named as amino acid-DNA affinity chromatography, to purify pDNA (Sousa *et al.* 2008b). The screening of the ability of histidine and arginine to isolate sc pDNA has shown that both amino acids ligands promote a specific interaction with pDNA. Extending beyond the possibility of histidine- and arginine-based affinity matrixes to isolate the major plasmid isoforms (oc and sc pDNA) (Sousa *et al.* 2005; Sousa *et al.* 2008a), the applicability of these matrices for the efficiently purification of sc pDNA from host impurities present in a clarified *E. coli* lysate (Sousa *et al.* 2006; Sousa *et al.* 2009c) has been further demonstrated by those researchers. In the case of histidine-chromatography, the application of an ammonium sulphate gradient appears to allow the specific recognition of sc pDNA by the histidine ligands, whereas no interaction has been found with oc pDNA and gDNA. Since the interaction of histidine with the DNA bases may include hydrogen bonding, ring stacking/hydrophobic interactions and water mediated H-bonds, the mechanism behind the specific interaction with the bases of sc pDNA was explained by those researchers as a consequence of deformations induced by torsional strain. Thus, the higher exposure degree of sc bases favours the interaction with the histidine ligand. A fundamental additional study performed with synthetic oligonucleotides to explain the retention mechanism in the histidine-support, also corroborates this hypothesis since the presence of secondary structures on polyA and polyG oligonucleotides has a significant influence on retention (Sousa *et al.* 2009a). In addition, it was verified that histidine interacts preferentially with the guanine base (Sousa *et al.* 2009a). *Figure 2* represents the specific recognition of guanine by the histidine support, by performing two hydrogen bonds.

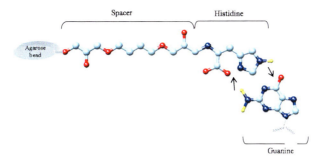

Figure 2. Schematic representation of an affinity interaction between a histidine-agarose support and guanine. Oxygen atoms are represented in red, carbon in light blue, nitrogen in dark blue and hydrogen in yellow. The zigzag line represents the nucleotide structure and the arrows indicate the hydrogen bonding established between the atoms involved and point from the donor to the acceptor atoms.

The quality control of the sc pDNA product showed that all the contaminants were eliminated or reduced to accepted levels, recovery was yielded in 45% and the 50% of transfection efficiency confirmed an efficient expression of the target gene in eukaryotic cells (Sousa *et al.* 2006), which is an interesting result concerning the therapeutic applications.

Otherwise, a study of the arginine-based matrix has revealed that after pDNA binding, due to the specific interaction, an elution can be performed using milder conditions. The application of an increased sodium chloride gradient or the addition of an arginine supplemented buffer has been shown to be equally efficient to purify sc pDNA, maintaining its stability and efficiency (Sousa *et al.* 2008c). Although, the interaction occurring between pDNA and the arginine support was supposed to be rather complex, the characteristics of arginine, namely its ability to interact in different conformations, the length of its side chain and the ability to produce good hydrogen bonding geometries also points to the possibility of specific recognition mechanisms. The investigation of the retention mechanism performed with synthetic oligonucleotides, has shown that although the electrostatic interaction plays an important role with regards the retention of single-stranded oligonucleotides, the interaction of double-stranded oligonucleotides onto the arginine support significantly decreased, as a result of the diminished exposure degree of bases. The underlying mechanism in this study involved phenomenological interactions like biorecognition which are themselves made up of elementary interaction forces such as (multiple) H-bond, electrostatic, hydrophobic interactions, dipole-dipole forces and cation-π interactions (Sousa *et al.* 2009b). Furthermore, the multiple interactions that an arginine-based matrix was able to promote has permitted the differential recognition of the biomolecules present in *E. coli* lysates, representing an important insight into the pDNA purification process. Using this simplified sc pDNA purification process, the majority of the contaminants were found to be removable and a 79% yield was achieved: sc pDNA was thereby purified under mild conditions and the process shown to be extremely efficient (62%) on cell transfection (Sousa *et al.* 2009c).

In spite of the efficacy to purify sc pDNA, these matrices present relatively low binding capacities. Hence, the application of large pore matrices in order to further increase capacity and open the way to process scale applications could be a great advantage for affinity chromatography.

Conclusions and future trends

Gene therapy and DNA vaccination are therapeutic strategies that are showing considerable potential with regards becoming an accepted and realistic therapeutic option for the treatment of several diseases and pandemic occurrences. As concerns are becoming dissipated and the success of these therapies is becoming increasingly apparent, the research on the implementation of appropriate enabling strategies for the production and purification of non-viral vectors, such as pDNA, has become intensified over the last few years and this review we have looked at the manufacturing processes.

The importance of pDNA and the processes required to obtain sufficient quantities of this molecule in pure and stable form has been challenging. A deep knowledge of the target molecule and associated impurities is required in order to improve the performance of the available unit operations for plasmid processing. An efficient high-scale production of pharmaceutical grade pDNA constitutes an engineering task demanding experience and a high technical level. In this regard, numerous chromatographic methodologies have now been established and characterized for plasmid purification. Ongoing investigations are therefore mainly focused on improving the capacity, recovery yields and selectivity of the chromatographic matrices, by considering the design of the supports and taking advantage of new and more specific ligands. An improved understanding of all unit operations involved in the process and its integration will comprise a great benefit to the implementation of new platforms for pDNA biopharmaceutical production.

Production of a plasmid product for therapeutic application requires an accurate control of purity and biological activity, conforming to appropriate quality requirements. To achieve this it is crucial that an integrative understanding of all the areas and technologies involved: the design of the vector, the implementation of new production strategies and the control of the purification techniques. Application of recent and established analytical methodologies and procedures capable of fully characterizing the product and process performance is a crucial task that has to be considered during the design stage.

References

ALTON, E.W., STERN, M., FARLEY, R. *et al.* (1999). Cationic lipid-mediated CFTR gene transfer to the lungs and nose of patients with cystic fibrosis: a double-blind placebo-controlled trial. *Lancet* **353**, 947-54.

ANDERSON, R.J. and SCHNEIDER, J. (2007). Plasmid DNA and viral vector-based vaccines for the treatment of cancer. *Vaccine* **25**, B24-34.

BARBOSA, H., HINE, A.V., BROCCHINI, S., SLATER, N.K. and MARCOS, J.C. (2008). Affinity partitioning of plasmid DNA with a zinc finger protein. *Journal of Chromatography A* **1206,** 105-12.

BENCINA, M., PODGORNIK, A. and STRANCAR, A. (2004). Characterization of methacrylate monoliths for purification of DNA molecules. *Journal of Separation Science* **27,** 801-10.

BIRNBOIM, H.C. and DOLY, J. (1979). A rapid alkaline extraction procedure for screening recombinant plasmid DNA. *Nucleic Acids Research* **7,** 1513-23.

BRANNON-PEPPAS, L., GHOSN, B., ROY, K. and CORNETTA, K. (2007). Encapsulation of nucleic acids and opportunities for cancer treatment. *Pharmaceutical Research* **24,** 618-27.

BRANOVIC, K., FORCIC, D., IVANCIC, J. *ET AL.* (2004). Application of short monolithic columns for fast purification of plasmid DNA. *Journal of Chromatography B* **801,** 331-7.

BRAUN, S. (2008). Muscular gene transfer using nonviral vectors. *Current Gene Therapy* **8,** 391-405.

BRAVE, A., LJUNGBERG, K., WAHREN, B. and LIU, M.A. (2007). Vaccine delivery methods using viral vectors. *Molecular Pharmaceutics* **4,** 18-32.

BYWATER, M., BYWATER, R. and HELLMAN, L. (1983). A novel chromatographic procedure for purification of bacterial plasmids. *Analytical Biochemistry* **132,** 219-24.

CANO, T., MURPHY, J.C., FOX, G.E. and WILLSON, R.C. (2005). Separation of genomic DNA from plasmid DNA by selective renaturation with immobilized metal affinity capture. *Biotechnology Progress* **21,** 1472-7.

CARNES, A.E. (2007). Fermentation process for continuous plasmid production. WO2007/089873.

CARNES, A.E., HODGSON, C.P. and WILLIAMS, J.A. (2006). Inducible Escherichia coli fermentation for increased plasmid DNA production. *Biotechnology and Applied Biochemistry* **45,** 155-66.

CARNES, A.E. and WILLIAMS, J.A. (2006). Media and process for plasmid DNA fermentation. PCT/US05/029238.

CARNES, A.E. and WILLIAMS, J.A. (2007). Plasmid DNA manufacturing technology. *Recent Patents on Biotechnology* **1,** 151-66.

CHAN, G., BOOTH, A.J., MANNWEILER, K. and HOARE, M. (2006). Ultra scale-down studies of the effect of flow and impact conditions during E. coli cell processing. *Biotechnology and Bioengineering* **95,** 671-83.

CHANDRA, G., PATEL, P., KOST, T.A. and GRAY, J.G. (1992). Large-scale purification of plasmid DNA by fast protein liquid chromatography using a Hi-Load Q Sepharose column. *Analytical Biochemistry* **203,** 169-72.

CHANG, C.W., CHRISTENSEN, L.V., LEE, M. and KIM, S.W. (2008). Efficient expression of vascular endothelial growth factor using minicircle DNA for angiogenic gene therapy. *Journal of Controlled Release* **125,** 155-63.

CHARTRAIN, M., KIZER, L.E., KRULEWICZ, B.A. *ET AL.* (2005). Process for large-scale production of plasmid DNA by E. coli fermentation. WO/2005/078115.

CHERNG, J.Y., SCHUURMANS-NIEUWENBROEK, N.M., JISKOOT, W. *ET AL.* (1999). Effect of DNA topology on the transfection efficiency of poly((2-dimethylamino)ethyl methacrylate)-plasmid complexes. *Journal of Controlled Release* **60,** 343-53.

CLARK, D.P. (2005). Molecular Biology - Understanding the Genetic Revolution: Elsevier, Academic Press.

CLEMSON, M. and KELLY, W.J. (2003). Optimizing alkaline lysis for DNA plasmid recovery. *Biotechnology and Applied Biochemistry* **37**, 235-44.

COOKE, J.R., MCKIE, E.A., WARD, J.M. and KESHAVARZ-MOORE, E. (2004). Impact of intrinsic DNA structure on processing of plasmids for gene therapy and DNA vaccines. *Journal of Biotechnology* **114**, 239-54.

CUPILLARD, L., JUILLARD, V., LATOUR, S. *et al.* (2005). Impact of plasmid supercoiling on the efficacy of a rabies DNA vaccine to protect cats. *Vaccine* **23**, 1910-6.

DARBY, R.A., FORDE, G.M., SLATER, N.K. and HINE, A.V. (2007). Affinity purification of plasmid DNA directly from crude bacterial cell lysates. *Biotechnology and Bioengineering* **98**, 1103-8.

DESHMUKH, N.R. and LALI, A.M. (2005). Adsorptive purification of pDNA on superporous rigid cross-linked cellulose matrix. *Journal of Chromatography B* **818**, 5-10.

DIOGO, M.M., QUEIROZ, J.A., MONTEIRO, G.A. and PRAZERES, D.M.F. (1999). Separation and analysis of plasmid denatured forms using hydrophobic interaction chromatography. *Analytical Biochemistry* **275**, 122-4.

DIOGO, M.M., QUEIROZ, J.A., MONTEIRO, G.A. *et al.* (2000). Purification of a cystic fibrosis plasmid vector for gene therapy using hydrophobic interaction chromatography. *Biotechnology and Bioengineering* **68**, 576-83.

DIOGO, M.M., QUEIROZ, J.A. and PRAZERES, D.M.F. (2001a). Studies on the retention of plasmid DNA and Escherichia coli nucleic acids by hydrophobic interaction chromatography. *Bioseparation* **10**, 211-20.

DIOGO, M.M., RIBEIRO, S.C., QUEIROZ, J.A. *et al.* (2001b). Production, purification and analysis of an experimental DNA vaccine against rabies. *Journal of Gene Medicine* **3**, 577-84.

DIOGO, M.M., QUEIROZ, J.A. and PRAZERES, D.M.F. (2003). Assessment of purity and quantification of plasmid DNA in process solutions using high-performance hydrophobic interaction chromatography. *Journal of Chromatography A* **998**, 109-17.

DIOGO, M.M., QUEIROZ, J.A. and PRAZERES, D.M.F. (2005). Chromatography of plasmid DNA. *Journal of Chromatography A* **1069**, 3-22.

DRAPE, R.J., MACKLIN, M.D., BARR, L.J. *et al.* (2006). Epidermal DNA vaccine for influenza is immunogenic in humans. *Vaccine* **24**, 4475-81.

EON-DUVAL, A., MACDUFF, R.H., FISHER, C.A., HARRIS, M.J. and BROOK, C. (2003). Removal of RNA impurities by tangential flow filtration in an RNase-free plasmid DNA purification process. *Analytical Biochemistry* **316**, 66-73.

EON-DUVAL, A. and BURKE, G. (2004). Purification of pharmaceutical-grade plasmid DNA by anion-exchange chromatography in an RNase-free process. *Journal of Chromatography B* **804**, 327-35.

FELICIELLO, I. and CHINALI, G. (1993). A modified alkaline lysis method for the preparation of highly purified plasmid DNA from Escherichia coli. *Analytical Biochemistry* **212**, 394-401.

FERREIRA, G.N., CABRAL, J.M. and PRAZERES, D.M.F. (1997). A comparison of gel filtration chromatographic supports for plasmid purification. *Biotechnology Techniques* **11**, 417-420.

FERREIRA, G.N., CABRAL, J.M. and PRAZERES, D.M.F. (1999). Development of process flow sheets for the purification of supercoiled plasmids for gene therapy applications. *Biotechnology Progress* **15**, 725-31.

FERREIRA, G.N., MONTEIRO, G.A., PRAZERES, D.M.F. and CABRAL, J.M. (2000). Downstream processing of plasmid DNA for gene therapy and DNA vaccine applications. *Trends in Biotechnology* **18**, 380-8.

FLORES, S., DE ANDA-HERRERA, R., GOSSET, G. and BOLIVAR, F.G. (2004). Growth-rate recovery of Escherichia coli cultures carrying a multicopy plasmid, by engineering of the pentose-phosphate pathway. *Biotechnology and Bioengineering* **87**, 485-94.

FORDE, G.M., GHOSE, S., SLATER, N.K. *et al.* (2006). LacO-LacI interaction in affinity adsorption of plasmid DNA. *Biotechnology and Bioengineering* **95**, 67-75.

GIOVANNINI, R., FREITAG, R. and TENNIKOVA, T.B. (1998). High-performance membrane chromatography of supercoiled plasmid DNA. *Analytical Chemistry* **70**, 3348-54.

GUO, Z.S., LI, Q., BARTLETT, D.L., YANG, J.Y. and FANG, B. (2008). Gene transfer: the challenge of regulated gene expression. *Trends in Molecular Medicine* **14**, 410-8.

GURUNATHAN, S., KLINMAN, D.M. and SEDER, R.A. (2000). DNA vaccines: immunology, application, and optimization. *Annual Review of Immunology* **18**, 27-74.

GUSTAVSSON, P.E., AXELSSON, A. and LARSSON, P.O. (1999). Superporous agarose beads as a hydrophobic interaction chromatography support. *Journal of Chromatography A* **830**, 275-284.

GUSTAVSSON, P.E., LEMMENS, R., NYHAMMAR, T., BUSSON, P. and LARSSON, P.O. (2004). Purification of plasmid DNA with a new type of anion-exchange beads having a non-charged surface. *Journal of Chromatography A* **1038**, 131-40.

HAMANN, C.J., NIELSEN, J. and INGERSLEV, E. (2000). Novel plasmids for use in medicine and method of producing same. WO 00/28048.

HAN, Y. and FORDE, G.M. (2008). Single step purification of plasmid DNA using peptide ligand affinity chromatography. *Journal of Chromatography B* **874**, 21-6.

HASCHE, A. and VOSS, C. (2005). Immobilisation of a repressor protein for binding of plasmid DNA. *Journal of Chromatography A* **1080**, 76-82.

HILBRIG, F., FREITAG, R. and SCHMACHER, I. (2004). Method for treating biomass for producing cell lysate containing plasmid DNA. US2004/0014098.

HOARE, M., LEVY, M.S., BRACEWELL, D.G. *et al.* (2005). Bioprocess engineering issues that would be faced in producing a DNA vaccine at up to 100 m3 fermentation scale for an influenza pandemic. *Biotechnology Progress* **21**, 1577-92.

HOLMES, D.S. and QUIGLEY, M. (1981). A rapid boiling method for the preparation of bacterial plasmids. *Analytical Biochemistry* **114**, 193-7.

HORN, N.A., MEEK, J.A., BUDAHAZI, G. and MARQUET, M. (1995). Cancer gene therapy using plasmid DNA: purification of DNA for human clinical trials. *Human Gene Therapy* **6**, 565-73.

HU, X., MA, Q. and ZHANG, S. (2006). Biopharmaceuticals in China. *Biotechnology Journal* **1**, 1215-24.

HUBER, H., WEIGL, G. and BUCHINGER, W. (2005). WO2005097990

IULIANO, S., FISHER, J.R., CHEN, M. and KELLY, W.J. (2002). Rapid analysis of a plasmid by hydrophobic-interaction chromatography with a non-porous resin. *Journal of Chromatography A* **972**, 77-86.

JUNGBAUER, A. and HAHN, R. (2004). Monoliths for fast bioseparation and bioconversion and their applications in biotechnology. *Journal of Separation Science* **27**, 767-78.

JUNGBAUER, A. and HAHN, R. (2008). Polymethacrylate monoliths for preparative and industrial separation of biomolecular assemblies. *Journal of Chromatography A* **1184**, 62-79.

JUNGBAUER, A., MACHOLD, C. and HAHN, R. (2005). Hydrophobic interaction chromatography of proteins. III. Unfolding of proteins upon adsorption. *Journal of Chromatography A* **1079**, 221-8.

KASLOW DC. (2004). A potential disruptive technology in vaccine development: gene-based vaccines and their application to infectious diseases. *Transactions of the Royal Society of Tropical Medicine and Hygiene* **98**, 593-601.

KODIHALLI, S., GOTO, H., KOBASA, D.L. *et al.* (1999). DNA vaccine encoding hemagglutinin provides protective immunity against H5N1 influenza virus infection in mice. *The Journal of Virology* **73**, 2094-8.

KONG, S., ROCK, C.F., BOOTH, A. *et al.* (2008). Large-scale plasmid DNA processing: evidence that cell harvesting and storage methods affect yield of supercoiled plasmid DNA. *Biotechnology and Applied Biochemistry* **51**, 43-51.

LAHIJANI, R., HULLEY, G., SORIANO, G., HORN, N.A. and MARQUET, M. (1996). High-yield production of pBR322-derived plasmids intended for human gene therapy by employing a temperature-controllable point mutation. *Human Gene Therapy* **7**, 1971-80.

LEE, A.L. and SAGAR, S. (1999). A method for large scale plasmid purification. WO96/36706.

LEE, T.W., MATTHEWS, D.A. and BLAIR, G.E. (2005). Novel molecular approaches to cystic fibrosis gene therapy. *Biochemical Journal* **387**, 1-15.

LEVY, M.S., O'KENNEDY, R.D., AYAZI-SHAMLOU, P. and DUNNILL, P. (2000). Biochemical engineering approaches to the challenges of producing pure plasmid DNA. *Trends in Biotechnology* **18**, 296-305.

LI, L.Z., LIU, Y., SUN, M.S. and SHAO. YM. (2007). Effect of salt on purification of plasmid DNA using size-exclusion chromatography. *Journal of Chromatography A* **1139**, 228-35.

LI, Y., DONG, X-Y. and SUN, Y. (2005). High-speed chromatographic purification of plasmid DNA with a customized biporous hydrophobic adsorbent. *Biochemical Engineering Journal* **27**, 33-9.

LIN-CHAO, S. and BREMER, H. (1986). Effect of the bacterial growth rate on replication control of plasmid pBR322 in Escherichia coli. *Molecular Genomics and Genetics* **203**, 143-9.

LISTNER, K., BENTLEY, L., OKONKOWSKI, J. *et al.* (2006). Development of a highly productive and scalable plasmid DNA production platform. *Biotechnology Progress* **22**, 1335-45.

LIU, M.A. (2003). DNA vaccines: a review. *Journal of Internal Medicine* **253**, 402-10.

LJUNGLOF, A., BERGVALL, P., BHIKHABHAI, R. and HJORTH, R. (1999). Direct visualisation of plasmid DNA in individual chromatography adsorbent particles by confocal scanning laser microscopy. *Journal of Chromatography A* **844**, 129-35.

LOWE, C.R., LOWE, A.R. and GUPTA, G. (2001). New developments in affinity chromatography with potential application in the production of biopharmaceuticals. *Journal of Biochemical and Biophysical Methods* **49**, 561-74.

MALLIK, R. and HAGE, D.S. (2006). Affinity monolith chromatography. *Journal of Separation Science* **29**, 1686-704.

MANTHORPE, M., HOBART, P., HERMANSON, G. *et al.* (2005). Plasmid vaccines and therapeutics: from design to applications. *Advances in Biochemical Engineering Biotechnology* **99**, 41-92.

MARTIN, J.E., SULLIVAN, N.J., ENAMA, M.E. *et al.* (2006). A DNA vaccine for Ebola virus is safe and immunogenic in a phase I clinical trial. *Clinical and Vaccine Immunology* **13**, 1267-77.

MCCARTHY, S.M., GILAR, M. and GEBLER, J. (2009). Reversed-phase ion-pair liquid chromatography analysis and purification of small interfering RNA. *Analytical Biochemistry. In press.*

MCDONNELL, W.M. and ASKARI, F.K. (1996). DNA vaccines. *The New England Journal of Medicine* **334**, 42-5.

MICARD, D., SOBRIER, M.L., COUDERC, J.L. and DASTUGUE, B. (1985). Purification of RNA-free plasmid DNA using alkaline extraction followed by Ultrogel A2 column chromatography. *Analytical Biochemistry* **148**, 121-6.

MORISHITA, R., AOKI, M., HASHIYA, N. *et al.* (2004). Safety evaluation of clinical gene therapy using hepatocyte growth factor to treat peripheral arterial disease. *Hypertension* **44**, 203-9.

MOUNTAIN, A. (2000). Gene therapy: the first decade. *Trends in Biotechnology* **18**, 119-28.

MURPHY, J.C., WIBBENMEYER, J.A., FOX, G.E. and WILLSON, R.C. (1999). Purification of plasmid DNA using selective precipitation by compaction agents. *Nature Biotechnology* **17**, 822-3.

MURPHY, J.C., FOX, G.E. and WILLSON, R.C. (2003a). Enhancement of anion-exchange chromatography of DNA using compaction agents. *Journal of Chromatography A* **984**, 215-21.

MURPHY, J.C., JEWELL, D.L., WHITE, K.I., FOX, G.E. and WILSSON, R.C. (2003b). Nucleic acid separations utilizing immobilized metal affinity chromatography. *Biotechnology Progress* **19**, 982-6.

MWAU, M., CEBERE, I., SUTTON, J. *et al.* (2004). A human immunodeficiency virus 1 (HIV-1) clade A vaccine in clinical trials: stimulation of HIV-specific T-cell responses by DNA and recombinant modified vaccinia virus Ankara (MVA) vaccines in humans. *Journal of General Virology* **85**, 911-9.

O'KENNEDY, R.D., BALDWIN, C. and KESHAVARZ-MOORE, E. (2000). Effects of growth medium selection on plasmid DNA production and initial processing steps. *Journal of Biotechnology* **76**, 175-83.

O'MAHONY, K., FREITAG, R., HILBRIG, F., MULLER, P. and SCHUMACHER, I. (2005). Proposal for a better integration of bacterial lysis into the production of plasmid DNA at large scale. *Journal of Biotechnology* **119**, 118-32.

O'MAHONY, K., FREITAG, R., HILBRIG, F., SCHUMACHER, I. and MULLER, P. (2007). Integration of bacteria capture via filtration and in situ lysis for recovery of plasmid DNA under industry-compatible conditions. *Biotechnology Progress*

23, 895-03.

PALECEK, E., BRAZDA, V., JAGELSKA, E. *et al.* (2004). Enhancement of p53 sequence-specific binding by DNA supercoiling. *Oncogene* **23**, 2119-27.

PASSARINHA, L.A., DIOGO, M.M., QUEIROZ, J.A. *et al.* (2006). Production of ColE1 type plasmid by Escherichia coli DH5 alpha cultured under nonselective conditions. *Journal of Microbiology and Biotechnology* **16**, 20-4.

PATIL, S.D., RHODES, D.G. and BURGESS, D.J. (2005). DNA-based therapeutics and DNA delivery systems: a comprehensive review. *The AAPS Journal* **7**, E61-77.

PHUE, J.N., LEE, S.J., TRINH, L. and SHILOACH, J. (2008). Modified Escherichia coli B (BL21), a superior producer of plasmid DNA compared with Escherichia coli K (DH5alpha). *Biotechnology Bioengineering* **101**, 831-6.

PLATONOVA, G.A. and Tennikova TB. (2005). Chromatographic investigation of macromolecular affinity interactions. *Journal of Chromatography A* **1065**, 75-81.

POWELL, R.J., SIMONS, M., MENDELSOHN, F.O. *et al.* (2008). Results of a double-blind, placebo-controlled study to assess the safety of intramuscular injection of hepatocyte growth factor plasmid to improve limb perfusion in patients with critical limb ischemia. *Circulation* **118**, 58-65.

PRATHER, K.J., SAGAR, S., MURPHY, J. and CHARTRAIN, M. (2003). Industrial scale production of plasmid DNA for vaccine and gene therapy: plasmid design, production, and purification. *Enzyme and Microbial Technology* **33**, 865-83.

PRAZERES, D.M.F., SCHLUEP, T. and COONEY, C. (1998). Preparative purification of supercoiled plasmid DNA using anion-exchange chromatography. *Journal of Chromatography A* **806**, 31-45.

PRAZERES, D.M.F., FERREIRA, G.N., MONTEIRO, G.A., COONEY, C.L. and CABRAL, J.M. (1999). Large-scale production of pharmaceutical-grade plasmid DNA for gene therapy: problems and bottlenecks. *Trends in Biotechnology* **17**, 169-74.

PRAZERES, D.M.F. and FERREIRA, G.N.M. (2004). Design of flowsheets for the recovery and purification of plasmids for gene therapy and DNA vaccination. *Chemical Engineering Process* **43**, 615-30.

PRZYBYLOWSKI, M., BARTIDO, S., BORQUEZ-OJEDA, O., SADELAIN, M. and RIVIERE, I. (2007). Production of clinical-grade plasmid DNA for human Phase I clinical trials and large animal clinical studies. *Vaccine* **25**, 5013-24.

QUEIROZ, J.A., TOMAZ, C.T. and CABRAL, J.M. (2001). Hydrophobic interaction chromatography of proteins. *Journal of Biotechnology* **87**, 143-59.

RAYMOND, G.J., BRYANT, P.K., NELSON, A. and JOHNSON, J.D. (1988). Large-scale isolation of covalently closed circular DNA using gel filtration chromatography. *Analytical Biochemistry* **173**, 125-33.

ROLLAND, A. (2005). Gene medicines: the end of the beginning? *Advanced Drug Delivery Reviews* **57**, 669-73.

ROSENBERG, S.A., YANG, J.C., SHERRY, R.M. *et al.* (2003). Inability to immunize patients with metastatic melanoma using plasmid DNA encoding the gp100 melanoma-melanocyte antigen. Human Gene Therapy **14**, 709-14.

ROZKOV, A., AVIGNONE-ROSSA, C.A., ERTL, P.F. *et al.* (2004). Characterization of the metabolic burden on Escherichia coli DH1 cells imposed by the presence of a plasmid containing a gene therapy sequence. *Biotechnology and Bioengineering* **88**, 909-15.

SCHIEL, J.E., MALLIK, R., SOMAN, S., JOSEPH, K.S. and HAGE, D.S. (2006). Applications of silica supports in affinity chromatography. *Journal of Separation Science* **29**, 719-37.

SCHLEEF, M. and SCHMIDT, T. (2004). Animal-free production of ccc-supercoiled plasmids for research and clinical applications. *Journal of Gene Medicine* **6**, S45-53.

SCHLUEP, T. and COONEY, C.L. (1998). Purification of plasmids by triplex affinity interaction. *Nucleic Acids Research* **26**, 4524-8.

SCHMIDT, T., SCHLEEF, M., FRIEHS, K. and FLASCHEL, E. (2002). Method for producing homogeneus nucleic acid multimers for use in pharmaceutical compositions. WO02/062987.

SEBESTYEN, M.G., HEGGE, J.O., NOBLE, M.A. *et al.* (2007). Progress toward a nonviral gene therapy protocol for the treatment of anemia. *Human Gene Therapy* **18**, 269-85.

SEO, J.H. and BAILEY, J.E. (1985). Effects of recombinant plasmid content on growth properties and cloned gene product formation in Escherichia coli. *Biotechnology and Bioengineering* **27**, 1668-74.

SHEETS, E.E., URBAN, R.G., CRUM, C.P. *et al.* (2003). Immunotherapy of human cervical high-grade cervical intraepithelial neoplasia with microparticle-delivered human papillomavirus 16 E7 plasmid DNA. *American Journal of Obstetrics and Gynecology* **188**, 916-26.

SINGER, A., EITEMAN, M.A. and ALTMAN, E. (2009). DNA plasmid production in different host strains of Escherichia coli. *Journal of Industrial Microbiology and Biotechnology* **36**, 521-30.

SOFER, G. and HAGER, L. Handbook of process chromatography: A guide to optimization, scale-up and validation.

SOUSA, A., SOUSA, F., PRAZERES, D.M.F. and QUEIROZ, J.A. (2009a). Histidine affinity chromatography of homo-oligonucleotides. Role of multiple interactions on retention. *Biomedical Chromatography* **23**, 745-53.

SOUSA, A., SOUSA, F. and QUEIROZ, J.A. (2009b). Selectivity of arginine chromatography in promoting different interactions using synthetic oligonucleotides as model. *Journal of Separation Science* **32**, 1665-72.

SOUSA, F., TOMAZ, C.T., PRAZERES, D.M.F. and QUEIROZ, J.A. (2005). Separation of supercoiled and open circular plasmid DNA isoforms by chromatography with a histidine-agarose support. *Analytical Biochemistry* **343**, 183-5.

SOUSA, F., FREITAS, S., AZZONI, A.R., PRAZERES, D.M.F. and QUEIROZ, J.A. (2006). Selective purification of supercoiled plasmid DNA from clarified cell lysates with a single histidine-agarose chromatography step. *Biotechnology and Applied Biochemistry* **45**, 131-40.

SOUSA, F., PRAZERES, D.M.F. and QUEIROZ, J.A. (2007). Circular dichroism investigation of the effect of plasmid DNA structure on retention in histidine chromatography. *Archives of Biochemistry and Biophysics* **467**, 154-62.

SOUSA, F., MATOS, T., PRAZERES, D.M.F. and QUEIROZ, J.A. (2008a). Specific recognition of supercoiled plasmid DNA in arginine affinity chromatography. *Analytical Biochemistry* **374**, 432-4.

SOUSA, F., PRAZERES, D.M.F. and QUEIROZ, J.A. (2008b). Affinity chromatography approaches to overcome the challenges of purifying plasmid DNA. *Trends in*

Biotechnology **26**, 518-25.

SOUSA, F., PRAZERES, D.M.F. and QUEIROZ, J.A. (2008c). Binding and elution strategy for improved performance of arginine affinity chromatography in supercoiled plasmid DNA purification. *Biomedical Chromatography* **23**, 160-5.

SOUSA, F., PRAZERES, D.M.F. and QUEIROZ, J.A. (2009c). Improvement of transfection efficiency by using supercoiled plasmid DNA purified with arginine affinity chromatography. *Journal of Gene Medicine* **11**, 79-88.

STADLER, J., LEMMENS, R. and NYHAMMAR, T. (2004). Plasmid DNA purification. *Journal of Gene Medicine* **6**, S54-66.

SUBBARAO, K., MURPHY, B.R. and FAUCI, A.S. (2006). Development of effective vaccines against pandemic influenza. *Immunity* **24**, 5-9.

TAN, L., LAI, W.B., LEE, C.T., KIM, D.S. and CHOE, W.S. (2007). Differential interactions of plasmid DNA, RNA and endotoxin with immobilised and free metal ions. *Journal of Chromatography A* **1141**, 226-34.

TANGNEY, M., CASEY, G., LARKIN, J.O. *et al.* (2006). Non-viral in vivo immune gene therapy of cancer: combined strategies for treatment of systemic disease. *Cancer Immunology Immunotherapy* **55**, 1443-50.

TEETERS, M.A., CONRARDY, S.E., THOMAS, B.L., ROOT, T.W. and LIGHTFOOT, E.N. (2003). Adsorptive membrane chromatography for purification of plasmid DNA. *Journal of Chromatography A* **989**, 165-73.

TEJEDA-MANSIR, A. and MONTESINOS, R.M. (2008). Upstream processing of plasmid DNA for vaccine and gene therapy applications. *Recent Patents on Biotechnology* **2**, 156-72.

THATCHER, D.R., HITCHCOCK, A.G., HANAK, J.A. and VARLEY, D. (1997). Method of plasmid DNA production and purification patent. WO97/29190.

THEODOSSIOU, I., SONDERGAARD, M. and THOMAS, O.R. (2001). Design of expanded bed supports for the recovery of plasmid DNA by anion exchange adsorption. *Bioseparation* **10**, 31-44.

THEODOSSIOU, I. and THOMAS, O.R. (2002). DNA-induced inter-particle cross-linking during expanded bed adsorption chromatography. Impact on future support design. *Journal of Chromatography A* **971**, 73-86.

THWAITES, E., BURTON, S.C. and LYDDIATT, A. (2002). Impact of the physical and topographical characteristics of adsorbent solid-phases upon the fluidised bed recovery of plasmid DNA from Escherichia coli lysates. *Journal of Chromatography A* **943**, 77-90.

TIAINEN, P., GALAEV, I. and LARSSON, P.O. (2007a). Plasmid adsorption to anion-exchange matrices: Comments on plasmid recovery. *Biotechnology Journal* **2**, 726-35.

TIAINEN, P., GUSTAVSSON, P.E., LJUNGLOF, A. and LARSSON, P.O. (2007b). Superporous agarose anion exchangers for plasmid isolation. *Journal of Chromatography A* **1138**, 84-94.

TIAINEN, P., GUSTAVSSON, P.E., MANSSON, M.O. and LARSSON, P.O. (2007c). Plasmid purification using non-porous anion-exchange silica fibres. *Journal of Chromatography A* **1149**, 158-68.

UNSAL, E., BAHAR, T., TUNCEL, M. and TUNCEL, A. (2000). DNA adsorption onto polyethylenimine-attached poly(p-chloromethylstyrene) beads. *Journal of Chromatography A* **898**, 167-77.

URTHALER, J., SCHLEGL, R., PODGORNIK, A. *et al.* (2005). Application of monoliths for plasmid DNA purification development and transfer to production. *Journal of Chromatography A* **1065**, 93-106

VO-QUANG, T., MALPIECE, Y., BUFFARD, D. *et al.* (1985). Rapid large-scale purification of plasmid DNA by medium or low pressure gel filtration. Application: construction of thermoamplifiable expression vectors. *Bioscience Reports* **5**, 101-11.

WANG, R., DOOLAN, D.L., LE, T.P. *et al.* (1998). Induction of antigen-specific cytotoxic T lymphocytes in humans by a malaria DNA vaccine. *Science* **282**, 476-80.

WEIDE, B., GARBE, C., RAMMENSEE, H.G. and PASCOLO, S. (2008). Plasmid DNA- and messenger RNA-based anti-cancer vaccination. *Immunology Letters* **115**, 33-42.

WICKS, I.P., HOWELL, M.L., HANCOCK, T. *et al.* (1995). Bacterial lipopolysaccharide copurifies with plasmid DNA: implications for animal models and human gene therapy. *Human Gene Therapy* **6**, 317-23.

WILS, P., ESCRIOU, V., WARNERY, A. *et al.* (1997). Efficient purification of plasmid DNA for gene transfer using triple-helix affinity chromatography. *Gene Therapy* **4**, 323-30.

WOLFF, J.A., MALONE, R.W., WILLIAMS, P. *et al.* (1990). Direct gene transfer into mouse muscle in vivo. *Science* **247**, 1465-8.

WOODGATE, J., PALFREY, D., NAGEL, D.A., HINE, A.V. and SLATER, N.K. (2002). Protein-mediated isolation of plasmid DNA by a zinc finger-glutathione S-transferase affinity linker. *Biotechnology and Bioengineering* **79**, 450-6.

XIAO, Y., RATHORE, A., O'CONNELL, J.P. and FERNANDEZ, E.J. (2007). Generalizing a two-conformation model for describing salt and temperature effects on protein retention and stability in hydrophobic interaction chromatography. *Journal of Chromatography A* **1157**, 197-206.

XU, Z.N., SHEN, W.H., CHEN, H. and CEN, P.L. (2005). Effects of medium composition on the production of plasmid DNA vector potentially for human gene therapy. *Journal of Zhejiang University Science B* **6**, 396-400.

YAU, S.Y., KESHAVARZ-MOORE, E. and WARD, J. (2008). Host strain influences on supercoiled plasmid DNA production in Escherichia coli: Implications for efficient design of large-scale processes. *Biotechnology and Bioengineering* **101**, 529-44.

YONENAGA, Y., MORI, A., FUJIMOTO, A. *et al.* (2007). The administration of naked plasmid DNA into the liver induces antitumor innate immunity in a murine liver metastasis model. *Journal of Gene Medicine* **9**, 299-307.

ZANIN, M.P., WEBSTER, D.E. and WESSELINGH, S.L. (2007). A DNA prime, orally delivered protein boost vaccination strategy against viral encephalitis. *Journal of Neurovirology* **13**, 284-9.

ZHU, K., JIN, H., MA, Y. *et al.* (2005). A continuous thermal lysis procedure for the large-scale preparation of plasmid DNA. *Journal of Biotechnology* **118**, 257-64.

Biotechnology and Genetic Engineering Reviews - Vol. 26, 117-138 (2009)

Three-Dimensional Cell Cultures in Toxicology

FRANCESCO PAMPALONI[1*] AND ERNST H. K. STELZER[1]

[1]*Cell Biology & Biophysics Unit, European Molecular Biology Laboratory (EMBL), Meyerhofstrasse 1, D-69117, Heidelberg, Germany*

Abstract

Toxicity testing with animals is expensive, ethically controversial, and not always predictive of the human response. Cell-based assays are regarded as an alternative. However, conventional two-dimensional cell cultures do not reproduce the tissue architecture *in vivo*, and do not forecast organ-specific toxicity. On the other hand, three-dimensional cultures emulate the biochemistry and mechanics of the microenvironment in tissues more closely. Therefore, they address the limitations of both animals and two-dimensional cultures, and provide more accurate data on the effects of short- and long-term exposure to toxicants. We provide an up-to-date overview on the use of three-dimensional cell cultures in toxicology. We anticipate that three-dimensional cultures will become invaluable to accomplish the 3R agenda (refinement, reduction, and replacement) for animal-based toxicity testing and will play a major role for the Registration, Evaluation and Authorisation of Chemicals in the European Union (REACH legislation).

---

* To whom correspondence may be addressed (francesco.pampaloni@embl.de)

Abbreviations: 3R, Refinement, Reduction, and Replacement; REACH, Registration, Evaluation, and Authorization of Chemicals; ADME, Absorption, Distribution, Metabolism, Elimination; 3D, three-dimensional; 2D, two-dimensional; ECM, Extracellular Matrix; MDCK, Madin-Darby Canine Kidney; EHS, Engelbreth-Holm-Swarm; GOT, glutamic oxaloacetic transaminase; GPT, glutamic-pyruvic transaminase; LDH, lactate dehydrogenase; gamma-GT, gamma-glutamyl transferase; ISC, intestinal stem cells; EC-VAM, European Center for the Validation of Alternative Methods; CYP, cytochrom P450; RWV, rotating well vessel; HCS, high content screening; ESEM, environmental scanning electron microscope; LSFM, light-sheet-based fluorescence microscope; SPIM, single plane illumination microscope; DSLM, digital scanned light-sheet microscope.

Introduction

The toxicity of a substance is usually tested using animals. The tests are time-consuming and expensive, as well as ethically controversial, and in general of limited reliability. A comprehensive survey of drug screening tests has shown that 57% of the toxicological data derived from experiments with rodents do not correlate with the results of human trials (Olson *et al.*, 2000). In many cases, drugs have failed during clinical trials due to their adverse toxicity. Pre-clinical toxicity tests with multiple animal species are also poorly predictive (de Boo & Hendriksen, 2005; Knight, 2007a; Knight, 2007b). These facts motivate governments and authorities to support alternative methods. The replacement of animal models for testing drugs and chemical substances is one of the three Rs of the Replacement, Reduction, and Refinement (3R) agenda for the humane handling of laboratory animals (de Boo & Hendriksen, 2005). The 3R compliancy is required as a good laboratory animal practice. Important regulatory initiatives, such as the European REACH (Registration, Evaluation and Authorisation of Chemicals) (Foth & Hayes, 2008; Lilienblum *et al.*, 2008), require the application of alternatives to animal-based tests. To date, several basic cytotoxicity and ADME-Tox (drug Absorption, Distribution, Metabolism, Elimination and Toxicity) tests rely on cell cultures (Hasspieler *et al.*, 2006; Li, 2001; Lin *et al.*, 2003; Rausch, 2006). However, conventional two-dimensional cell cultures fail to detect organ-specific toxicity (Mazzoleni, Di Lorenzo & Steimberg, 2009). The reason seems to be that the flat culture plastic substrates provide a non-physiological environment to the cell (Mazzoleni *et al.*, 2009; Pampaloni, Reynaud & Stelzer, 2007). Plastic substrates are two-dimensional (2D) and fairly stiff. In contrast, real tissues have a three-dimensional (3D) geometry, gel-like (soft) stiffness, and a specific biochemistry determined by the proteins of the extra-cellular matrix (ECM). Examples of ECM proteins are collagen, laminin, and fibronectin. Although a 2D substrate can be coated with a thin layer of ECM proteins (usually collagen I or fibronectin) this does not seem to reproduce the complexity of the ECM *in vivo*. Due to the non-physiological microenvironment, cells mostly proliferate and de-differentiate in 2D cultures (Bhadriraju & Chen, 2002). Fibroblasts cultured in 2D have a flat shape strikingly dissimilar from the bipolar/stellate shape found in tissues (Beningo, Dembo & Wang, 2004; Rhee & Grinnell, 2007; Rhee *et al.*, 2007). A comparison of the gene expression profile of melanoma cells showed that 173 genes differ between the same cells cultured in 3D and 2D (Birgersdotter, Sandberg & Ernberg, 2005). Most of the genes strongly up-regulated in 3D are chemokines, as well as laminin and c-Jun (Ghosh *et al.*, 2005). Expression profiles of vascular smooth muscle cells showed evidence for ~100 genes, which are differently up-regulated in 2D compared to the situation in 3D (Li *et al.*, 2003). Primary hepatocytes plated in 2D systems lose liver-specific functions after a few passages. The first function lost is the biosynthesis of the drug metabolizing enzymes, which are essential for testing toxicity (Gomez-Lechon *et al.*, 1998; Pampaloni *et al.*, 2007). On the other end, culturing hepatocytes in 3D collagen or purified basement membrane maintains liver-specific functions for several weeks. An extended liver-specific functionality is also obtained by aggregating hepatocytes into "spheroids" with a diameter of several hundreds of micrometers (Semino *et al.*, 2003). Fibroblasts cultured in 3D collagen show a more typical *in vivo* phenotype (Cukierman *et al.*, 2001). Madin-Darby canine kidney

(MDCK) cells generate kidney organoids (spherical hollow spheres called *cysts*) in 3D collagen (Figure 1a) (Montesano, 1986). This phenotype is not observed in 2D cultures of MDCK cells.

These selected examples show that establishing 3D cell-cell contacts and embedding cells in 3D gels that emulate the natural ECM reduces the gap between cell culture and real tissue. Here, we review the application of 3D cell cultures to toxicity screening. Properly standardized and validated toxicity assays based on 3D cultures will allow us to predict the effect of toxicants on humans with higher accuracy than both 2D cultures and animal models. 3D cultures could greatly improve the toxicity screening of industrial chemicals as well as eliminate the toxic substances at an early stage of the drug discovery pipeline. The available 3D systems still reproduce at rudimental level the microstructure and the function of live tissues. Substantial effort is currently devoted to optimize 3D cell cultures for toxicity screening purposes. This requires the introduction of new synthetic ECM-like scaffolds, bioreactors, as well as advanced imaging technologies.

Methods for 3D cell culture

3D HYDROGEL-BASED CULTURES

The natural ECM in tissues consists of a tight network of fibrous proteins filled with glycosaminoglycan hydrogel (Lutolf & Hubbell, 2005). ECM hydrogels for 3D cell cultures mimic the biochemical and mechanical properties of tissues. Cells are confronted with a more physiological condition compared to 2D systems. Therefore, the relationship between cell function and tissue architecture can be isolated and addressed. ECM hydrogels for 3D cell culture are either of animal origin or synthetic. The most often employed hydrogels in a 3D cell culture are collagen type I and basement membrane extract, both of animal origin.

COLLAGEN TYPE I

Collagen I is the prevalent ECM component of the connective tissue (stroma). Collagen is extracted from animal tendons and is commercially available as an acidic solution. A 3D collagen fibrillar hydrogel can be easily reconstituted *in vitro* by neutralizing the solution. A highly ordered fibrillar architecture, similar to the one *in vivo*, can obtained by controlling the collagen concentration and sonicating the solution before gelation (Bessea *et al.*, 2002). Collagen hydrogels, sponges, microspheres, and membranes can be employed for 3D cultures (Chevallay & Herbage, 2000).

RECONSTITUTED BASAL MEMBRANE

The basement membrane is a fibrous sheet underlying the epithelia. It is composed of laminin 1, collagen III-IV, and heparin sulphate proteoglycans. The basement membrane hydrogel (commercial name "Matrigel" from BD Biosciences) is derived from the Engelbreth-Holm-Swarm (EHS) mouse sarcoma (Kleinman & Martin, 1989) and can be reconstituted as a 3D gel under physiological pH and temperature. 3D cell cultures in Matrigel and collagen I have been intensively and very successfully applied in breast cancer research (Bissell & Radisky, 2001; Cukierman *et al.*, 2001; Petersen *et al.*, 2001; Radisky, Hagios & Bissell, 2001), and to elucidate the steps leading to the establishment of epithelial cell polarity and epithelial morphogenesis (Montesano, 1986; Mostov, Su & ter Beest, 2003; O'Brien, Zegers & Mostov, 2002). These studies show very clearly that both morphology and gene expression adapt to the cell microenvironment.

The use of 3D cell cultures in hydrogels for toxicity screening requires standardized and biochemically well-defined matrices, which should suffer from minimal batch-to-batch variations. A significant improvement is the introduction of novel synthetic gels alternative to the "classical" extracted collagen I and to Matrigel, which is discussed in the last paragraph.

CELLULAR SPHEROIDS

Cellular spheroids are large (hundreds of micrometers in diameter) spheres composed of several hundreds to thousands of cells (*Figure 1b*). They form by aggregation of isolated cells. Cellular spheroids are among the first 3D cell culture models adopted in clinical pharmacology (Mueller-Klieser, 1997). An exogenous scaffold or matrix is not required to support the cells, since cell aggregation is spontaneous and facilitated by buoyancy or stirring. Spheroids can be obtained by the hanging drop technique (Kelm *et al.*, 2003; Korff & Augustin, 1998; Timmins, Dietmair & Nielsen, 2004), by seeding cells on non-adhesive surfaces, such as 3D alginate porous scaffolds (Glicklis *et al.*, 2000), or by employing rotating well vessel cultures (Bilodeau & Mantovani, 2006). Buoyancy is exploited in the "hanging-drop" method (*Figure 1b-1*). In this method, droplets of culture medium containing isolated cells are suspended from a Petri dish lid. After three to seven days of growth, large spheroids can be harvested (Kelm *et al.*, 2003; Timmins *et al.*, 2004). Alternatively, rotational stirrers can be employed for spheroids formation (Bilodeau & Mantovani, 2006; Moscona, 1961) (*Figure 1b-2*). The sedimentation of the cells within the vessel is offset by the rotating fluid, which keeps the cells continuously suspended in the culture medium. In a low-shear and low-turbulence regime, the rotating vessel bioreactors minimize the mechanical damage of cells and provide adequate nutrition and oxygenation.

SPHEROID TYPES

Many common cell lines can aggregate to spheroids, including MCF-10a (human mammary cell line), Caco2 (intestinal cell line), and HepG2 (human liver cell line) (Kelm *et al.*, 2003). Spheroids from human teratocarcinoma cell line Ntera2 (NT2) are a useful model system for biomedical studies and toxicity assays on the nervous system (Podrygajlo *et al.*, 2009). Podrygajlo *et al.* have shown that the differentiation time of the to mature post-mitotic neurons can be drastically reduced from two to one month by aggregation in spheroids (Podrygajlo *et al.*, 2009). Spheroids obtained from primary liver cells are particularly important for toxicity testing. Rat hepatocyte spheroids can be easily obtained by following standard procedures (Xu, Ma & Purcell, 2003a; Xu, Ma & Purcell, 2003b). During the transitional phase from

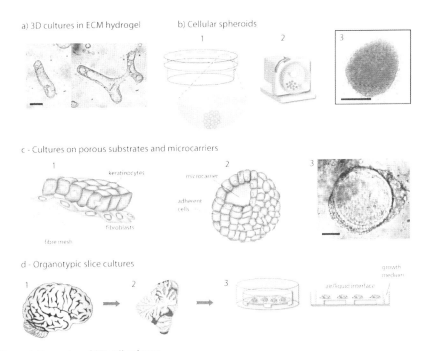

Figure 1. Examples of 3D cell cultures.

a. *3 D cultures in ECM hydrogels cultures.* An example is a 3D culture of MDCK cells (a kidney cell line) in hydrated collagen gel. The gel forms thick fibres, which support tissue-like cell growth. MDCK cells yields polarized epithelia resembling kidney tubules. The basal surface is in contact with the gel, and the apical side faces the fluid-filled internal cavity. The phase contrast micrograph shows MDCK tubes obtained by mechanically stretching the collagen fibers (scale bar 50 μm).

b. *Cellular spheroids.* Cells can re-establish mutual contacts and specific microenvironments that allow them to express a tissue-like phenotype by aggregating into large (several hundreds of micrometers) spheroids. These can be obtained at specific cellular concentrations within a "hanging drop" (1) or in rotating-wall vessels (2). In both methods, the cells cluster by gravity and then aggregate. (3) The phase contrast micrograph shows a spheroid obtained from the pancreas tumor cell line BxPC-3.

c. *Cultures on porous substrates and microcarriers.* (1) Culture on fibre mesh. For example, primary fibroblasts are cultured in Petri dishes. The cells are subsequently seeded onto a biodegradable fibre mesh. After several weeks in culture, keratinocytes, e.g. extracted from the foreskin, are placed onto the new dermal tissue and form an epidermal layer. (2-3) Microscaled materials beads derived from dextran, gelatine, glycosaminoglycans and other porous polymers can be used as a three-dimensional support for the culture of anchorage-dependent animal cell lines.

d. *Organotypic cultures.* Dissected organ slices, such as brain, are placed on porous substrates, supported by a metal grid (1, 2) and cultured at the air-growth medium interface (3).

isolated cells to mature spheroid (1-5 days) a drastic biochemical rearrangement has been observed in hepatocytes. Initially, glucose secretion and cellular activity of GOT and GPT increased (day 1-6). In contrast, LDH and γ-GT activity were undetectable during spheroid formation. Albumin secretion decreased rapidly during the first two days. Arginine uptake as well as urea and nitric oxide synthesis increased. The physiological liver-specific functions recovered with spheroid maturation (after day 6), and remained stable at least from day 6 until day 15. This nine day long time window is recommended by the authors as the best suited interval to perform toxicity assays (Xu *et al.*, 2003a; Xu *et al.*, 2003b). In another study, the liver-like functionality of porcine hepatocytes has been extended to over 21 days by embedding them in 3D collagen (Lazar *et al.*, 1995). Embryonic cells, such as retina cells (Layer *et al.*, 2002), and adult multipotent neuronal stem cells (Jessberger, Clemenson & Gage, 2007; Wang *et al.*, 2006) can also form spheroids (neuronal cells spheroids are termed "neurospheres"). Spheroids from embryonic retinal cells had been obtained by a static culture in a microscaffold cell array. Each microscaffold was a square with a lateral length of 300 μm. Each array contained 506 microcontainers. The cells produced mature spheroids within ten days (Rieke *et al.*, 2008). This microarray format is promising for large-scale assays.

ADVANTAGES OF CELLULAR SPHEROIDS

Since the addition of exogenous extra-cellular scaffolds is not required, spheroid aggregation can be easily automated. Therefore, scaling-up cellular spheroids-based assays to high-throughput analysis is feasible (Friedrich *et al.*, 2009; Ivascu & Kubbies, 2006; Kunz-Schughart *et al.*, 2004; Zhang, Gelain & Zhao, 2005). Also, the relatively easy handling allows a precise patterning of the spheroids into 3D shapes. Jakab *et al.* positioned ten CHO cell spheroids within 3D collagen, patterning a circle. After ~5 days in culture, the single spheroids along the circle merged, forming a continuous tissue of toroidal shape. This work shows that "organ printing" with cellular spheroids can establish a precise 3D geometry and mimic basic organ architectures (Jakab *et al.*, 2004a; Jakab *et al.*, 2004b).

In fact, a well-defined geometry allows the modelling of dynamic processes, such as organ formation and growth (Glicklis, Merchuk & Cohen, 2004), diffusion of drugs, cell invasion, as well as angiogenesis (Jiang *et al.*, 2005; Stein *et al.*, 2007; Tabatabai, Williams & Bursac, 2005). Spheroids are already well-established in clinical research, especially as models of small solid tumors (Sutherland, 1988; Sutherland, McCredie & Inch, 1971). They are the system of choice for therapeutically-oriented biomedical studies (Mueller-Klieser, 1997; Sutherland, 1988; Sutherland *et al.*, 1971), and have been applied in biotechnology, e.g. in tissue engineering of human bone (Kale *et al.*, 2000). With current advances in automation, microscopy imaging, and processing of large amount of data, cellular spheroids will become increasingly important for *in vitro* drug discovery screening and toxicity assays.

ORGANOTYPIC CULTURES

Organotypic cultures allow testing the functions of an organ exposed to toxicants by employing the actual organ itself grown *in vitro* under tissue culture conditions (Holopainen, 2005). This allows the maintenance of the tissue's architecture. Organotypic conditions are obtained by culturing organ slices of microscopic thickness (*Figure 1d*). Slices are cultured on semiporous membranes at the air-liquid interface, on collagen-coated substrates, or within 3D collagen. Each cell in the slice is less than a few hundred micrometers away from the media and oxygen supply. Thus, the tissue slice remains viable for many weeks or months even in serum-free media.

Organotypic lung, skin, and brain cultures have been developed and are widespread in drug discovery and toxicology. Organotypic brain slice cultures are well established for neurotoxicological screening and toxicological mechanistic research (Noraberg, 2004). Ootani *et al.* developed a promising combination of 3D culture in collagen hydrogel and organotypic culture. They realized a 3D intestinal model with murine intestinal stem cells (ISC), employing explants as starting material. Intestine explants were cultured within 3D collagen I hydrogel maintained at the air-liquid interface. On day seven, cystic structures (termed "intestinal spheres") appeared. Ootani *et al.* observed that the intestinal sphere is a polarized epithelial monolayer expressing the markers of intestine epithelium. They have shown that ISC in 3D culture differentiated to absorptive enterocytes and Goblet cells. The myofibroblasts present in the explants (the candidate ISC niche) supported long-term (i.e. for over 350 days) proliferation and differentiation of the tissue. This intestinal organotypic culture could have a significant impact on drug and toxicity assays.

In summary, the major advantages of whole organ slices are that the basic organ architecture is retained, and that the inter-individual variability is maintained. The drawbacks are the need of numerous biopsies and the difficult standardization of the assays.

3D CULTURES ON POROUS SUBSTRATES AND MICROCARRIERS

CULTURES ON POROUS MEMBRANES

Epithelial structures such as human skin models have been developed as 3D cultures on porous membrane ("filter-well inserts", e.g. Millicell from Millipore, Billerica, MA, USA or Transwell from Nunc, Rochester, NY, USA) or on polymeric fibre mesh (*Figure 1c-1*). The epithelial cells are cultured to confluence at the air-liquid interface. In this configuration, cells differentiate to polarized epithelial sheets (Justice, Badr & Felder, 2009). Sun *et al.* developed a 3D skin model by culturing the human keratinocyte cell line HaCaT on a biologically inert non-hydrolysable commercial scaffold (non-woven viscose rayon, trade name Azowipes™) (Sun *et al.*, 2006). The ability of the skin cells to respond to toxic xenobiotics (hydrogen peroxide and silver nitrate) in 3D and 2D was compared. The results showed that the concentration of H_2O_2 and $AgNO_3$ necessary to produce a 50% loss of viability (IC_{50}) was doubled in 3D compared to 2D. Consistent results were obtained by comparing the IC_{50} of human dermal fibroblasts and endothelial cells 3D and 2D cultures. The improved viability of 3D cultures vs. 2D ("multicellular resistance") has been observed by other researchers that employed 3D macroporous hydrogel as a substrate (Dainiak *et al.*, 2008), and is consistent with the higher resistance to cytotoxic drugs observed *in vivo* (Desoize & Jardillier, 2000). These data suggest that the outcome of toxicological assays obtained with 2D cultures needs to be critically reassessed (Sun *et al.*, 2006). Human colon carcinoma cells (Caco2) cultured for 14-20 days on a porous membrane establish intestinal cell polarity, tight junction, and transport properties. These differentiated Caco2 cells are morphologically and functionally similar to intestinal enterocytes. This is a valuable in vitro system to screen the intestinal absorption of drugs and toxicants (Bohets *et al.*, 2001). A drawback is that Caco2 cells represent enterocytes only. A more realistic intestinal model includes the adenocarcinoma cells

HT29. These cells differentiate into mature Goblet cells under the influence of the drug methotrexate (Behrens *et al.*, 2001). The mucus-secreting Goblet cells represent the second major cell type in the intestine.

Some 3D cell cultures on porous membrane mimicking epidermis and other epithelial organs have been fully validated for toxicity testing and will be extensively discussed in the paragraph "3D cultures in toxicology".

CULTURES ON MICROCARRIERS

Differentiated epithelia can also be obtained with microcarriers (*Figure 1c-2,3*). Microcarriers are polymeric beads with a diameter of 300-500 µm. Microcarriers provide an enourmous surface to volume ratio and support cell differentiation similarly to porous membranes. A commercially available microcarrier is the gelatin-coated Cytodex 3 (GE Healthcare, Chalfont St. Giles, UK). The cell-seeded microcarriers can be cultured in a rocked dish or in a rotating well vessel (see following paragraph). An advantage is that cells on microcarriers are maintained in homogeneous liquid medium during culture. This minimizes sample-to-sample variability. A further advantage is that the microcarrier system can be easily scaled-up to a high-throughput format (Justice *et al.*, 2009).

3D CELL CULTURES IN TOXICOLOGY

The European REACH-agenda will re-evaluate tens of thousands of chemical substances. This will be impossible without the use of *in vitro* assays. Moreover, EU regulation has already banned the use of animals for toxicological testing in the cosmetic industry since 2009 (76/768/EEC, February 2003). Assays based on 3D culture will have a pivotal role in replacing animals (Bhogal *et al.*, 2005). However, the validation of new toxicity assays for regulatory purposes is a technically demanding process, very expensive and very time-consuming. Therefore, it is no surprise that just few 3D cell models have been validated so far by authorities, such as the European Center for the Validation of Alternative Methods (ECVAM, http://ecvam.jrc.it/).

ASSAYS WITH 3D EPIDERMAL MODELS

Three commercial epidermal models have been fully validated, namely EPISKIN™ (L'Oréal, www.invitroskin.com), EpiDerm™ (MatTek, www.mattek.com) and SkinEthic™ (www.skinethic.com) (Bhogal *et al.*, 2005; Netzlaff *et al.*, 2005). In all the three models, several stratified layers of epidermal cells are cultured at the air-liquid interface on microporous filter inserts (such as MillicellCM by Millipore or Nunc polycarbonate inserts). Primary normal human keratinocytes are employed. The cells are obtained from donor specimens and expanded in monolayer culture to provide large pools. The general morphology, lipid composition, and biochemical markers of the 3D epidermis models are close to that of human skin. A common problem of these models is the higher permeability compared to human skin (Netzlaff *et al.*, 2005). Protocols to discriminate between corrosive and non-corrosive substances on skin based on EPISKIN and EpiDerm have been validated by ECVAM, and are now accepted as EU test guidelines (Bhogal *et al.*, 2005; Kandarova *et al.*, 2004; Kandarova *et al.*, 2005). 3D models of human corneal epithelium, tracheal/bronchial epithelium, buccal and gingival mucosa, and vaginal cervical mucosa are also commercially available (MakTek, SkinEthic). All the systems are based on primary normal human cells.

ASSAYS WITH LIVER MODELS

While 3D epidermis models provide essential data on skin corrosivity and phototoxicity, they are not suitable for metabolite toxicity or organ-specific toxicity, such as liver toxicity. Liver toxicity is the first cause of failing of a drug candidate during the clinical phase, and of withdrawal of a drug from the market (Kaplowitz, 2005). Liver models suitable for toxicity testing should maintain the liver functionality for up to several weeks. Particularly, the expression of cytochrome P450-mixed function monooxygenases (CYPs, the most important group of chemicals-metabolizing enzymes in the liver) must be conserved. Liver models based on 3D cell cultures have been intensively investigated. A broad range of 3D cell culture techniques has been employed to develop liver models, including culture in hydrogels, porous membrane, and cellular spheroids.

LIVER MODELS WITH 3D HYDROGELS

Lee *et al.* developed a microarray ("DataChip") for high-throughput toxicity assay based on Hep3B cells (a human hepatocellular carcinoma cell line that partially retains liver-specific functions) embedded in 3D hydrogels (Lee *et al.*, 2008). The cells were encapsulated in microdroplets of alginate gel. Each microdroplets contained ~60 cells. The droplets were spotted onto glass slides producing a 1080-spots array. Alginate is not degraded by matrix metalloproteinases produced by cells. Thus, the droplets were stable over several days, and their volume was as small as 20 nanoliters. This allowed increasing the density of the spots in the array. The toxicity of CYPs-metabolites was tested with the 3D cell microarray, as a proof of principle for toxicity screening of drugs and drugs metabolites. Dose-response curves were measured on 27 CYPs-metabolized substances, including digoxin, doxorubicin, acetaminophen, and ketokonazole. The assay was able to predict the influence of CYPs metabolism on the toxicity of this diverse range of xenobiotics.

LIVER MODELS WITH POROUS MEMBRANES

A sandwich culture of rat primary hepatocytes employing synthetic ECM was developed by Du *et al.* (Du *et al.*, 2008). The culture was maintained in a commercial small-sized perfusion bioreactor (Minusheet carriers, Minucell GmbH, Germany), suitable for metabolism/toxicity assays. The synthetic ECM consisted of two porous membranes sandwiching the hepatocytes. A galactosilated polyethylene-terephtalate (PET-Gal) was employed as bottom (basal) substrate for the hepatocytes. A polyethylene porous membrane coated with the oligopeptide Gly-Arg-Gly-Asp-Ser (PET-GRGDS) was employed as top (apical) support. The two synthetic basal and apical membranes mimic the space of Disse in liver. The authors compared the performance of the PET-Gal/PET-GRGDS system and the collagen sandwich system. The parameters examined in the comparison were the establishment of *in vivo*-like cell polarity, cell-cell interaction, biliary excretion, differentiated liver function, and mass transfer of medium and metabolites. The authors observed that PET-Gal/PET-GRGDS sandwich has a similar or improved performance compared to the collagen sandwich system. Hepatocyte polarity was established in both systems, as confirmed by the cortical rearrangement of the F-actin cytoskeleton and formation of bile canaliculi. Also the biliary excretion measured with fluorescein diacetate was comparable. Interestingly, liver functionality parameters, i.e. albumin and urea production, as well as cytochrome P540 1A activity were significantly higher in the PET-Gal/PET-GRGDS sandwich than in the collagen sandwich for over 14 days. The improvement of *in vivo*-like liver functionality in the PET-Gal/PET-GRGDS system could be due to better cell-cell contacts and more efficient diffusion of nutrients and waste removal.

SPHEROID-BASED LIVER MODELS

Hepatocyte spheroids have been produced by (Brown *et al.*, 2003) and maintained in a differentiated state for several weeks by culturing in a rotating well vessel (RWV, www.synthecon.com). The spheroids were obtained by culturing primary rat hepatocytes in 100-mm Petri dishes constantly shacked with an orbital shaker. Then, the spheroids were further cultured in a RWV. The hepatocyte spheroids remained viable for several weeks in the RWV and maintained specific liver functions. Phase 1 and phase 2 CYPs xenobiotics metabolism was confirmed by measuring the activity of CYP 2B1/2, CYP 2E1, CYP 2D and CYP 3A. Prolonged albumin secretion was also observed. Thus, RWV is well suited for long-term culture of liver cell spheroids, with good viability and maintenance of liver-specific functions. With this system, toxicological and pharmacological assays are possible.

   Xu *et al.* compared primary primary rat hepatocytes spheroids and HepG2 (a human hepatocellular carcinoma cell line) spheroids as *in vitro* models for toxicity assays (Xu, Ma & Purcell, 2003c). The authors claim that HepG2 spheroids are suitable for *in vitro* toxicity studies, and produce results comparable with that obtained with primary hepatocytes.

FURTHER SPHEROID-BASED ASSAYS

Kloss *et al.* developed a spheroids biochip for toxicological screening based on impedance spectroscopy (Kloss *et al.*, 2008). The frequency-dependent impedance spectrum provides information on the dielectric and structural properties of the spheroids. Single spheroids were trapped into pyramidal cavities with a size of 300 µm etched on a silicon wafer. Each cavity contained four gold microelectrodes to measure the spheroid's impedance. Each biochip contained 25 cavities. The biochip was used to assess the effect of drugs on cell packaging density, surface structure, and ECM architecture of the spheroids. Spheroids obtained from human melanoma cells (Bro), african green monkey kidney (Cos-7), chinese hamster ovary cells (CHO), and chicken retina cells were tested. The most compact spheroids (Bro) showed low impedance, while less compact and irregular chicken retina cell spheroids displayed the highest impedance. Human melanoma (Bro) spheroids were employed for a screening assay with cytotoxic drugs. After eight hours of exposition, the drugs forskolin, staurosporine and campthotecin caused an increase of the impedance, while doxorubicin and tamoxifen induced a decrease of the impedance. Correlating the change of the impedance with specific drug-induced effect requires additional investigation, e.g. with light microscopy. Previous work on spheroid impedance spectra suggests that changes in the spheroid's compactness (Kloss *et al.*, 2008) as well as disintegration of the cell's membrane (Thielecke, Mack & Robitzki, 2001) are the parameters with the largest influence on the impedance.

Further works have shown that impedance assays on embryonic chicken heart muscle spheroids are suitable for drug and toxicity testing (Reininger-Mack, Thielecke & Robitzki, 2002). Thus, spheroid-based biochips with impedance/potential recording are promising for toxicity assay for food quality control, environmental control, pharmaceutical and cosmetic industry.

In summary, several toxicity assays based on 3D cell models have been developed and are promising in their ability to mimic specific organ response to toxicants. However, the validation of these assays from the research stage to real-world application is a long and cumbersome process. This is evidenced by the fact that just three skin models have been fully validated so far to replace animal testing. A close cooperation between regulatory authorities, industry, and academia is required in order to speed up the validation process.

Conclusions and prospects

NOVEL EXTRACELLULAR MATRICES

The successful introduction of toxicity screenings based on 3D cell cultures strongly depends on the availability of 3D matrices that are easy to handle and with low price. Reproducible and standardized matrices designed to mimic specific tissue microenvironments are required. A broad range of synthetic scaffolds developed for tissue engineering is available (Kim & Mooney, 1998). Substantial efforts have been devoted to increase tissue-like specificity and biocompatibility of 3D hydrogels. In the near future, novel 3D matrices will contribute to the wider adoption of 3D cell cultures for large-scale toxicity screenings.

EXTRACTED ECM HYDROGELS

Collagen I and basement membrane extract alone do not accurately mimic the composition of the different ECM types found in tissues (Uriel *et al.*, 2008). Laminin-1 represents a minor fraction of the ECM proteins found in most adult tissues. Further laminin isoforms or basement membrane components are essential to recapitulate specific property of tissues *in vitro* (Uriel *et al.*, 2008). A procedure to extract ECM proteins from selected tissues has been elaborated by (Uriel *et al.*, 2008) and (Brey *et al.*, 2007). The source tissue is decellularized by treating with EDTA, with dispase, or by grinding. Then, the basement membrane is homogenized in a high-salt solution containing a protease inhibitor. Finally, the basement membrane proteins are extracted by soaking in presence of urea. The extracted ECM proteins can be gelled by rising pH or temperature. The resulting gel is a highly specific duplicate of the tissue's ECM.

SYNTHETIC ECM HYDROGELS

Extracted hydrogels recapitulates many *in vivo* characteristic of the ECM. However, they may contain non quantified impurities, such as growth factors and intracellular proteins. Major issues are batch-to-batch variations and limited availability. Fully synthetic hydrogels for 3D cell-culture would reduce costs and improve the reproducibility of results. Zhang *et al.* introduced nanofibers hydrogels based on self-assembling oligopeptides with repeating motifs Glu-Ala-Lys or Arg-Ala-Asp (Zhang *et al.*, 2005). The oligopeptide is available on the market with the trade name "Puramatrix" (BD Biosciences) (Zhang *et al.*, 1997). Nanofiber self-assembly is triggered by mixing the peptide solution with culture medium at physiological pH (~7). The nanofibers have a diameter of about 10 nm and the gel pore size is between 5 and 200 nm (Zhang *et al.*, 2005). The nanofiber gel surrounds the cells similarly to the natural ECM. Ulijn *et al.* introduced short oligopeptides with the motifs Phe-Phe, Fmoc-Leu-Leu-Leu, as well as Fmoc-X-Phe-Phe (X=Ala, Val, Leu, Phe) (Ulijn *et al.*, 2007). These oligopeptides form stable hydrogels that support cell growth in 3D. The drawback of fully synthetic hydrogels is that they lack cell-specific adhesion-promoting sites. Thus, ECM proteins such as laminin-1 or fibronectin must be added in order to promote cell growth and differentiation. Semi-synthetic hydrogels consisting of cross-linked derivatives of hyaluronan and gelatine are a promising alternative to both animal-derived and fully synthetic hydrogels (Serban, Scott & Prestwich, 2008). Hyaluronan is a major constituent of natural ECM, while gelatine is denatured collagen. In the semisynthetic hydrogel, both hyaluronan and gelatine are modified with reactive thiol groups. Polyethylene glycol diacrylate is employed as cross-linker. Studies have shown that primary hepatocytes cultured in the 3D hyaluronan-gelatine matrix maintain liver-specific functionality longer than 2D-cultured hepatocytes (Prestwich *et al.*, 2007). A hyaluronan semi-synthetic matrix is available commercially (trade name "Extracel", Glycosan Biosystems).

Alginate hydrogels are also widely employed in tissue engineering (Lee & Mooney, 2001). Gelatine-coated alginate microcarriers are emerging as a promising matrix for large-scale 3D cell cultures (Justice *et al.*, 2009).

Recently, macroporous hydrogels (cryogels) have been applied to 3D cell cultures (Plieva *et al.*, 2008a). Cryogels are prepared by gelation at sub-zero temperature. They have a well-controlled porosity, determined by the size of the growing ice crystals. The cryogel surface can be easily functionalized with ECM molecules, such as collagen I or RGD peptides (Dainiak *et al.*, 2008). The porous structure of 3D cryogels, as well as tissue-like elasticity, allows for a physiological cell microenvironment (Plieva *et al.*, 2008b). 3D cryogel scaffolds have been adapted to standard 96-well format. A drug toxicity test performed with this system has shown that human colon cancer cells (HCT116) are 3.5-fold less sensitive to cisplatin than the same cells cultured in 2D (Dainiak *et al.*, 2008).

Photopolymerizable polyethylene glycol diacrylate (PEG-DA) hydrogels supplemented with ECM molecules have been employed to fabricate 3D liver cell constructs. PEG-based hydrogels are biocompatible, hydrophilic, and can easily be functionalized. By using micro-photolithography, Liu Tsang *et al.* patterned a three-layer hexagonal multilayer structure with PEG-DA embedding hepatocytes (Liu Tsang *et al.*, 2007). The hexagonal geometry mimics the branching architecture of liver *in vivo* and optimizes the exchange of nutrients and waste in the perfusion system. The study showed that hepatocytes can be maintained for two weeks in the perfused PEG-DA. The fabrication of photopatterned multilayer 3D cultures could be straightforwardly extended to other cell types, realizing miniaturized "printed organs" suitable for toxicology screening.

LIGHT-SHEET BASED FLUORESCENCE MICROSCOPY

The possibility to perform high-content screening (HCS) of 3D cell cultures is essential to accomplish efficient and predictive toxicity assays and replace testing on animals. HCS harnesses fluorescence images of cells with algorithms that track the spatiotemporal distribution of target proteins. HCS can probe events in live cells (e.g. sub-cellular protein translocation) that are not accessible in conventional high throughput screening, and is establishing itself as a core technology in cell-based toxicity assays (Rausch, 2006). However, 3D cell cultures are challenging for optical microscopy. Wide-field epifluorescence microscopy and confocal microscopy work optimally on flat specimens such as 2D cultured cells, but meet their limits when applied to 3D cell cultures. This is due to large size (from hundreds of micrometers to few millimeters), strong light scattering due to the ECM fibrous network, and high cell density. Further serious issues are phototoxicity as well as photobleaching due to high-intensity illumination of the specimen. Other imaging techniques, such as the environmental scanning electron microscope (ESEM), have been successfully applied to 3D cultures (Uroukov & Patton, 2008), but they are not suitable for large-scale screening due to slow specimen preparation. Recent advances in imaging technology are removing some of the obstacles that have prevented the observation of live cells that grow in a more natural, physiological three-dimensional environment. The Light Sheet-based Fluorescence Microscope (LSFM) (*Figure 2A*) is optimally suited for imaging of 3D cell cultures. Two implementations of LSFM exist: Single Plane Illumination Microscope (SPIM) (Greger, Swoger & Stelzer, 2007; Huisken *et al.*, 2004; Pampaloni *et al.*, 2007), and Digital Scanned Light-sheet Microscope (DSLM) (Keller *et al.*, 2008). In LSFM, the specimen is selectively illuminated in the focal plane with a laser light-sheet. Thus, fluorescence emission is excited exclusively in a thin plane in the specimen. The out-of-plane regions are not illuminated. The fluorescence from the illuminated plane is collected with an objective lens placed perpendicularly to the light-sheet and imaged with a sensitive CCD camera. The objective lens focal plane and the illuminated plane overlap, so that only the plane that is illuminated is imaged. A 3D image stack is obtained by moving the specimen through the light-sheet. Adjacent planes can be as close as 0.5 micrometers. Additionally, the specimen can be axially rotated. This allows recording 3D image stacks of the same specimen from different angles. By using high-numerical aperture water-dipping objective lenses (e.g. with NA=0.85 or NA=1.0), a resolution at sub-cellular level can be achieved. LSFM has a large penetration depth also in strongly scattering specimens, very low photobleaching, and high recording speed. Typical specimens for LSFM are 3D cell cultures in ECM gel, cellular spheroids (*Figure 2B-C, Figure 3*), zebrafish and medaka fish embryos, *Drosophila* embryos, as well as tissue explants. The specimen handling and preparation for 3D imaging is straightforward for most types of specimens. Cellular spheroids are just embedded in 0.5% agarose before imaging (*Figure 2B*). Simple mounting procedures, fast recording speed, and high-resolution multi-channel fluorescence imaging allow the fast screening of a large number of spheroids, tissue explants, and cell cultures. We anticipate that LSFM will greatly advance toxicity assays, allowing HCS and HCA of size, shape, texture, and intensity at individual cell level in 3D cell cultures.

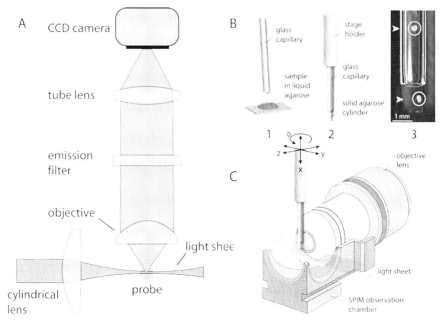

Figure 2. LSFM set-up. (A) In an LSFM, a laser light sheet is fed into the specimen from the side and observed at an angle of 90° to the illumination optical axis. In this implementation (Single Plane Illumination Microscope, SPIM (Greger *et al.*, 2007; Huisken *et al.*, 2004)), the light sheet is produced by a cylindrical lens. The optical sectioning arises from the overlap of the focal plane of the fluorescence detection system with the central plane of the excitation light sheet. Light-sheet-based fluorescence microscopes perform particularly well with long working distance lenses. Since the numerical aperture of the illumination system is much smaller than that of the detection system, light-sheet-based fluorescence microscopes have a good penetration depth. Millimetre-sized specimens can be observed in their totality. A further increase in resolution and information content can be obtained by observing the same specimen multiple times but along different directions. Parts of the sample that would otherwise be hidden or obscured along one direction now become visible. In a further step, the images stacks that are independently recorded along different angles are combined to yield a single fused images stack. (B1) The spheroid is inside a droplet of liquid agarose and sucked into a glass capillary. (B2) As soon as the agarose has formed a gel, the sample is pushed out of the capillary and imaged. (B3) A photograph of the specimen with the spheroid marked by the circle at the bottom. A second spheroid within the capillary is found further up (upper circle). The white arrow points at the boundary of the agarose. The yellow arrow indicates the boundary of the capillary.

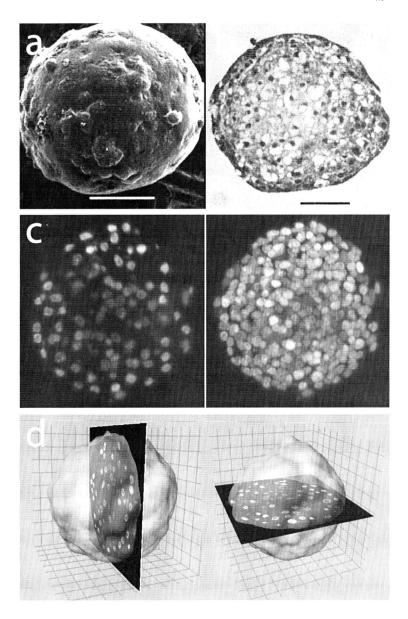

Figure 3. (A) An image of a liver spheroid recorded by scanning electron microscope. The bar represents 50 μm. (B) The histological section of a six-days old spheroid. Scale bar 50 μm. (C) Fluorescence image of a cellular pancreas tumour cells (BxCP-3) spheroid. The image was recorded with Light-sheet based fluorescence microscopy (SPIM). The image is calculated by combining twelve views recorded along different angles. The cell nuclei are labelled with the fluorescent dye Draq5™. The spheroid diameter is ~100 μm. (C, left) A three-dimensional maximum projection of about 100 single slices spaced 0.5 μm apart. (C, right) A single slice showing the central region of the spheroid. (D) The same spheroid rendered in three dimensions with the image processing software Imaris™.

Acknowledgements

FP and EHK Stelzer acknowledge Ruth Kroschewski and Ulrich Berge (ETH Zürich), Marco Marcello (University of Liverpool), and Emmanuel Reynaud (University College Dublin) for many interesting discussions and support.

References

Behrens, I., Stenberg, P., Artursson, P. & Kissel, T. (2001). Transport of lipophilic drug molecules in a new mucus-secreting cell culture model based on HT29-MTX cells. *Pharm. Res.* 18, 1138-45.

Beningo, K. A., Dembo, M. & Wang, Y. L. (2004). Responses of fibroblasts to anchorage of dorsal extracellular matrix receptors. *Proc. Natl. Acad. Sci. USA* **101,** 18024-9.

Bessea, L., Coulomb, B., Lebreton-Decoster, C. & Giraud-Guille, M. M. (2002). Production of ordered collagen matrices for three-dimensional cell culture. *Biomaterials* **23,** 27-36.

Bhadriraju, K. & Chen, C. S. (2002). Engineering cellular microenvironments to improve cell-based drug testing. *Drug Discov. Today* **7,** 612-20.

Bhogal, N., Grindon, C., Combes, R. & Balls, M. (2005). Toxicity testing: creating a revolution based on new technologies. *Trends Biotechnol.* **23,** 299-307.

Bilodeau, K. & Mantovani, D. (2006). Bioreactors for tissue engineering: focus on mechanical constraints. A comparative review. *Tissue Eng.* **12,** 2367-2383.

Birgersdotter, A., Sandberg, R. & Ernberg, I. (2005). Gene expression perturbation in vitro--a growing case for three-dimensional (3D) culture systems. *Semin. Cancer Biol.* **15,** 405-12.

Bissell, M. J. & Radisky, D. (2001). Putting tumours in context. *Nat. Rev. Cancer* **1,** 46-54.

Bohets, H., Annaert, P., Mannens, G., Van Beijsterveldt, L., Anciaux, K., Verboven, P., Meuldermans, W. & Lavrijsen, K. (2001). Strategies for absorption screening in drug discovery and development. *Curr. Top. Med. Chem.* **1,** 367-83.

Brey, E. M., Uriel, S., Labay, E. & Weichselbaum, R. R. (2007). Tissue-specific basement membrane gels. US Patent App. 11/585,353, 2006.

Brown, L. A., Arterburn, L. M., Miller, A. P., Cowger, N. L., Hartley, S. M., Andrews, A., Silber, P. M. & Li, A. P. (2003). Maintenance of liver functions in rat hepatocytes cultured as spheroids in a rotating wall vessel. *In vitro Cell Dev. Biol. Anim.* **39,** 13-20.

Chevallay, B. & Herbage, D. (2000). Collagen-based biomaterials as 3D scaffold for cell cultures: applications for tissue engineering and gene therapy. *Med. Biol. Eng. Comput.* **38,** 211-8.

Cukierman, E., Pankov, R., Stevens, D. R. & Yamada, K. M. (2001). Taking cell-matrix adhesions to the third dimension. *Science* **294,** 1708-1712.

Dainiak, M. B., Savina, I. N., Musolino, I., Kumar, A., Mattiasson, B. & Galaev, I. Y. (2008). Biomimetic macroporous hydrogel scaffolds in a high-throughput screening format for cell-based assays. *Biotechnol. Prog.* **24,** 1373-83.

de Boo, J. & Hendriksen, C. (2005). Reduction strategies in animal research: a review of scientific approaches at the intra-experimental, supra-experimental and extra-experimental levels. *Altern. Lab. Anim.* **33,** 369-77.

DESOIZE, B. & JARDILLIER, J. (2000). Multicellular resistance: a paradigm for clinical resistance? *Crit Rev. Oncol. Hematol.* **36**, 193-207.

DU, Y., HAN, R., WEN, F., NG SAN SAN, S., XIA, L., WOHLAND, T., LEO, H. L. & YU, H. (2008). Synthetic sandwich culture of 3D hepatocyte monolayer. *Biomaterials* **29**, 290-301.

FOTH, H. & HAYES, A. (2008). Concept of REACH and impact on evaluation of chemicals. *Hum. Exp. Toxicol.* **27**, 5-21.

FRIEDRICH, J., SEIDEL, C., EBNER, R. & KUNZ-SCHUGHART, L. A. (2009). Spheroid-based drug screen: considerations and practical approach. *Nat. Protoc.* **4**, 309-24.

GHOSH, S., SPAGNOLI, G. C., MARTIN, I., PLOEGERT, S., DEMOUGIN, P., HEBERER, M. & RESCHNER, A. (2005). Three-dimensional culture of melanoma cells profoundly affects gene expression profile: a high density oligonucleotide array study. *J. Cell Physiol.* **204**, 522-31.

GLICKLIS, R., MERCHUK, J. C. & COHEN, S. (2004). Modeling mass transfer in hepatocyte spheroids via cell viability, spheroid size, and hepatocellular functions. *Biotechnol. Bioeng.* **86**, 672-80.

GLICKLIS, R., SHAPIRO, L., AGBARIA, R., MERCHUK, J. C. & COHEN, S. (2000). Hepatocyte behavior within three-dimensional porous alginate scaffolds. *Biotechnol. Bioeng.* **67**, 344-53.

GOMEZ-LECHON, M. J., JOVER, R., DONATO, T., PONSODA, X., RODRIGUEZ, C., STENZEL, K. G., KLOCKE, R., PAUL, D., GUILLEN, I., BORT, R. & CASTELL, J. V. (1998). Long-term expression of differentiated functions in hepatocytes cultured in three-dimensional collagen matrix. *J. Cell Physiol.* **177**, 553-562.

GREGER, K., SWOGER, J. & STELZER, E. H. (2007). Basic building units and properties of a fluorescence single plane illumination microscope. *Rev. Sci. Instrum.* **78**, 023705.

HASSPIELER, B., HAFFNER, D., STELLJES, M. & ADELI, K. (2006). Toxicological assessment of industrial solvents using human cell bioassays: assessment of short-term cytotoxicity and long-term genotoxicity potential. *Toxicol. Ind. Health* **22**, 301-15.

HOLOPAINEN, I. E. (2005). Organotypic hippocampal slice cultures: a model system to study basic cellular and molecular mechanisms of neuronal cell death, neuroprotection, and synaptic plasticity. *Neurochem. Res.* **30**, 1521-8.

HUISKEN, J., SWOGER, J., DEL BENE, F., WITTBRODT, J. & STELZER, E. H. (2004). Optical sectioning deep inside live embryos by selective plane illumination microscopy. *Science* **305**, 1007-9.

IVASCU, A. & KUBBIES, M. (2006). Rapid generation of single-tumor spheroids for high-throughput cell function and toxicity analysis. *J. Biomol. Screen.* **11**, 922-932.

JAKAB, K., NEAGU, A., MIRONOV, V. & FORGACS, G. (2004a). Organ printing: fiction or science. *Biorheology* **41**, 371-5.

JAKAB, K., NEAGU, A., MIRONOV, V., MARKWALD, R. R. & FORGACS, G. (2004b). Engineering biological structures of prescribed shape using self-assembling multicellular systems. *Proc. Natl. Acad. Sci. USA* **101**, 2864-9.

JESSBERGER, S., CLEMENSON, G. D., JR. & GAGE, F. H. (2007). Spontaneous fusion and nonclonal growth of adult neural stem cells. *Stem Cells* **25**, 871-4.

JIANG, Y., PJESIVAC-GRBOVIC, J., CANTRELL, C. & FREYER, J. P. (2005). A multiscale model for avascular tumor growth. *Biophysical Journal* **89**, 3884-3894.

JUSTICE, B. A., BADR, N. A. & FELDER, R. A. (2009). 3D cell culture opens new dimensions in cell-based assays. *Drug Discov. Today* **14**, 102-7.

Kale, S., Biermann, S., Edwards, C., Tarnowski, C., Morris, M. & Long, M. W. (2000). Three-dimensional cellular development is essential for ex vivo formation of human bone. *Nat.Biotechnol.* **18,** 954-8.

Kandarova, H., Liebsch, M., Genschow, E., Gerner, I., Traue, D., Slawik, B. & Spielmann, H. (2004). Optimisation of the EpiDerm test protocol for the upcoming ECVAM validation study on in vitro skin irritation tests. *Altex* **21,** 107-14.

Kandarova, H., Liebsch, M., Gerner, I., Schmidt, E., Genschow, E., Traue, D. & Spielmann, H. (2005). The EpiDerm test protocol for the upcoming ECVAM validation study on in vitro skin irritation tests--an assessment of the performance of the optimised test. *Altern. Lab. Anim.* **33,** 351-67.

Kaplowitz, N. (2005). Idiosyncratic drug hepatotoxicity. *Nat. Rev. Drug Discov.* **4,** 489-99.

Keller, P. J., Schmidt, A. D., Wittbrodt, J. & Stelzer, E. H. (2008). Reconstruction of zebrafish early embryonic development by scanned light sheet microscopy. *Science* **322,** 1065-9.

Kelm, J. M., Timmins, N. E., Brown, C. J., Fussenegger, M. & Nielsen, L. K. (2003). Method for generation of homogeneous multicellular tumor spheroids applicable to a wide variety of cell types. *Biotechnol. Bioeng.* **83,** 173-180.

Kim, B. S. & Mooney, D. J. (1998). Development of biocompatible synthetic extracellular matrices for tissue engineering. *Trends Biotechnol.* **16,** 224-30.

Kleinman, H. K. & Martin, G. R. (1989). Reconstituted basement membrane complex with biological activity. US Patent 4,829,000, 1989

Kloss, D., Fischer, M., Rothermel, A., Simon, J. C. & Robitzki, A. A. (2008). Drug testing on 3D in vitro tissues trapped on a microcavity chip. *Lab Chip* **8,** 879-84.

Knight, A. (2007a). Animal experiments scrutinised: systematic reviews demonstrate poor human clinical and toxicological utility. *Altex* **24,** 320-5.

Knight, A. (2007b). Systematic reviews of animal experiments demonstrate poor human clinical and toxicological utility. *Altern. Lab. Anim.* **35,** 641-59.

Korff, T. & Augustin, H. G. (1998). Integration of endothelial cells in multicellular spheroids prevents apoptosis and induces differentiation. *J. Cell Biol.* **143,** 1341-52.

Kunz-Schughart, L. A., Freyer, J. P., Hofstaedter, F. & Ebner, R. (2004). The use of 3-D cultures for high-throughput screening: the multicellular spheroid model. *J. Biomol. Screen* **9,** 273-85.

Layer, P. G., Robitzki, A., Rothermel, A. & Willbold, E. (2002). Of layers and spheres: the reaggregate approach in tissue engineering. *Trends Neurosci.* **25,** 131-4.

Lazar, A., Mann, H. J., Remmel, R. P., Shatford, R. A., Cerra, F. B. & Hu, W. S. (1995). Extended liver-specific functions of porcine hepatocyte spheroids entrapped in collagen gel. *In Vitro Cell Dev. Biol. Anim.* **31,** 340-6.

Lee, K. Y. & Mooney, D. J. (2001). Hydrogels for tissue engineering. *Chem. Rev* .**101,** 1869-79.

Lee, M. Y., Kumar, R. A., Sukumaran, S. M., Hogg, M. G., Clark, D. S. & Dordick, J. S. (2008). Three-dimensional cellular microarray for high-throughput toxicology assays. *Proc. Natl. Acad. Sci. USA* **105,** 59-63.

Li, A. P. (2001). Screening for human ADME/Tox drug properties in drug discovery. *Drug Discov. Today* **6,** 357-366.

Li, S., Lao, J., Chen, B. P., Li, Y. S., Zhao, Y., Chu, J., Chen, K. D., Tsou, T. C., Peck,

K. & CHIEN, S. (2003). Genomic analysis of smooth muscle cells in 3-dimensional collagen matrix. *Faseb. J.* **17**, 97-9.

LILIENBLUM, W., DEKANT, W., FOTH, H., GEBEL, T., HENGSTLER, J. G., KAHL, R., KRAMER, P. J., SCHWEINFURTH, H. & WOLLIN, K. M. (2008). Alternative methods to safety studies in experimental animals: role in the risk assessment of chemicals under the new European Chemicals Legislation (REACH). *Arch. Toxicol.* **82**, 211-36.

LIN, J., SAHAKIAN, D. C., DE MORAIS, S. M., XU, J. J., POLZER, R. J. & WINTER, S. M. (2003). The role of absorption, distribution, metabolism, excretion and toxicity in drug discovery. *Curr. Top. Med. Chem.* **3**, 1125-54.

LIU TSANG, V., CHEN, A. A., CHO, L. M., JADIN, K. D., SAH, R. L., DELONG, S., WEST, J. L. & BHATIA, S. N. (2007). Fabrication of 3D hepatic tissues by additive photopatterning of cellular hydrogels. *Faseb J* .**21**, 790-801.

LUTOLF, M. P. & HUBBELL, J. A. (2005). Synthetic biomaterials as instructive extracellular microenvironments for morphogenesis in tissue engineering. *Nat. Biotechnol.* **23**, 47-55.

MAZZOLENI, G., DI LORENZO, D. & STEIMBERG, N. (2009). Modelling tissues in 3D: the next future of pharmaco-toxicology and food research? *Genes Nutr.* **4**, 13-22.

MONTESANO, R. (1986). Cell-extracellular matrix interactions in morphogenesis: an in vitro approach. *Experientia* **42**, 977-85.

MOSCONA, A. (1961). Rotation-mediated histogenetic aggregation of dissociated cells. A quantifiable approach to cell interactions in vitro. *Exp. Cell Res.* **22**, 455-475.

MOSTOV, K., SU, T. & TER BEEST, M. (2003). Polarized epithelial membrane traffic: conservation and plasticity. *Nat. Cell Biol.* **5**, 287-293.

MUELLER-KLIESER, W. (1997). Three-dimensional cell cultures: from molecular mechanisms to clinical applications. *Am. J. Physiol.* **273**, C1109-23.

NETZLAFF, F., LEHR, C. M., WERTZ, P. W. & SCHAEFER, U. F. (2005). The human epidermis models EpiSkin, SkinEthic and EpiDerm: an evaluation of morphology and their suitability for testing phototoxicity, irritancy, corrosivity, and substance transport. *Eur. J. Pharm. Biopharm.* **60**, 167-78.

NORABERG, J. (2004). Organotypic brain slice cultures: an efficient and reliable method for neurotoxicological screening and mechanistic studies. *Altern. Lab. Anim.* **32**, 329-37.

O'BRIEN, L. E., ZEGERS, M. M. & MOSTOV, K. E. (2002). Opinion: Building epithelial architecture: insights from three-dimensional culture models. *Nat. Rev. Mol. Cell Biol.* **3**, 531-537.

OLSON, H., BETTON, G., ROBINSON, D., THOMAS, K., MONRO, A., KOLAJA, G., LILLY, P., SANDERS, J., SIPES, G., BRACKEN, W., DORATO, M., VAN DEUN, K., SMITH, P., BERGER, B. & HELLER, A. (2000). Concordance of the toxicity of pharmaceuticals in humans and in animals. *Regul. Toxicol. Pharmacol.* **32**, 56-67.

PAMPALONI, F., REYNAUD, E. G. & STELZER, E. H. (2007). The third dimension bridges the gap between cell culture and live tissue. *Nat. Rev. Mol. Cell Biol.* **8**, 839-845.

PETERSEN, O. W., LIND NIELSEN, H., GUDJONSSON, T., VILLADSEN, R., RONNOV-JESSEN, L. & BISSELL, M. J. (2001). The plasticity of human breast carcinoma cells is more than epithelial to mesenchymal conversion. *Breast Cancer Res.* **3**, 213-7.

PLIEVA, F. M., GALAEV, I. Y., NOPPE, W. & MATTIASSON, B. (2008a). Cryogel applications in microbiology. *Trends Microbiol.* **16**, 543-51.

PLIEVA, F. M., OKNIANSKA, A., DEGERMAN, E. & MATTIASSON, B. (2008b). Macroporous gel particles as robust macroporous matrices for cell immobilization. *Biotechnol. J.* **3,** 410-7.

PODRYGAJLO, G., TEGENGE, M. A., GIERSE, A., PAQUET-DURAND, F., TAN, S., BICKER, G. & STERN, M. (2009). Cellular phenotypes of human model neurons (NT2) after differentiation in aggregate culture. *Cell Tissue Res.* **336,** 439-52.

PRESTWICH, G. D., LIU, Y., YU, B., SHU, X. Z. & SCOTT, A. (2007). 3-D culture in synthetic extracellular matrices: new tissue models for drug toxicology and cancer drug discovery. *Adv Enzyme Regul.* **47,** 196-207.

RADISKY, D., HAGIOS, C. & BISSELL, M. J. (2001). Tumors are unique organs defined by abnormal signaling and context. *Semin. Cancer Biol.* **11,** 87-95.

RAUSCH, O. (2006). High content cellular screening. *Curr Opin. Chem. Biol.* **10,** 316-20.

REININGER-MACK, A., THIELECKE, H. & ROBITZKI, A. A. (2002). 3D-biohybrid systems: applications in drug screening. *Trends Biotechnol.* **20,** 56-61.

RHEE, S. & GRINNELL, F. (2007). Fibroblast mechanics in 3D collagen matrices. *Adv. Drug Deliv. Rev.* **59,** 1299-305.

RHEE, S., JIANG, H., HO, C. H. & GRINNELL, F. (2007). Microtubule function in fibroblast spreading is modulated according to the tension state of cell-matrix interactions. *Proc. Natl. Acad. Sci. USA* **104,** 5425-30.

RIEKE, M., GOTTWALD, E., WEIBEZAHN, K. F. & LAYER, P. G. (2008). Tissue reconstruction in 3D-spheroids from rodent retina in a motion-free, bioreactor-based microstructure. *Lab Chip* **8,** 2206-13.

SEMINO, C. E., MEROK, J. R., CRANE, G. G., PANAGIOTAKOS, G. & ZHANG, S. (2003). Functional differentiation of hepatocyte-like spheroid structures from putative liver progenitor cells in three-dimensional peptide scaffolds. *Differentiation* **71,** 262-70.

SERBAN, M. A., SCOTT, A. & PRESTWICH, G. D. (2008). Use of hyaluronan-derived hydrogels for three-dimensional cell culture and tumor xenografts. *Curr. Protoc. Cell Biol.* **Chapter 10,** Unit 10 14.

STEIN, A. M., DEMUTH, T., MOBLEY, D., BERENS, M. & SANDER, L. M. (2007). A mathematical model of glioblastoma tumor spheroid invasion in a three-dimensional in vitro experiment. *Biophys. J.* **92,** 356-365.

SUN, T., JACKSON, S., HAYCOCK, J. W. & MACNEIL, S. (2006). Culture of skin cells in 3D rather than 2D improves their ability to survive exposure to cytotoxic agents. *J. Biotechnol.* **122,** 372-81.

SUTHERLAND, R. M. (1988). Cell and environment interactions in tumor microregions: the multicell spheroid model. *Science* **240,** 177-184.

SUTHERLAND, R. M., MCCREDIE, J. A. & INCH, W. R. (1971). Growth of multicell spheroids in tissue culture as a model of nodular carcinomas. *J. Natl. Cancer Inst.* **46,** 113-120.

TABATABAI, M., WILLIAMS, D. K. & BURSAC, Z. (2005). Hyperbolastic growth models: theory and application. *Theor. Biol. Med. Model.* **2,** 14.

THIELECKE, H., MACK, A. & ROBITZKI, A. (2001). Biohybrid microarrays--impedimetric biosensors with 3D in vitro tissues for toxicological and biomedical screening. *Fresenius J. Anal. Chem.* **369,** 23-9.

TIMMINS, N. E., DIETMAIR, S. & NIELSEN, L. K. (2004). Hanging-drop multicellular spheroids as a model of tumour angiogenesis. *Angiogenesis* **7,** 97-103.

ULIJN, R. V., JAYAWARNA, V., SMITH, A. & GOUGH, J. E. (2007). Hydrogel compositions. US Patent App. 11/470,962, 2006.

URIEL, S., LABAY, E., FRANCIS-SEDLAK, M., MOYA, M. L., WEICHSELBAUM, R. R., ERVIN, N., CANKOVA, Z. & BREY, E. M. (2008). Extraction and Assembly of Tissue-Derived Gels for Cell Culture and Tissue Engineering. *Tissue Eng. Part C Methods.*

UROUKOV, I. S. & PATTON, D. (2008). Optimizing environmental scanning electron microscopy of spheroidal reaggregated neuronal cultures. *Microsc. Res. Tech.* **71**, 792-801.

WANG, T. Y., SEN, A., BEHIE, L. A. & KALLOS, M. S. (2006). Dynamic behavior of cells within neurospheres in expanding populations of neural precursors. *Brain Res.* **1107**, 82-96.

XU, J., MA, M. & PURCELL, W. M. (2003a). Biochemical and functional changes of rat liver spheroids during spheroid formation and maintenance in culture: II. nitric oxide synthesis and related changes. *J. Cell Biochem.* **90**, 1176-85.

XU, J., MA, M. & PURCELL, W. M. (2003b). Characterisation of some cytotoxic endpoints using rat liver and HepG2 spheroids as in vitro models and their application in hepatotoxicity studies. I. Glucose metabolism and enzyme release as cytotoxic markers. *Toxicol. Appl. Pharmacol.* **189**, 100-11.

XU, J., MA, M. & PURCELL, W. M. (2003c). Characterisation of some cytotoxic endpoints using rat liver and HepG2 spheroids as in vitro models and their application in hepatotoxicity studies. II. Spheroid cell spreading inhibition as a new cytotoxic marker. *Toxicol. Appl. Pharmacol.* **189**, 112-9.

ZHANG, S., GELAIN, F. & ZHAO, X. (2005). Designer self-assembling peptide nanofiber scaffolds for 3D tissue cell cultures. *Semin. Cancer Biol.* **15**, 413-420.

ZHANG, S., LOCKSHIN, C., RICH, A. & HOLMES, T. (1997). Stable macroscopic membranes formed by self-assembly of amphiphilic peptides and uses therefore. US Patent 5,670,483, 1997.

Biotechnology and Genetic Engineering Reviews - Vol. 26, 139-162 (2009)

Biosynthesis and Engineering of Carotenoids and Apocarotenoids in Plants: State of the Art and Future Prospects

CARLO ROSATI[1*], GIANFRANCO DIRETTO[2], GIOVANNI GIULIANO[2]

*Ente per le Nuove tecnologie, l'Energia e l'Ambiente (ENEA), Trisaia Research Center, S.S. 106 km 419.5, 75026 Rotondella (Matera), Italy and *ENEA, Casaccia Research Center, PO Box 2400, 00123 S.M. di Galeria (Roma), Italy*

Abstract

Carotenoids and their apocarotenoid derivatives are isoprenoid molecules important for the primary and secondary metabolisms of plants and other living organisms, displaying also key health-related roles in humans and animals. Progress in the knowledge of the carotenoid pathway at the genetic, biochemical and molecular level, supported by successful genetic engineering examples for an increasing number of important plant crops have paved the way for precise molecular breeding of carotenoids. In this review, following a description of the general carotenoid pathway, select examples of plant species able to produce specialty carotenoids and apocarotenoids are illustrated.

* To whom correspondence may be addressed (carlo.rosati@enea.it)

Abbreviations:. ABA, abscisic acid; ADH, aldehyde dehydrogenase; BADH, bixin ADH; BETA, chromoplast-specific lycopene β-cyclase; BMT, norBixin methyltransferase; CCD, carotenoid cleavage dioxygenase; CCS, capsanthin/capsorubin synthase; CHY, non-heme carotene hydroxylase; CMK, 4-(cytidine 5'-diphospho)-2-C-methyl-D-erythritol kinase; CrtB, bacterial phytoene synthase; CrtI, bacterial carotenoid desaturase/isomerase; CRTISO, carotenoid isomerase; CrtW, bacterial carotene ketolase; CrtY, bacterial lycopene β-cyclase; CrtZ, bacterial CHY; CYP97A and CYP97C, cytochrome P450 carotene hydroxylases; DXR, 1-deoxy-D-xylulose 5-phosphate reductoisomerase; DXS, 1-deoxy-D-xylulose 5-phosphate synthase; HDR, 4-hydroxy-3-methylbut-2-enyl diphosphate reductase; HDS, 4-hydroxy-3-methylbut-2-enyl diphosphate synthase; LCY-b, lycopene β-cyclase; LCY-e, lycopene ε-cyclase; MCT, 2-C-methyl-D-erythritol 4-phosphate cytidylyltransferase; MDS, 2-C-methyl-D-erythritol 2,4-cyclodiphosphate synthase; MVA, mevalonic acid / mevalonate; MVAP, mevalonate-5-phosphate; MVAPP, mevalonate-5-pyrophosphate; NXS, neoxanthin synthase; PDS, phytoene desaturase; PSY, phytoene synthase; VDE, violaxanthin de-epoxidase; ZCD, zeaxanthin-specific cleavage dioxygenase; ZDS, ζ-carotene desaturase; ZEP, zeaxanthin epoxidase; Z-ISO, ζ-carotene isomerase.

An update on plant carotenoid engineering is also provided for non-solanaceous crops and members of the Solanaceae family, by means of different strategies and making use of plant and bacterial genes.

Introduction

The visual properties and health benefits of carotenoid pigments have fostered fundamental research and industrial applications for decades. Carotenoids are isoprenoid molecules synthesized by plants as well as by bacteria, fungi, yeast and algae. In plants, the presence of carotenoids is mandatory for photosynthesis, and determines the pigmentation of flower organs, fruits and seeds in many important crops. Carotenoids cover a wide range of functions in plants and animals such as energy transfer and protection against photooxidative damage during photosynthesis, attraction of insects and animals for pollination and fruit/seed dispersal, contribution to plant cross-talk with pathogens, pests, and symbiotic organisms, reduction of degenerative disease risks and prevention of cardiovascular accidents, and provide a source of pro-vitamin A (Gann *et al.*, 1999; Hirschberg, 2001 and references therein). To date, hundreds of carotenoid and apocarotenoid derivatives have been described (Britton *et al.*, 2003; Winterhalter and Rouseff, 2001). This figure is most likely to increase owing to the continuous progress in analytical technologies and the advances in the knowledge of the pathway.

The existing scientific literature on the carotenoid pathway is vast and exceeds the scope of the present article, which will focus on the most recent publications after the latest reviews (Giuliano *et al.*, 2008; Fraser *et al.*, 2009) and on the exploitation of plant systems for the spontaneous or engineered production of high-value carotenoid and apocarotenoid compounds.

The carotenoid pathway in plants

CAROTENOID BIOSYNTHESIS

Plant carotenoids are isoprenoid-derived molecules generally synthesized and located in plastids, and their production is driven by nuclear-encoded enzymes (Hirschberg, 2001). In plants, isoprenoids are produced from the distinct but cross-talking cytosolic mevalonate (MVA) and plastidial 2-C-methyl-D-erythritol 4-phosphate (MEP) routes, to give rise to isopentenyl phosphate (IPP), the C5 building block for the synthesis of carotenoids and other compounds (*Figure 1*). Phytoene, the first carotenoid, is a C40 molecule synthesized from the condensation of two C20 geranylgeranyl diphosphate units by phytoene synthase (PSY). Subsequent desaturation steps, catalyzed by the phytoene desaturase (PDS) and ζ-carotene desaturase (ZDS) enzymes, take to the synthesis of *cis*-lycopene, which is converted into *all-trans*-lycopene either by the carotenoid isomerase (CRTISO) enzyme in non-photosynthetic tissues or spontaneously by light in chlorophyll-containing ones (Giuliano *et al.*, 2002; Isaacson *et al.*, 2004; Breitenbach and Sandmann, 2005). A recent study characterized the *Y9* locus of maize to code for a ζ-carotene isomerase (Z-ISO), and put forward the existence of such an activity across higher plants (Li *et al.*, 2007). The pathway then bifurcates

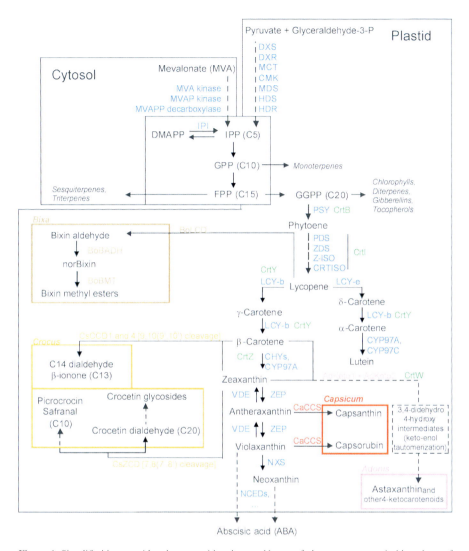

Figure 1. Simplified isoprenoid and carotenoid pathways. Names of plant enzymes are in blue, those of bacterial orthologs in green, while those of plants able to produce particular carotenoids and apocarotenoids are highlighted in brown (*Bixa orellana*), orange (*Crocus sativus*), red (*Capsicum annuum*) and magenta (*Adonis aestivalis*). Steps involving multiple enzymes are indicated with dashed arrows.

into two branches through the action of lycopene β- (LCY-b and chromoplast-specific BETA; Ronen *et al.*, 2000) and ε- (LCY-e; Cunningham and Gantt, 2001) cyclases. Two kinds of carotene hydroxylases – cytochrome P450 (CYP97A and C) and non-heme (CHY) enzymes – hydroxylate the β- and ε-rings to produce xanthophylls (DellaPenna and Pogson, 2006). The β-ε-ring lutein is the most abundant natural carotenoid present in photosynthetic tissues, while zeaxanthin is the first xanthophyll of the β-β-ring branch leading to the synthesis of abscisic acid (ABA). The xanthophyll cycle enzymes zeaxanthin epoxidase (ZEP) and violaxanthin de-epoxidase (VDE) regulate the amounts of such compounds during day and night: levels of violaxanthin are high at night, while those of zeaxanthin, a better quencher of singlet chlorophyll as well as a ROS scavenger, increase during the day when low lumen pH conditions induce the VDE enzyme activity. The late biosynthetic steps to neoxanthin and ABA have to be fully clarified yet. Neoxanthin synthase (NXS) genes, encoding paralogs of BETA β-cyclase in tomato, have been identified to date only in tomato and potato (Al-Babili *et al.*, 2000; Bouvier *et al.*, 2000). Nevertheless, tomato *beta* mutants are not impaired in neoxanthin biosythesis (Ronen *et al.*, 2000). *Arabidopsis* is able to produce ABA, although it lacks a *BETA* ortholog in its genome. The AtABA4 protein is directly involved in neoxanthin biosynthesis, but analysis of *ABA4* mutants and overexpressors could not clarify whether it coincides with NXS or it is part of a NXS multienzyme complex (North *et al.*, 2007). The possible multiple routes to ABA formation, involving not only enzymatic cleavage (see below), also await further clarification. Last but not least, the observed redundancy of carotenoid ortholog and paralog genes/enzymes is critical for the spatial and developmental regulation of carotenogenesis among and within species (e.g., Galpaz *et al.*, 2006; Li *et al.*, 2008).

CAROTENOID–SPECIFIC CLEAVAGE DIOXYGENASES PRODUCE APOCAROTENOID VOLATILES AND ABA PRECURSORS

Plants emit volatile organic compounds (VOCs) with crucial roles in plant-environment crosstalk, e.g. attraction of pollinators, herbivore deterrence and defense against pathogens. The structure similarity of some flower and spice isoprenoid VOCs (e.g., β-ionone and safranal) to carotenoids recently led to the discovery of carotenoid-specific cleavage dioxygenases (CCDs), which belong to a multienzyme family counting nine members in *Arabidopsis* (Tan *et al.*, 2003). Biochemical characterization established that most CCDs have wide substrate promiscuity, but are highly specific for the position of the double bond that they cleave (reviewed by Auldridge *et al.*, 2006). CCDs also differ for their subcellular localization: some, like CCD1s, are predicted to be cytosolic; others possess transit peptides for plastid or plastoglobule targeting (e.g., Tan *et al.*, 2003; Rubio *et al.*, 2008). CCDs include 9-*cis*-epoxycarotenoid dioxygenases (NCEDs), which cleave epoxy-xanthophylls to produce xanthoxin, a precursor of ABA (Schwartz *et al.*, 1997). Among CCDs, CCD7 and CCD8 are strikingly divergent: they cleave carotenoids and apocarotenoids asymmetrically, and were shown to be involved in the synthesis of a hormone inhibiting lateral branching, as confirmed by selective chemical inhibition, as well as controlling leaf senescence, root growth and flower development (Schwartz *et al.*, 2004; Snowden *et al.*, 2005; Sergeant *et al.*, 2009). CCD genes and enzymes have been characterized in a growing number of species, including fruit, vegetable, flower and field crops (e.g., Simkin *et al.*, 2004a and 2004b; Ibdah *et al.*, 2006; Garcia-Limones *et al.*, 2008; Huang *et al.*, 2009; Ilg *et al.*, 2009). Because of their economic value, the CCD-derived apocarotenoid products of *Crocus sativus* and *Bixa orellana* will be dealt more in detail in the following section.

Selected plant species produce specialty carotenoids and apocarotenoids

ACHIOTE (*BIXA ORELLANA* L.)

Annatto (bixin; European colorant code E160b) is a pigment extracted from achiote (*Bixa orellana*) seeds, used in food and cosmetic industry. Structure similarity between bixin and the *Crocus* pigment crocin, as well as the mixture of bixin aldehyde, carboxylated and methylated forms in annatto, indicated the possible involvement of CCD, aldehyde dehydrogenase (ADH) and methyltransferase enzymes in the minipathway (*Figure 1*). The corresponding genes were isolated from a seed cDNA library of *B. orellana* and characterized (Bouvier *et al.*, 2003a). The lycopene cleavage dioxygenase (BoLCD) is peculiar in that it cleaves lycopene at the 5,6(5',6') double bonds, but neither β-carotene nor zeaxanthin substrates. The bixin ADH (BoBADH) clustered with cytosolic ADHs, consistently with its putative localization, while the (nor)bixin MT (BonBMT) shared similar function (methylation of carboxyl groups) with jasmonic and salicylic acid MTs (Giuliano *et al.*, 2003). Expression of the *Bixa* minipathway enzymes in tomato might represent a biotechnological alternative for annatto production.

SAFFRON CROCUS (*CROCUS SATIVUS* L.)

Saffron is an expensive spice obtained from the dried styles of saffron crocus (*Crocus sativus*), whose distinctive flavor and color are due to picrocrocin, safranal and crocetin-derived apocarotenoids. The biosynthesis starts from the cleavage of zeaxanthin, to produce cyclic carotenoid VOCs (picrocrocin and safranal) and crocetin, which is eventually glycosylated to crocin (*Figure 1*). Some of the genes and enzymes involved in these steps have been studied. The *Crocus* CCDs so far characterized are similar to CCD1 and CCD4 enzymes, are differentially expressed in flower organs and CCD4s only contain predicted transit peptides for plastid (plastoglobule) localization (Rubio *et al.*, 2008). The CCD1-like generic CsCCDs possess 9,10 (9',10') cleavage activity on various carotenoid substrates (Bouvier *et al.*, 2003b; Rubio *et al.*, 2008). CCD4-like CsCCD4a and CsCCD4b proteins (Rubio *et al.*, 2008) are very similar (98-100% homology) to the CsZCD enzyme, previously reported to cleave zeaxanthin at the 7,8(7',8') positions and giving rise to crocetin dialdehyde (Bouvier et al., 2003b). CsCCD4 enzymes are longer than CsZCD and carry a plastid transit peptide. They perform a 9,10(9',10') cleavage, and are also able to cleave zeaxanthin, although the expected apocarotenoids could not be detected by neither LC nor GC (Rubio *et al.*, 2008). Further investigation is needed to clarify the exact number, protein structure, and enzymatic activity of *Crocus* CCD4/ZCD enzymes. Besides apocarotenoid volatiles, water-soluble crocetin glycosides are presumed to accumulate in vacuoles to confer pigmentation. A UDP-glucose crocetin 8-8'-glycosyltransferase enzyme was purified from cell suspensions and characterized (Côté *et al.*, 2000), and the product of the stigma-expressed *UGTCs2* gene was shown to glucosylate crocetin aglycones and glycosides *in vitro* (Rubio Moraga *et al.*, 2004).

PHEASANT'S EYE (*ADONIS SPP.*)

The ability to synthesize ketocarotenoids is usually exclusive in some bacteria and marine organisms. Few plants are known to produce ketocarotenoids, but only two of them accumulate these pigments at high levels: *Adonis aestivalis* and *Adonis annua*, whose ketocarotenoid content reaches 1% of dry weight in flower petals of the latter (Renstrøm *et* al., 1981). Astaxanthin, the high-value ketocarotenoid widely used as a feed supplement in aquaculture and poultry farming, is the major ketocarotenoid accumulated in *Adonis* flowers: *AdKeto* genes involved in ketocarotenoid formation have been isolated and characterized (Cunningham and Gantt, 2005). Interestingly, both AdKeto1 and AdKeto2 enzymes display high identity (>90%) to CHY enzymes, but, contrary to the latter, do not possess 3-hydroxylase activity, nor directly synthesize 4-keto carotenoids. Surprisingly, the Keto enzymes were shown to act as either 3,4-desaturase or 4-hydroxylase: in this context, a keto-enol tautomerization mechanism would rather be responsible for ketocarotenoid synthesis. More recently, a novel approach based on the use of several carotenoid-producing *E. coli* strains has been developed in *Adonis aestivalis* to characterize carotenoid and isoprenoid biosynthesis enzymes (Cunningham and Gantt, 2007).

PEPPERS (*CAPSICUM SPP.*)

Pepper and its hot chili varieties have long been cultivated and widely used as food and additives, also owing to the production of several peculiar carotenoids. Capsanthin, capsorubin and capsanthin 5,6-epoxide are the red carotenoids exclusively accumulating in *Capsicum spp.* fruits (Deli *et al.*, 2001). The gene encoding capsanthin-capsorubin synthase (CCS), involved in the synthesis of these pigments, has been isolated and characterized: CCS is a bifunctional cyclase enzyme able to use both antheraxanthin and violaxanthin as precursors and is specifically expressed during chromoplast development in fruits accumulating ketocarotenoids (Bouvier *et al.*, 1994). Pepper varieties can be classified according to fruit color: the two main isochromic families include red varieties, synthesizing capsanthin and capsorubin pigments, and yellow ones accumulate a carotenoid pool comprising antheraxanthin, violaxanthin (the precursors of capsanthin and capsorubin) as well as zeaxanthin, α-cryptoxanthin, α-carotene, and cucurbitaxanthin A (Guil-Guerrero *et al.*, 2006).

Genetically, three independent loci (*y*, *c1*, and *c2*) have been proposed to control fruit pigmentation. The *Ccs* gene was determined to be tightly associated with or to be encoded by the *y* locus, and yellow fruit phenotypes suggested to result from deletions of this gene (Lefebvre *et al.*, 1998 and references therein). A more complex scenario concerns orange-fruited varieties, with two possible genotypes that are responsible for this colour (Popovsky and Paran, 2000). From the molecular and biochemical viewpoint, red-fruited varieties display high *Ccs* and other carotenoid gene transcript levels, and high red-to-yellow (*R/Y*) isochromic pigment fraction and capsanthin-to-zeaxanthin (Caps/Zeax) ratios. Such ratios are very important to select varieties for paprika production, and are inversely related to the total carotenoid content of pepper fruit (Hornero-Méndez *et al.*, 2000). In contrast, yellow varieties are generally characterized by low carotenoid gene transcript levels, reduced carotenoid content, and absence, impaired expression or mutations of the *Ccs* gene (Bouvier *et al.*, 1994; Popovsky and Paran, 2000; Ha *et al.*, 2007). Recently, the analysis of yellow-fruited varieties, containing *Ccs* genes with premature stop-codons or frame-shifts in their coding sequences, indicated that post-transcriptional silencing mechanisms should also be taken into account in the determination of fruit yellow color (Ha *et al.*, 2007).

Engineering carotenoids and apocarotenoids in plants

Because of their pro-vitamin A and antioxidant properties, carotenoid biofortification of major food crops *via* either conventional or molecular breeding is an important issue especially in developing countries, where vitamin A deficiency (VAD) poses serious public health concerns and is responsible for various diseases and mortality (Al-Babili and Beyer, 2005 and references therein). For two of the three major cereal crops (in decreasing order of production: maize, rice and wheat), successful carotenoid engineering has been achieved: by *de novo* synthesis in the case of rice, or by increasing/optimizing the carotenoid profile in maize (see below). Novel approaches combining genetic, molecular and bioinformatics analyses open the way to marker-assisted breeding for improved carotenoid content in maize and wheat (Harjes *et al.*, 2008; Singh *et al.*, 2009; Vallabhaneni and Wurtzel, 2009). The Solanaceae family has been the target of genetic engineering since it includes economically important crops, among which tobacco, tomato and potato are easily amenable to transformation and have been engineered for carotenoid content with diverse constructs and strategies. Solanaceous crops are also being targeted by the SOL genomics network (http://sgn.cornell.edu/), which is expected to unravel the complexity of genomic information in the coming years and to provide new molecular tools for crop breeding. Other major crops engineered for carotenoid production are canola and carrot. Altogether, the six mentioned engineered crops (rice, maize, tomato, potato, canola and carrot) account for almost two billion tons production and 370 million hectares surface worldwide (*Table 1*). These data demonstrate how plant-based carotenoid molecular breeding could provide economic advantages and be a feasible alternative to chemical synthesis.

Table 1. World 2007 production and area harvested data of the six major crops already engineered for carotenoid content (Source: FAO Statistics; http://faostat.fao.org).

	Production (million metric tons)	Surface (million hectares)
Maize	785	157.8
Rice	651	157.0
Potatoes	321	19.3
Tomatoes	126	4.6
Rapeseed	50	30.2
Carrots[a]	27	1.2
TOTAL	1960	370.1

[a] includes turnips

Plant carotenoids have been successfully engineered with either plant or bacterial genes – or combinations of genes from the two sources. Because they display some particular features, bacterial genes have been used to engineer both early (phytoene synthesis, desaturation and isomerization) and late (lycopene cyclization, ketocarotenoid biosynthesis) biosynthetic steps and deserve a particular mention. Engineering of early steps using the *CrtB* and/or *CrtI* genes, encoding respectively PSY and PDS (*Figure 1*), has been obtained in tobacco leaves, tomato fruits, potato tubers, and seeds of canola, flax, maize and rice (for details and references, see specific sections below). Overexpression of *CrtB* results, depending on the plant species, in an increase of phytoene as well as of compounds further downstream: β- and α-carotene and lutein. Overexpression of *CrtI* results, rather than in an increase of its product lycopene, in a re-direction of synthesis towards the beta-branch (Misawa *et al.*, 1994; Römer *et al.*, 2000).

Carotenoid biosynthesis enzymes of bacterial origin lack a transit peptide, and when expressed from the nucleus must be fused to an artificial transit peptide, usually derived from the *RbcS* gene, for correct localization in the plastid compartment (Ye *et al.*, 2000; Fraser *et al.*, 2002; Ducreux *et al.*, 2005; Diretto *et al.*, 2007a). Alternatively, bacterial enzymes can be expressed directly from the plastid genome. A lycopene cyclase from *Erwinia* has been expressed from the tomato plastid genome, resulting in an elevation of fruit β-carotene levels (Wurbs *et al.*, 2007). The polycistronic nature of the plastid transcripts allows also the expression of multiple genes under the control of a single promoter sequence. The *Brevundimonas CrtZ* and *CrtW* genes, encoding respectively β-carotene hydroxylase and ketolase enzymes, have been expressed as a single operon under the control of a single tobacco plastid promoter, for astaxanthin production in tobacco leaves (Hasunuma *et al.*, 2008).

Another possibility is the overexpression of bacterial mini-pathways from the nuclear genome. The *CrtI* gene from *Erwinia* is often used in these mini-pathways, since it contains, in a single enzyme, activities corresponding to plant PDS, ZDS, CrtISO and, possibly, Z-ISO (*Figure 1*). This multifunctional bacterial enzyme has been used, in metabolic engineering efforts, to surrogate the plant enzymes in the phytoene desaturation/isomerization pathway (Misawa *et al.*, 1994; Ye *et al.*, 2000; Paine *et al.*, 2005; Römer *et al.*, 2000; Ravanello *et al.*, 2003; Diretto *et al.*, 2007a). As a consequence, we know very little on the consequences of the overexpression of the latter enzymes. Examples of mini-pathway expression are the *PSY-CrtI-CrtY* mini-pathway overexpressed in rice (Ye *et al.*, 2000), and the *CrtB-CrtI-CrtY* minipathway overepxressed in canola and potato (Ravanello *et al.*, 2003; Diretto *et al.*, 2007a). In some cases, *CrtY*, which encodes a lycopene β-cyclase, has been shown to be dispensable for β-carotene biosynthesis, probably because the endogenous LCY-b enzymes are sufficiently expressed. In an engineering *tour de force*, seven genes involved in ketocarotenoid formation from three different bacterial genera were introduced into canola (Fujisawa *et al.*, 2009). Recent combinatorial approaches, based on biolistic transformation and introducing up to five of plant and bacterial gene constructs of interest, produced repertoires of plants with several transgene combinations in maize, showing the potential for pathway investigation and vitamin biofortification (Zhu *et al.*, 2008; Naqvi *et al.*, 2009).

NON-SOLANACEOUS PLANTS

RICE (ORYZA SATIVA)

Rice dehusked grain is devoid of carotenoids. Overexpression of a *PSY* gene alone produced the accumulation of phytoene but not that of downstream carotenoids (Burkhardt *et al.*, 1997). One of the major biotechnological breakthroughs of the last decade has been the molecular breeding of golden rice (GR) in both japonica and indica backgrounds, whose grain accumulate β-carotene (pro-vitamin A) through the expression of a minipathway composed of a plant *PSY* and a bacterial desaturase/isomerase *CrtI* under different promoters (Ye *et al.*, 2000; Datta *et al.*, 2003; Hoa *et al.*, 2003). The synthesis of β-carotene and xanthophylls was made possible likely by the sufficient expression of endogenous carotene cyclase and hydroxylase genes in the endosperm (Schaub *et al.*, 2005). In second-generation GR (GR2), the use of a maize optimized *PSY* coding sequence increased up to 17-fold and 23-fold the synthesis of β-carotene and total carotenoids, respectively (Paine *et al.*, 2005). Such levels allow to efficiently fight VAD with diet-compatible amounts of GR2 (Giuliano *et al.*, 2008).

MAIZE (ZEA MAYS)

Maize kernels naturally contain β-carotene and other xanthophylls, albeit at levels usually insufficient to cover pro-vitamin A recommended daily allowance (RDA) values. As an alternative to conventional breeding, transgenic maize plants containing the bacterial *CrtB* and *CrtI* genes, controlled by either "normal" or "super" γ-zein promoters, showed increased total carotenoid levels (especially β-carotene) up to 34-fold, contributing the first transgenic maize designed to alleviate VAD in third world countries (Aluru *et al.*, 2008). Two recent works showed the dramatic combinatorial potential of biolistic transformation for fast production of repertoires of multitransgenic plants in maize biofortified in carotenoids and vitamins. In the first work, Zhu *et al.* (2008) introduced up to five carotenoid genes (plant *Psy1*, *Lcy-b* and *Chy*; bacterial *CrtI* and *CrtW*), each with a different endosperm-specific promoter, to produce an array of transformants with different combinations of transgenes. Transgenic kernels displayed several phenotypes consistent with the combinations of transgenes integrated. In the second work, Naqvi *et al.* (2009) transformed genes for three independent vitamin pathways (carotenoid, ascorbate and folate) to produce transformants with increased β-carotene (169-fold), ascorbic acid (sixfold) and folate (twofold).

CANOLA (BRASSICA NAPUS)

Overexpression of a bacterial *CrtB* (*PSY*) gene in oilseed rape directed by a seed-specific napin promoter and a plastid target peptide was sufficient to increase total carotenoids 50-fold, mostly α- and β-carotene (Shewmaker *et al.*, 1999). Overexpression of *CrtB* with different combinations of other genes (*CrtI, CrtY* and a plant LCY-b) increased β-carotene levels and β- to α-carotene ratios, suggesting the existence of carotenoid multienzyme complexes (Ravanello *et al.*, 2003). Downregulation of a LCY-e diverted the carotenoid flux and increased not only the content of β-carotene and β-β xanthophylls, but unexpectedly also that of lutein (Yu *et al.*, 2008). More recently, transformation of seven genes for ketocarotenoid biosynthesis from *Erwinia*, *Brevundimonas* and *Paracoccus*, each with its own promoter, terminator and transit peptide, was achieved in canola. Approx. 80% of the regenerated plants retained all seven genes, and formed orange- or pinkish orange-colored seeds. The total amount of carotenoids increased up to 30-fold, with β-carotene and ketocarotenoids making up a significant proportion (Fujisawa *et al.*, 2009).

CARROT (DAUCUS CAROTA)

Despite being a traditional and rich source of carotenoids in the diet, molecular research on carrot carotenoid pathway was reported only recently (Just *et al.*, 2007). Besides orange-rooted types, carrot germplasm display purple, red, yellow and white root color phenotypes, combinations of them in root tissues, and tissue-specific accumulation of the different carotenoids (Surles *et al.*, 2004; Baranska *et al.*, 2006). Ketocarotenoid engineering was accomplished by overexpressing an algal β-carotene ketolase *CrtO* gene in an orange-rooted genotype using three different promoters (Jayarai *et al.*, 2008). Transgenic roots produced several β-ketocarotenoids but not ketolutein, in quantities as high as 70% the total carotenoid pool (up to 2.4 mg/g root dry weight). Transformants were also more resistant to UV-B radiation and oxidative stress than wt ones (Jayarai and Punja, 2008).

ARABIDOPSIS (ARABIDOPSIS THALIANA)

Arabidopsis has been used in fundamental and proof-of-concept studies, also for carotenoid engineering. Carotenoid engineering in *Arabidopsis* seeds was achieved by up-regulation of an endogenous *PSY* gene, which increased the total carotenoid levels, namely β-carotene, lutein and violaxanthin (Lindgren *et al.*, 2003). Seeds also had increased chlorophyll and ABA content, which delayed germination. Comparison of *Arabidopsis* with canola and flax seed-targeted transgenic systems, using different promoters and *PSY* genes, revealed differences in metabolite flux. In *Arabidopsis*, the use of an endogenous *AtPSY* coding sequence directed the flux towards xanthophyll end products lutein and violaxanthin. In the other two systems, bacterial *CrtB* genes preferentially increased the levels of α- and β-carotene (Shewmaker *et al.*, 1999; Lindgren *et al.*, 2003; Fujisawa *et al.*, 2008). Transformed *Arabidopsis* plants more resistant to elevated light and temperature conditions were produced by up-regulating an endogenous *AtCHY* hydroxylase (Davison *et al.*, 2002). This increased two-fold the xanthophyll cycle pool and levels of zeaxanthin, which has a photoprotective role in photosynthesis. *De novo* synthesis of ketocarotenoids in *Arabidopsis* seeds was brought by the expression of an oxygenase (ketolase) gene from *Haematococcus* (Stålberg *et al.*, 2003). Crossing with *PSY*-overexpressing lines increased total carotenoid levels by almost 5-fold and those of major ketocarotenoids 13-fold, demonstrating the pivotal role of PSY in the governing the metabolite flux in seeds.

SOLANACEAE CROPS

TOMATO (SOLANUM LYCOPERSICUM)

Tomato is the best-investigated species within Solanaceae family, due to its impor-tance as food crop and the nutritional value of fruits accumulating lycopene. Parallel holistic approaches are currently in progress to sequence the gene-rich region of its genome (Mueller *et al.*, 2009) and characterize the berry ripening process with transcriptomic (Alba *et* al., 2004), proteomic (Rocco *et* al., 2006) and metabolomic (Moco *et* al., 2006) approaches.

Carotenoid studies in tomato fruits, both using both transgenic plants and natural mutants, have been carried out on several isoprenoid and carotenoid pathway steps. Within the former, tomato transgenic plants overexpressing a plant 3-hydroxy-3-methylglutaryl CoA reductase (HMGR) showed no perturbation of the pathway, while those up-regulated for a bacterial 1-deoxy-D-xylulose 5-phosphate synthase (DXS) had enhanced phytoene and β-carotene contents (Enfissi *et al.*, 2005).

In the latter, PSY is the key step in carotenoid biosynthesis, whose modification affects also other pathways. For these reasons, PSY has been extensively investigated in tomato: the first example is the antisense silencing of *PSY1* (preferentially expressed in reproductive organs), resulting in a great reduction in flower and fruit carotenogenesis (Bird *et al.*, 1991). In an opposite trial, overexpression of the same gene in a constitutive fashion produced fruits with earlier lycopene production along development, but showing a lower content of this metabolite at maturity (Fray *et al.*, 1995). Interestingly, these plants displayed several pleiotropic effects, including albino fruit production, and reduced plant height and chlorophyll content; these phenotypes find an explanation in both endogenous *PSY1* gene silencing and decrease in gibberellin (GA) content, derived from the competition of the two pathways for the geranylgeranyl diphosphate pool. Fruit-specific expression of a bacterial *PSY* gene (*CrtB* from *Erwinia*) produced fruits with higher phytoene, β-carotene and total carotenoid levels, but did not increase lycopene content (Fraser *et al.*, 2002). In a more recent attempt, it has been shown that overexpression of endogenous *PSY1* is able to perturb not only carotenoid gene transcripts and metabolites, but also plastid type and development and the accumulation of several metabolites (Fraser *et al.*, 2007).

Subsequent carotenoid biosynthetic steps have also been altered through the ectopic expression of a bacterial *CrtI* gene: unexpectedly, β-carotene and xanthophyll levels in transgenic fruits increased (phenotype partially explained by endogenous *LCY-b* up-regulation), while lycopene and total carotenoids were unchanged (Römer *el al.*, 2000). Large increases in fruit β-carotene and total carotenoids were achieved manipulating expression of lycopene β-cyclase genes: overexpressing *Arabidopsis/* tomato *LCY-b* genes under the control of chromoplast-specific promoters resulted in higher β-carotene level up to 7- (Rosati *et al.*, 2000) and 32-fold (D'Ambrosio *et al.*, 2004), respectively. In a more recent study, an *Erwinia CrtY* gene was expressed at the chloroplast level leading to a maximum β-carotene accumulation of 4-fold in transplastomic fruits (Wurbs *et al.*, 2007). With a different aim, *LCY-b* was also silenced using an antisense approach in order to obtain, as expected, higher lycopene levels in fruits (Rosati *et al.*, 2000).

Tomato fruit has also been investigated as potential target for collecting carotenoid down-stream compounds. Simultaneous and fruit-specific overexpression of *Arabidopsis LCY-b* and pepper *CHY* increased both β-carotene and total xanthophylls, namely zeaxanthin and β-cryptoxanthin, virtually absent in wild-type fruits (Dharmapuri *et al.*, 2002). In another attempt, the *CrtW* and *CrtZ* genes from *Paracoccus* encoding, respectively, a 4,4' ketolase and a 3,3' hydroxylase, were assembled as a polyprotein and expressed in a constitutive manner: transgenic leaves accumulated ketocarotenoids, including canthaxanthin, astaxanthin and ketozeaxanthin, while no alteration in ketocarotenoid patterns was revealed in tomato fruits (Ralley *et al.*, 2004).

Unintended lycopene accumulation has also been obtained as "secondary effect" in different studies: ectopic expression of pepper fibrillin gene in tomato fruits did not lead to fibril or fibril-like arising but, contrary to the expectations, total carotenoids and volatiles, lycopene and lycopene-derived apocarotenoids and, in a slighter way, β-carotene increased in transgenic fruits (Simkin *et al.*, 2007). The effects derived by the perturbation of pathways not apparently linked to carotenoid biosynthesis are intriguing: for example, modification of polyamine pathway, by the expression of a yeast S-adenosylmethionine decarboxylase (*ySAMdc*) gene (Mehta *et al.*, 2002) led to an increase in lycopene content up to 2-fold. A strong cross-link between carotenoid biosynthesis and light perception and signaling has been shown: silencing the expression of *LeDet1* and *LeCOP1LIKE* photomorphogenic repressors (Davuluri *et al.*, 2005) or overexpressing the blue light photoreceptor cryptochrome 2 (CRY2) (Giliberto *et al.*, 2005) resulted in an enhanced lycopene level, with a maximum of 2-fold with respect to the control. An opposite phenotype (decrease of lycopene content) was reported through silencing of *LeHY5* gene (Liu *et al.*, 2004). Very recently, the overexpression of a grape MYB-type transcription factor (TF), producing an increase in fruit β-carotene content, has opened the way for carotenoid engineering through the use of TFs (Mahjoub *et al.*, 2009).

An interesting approach in metabolic engineering of carotenoids is represented by the opportunity to modify the apocarotenoid pattern in tomato fruits: antisense downregulation of the ripening-induced *LeCCD1b* has been shown to reduce the production of both β-ionone (derived from cleavage of α- and β-carotene) and geranylacetone (derived from cleavage of lycopene precursors) in fruits (Simkin *et al.*, 2004a). In a different study, diversion of apocarotenoid and volatile profiles has been obtained expressing, in a fruit-specific fashion, the geraniol synthase gene (*GES*) from basil (Davidovich-Rikanati *et al.*, 2007): transgenic fruits showed an increase of monoterpenes (geraniol, nerol, citronellol and citronellic acid) at the expense of lycopene content. These changes deeply altered the flavor of transgenic fruits over control ones.

POTATO (SOLANUM TUBEROSUM)

Potato is the fourth worldwide crop for production and consumption: its tubers are rich in minerals/micronutrients and vitamin C, but not in pro-vitamin A (Brown, 2005); moreover, and unfortunately, there is no available source of β-carotene within potato germplasm, including both current commercial potato cultivars and wild species (*Solanum phureja*, one of the few yellow-flesh wild species, show a high content in total carotenoids, but not in β-carotene) (Morris *et al.*, 2004).

Several attempts have been carried out during last decade to increase the carotenoid potential in potato. Basically, these studies pointed on either "push" or "block" strategies, respectively through the overexpression of heterologous constructs or silencing of endogenous genes. As to "push" approaches, similarly to tomato, constitutive or tuber-specific expression of bacterial genes involved in early steps in the isoprenoid/carotenoid pathway has been attempted: overexpression of a bacterial *DXS* caused an increase in phytoene levels, causing enhanced total carotenoid content, and an early sprouting phenotype, probably derived from concurrent enhanced levels of *trans*-zeatin riboside in tubers (Morris *et al.*, 2006b). In a different study, a bacterial phytoene synthase gene (*CrtB*) under the control of *patatin* promoter strongly increased β-carotene and total carotenoid levels in *S. tuberosum* and *S. phureja* tubers (Ducreux *et al.*, 2005). Interestingly, the accumulation of these metabolites is more pronounced in *S. tuberosum* than in *S. phureja* tubers, probably due to the fact that the endogenous carotenoid genes of the latter species are expressed at high levels. In a more extensive approach, a systematic investigation on modulation of potato leaf and tuber carotenogenesis has been performed using three bacterial genes (*CrtB*, phytoene desaturase *CrtI* and lycopene β-cyclase *CrtY*) expressed using different gene/promoter combinations (Diretto *et al.*, 2007a). Leaves of all transgenic lines displayed, in an additive way, a strong interference on endogenous carotenogenesis, both at transcript and metabolite levels. Quite the opposite, tuber total carotenoid content increased in all the transgenic plants, but only the simultaneous expression of all three genes in a tuber-specific manner led to a 20-fold increase in transgenic tubers. Furthermore, these tubers showed a great enhancement of β-carotene up to 3600-fold (47μg/gm of dry weight), corresponding to the highest content ever reported for any of the four major crops. These tubers showed a deep-yellow (*Golden*) flesh and also accumulated higher xanthophylls levels, as well as phytoene and α-carotene, which are normally undetected in wild-type tubers.

Within "block" approaches, silencing of genes involved in both ε-β- and β-β- carotenoid branching has been described. Reducing *LCY-e* (Diretto *et al.*, 2006) or *CHY* (Diretto *et al.*, 2007b) expression in a tuber-specific way increased β-carotene and total carotenoid levels, albeit to a smaller extent than in "*Golden*" tubers. Contrary to expectations, silencing of *ZEP* gene resulted in a dramatic increase in zeaxanthin and total carotenoids, but not in β-carotene (Römer *et al.*, 2002). Recently, it has been reported that zeaxanthin from such transgenic tubers is bioavailable to humans. Following ingestion of *ZEP*-silenced tubers, the zeaxanthin concentration incorporated into chylomicrons, the lipoprotein bodies that transport hydrophobic compounds through the body, was significantly increased (Bub *et al.*, 2008).

A third and more recent approach is the "sink" strategy, through the ectopic expression of plastid-associated proteins able to increase the total carotenoid pool: a notable example of this kind of manipulation is the overexpression of the *Orange (Or)* locus from cauliflower (Lu *et al*, 2006) which produced yellow-fleshed tubers with dramatically high levels of β-carotene (up to 1600-fold) and total carotenoids (up to 7-fold) with respect to the control. Interestingly, *Or* tubers were characterized by several novel features: they synthesize, *de novo*, phytoene, phytofluene and ζ-carotene, and chromoplasts containing carotenoid-sequestering structures were observed; moreover, as positive trait, they continued to enhance β-carotene and total carotenoids during cold-storage (Lopez *et al.*, 2008) according to a not yet elucidated mechanism.

Potato tubers have also been engineered to produce high-value ketocarotenoids. Gerjets and Sandmann (2006) constitutively expressed a β-carotene ketolase gene from *Synechocystis* in *ZEP*-silenced tubers to synthesize ketocarotenoids in both leaves (echinenone, hydroxyechinenone and ketozeaxanthin) and tubers (hydroxyechinenone, ketozeaxanthin and astaxanthin). In a second study, astaxanthin and ketolutein accumulation was achieved through the overexpression of a green alga β-carotene ketolase gene (Morris *et al.*, 2006a).

TOBACCO (NICOTIANA SPP.)

The possibility to manipulate the carotenogenesis in several tobacco (*N. tabacum* and *N. benthamiana*) organs has been intensively explored. Expressing a bacterial *DXR* gene in a transplastomic manner led to a general increase in various isoprenoids such as chlorophyll a, β-carotene, lutein, solanesol and β-sitosterol with no pleiotropic effects on vegetative growth (Hasunuma *et al.*, 2008). *CrtI* overexpression was shown to increase β-carotene and β-ring-xanthophyll content, as well as resistance to norflurazon and other bleaching herbicides (Misawa *et al.*, 1994).

Ketocarotenoid induction has been the main goal of metabolic engineering of tobacco carotenoids. In a first approach, the pepper *Ccs*, introduced *via* a viral vector, boosted the *de novo* accumulation of capsanthin up to around 36% of total leaf carotenoids in *N. benthamiana*, accompanied by a strong and unexpected increase in light-harvesting processes (Kumagai *et al.*, 1998). Another breakthrough is represented by the overexpression of β-carotene ketolase gene from *Haematococcus*, using a chromoplast-specific promoter, which induced a massive ketocarotenoid synthesis in *N. tabacum* nectaries (Mann *et al.*, 2000). Subsequent attempts encompassed the simultaneous expression of a 4,4' ketolase (*CrtW*) and a 3,3' hydroxylase (*CrtZ*) from *Paracoccus*, which produced several ketocarotenoids (canthaxanthin, astaxanthin and ketozeaxanthin) in both tobacco leaves and nectaries at less than 5% of the total carotenoids (Ralley *et al.*, 2004). Through the use of two ketolase (*CrtO* and *CrtW*) and a β-carotene hydroxylase (*CrtZ*) genes from, respectively, *Synechocystis*, *Nostoc* and *Erwinia* in different combinations, *Nicotiana glauca* and *Nicotiana tabacum* accumulated ketocarotenoids in petal flowers, nectaries and, in a minor fraction, in leaves but without any significant increase in astaxanthin content (Gerjets *et al.*, 2007; Zhu *et al.*, 2007a). A transplastomic approach was also devised with the *CrtZ* and *CrtW* genes, but from a different source (the marine bacterium *Brevundimonas*): transgenic plants exhibited a reddish leaf color and accumulated extremely high amounts of astaxanthin (>70% of total carotenoids, >0.5% on a dry weight basis) with no adverse effect on vegetative growth or photosynthesis (Hasunuma *et al.*, 2008). Interestingly, the new ketocarotenoid 4-ketoantheraxanthin was detected in transgenic leaves.

Tobacco leaves and reproductive organs have also been investigated for the possibility to modify plastid structures: with this aim, pepper fibrillin was constitutively introduced in tobacco and transgenic chloroplasts/leucoplasts showed an increased number of plastoglobules organized in clusters (Rey *et al.*, 2000). Fibrillin overexpressors displayed no alteration in vegetative growth under low light conditions, while longer main stem, higher number of lateral stems and faster floral development were observed under higher light intensities, suggesting a role of fibrillin in regulating plant development *vis-à-vis* environmental stresses. The relation between carotenoid pathway and abiotic stress response has been investigated through the overexpression of bacterial β-carotene hydroxylase *CrtZ*, which led to a higher UV tolerance, probably due to the involvement of carotenoid metabolites in the non-photochemical quenching process (Götz *et al.*, 2002).

Tobacco has also been used as a model for investigating the relation between carotenoid and ABA metabolism: transgenic plants expressing two *NCED* genes from *Gentiana lutea* were characterized for several ABA-related processes: contrary to expectations, delayed radicle formation and cotyledon appearance were the only phenotypes detected in these plants, albeit no data were reported on carotenoid composition (Zhu *et al.*, 2007b).

Conclusions and perspectives

Carotenoid engineering has proven beneficial for improving the nutritional value of important crops. The integration of genetic, molecular and biochemical approaches, and the current genomic initiatives are likely to provide further momentum to the understanding of the (apo)carotenoid pathways in the coming years. Successful carotenoid engineering through diverse transformation (nuclear, plastid and transient) approaches has been achieved by acting on most structural genes of the pathway. The recent modification of carotenoid composition in mutants and transformants for TFs, light signal transduction factors and plastid-associated proteins (e.g., Davuluri *et al.*, 2005; Lu *et al.*, 2006; Simkin *et al.*, 2007; Mahjoub *et al.*, 2009) enlarge the palette of tools for molecular breeding. The development of transformation procedures for other carotenoid-accumulating species, such as pepper (Lee *et al.*, 2009), potentially expands the number of target species for (apo)carotenoid engineering.

Acknowledgements

Work supported by the EC (projects EU-SOL and Develonutri) and by the Italian Ministry of Research (Special funds for Fundamental Research and for Industrial research).

References

AL-BABILI, S. AND BEYER, P. (2005) Golden Rice – five years on the road – five years to go?. *Trends Plant Sci.* **10**. 565-573

AL-BABILI, S., HUGUENEY, P., SCHLEDZ, M., WELSCH, R., FROHNMEYER, H., LAULE, O. AND BEYER, P. (2000) Identification of a novel gene coding for neoxanthin synthase from *Solanum tuberosum*. *FEBS Lett.* **485**. 168-172

ALBA, R., FEI, Z., PAYTON, P., LIU, Y., MOORE, S. L., DEBBIE, P., COHN, J., D'ASCENZO, M., GORDON, J. S., ROSE, J. K. C., MARTIN, G., TANKSLEY, S. D., BOUZAYEN, M., JAHN, M. M. AND GIOVANNONI, J. (2004) ESTs, cDNA microarrays, and gene expression profiling: tools for dissecting plant physiology and development. *Plant J.* **39**. 697-714

ALURU, M., XU, Y., GUO, R., WANG, Z., LI, S., WHITE, W., WANG, K. AND RODERMEL, S. (2008) Generation of transgenic maize with enhanced pro-vitamin A content. *J Exp Bot.* **59**. 3551-3562

AULDRIDGE, M. E., MCCARTY, D. R. AND KLEE, H. J. (2006) Plant carotenoid cleavage oxygenases and their apocarotenoid products. *Curr Opin Plant Biol.* **9**. 1-7

BIRD, C. R., RAY, J. A., FLETCHER, J. D., BONIWELL, J. M., BIRD, A. S., TEULIERET, C., BRAMLEY, P. M. AND SCHUCH, W. (1991) Using antisense RNA to study gene function: inhibition of carotenoid biosynthesis in transgenic tomatoes. *Bio/Technol.* **7**. 635-639

BOUVIER, F., HUGUENEY, P., D'HARLINGUE, A., KUNTZ, M. AND CAMARA, B. (1994) Xanthophyll biosynthesis in chromoplasts: isolation and molecular cloning of an enzyme catalyzing the conversion of 5,6-epoxycarotenoid into ketocarotenoid. *Plant J.* **6**. 45-54

BOUVIER, F., D'HARLINGUE, A., BACKHAUS, R. A., KUMAGAI, M. H. AND CAMARA, B.

(2000) Identification of neoxanthin synthase as a carotenoid cyclase paralog. *Eur J Biochem.* **267**. 6346-6352

Bouvier, F., Dogbo, O. and Camara, B. (2003a) Biosynthesis of the food and cosmetic plant pigment bixin (annatto). *Science* **300**. 2089-2091

Bouvier, F., Suire, C., Mutterer, J. and Camara, B. (2003b) Oxidative remodeling of chromoplast carotenoids: identification of the carotenoid dioxygenase *CsCCD* and *CsZCD* genes involved in *Crocus* secondary metabolite biogenesis. *Plant Cell.* **15**. 47-62

Breitenbach, J. and Sandmann, G. (2005) ζ-carotene *cis* isomers as products and substrates in the plant poly-*cis* carotenoid biosynthetic pathway to lycopene. *Planta.* **220**. 785-793

Britton, G., Liaaen-Jensen, S. and Pfander, H., eds. (2003). *Carotenoid Handbook.* Basel: Birkhäuser.

Brown, C. R. (2005) Antioxidants in potato. Am J Potato Res. 82. 163-172

Bub, A., Möseneder, J., Wenzel, G., Rechkemmer, G. and Briviba, K. (2008) Zeaxanthin is bioavailable from genetically modified zeaxanthin-rich potatoes. *Eur J Nutr.* **47**. 99-103

Burkhardt, P. K., Beyer, P., Wünn, J., Klöti, A., Armstrong, G. A., Schledz, M., Von Lintig, J. and Potrykus, I. (1997) Transgenic rice (*Oryza sativa*) endosperm expressing daffodil (*Narcissus psuedonarcissus*) phytoene synthase accumulates phytoene, a key intermediate of pro-vitamin A biosynthesis. *Plant J.* **11**. 1071-1078

Côté, F., Cormier, F., Dufresne, C. and Willemot, C. (2000) Properties of a glucosyltransferase involved in crocin synthesis. *Plant Sci.* **153**. 55-63

Cunningham, F. X. Jr. and Gantt, E. (2001) One ring or two? Determination of ring number in carotenoids by lycopene epsilon-cyclases. *Proc Natl Acad Sci USA.* **98**. 2905-2910

Cunningham, F. X. Jr. and Gantt, E. (2005) A study in scarlet: enzymes of ketocarotenoid biosynthesis in the flowers of *Adonis aestivalis*. *Plant J.* **41**. 478-492

Cunningham, F. X. Jr. and Gantt, E. (2007) A portfolio of plasmids for identification and analysis of carotenoid pathway enzymes: *Adonis aestivalis* as a case study. *Photosynth Res.* **92**. 245-259

D'ambrosio, C., Giorio, G., Marino, I., Merendino, A., Petrozza, A., Salfi, L., Stigliani, A. L. and Cellini, F. (2004) Virtually complete conversion of lycopene into β-carotene in fruits of tomato plants transformed with the tomato lycopene β-cyclase (t*lcy-b*) cDNA. *Plant Sci* **166**. 207-214

Datta, K., Baisakh, N., Oliva, N., Torrizo, L., Abrigo, E., Tan, J., Rai, M., Rehana, S., Al-Babili, S., Beyer, P. Potrykus, I. and Datta, S. K. (2000) Bioengineered '*golden*' indica rice cultivars with -carotene metabolism in the endosperm with hygromycin and mannose selection systems. *Plant Biotechnol J.* **1**. 81-90

Davidovich-Rikanati, R., Sitrit, Y., Tadmor, Y., Iijima, Y., Bilenko, N., Bar, E., Carmona, B., Fallik, E., Dudai, N., Simon, J. E., Pichersky, E. and Lewinsohn, E. (2007) Enrichment of tomato flavor by diversion of the early plastidial terpenoid pathway. *Nat Biotechnol.* **25**. 899-901.

Davison, P. A., Hunter, C. N. and Horton P. (2002) Overexpression of β-carotene hydroxylase enhances stress tolerance in *Arabidopsis*. *Nature*. **418**. 203-206

Davuluri, G. R., van Tuinen, A., Fraser, P. D., Manfredonia, A., Newman, R., Burgess, D., Brummell, D. A., King, S. R., Palys, J., Uhlig, J., Bramley, P. M., Pennings,

H. M. J. AND BOWLER, C. (2005). Fruit-specific RNAi-mediated suppression of *DET1* enhances carotenoid and flavonoid content in tomatoes. *Nat Biotechnol.* **23**. 890-895

DELI, J., MOLNÁR, P., MATUS, Z. AND TÓTH, G. (2001) Carotenoid composition in the fruits of red paprika (*Capsicum annuum* var. *lycopersiciforme rubrum*) during ripening; biosynthesis of carotenoids in red paprika. *J Agric Food Chem.* **49**. 1517-1523

DELLAPENNA, D. AND POGSON, B. J. (2006) Vitamin synthesis in plants: tocopherols and carotenoids. *Annu Rev Plant Biol.* **57**. 711-738

DHARMAPURI, S., ROSATI, C., PALLARA, P., AQUILANI, R., BOUVIER, F., CAMARA, B. AND GIULIANO, G. (2002) Metabolic engineering of xanthophyll content in tomato fruits. *FEBS Lett.* **519**. 30-34

DIRETTO, G., TAVAZZA, R., WELSCH, R., PIZZICHINI, D., MOURGUES, F., PAPACCHIOLI, V., BEYER, P. AND GIULIANO, G. (2006) Metabolic engineering of potato tuber carotenoids through tuber-specific silencing of lycopene epsilon cyclase. *BMC Plant Biol.* **6**:13.

DIRETTO, G., AL-BABILI, S., TAVAZZA, R., PAPACCHIOLI, V., BEYER, P. AND GIULIANO, G. (2007a) Metabolic engineering of potato carotenoid content through tuber-specific overexpression of a bacterial mini-pathway. *PLoS ONE.* **2**:e350.

DIRETTO, G., WELSCH, R., TAVAZZA, R., MOURGUES, F., PIZZICHINI, D., BEYER, P. AND GIULIANO, G. (2007b) Silencing of beta-carotene hydroxylase increases total carotenoid and beta-carotene levels in potato tubers. *BMC Plant Biol.* **7**:11.

DUCREUX, L. J., MORRIS, W. L., HEDLEY, P. E., SHEPHERD, T., DAVIES, H. V., MILLAM, S. AND TAYLOR, M. A. (2005) Metabolic engineering of high carotenoid potato tubers containing enhanced levels of beta-carotene and lutein. *J Exp Bot.* **56**. 81-89

ENFISSI, E. M. A., FRASER P. D., LOIS, L. M., BORONAT, A., SCHUCH, W. AND BRAMLEY, P. M. (2005) Metabolic engineering of the mevalonate and nonmevalonate isopentenyl diphosphate-forming pathways for the production of health-promoting isoprenoids in tomato. *Plant Biotechnol J.* **3**. 17-27

FRASER, P. D., RÖMER, S., SHIPTON, C. A., MILLS, P. B., KIANO, J. W., MISAWA, N., DRAKE, R. G., SCHUCH, W. AND BRAMLEY, P. M. (2002) Evaluation of transgenic tomato plants expressing an additional phytoene synthase in a fruit-specific manner. *Proc Natl Acad Sci USA.* **99**. 1092-1097

FRASER, P. D., ENFISSI, E. M., HALKET, J. M., TRUESDALE, M. R., YU, D., GERRISH, C. AND BRAMLEY, P. M. (2007) Manipulation of phytoene levels in tomato fruit: effects on isoprenoids, plastids, and intermediary metabolism. *Plant Cell.* **19**. 3194-3211

FRASER, P. D., ENFISSI, E. M. A. AND BRAMLEY, P. M. (2009) Genetic engineering of carotenoid formation in tomato fruit and the potential application of systems and synthetic biology approaches. *Arch Biochem Biophys.* **483**. 196-204

FRAY, R. G., WALLACE, A., FRASER, P. D., VALERO, D., HEDDEN, P., BRAMLEY, P. M. AND GRIERSON, D. (1995) Constitutive expression of a fruit phytoene synthase gene in transgenic tomatoes causes dwarfism by redirecting metabolites from the gibberellin pathway. *Plant J.* **8**. 693-701

FUJISAWA, M., WATANABE, M., CHOI, S. K., TERAMOTO, M., OHYAMA, K. AND MISAWA, N. (2008) Enrichment of carotenoids in flaxseed (*Linum usitatissimum*) by metabolic engineering with introduction of bacterial phytoene synthase gene *crtB*. *J Biosci Bioeng.* **105**. 636-641

FUJISAWA, M., TAKITA, E., HARADA, H., SAKURAI, N., SUZUKI, H., OHYAMA, K., SHIBATA,

D. AND MISAWA, N. (2009) Pathway engineering of *Brassica napus* seeds using multiple key enzyme genes involved in ketocarotenoid formation. *J Exp Bot.* **60**. 1319-1332

GALPAZ, N., RONEN, G., KHALFA, Z., ZAMIR, D. AND HIRSCHBERG, J. (2006) A chromoplast-specific carotenoid biosynthesis pathway is revealed by cloning of the tomato *white-flower* locus. *Plant Cell.* **18**. 1947-1960

GANN, P. H., MA, J., GIOVANNUCCI, E., WILLETT, W., SACKS, F. M., HENNEKENS, C. H. AND STAMPFER, M. J. (1999) Lower prostate cancer risk in men with elevated plasma lycopene levels: results of a prospective analysis. *Cancer Res.* **59**. 1225-1230

GARCIA-LIMONES, C., SCHNÄBELE, K., BLANCO-PORTALES, R., BELLIDO, M. L., CABALLERO, J. L., SCHWAB, W. AND MUÑOZ-BLANCO, J. (2008) Functional characterization of FaCCD1: a carotenoid cleavage dioxygenase from strawberry involved in lutein degradation during fruit ripening. *J. Agric. Food Chem.* **56**. 9277-9285

GERJETS, T. AND SANDMANN, G. (2006) Ketocarotenoid formation in transgenic potato. *J Exp Bot.* **57**. 3639-3645

GERJETS, T., SANDMANN, M., ZHU, C. AND SANDMANN, G. (2007) Metabolic engineering of ketocarotenoid biosynthesis in leaves and flowers of tobacco species. Biotechnol J. 2. 1263-1269

GILIBERTO, L., PERROTTA, G., PALLARA, P., WELLER, J. L., FRASER, P. D., BRAMLEY, P. M., FIORE, A., TAVAZZA, M. AND GIULIANO, G. (2005) Manipulation of the blue light photoreceptor cryptochrome 2 in tomato affects vegetative development, flowering time, and fruit antioxidant content. *Plant Physiol.* **137**. 199-208

GIULIANO, G., GILIBERTO, L. AND ROSATI, C. (2002) Carotenoid isomerase: a tale of light and isomers. *Trends Plant Sci.* **7**. 513-516

GIULIANO, G., ROSATI, C. AND BRAMLEY, P. M. (2003) To dye or not to dye: biochemistry of annatto unveiled. *Trends Biotechnol.* **21**. 427-429

GIULIANO, G., TAVAZZA, R., DIRETTO, G., BEYER, P. AND TAYLOR, M. A. (2008) Metabolic engineering of carotenoid biosynthesis in plants. *Trends Biotechnol.* **26**. 139-145

GÖTZ ,T., SANDMANN, G. AND RÖMER, S. (2002) Expression of a bacterial carotene hydroxylase gene (*crtZ*) enhances UV tolerance in tobacco. *Plant Mol Biol.* **50**. 129-142

GUIL-GUERRERO, J. L., MARTÍNEZ-GUIRADO, C., DEL MAR REBOLLOSO-FUENTES, M. AND CARRIQUE-PÉREZ, A. (2006) Nutrient composition and antioxidant activity of 10 pepper (*Capsicum annuum*) varieties. *Eur Food Res Technol.* **224**. 1-9

HA, S. H., KIM, J. B., PARK, J. S., LEE, S. W. AND CHO, K. J. (2007) A comparison of the carotenoid accumulation in *Capsicum* varieties that show different ripening colours: Deletion of the capsanthin-capsorubin synthase gene is not a prerequisite for the formation of a yellow pepper. *J Exp Bot.* **133**. 161-169

HARJES, C. E., ROCHEFORD, T. R., BAI, L., BRUTNELL, T. P., BERMUDEZ KANDIANIS, C., SOWINSKI, S. G., STAPLETON, A. E., VALLABHANENI, R., WILLIAMS, M., WURTZEL, E. T., YAN, J. AND BUCKLER, E. S. (2008) Natural genetic variation in *lycopene epsilon cyclase* tapped for maize biofortification. *Science.* **319**. 330-333

HASUNUMA, T., MIYAZAWA, S., YOSHIMURA, S., SHINZAKI, Y., TOMIZAWA, K., SHINDO, K., CHOI, S. K., MISAWA, N. AND MIYAKE, C. (2008) Biosynthesis of astaxanthin in tobacco leaves by transplastomic engineering. *Plant J.* **55**. 857-868

HIRSCHBERG, J. (2001) Carotenoid biosynthesis in flowering plants. *Curr Opin Plant Biol.* **4**. 210-218

HOA, T. T. C., AL-BABILI, S., SCHAUB, P., POTRYKUS, I. AND BEYER, P. (2003) Golden indica and japonica rice lines amenable to deregulation. *Plant Physiol.* **133**. 161-169

HORNERO-MÉNDEZ, D., DE GUEVARA, R. G. L. AND MÍNGUEZ-MOSQUERA, I. M. (2000) Carotenoid biosynthesis changes in five red pepper (*Capsicum annuum*) cultivars during ripening. Cultivar selection for breeding. *J Agric Food Chem.* **48**. 3857-3864

HUANG, F. C., HORVÁTH, G., MOLNÁR, P., TURCSI, E., DELI, J., SCHRADER, J., SANDMANN, G., SCHMIDT, H. AND SCHWAB, W. (2009) Substrate promiscuity of RdCCD1, a carotenoid cleavage oxygenase from *Rosa damascena. Phytochem.* **70**. 457-464

IBDAH, M., AZULAY, Y., PORTNOY, V., WASSERMAN, B., BAR, E., MEIR, A., BURGER, Y., HIRSCHBERG, J., SCHAFFER, A. A., KATZIR, N., TADMOR, Y. AND LEWINSOHN E. (2006) Functional characterization of CmCCD1, a carotenoid cleavage dioxygenase from melon. *Phytochem.* **67**. 1579-1589.

ILG, A., BEYER, P. AND AL-BABILI, S. (2009) Characterization of the rice carotenoid cleavage dioxygenase 1 reveals a novel route for geranial biosynthesis. *FEBS J.* **276**. 736-747

ISAACSON, T., OHAD, I., BEYER, P. AND HIRSCHBERG, J. (2004). Analysis in vitro of the enzyme CRTISO establishes a poly-*cis*-carotenoid biosynthesis pathway in plants. *Plant Physiol.* **136**. 4246-55

JAYARAJ, J. AND PUNJA, Z. K. (2008) Transgenic carrot plants accumulating ketocarotenoids show tolerance to UV and oxidative stress. *Plant Physiol Biochem.* **46**. 875-883

JAYARAJ, J., DEVLIN, R. AND PUNJA, Z. K. (2008) Metabolic engineering of novel ketocarotenoids production in carrot plants. *Transgenic Res.* **17**. 489-501

JUST, B. J., SANTOS, C. A., FONSECA, M. E., BOITEUX, L. S., OLOIZIA, B. B. AND SIMON, P. W. (2007). Carotenoid biosynthesis structural genes in carrot (*Daucus carota*): isolation, sequence-characterization, single nucleotide polymorphism (SNP) markers and genome mapping. *Theor Appl Genet.* **114**. 693-704

KUMAGAI, M. H., KELLER, Y., BOUVIER, F., CLARY, D. AND CAMARA, B. (1998) Functional integration of non-native carotenoids into chloroplasts by viral-derived expression of capsanthin-capsorubin synthase in *Nicotiana benthamiana. Plant J.* **14**. 305-315

LEE, Y. H., JUNG, M., SHIN, S. H., LEE, J. H., CHOI, S. H., HER, N. H., LEE, J. H., RYU, K. H., PAEK, K. Y. AND HARN, C. H. (2009) Transgenic peppers that are highly tolerant to a new CMV pathotype. *Plant Cell Rep.* **28**. 223-232

LEFEBVRE, V., KUNTZ, M., CAMARA, B. AND PALLOIX, A. (1998) The capsanthin-capsorubin synthase gene: a candidate locus for the *y* locus controlling the red fruit colour in pepper. *Plant Mol Biol.* **36**. 785-789

LI, F., MURILLO, C. AND WURTZEL, E. T. (2007) Maize *Y9* encodes a product essential for 15-cis-ζ-carotene isomerization. *Plant Physiol.* **144**. 1181-1189

LI, F., VALLABHANENI, R. AND WURTZEL, E. T. (2008) *PSY3*, a new member of the phytoene synthase gene family conserved in the Poaceae and regulator of abiotic stress-induced root carotenogenesis. *Plant Physiol.* **146**. 1333-1345

LINDGREN, L. O., STÅLBERG, K. G. AND HÖGLUND, A. S. (2003) Seed-specific overexpression of an endogenous *Arabidopsis* phytoene synthase gene results in delayed germination and increased levels of carotenoids, chlorophyll, and abscisic acid. *Plant Physiol.* **132**. 779-785

LIU, Y., ROOF, S., YE, Z., BARRY, C., VAN TUINEN, A., VREBALOV, J., BOWLER, C. AND GIOVANNONI, J. (2004) Manipulation of light signal transduction as a means of

modifying fruit nutritional quality in tomato. *Proc Natl Acad Sci USA.* **101**. 9897-9902

Lopez, A. B., Van Eck, J., Conlin, B. J., Paolillo, D. J., O'Neill, J. and Li, L. (2008) Effect of the cauliflower *Or* transgene on carotenoid accumulation and chromoplast formation in transgenic potato tubers. *J Exp Bot.* **59**. 213-223

Lu, S., Van Eck, J., Zhou, X., Lopez, A. B., O'Halloran, D. M., Cosman, K. M., Conlin, B. J., Paolillo, D. J., Garvin, D. F., Vrebalov, J., Kochian, L. V., Küpper, H., Earle, E. D., Cao, J. and Li, L. (2006) The cauliflower *Or* gene encodes a DnaJ cysteine-rich domain-containing protein that mediates high-levels of β-carotene accumulation. *Plant Cell.* **18**. 3594-3605

Mahjoub, A., Hernould, M., Joubes, J., Decendit, A., Mars, M., Barrieu, F., Hamdi, S. and Delrot, S. (2009) Overexpression of a grapevine R2R3-MYB factor in tomato affects vegetative development, flower morphology and flavonoid and terpenoid metabolism. *Plant Physiol Biochem.* **47**. 551-561

Mann, V., Harker, M., Pecker, I. and Hirschberg, J. (2000) Metabolic engineering of astaxanthin production in tobacco flowers. *Nat Biotechnol.* **18**. 888-892

Mehta, R. A., Cassol, T., Li, N., Ali, N., Handa, A. K. and Mattoo, A. K. (2002) Engineered polyamine accumulation in tomato enhances phytonutrient content, juice quality, and vine life. *Nat Biotechnol.* **20**. 613-618

Misawa, N., Kazumori, M., Hori, T., Ohtani, T., Böger P. and Sandmann, G. (1994) Expression of an *Erwinia* phytoene desaturase gene not only confers multiple resistance to herbicides interfering with carotenoid biosynthesis but also alters xanthophyll metabolism in transgenic plants. *Plant J.* **6**. 481-489

Moco, S., Bino, R. J., Vorst, O., Verhoeven, H. A., de Groot, J., van Beek, T. A., Vervoort, J. and de Vos, R. C. H. (2006) A liquid chromatography-mass spectrometry-based metabolome database for tomato. *Plant Physiol.* **141**. 1205-1218

Morris, W. L., Ducreux, L., Griffiths, D. W., Stewart, D., Davies, H. V. and Taylor, M. A. (2004) Carotenogenesis during tuber development and storage in potato. *J Exp Bot.* **55**. 975-982

Morris, W. L., Ducreux, L., Fraser, P. D., Millam S. and Taylor, M. A. (2006a) Engineering ketocarotenoid biosynthesis in potato tubers. *Metab Eng.* **8**. 253-63

Morris, W. L., Ducreux, L., Hedden, P., Millam, S. and Taylor, M. A. (2006b) Overexpression of a bacterial 1-deoxy-D-xylulose 5-phosphate synthase gene in potato tubers perturbs the isoprenoid metabolic network: implications for the control of the tuber life cycle. *J Exp Bot.* **57**. 3007-3018

Mueller, L. A., Lankhorst, R. K., Tanksley, S. D., Giovannoni, J. J., White, R., Vrebalov, J., Fei, Z., van Eck, J., Buels, R., Mills, A. A., Menda, N.,Tecle, I. Y., Bombarely, A., Stack, S., Royer, S. M., Chang, S., Shearer, L. A., Kim, B. D., Jo, S., Hur, C. G., Choi, D., Li, C., Zhao, J., Jiang, H., Geng, Y., Dai, Y., Fan, H., Chen, J., Lu, F., Shi, J., Sun, S., J., Chen, X., Yang, C., Lu, M., Chen, Z., Cheng, C., Li, H., Ling, Xue, Y., Wang, Y., Seymour, G. B., Bishop, G. J., Bryan, G., Rogers, J., Sims, S., Butcher, S., Buchan, D., Abbott, J., Beasley, H., Nicholson, C., Riddle, C., Humphray, S., McLaren, K., Mathur, S., Vyas, S., Solanke, A. U., Kumar, R., Gupta, V., Sharma, A. K., Khurana, P., Khurana, J. P., Tyagi, A., Sarita, Chowdhury, P., Shridhar, S., Chattopadhyay, D., Pandit, A., Singh, P., Kumar, A., Dixit, R., Singh, A., Praveen, S., Dalal, V., Yadav, M., Ghazi, I. A., Gaikwad, K., Sharma, T. R., Mohapatra,T., Singh, K. N., Szinay, D., de Jong, H., Peters, S.,

VAN STAVEREN, M., DATEMA, E., FIERS, M. W. E. J., VAN HAM, R. C. H. J., LINDHOUT, P., PHILIPPOT, M., FRASSE, P., REGAD, F., ZOUINE, M., BOUZAYEN, M., ASAMIZU, E., SATO, S., FUKUOKA, H., TABATA, S., SHIBATA, D., BOTELLA, M. A., PEREZ-ALONSO, M., FERNANDEZ-PEDROSA, V., OSORIO, S., MICO, A., GRANELL, A., ZHANG, Z., HE, J., HUANG, S., DU, Y., QU, D., LIU, L., LIU, D., WANG, J., YE, Z., YANG, W., WANG, G., VEZZI, A., TODESCO, S., VALLE, G., FALCONE, G., PIETRELLA, M., GIULIANO, G., GRANDILLO, S., TRAINI, A., D'AGOSTINO, N., CHIUSANO, M. L., ERCOLANO, M., BARONE, A., FRUSCIANTE, L., SCHOOF, H., JÖCKER, A., BRUGGMANN, R., SPANNAGL, M., MAYER, K. X.F., GUIGÓ, R., CAMARA, F., ROMBAUTS, S., FAWCETT, J. A., VAN DE PEER, Y., KNAPP, S., ZAMIR, D. AND STIEKEMA, W. (2009) A snapshot of the emerging tomato genome sequence. *Plant Genome*. **2**. 78–92.

NAQVI, S., ZHU, C., FARRE, G., RAMESSAR, K., BASSIE, L., BREITENBACH, J., PEREZ CONESA, D., ROS, G., SANDMANN, G., CAPELL, T. AND CHRISTOU, P. (2009) Transgenic multivitamin corn through biofortification of endosperm with three vitamins representing three distinct metabolic pathways. *Proc Natl Acad Sci USA*. **106**. 7762-7767

NORTH, H. M., DE ALMEIDA, A., BOUTIN, J. P., FREY, A., TO, A., BOTRAN, L., SOTTA, B. AND MARION-POLL, A. (2007) The *Arabidopsis* ABA-deficient mutant *aba4* demonstrates that the major route for stress-induced ABA accumulation is via neoxanthin isomers. *Plant J*. **50**. 810-824

PAINE, J. A., SHIPTON, C. A., CHAGGAR, S., HOWELLS, R. M., KENNEDY M. J., VERNON, G., WRIGHT, S. Y., HINCHLIFFE, E., ADAMS, J. L., SILVERSTONE, A. L. AND DRAKE, R. (2005) Improving the nutritional value of Golden Rice through increased pro-vitamin A content. *Nat Biotechnol*. **23**. 482-487

POPOVSKY, S. AND PARAN, I. (2000) Molecular genetics of the *y* locus in pepper: its relation to capsanthin-capsorubin synthase and to fruit color. *Theor Appl Genet*. **101**. 86-89

RALLEY, L., ENFISSI, E. M., MISAWA, N., SCHUCH, W., BRAMLEY, P. M. AND FRASER, P. D. (2004) Metabolic engineering of ketocarotenoid formation in higher plants. *Plant J*. **39**. 477-486

RAVANELLO, M. P., KE, D., ALVAREZ, J., HUANG, B. AND SHEWMAKER, C. K. (2003) Coordinate expression of multiple bacterial carotenoid genes in canola leading to altered carotenoid production. *Metab Eng*. **5**. 255-263

RENSTRØM, B., BERGER, H. AND LIAAEN-JENSEN, S. (1981) Esterified, optical pure (3*S*, 3'*S*)-astaxanthin from flowers of *Adonis annua*. *Biochem Syst Ecol*. **9**. 249-250

REY, P., GILLET, B., RÖMER, S., EYMERY, F., MASSIMINO, J., PELTIER, G. AND KUNTZ, M. (2000) Over-expression of a pepper plastid lipid-associated protein in tobacco leads to changes in plastid ultrastructure and plant development upon stress. *Plant J*. **21**. 483-494

ROCCO, M., D'AMBROSIO, C., ARENA, S., FAUROBERT, M., SCALONI, A. AND MARRA, M. (2006) Proteomic analysis of tomato fruits from two ecotypes during ripening. *Proteomics*. **6**. 3781-3791

RÖMER, S. FRASER, P. D., KIANO, J. W., SHIPTON, C. A., MISAWA, N., SCHUCH, W. AND BRAMLEY, P. M. (2000) Elevation of the pro-vitamin A content of transgenic tomato plants. *Nat Biotechnol*. **18**. 666-669

RÖMER, S., LUBECK, J., KAUDER, F., STEIGER, S., ADOMAT, C. AND SANDMANN, G. (2002) Genetic engineering of a zeaxanthin-rich potato by antisense inactivation and co-suppression of carotenoid epoxidation. *Metab Eng*. **4**. 263-272

Ronen, G., Carmel-Goren, L., Zamir, D., and Hirschberg, J. (2000) An alternative pathway to β-carotene formation in plant chromoplasts discovered by map-based cloning of *Beta* and *old-gold* color mutations in tomato. *Proc Natl Acad Sci USA.* **97**. 11102-11107

Rosati, C., Aquilani, R., Dharmapuri, S., Pallara, P., Marusic, C., Tavazza, R., Bouvier, F., Camara, B. and Giuliano, G. (2000) Metabolic engineering of beta-carotene and lycopene content in tomato fruit. *Plant J.* **24**. 413-420

Rubio, A., Rambla, J. L., Santaella, M., Gomez, M. D., Orzaez, D., Granell, A. and Gómez-Gómez, L. (2008) Cytosolic and plastoglobule-targeted carotenoid dioxygenases from *Crocus sativus* are both involved in beta-ionone release. *J Biol Chem.* **283**. 24816-24825

Rubio Moraga, A., Fernández Nohales, P., Fernández Pérez, J. A. and Gómez-Gómez, L. (2004) Cytosolic and plastoglobule-targeted carotenoid dioxygenases from *Crocus sativus* are both involved in beta-ionone release. *Planta.* **219**. 955-966

Schaub, P., Al-Babili, S., Drake, R. and Beyer, P. (2005) Why is golden rice golden (yellow) and not red? *Plant Physiol.* **138**. 441-450

Schwartz, S. H., Tan, B. C., Gage D. A., Zeevart J. A. and McCarty, D. R. (1997) Specific oxidative cleavage of carotenoids by VP14 of maize. *Science.* **276**. 1872-1874

Schwartz, S. H., Qin, X. and Loewen, M. C. (2004) The biochemical characterization of two carotenoid cleavage enzymes from *Arabidopsis* indicates that a carotenoid-derived compound inhibits lateral branching. *J Biol Chem.* **279**. 46940-46945

Sergeant, M. J., Li, J. J., Fox, C., Brookbank, N., Rea, D., Bugg, T. D. H. and Thompson, A. J. (2009) Selective inhibition of carotenoid cleavage dioxygenases: phenotypic effects on shoot branching. *J Biol Chem.* **284**. 5257-5264

Shewmaker, C. K., Sheehy, J. A., Daley, M., Colburn, S. and Ke, D. Y. (1999) Seed-specific overexpression of phytoene synthase: increase in carotenoids and other metabolic effects *Plant J.* **20**. 401-412

Simkin, A. J., Schwartz, S. H., Auldridge, M., Taylor, M. G. and Klee, H. J. (2004a) The tomato *carotenoid cleavage dioxygenase 1* genes contribute to the formation of the flavor volatiles β-ionone, pseudoionone, and geranylacetone. *Plant J.* **40**. 882-892

Simkin, A. J., Underwood, B. A., Auldridge, M., Loucas, H. M., Shibuya, K., Schmelz, E., Clark, D. G. and Klee, H. J. (2004b) Circadian regulation of the PhCCD1 carotenoid cleavage dioxygenase controls emission of β-ionone, a fragrance volatile of petunia flowers. *Plant Physiol.* **136**. 3504-3514

Simkin, A. J., Gaffe, J., Alcaraz, J. P., Carde, J. P., Bramley, P. M., Fraser, P. D. and Kuntz, M. (2007) Fibrillin influence on plastid ultrastructure and pigment content in tomato fruit. *Phytochem.* **68**. 1545-1556

Singh, A., Reimer, S., Pozniak, C. J., Clarke, F. R., Clarke, J. M., Knox, R. E. and Singh, A. K. (2009) Allelic variation at *Psy1-A1* and association with yellow pigment in durum wheat grain. *Theor Appl Genet.* 118. 1539-1548

Snowden, K. C., Simkin, A. J., Janssen, B. J., Templeton, K. R., Loucas, H. M., Simons, J. L., Karunairetnam, S., Gleave, A. P., Clark, D. G. and Klee, H. J. (2005) The Decreased apical dominance1/*Petunia hybrida* CAROTENOID CLEAVAGE DIOXYGENASE8 gene affects branch production and plays a role in leaf senescence, root growth, and flower development. *Plant Cell.* **17**. 746-759

STÅLBERG, K., LINDGREN, O., EK, B. AND HÖGLUND, A. S. (2003) Synthesis of ketocarotenoids in the seed of *Arabidopsis thaliana*. *Plant J*. **36**. 771-779

SURLES, R. L., WENG, N., SIMON, P. W. AND TANUMIHARDJO, S. A. (2004) Carotenoid profiles and consumer sensory evaluation of specialty carrots (*Daucus carota* L.) of various colors. *J Agric Food Chem*. **52**. 3417-3421.

TAN, B. C., JOSEPH, L. M., DENG, W. T., LIU, L., LI, Q. B., CLINE, K. AND MCCARTY, D. R. (2003) Molecular characterization of the *Arabidopsis* 9-cis-epoxycarotenoid dioxygenase gene family. Plant J. 35. 44-56

VALLABHANENI, R. AND WURTZEL, E. T. (2009) Timing and biosynthetic potential for carotenoid accumulation in genetically diverse germplasm of maize. *Plant Physiol*. 150. 562-572

WINTERHALTER, P. AND ROUSEFF, R. L., EDS. (2001). *Carotenoid-derived aroma compounds*. Washington DC: American Chemical Society.

WURBS, D., RUF, S. AND BOCK, R. (2007) Contained metabolic engineering in tomatoes by expression of carotenoid biosynthesis genes from the plastid genome. *Plant J*. **49**. 276-288

YE, X., AL-BABILI, S., ZHANG, J., LUCCA, P., BEYER, P. AND POTRYKUS, I. (2000) Engineering the pro-vitamin A (β-carotene) biosynthetic pathway into (carotenoid-free) rice endosperm. *Science*. **287**. 303-305

YU, B., LYDIATE D. J., YOUNG L. W., SCHÄFER, U. A. AND HANNOUFA, A. (2008) Enhancing the carotenoid content of *Brassica napus* seeds by downregulating lycopene epsilon cyclase. *Transgenic Res*. **17**. 573-585

ZHU, C., GERJETS, T. AND SANDMANN, G. (2007a) *Nicotiana glauca* engineered for the production of ketocarotenoids in flowers and leaves by expressing the cyanobacterial *crtO* ketolase gene. *Transgenic Res*. **16**. 813-821

ZHU, C. F., KAUDER, F., RÖMER, S. AND SANDMANN, G. (2007b) Cloning of two individual cDNAS encoding 9-cis-epoxycarotenoid dioxygenase from *Gentiana lutea*, their tissue-specific expression and physiological effect in transgenic tobacco. *J Plant Physiol*. **164**. 195-204

ZHU, C., NAQVI, S., BREITENBACH, J., SANDMANN, G., CHRISTOU, P. AND CAPELL, T. (2008) Combinatorial genetic transformation generates a library of metabolic phenotypes for the carotenoid pathway in maize. *Proc Natl Acad Sci USA*. **105**. 18232-18237

·

Biotechnology and Genetic Engineering Reviews - Vol. 26, 163-178 (2009)

Supply of Nutrients to Cells in Engineered Tissues

JEROEN ROUWKEMA*[1], BART F.J.M. KOOPMAN[1], CLEMENS A. VAN BLITTERSWIJK[2], WOUTER J.A. DHERT[3,4] AND JOS MALDA[3]

[1]Department of Biomechanical Engineering, Institute for Biomedical Technology, University of Twente, Enschede, The Netherlands. [2]Department of Tissue Regeneration, Institute for Biomedical Technology, University of Twente, Enschede, The Netherlands. [3]Department of Orthopaedics, University Medical Center Utrecht, Utrecht, The Netherlands. [4]Faculty of Veterinary Medicine, Utrecht University, The Netherlands

Abstract

A proper supply of nutrients to cells in engineered tissues is paramount for an optimal development and survival of these tissues. However, especially in tissues with clinically relevant sizes, the mass transport of nutrients into the tissue is often insufficient to sustain all the cells within the tissue. This is not only the case during *in vitro* culture. After implantation of an engineered tissue, a vascular network is not directly established. Therefore, the mass transport of nutrients is also critical during the initial period after implantation.

This review introduces the basics of mass transport, leading to the conclusion that three main concepts can be used to increase nutrient supply in tissue engineering. These are; increasing the overall diffusion coefficient, decreasing the diffusion distance, or increasing convective transport. Based on these concepts, the main strategies that have been developed to enhance the supply of nutrients to cells in engineered tissues will be discussed.

*To whom correspondence may be addressed (J.Rouwkema@ctw.utwente.nl)

Abbreviations: J_a, flux of component a; D_a, diffusion coefficient of component a; c_a, concentration of component a; x, distance; u, bulk convective velocity; L, characteristic length; Pe, Péclet number; N_a, rate of mass transfer of component a; A, area across which mass transfer occurs; PFC, perfluorocarbon.

Introduction

Cells in the human body need nutrients and oxygen to survive. Most tissues in the body rely on the active transport of blood to supply individual cells with nutrients and oxygen, with the exception of avascular tissues, such as articular cartilage. Almost every cell within our body is, therefore, close to a blood vessel. Generally, new blood vessel formation is required for a tissue to grow beyond 100-200 µm (the diffusion limit of oxygen)(Carmeliet *et al.* 2000). For tissues engineered in a laboratory, the same principles hold true. This poses a significant challenge, since *in vitro* created tissue-engineered constructs lack a vascular network and cells receive nutrients by diffusion over distances that are often on the order of millimetres. However, due to the consumption of nutrients by the cells and the relatively slow supply by diffusion, only cells up to a distance of approximately 200 µm have access to sufficient nutrients. For cells that are further away from the engineered tissue surface, this results in nutrient limitations, and thus reduced cell proliferation or non-optimal conditions for new matrix production (Lewis *et al.* 2005). In turn, this leads to inhomogeneous tissue formation, with the bulk of matrix in the periphery and little in the center (Malda *et al.* 2004; Wendt *et al.* 2006).

Apart from the supply of nutrients during *in vitro* culture, keeping a graft viable after *in vivo* implantation is an additional challenge. In order for implanted tissues beyond a size of several hundreds of micrometers to survive, the tissue has to be vascularized, meaning that a capillary network should be present in the tissue that delivers nutrients to the cells. After implantation of a graft, blood vessels from the host generally invade the graft tissue to form such a network, in part due to signals that are secreted by the implanted cells as a response to hypoxia. However, this spontaneous vascular ingrowth is often limited to several tenths of millimeters per day (Clark 1939), meaning that the time that is needed for complete vascularization of an implant of several millimeters is in the order of weeks. During this time, insufficient vascularization can lead to nutrient deficiencies and/or hypoxia deeper in the tissue. Moreover, nutrient and oxygen gradients will be present in the outer regions of the tissue, which could result in non-uniform cell differentiation and integration, and thus decreased tissue functionality (Malda, Rouwkema *et al.* 2004).

This review will therefore describe the issues that are associated with the supply of nutrients to the cells in engineered tissues, both during *in vitro* culture and after implantation, and the possible strategies that can be followed to cope with these problems.

Mass transport

Nutrients are supplied to the cells by the process of mass transport. Mass transport takes place in mixtures containing local concentration variations (Doran 2000), due to for example cellular consumption. During *in vitro* culture, mass transport can occur as the result of convection and diffusion. Convection is the relatively fast mass transfer as a result of bulk fluid motion, whereas diffusion is the relatively slow movement of component molecules in a mixture in the direction of the concentration gradient.

Mass transport of a nutrient within the medium can be described with the convection-diffusion equation.

$$J_a = -D_a \cdot \frac{dc_a}{dx} + u \cdot c_a$$

[Eq. 1]

where:

J_a is the flux of component a   [mol m^{-2} s^{-1}]
D_a is the diffusion coefficient of component a in the medium   [m^2 s^{-1}]
c_a is the concentration of component a   [mol m^{-3}]
x is the distance   [m]
u is the bulk convective velocity of the medium   [m s^{-1}]

Whether nutrient transport is dominated by diffusion or convection, depends on the relation between the diffusion coefficient of the nutrient (D) and the flow speed of the medium (u). This relation between the contribution of diffusion and convection to the overall transport is described in the dimensionless Péclet number.

$$Pe = \frac{L \cdot u}{D}$$

[Eq. 2]

where:

L is the characteristic length   [m]
(the distance over which diffusion takes place)
D is the diffusion coefficient of the nutrient in medium   [m^2 s^{-1}]
u is the bulk convective velocity of the medium   [m s^{-1}]

In free solution, convection dominates when Pe >> 1, and diffusion dominates when Pe << 1 (Friedman 2008). In tissue engineering, cultures are often static, meaning that no active perfusion is included in the culture regime. The minimal contribution of convection to the overall transport of nutrients within these cultures, results in a low Péclet number. Therefore, the contribution of convection to the overall transport of nutrients is often ignored in tissue engineered constructs (Swartz *et al.* 2007). In this case, the transport of nutrients can thus be simply described by *Fick's Law* of diffusion (Eq 3. and *Figure 1*).

$$J_a = \frac{N_a}{A} = -D_a \frac{dc_a}{dx}$$

[Eq. 3]

where:

J_a is the flux of component *a*   [mol m^{-2} s^{-1}]
N_a is the rate of mass transfer of component *a*   [mol s^{-1}]
A is the area across which mass transfer occurs   [m^2]
D_a is the diffusion coefficient of component *a*   [m^2 s^{-1}]
x is the distance   [m]

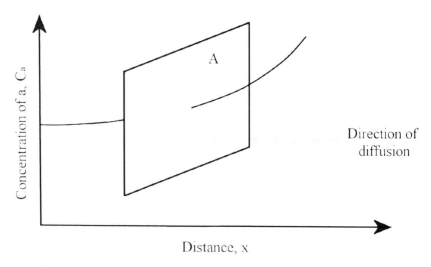

Figure 1. The concentration gradient of component "a" in a stagnant system (no convection) inducing mass transfer across area A. The diffusion occurs in the direction required to destroy the concentration gradient of the molecular components "a". The concentration of "a", varies as a function of the distance x and $C_{a2} > C_{a1}$. Adapted from Doran (Doran 2000) and Malda *et al.* (Malda *et al.* 2008) with permission of the publisher.

The mass transport of nutrients thus depends both on the diffusion coefficient (D) and the diffusion distance (x). Therefore, in cases where convective flow is not involved, strategies to improve mass transfer in tissue engineering should be focused on increasing the diffusion coefficient (by providing an open matrix through which nutrients can easily diffuse) or on decreasing the diffusion distance.

Cell Nutrition *in vitro*

Where diffusion and in some cases convection are the processes that take care of the transport of nutrients into a tissue, consumption by the cells is the main process that governs the removal of these nutrients. The challenge in tissue engineering is often to find a balance between these two processes. In cases where the consumption of a nutrient in a tissue is higher (for instance due to a high cellular density and/or cell activity) than the transport of this nutrient into the tissue (for instance due to the lack of convection and/or large diffusion distances), nutrient limitations will occur. This is translated into a concentration gradient of this nutrient, with a high concentration at the periphery and a low concentration at the centre (Malda, Rouwkema *et al.* 2004; Malda *et al.* 2004). As a result, cells at peripheral boundaries will experience completely different environmental conditions compared to cells located in the center. This can lead to tissue necrosis in the center and inhomogeneous tissue development throughout the construct (Ma *et al.* 2001; Griffith *et al.* 2006; Cioffi *et al.* 2008).

The occurrence of nutrient gradients is a process that plays a role for all nutrients. However, for the optimization of nutrient supply in tissue engineering, oxygen is often chosen as a model nutrient to describe gradient formation (Obradovic *et al.* 2000; Kellner *et al.* 2002; Malda, Rouwkema *et al.* 2004; Malda, Woodfield *et al.* 2004; Lewis, Macarthur *et al.* 2005; Radisic *et al.* 2006). Oxygen is required for cellular survival due to its central role as an acceptor of electrons in the mitochondrial respiratory chain and due to its influence on developmental processes. Due to high consumption rates combined with low solubility, oxygen is often regarded as the limiting nutrient.

DIFFERENCES IN NUTRIENT CONCENTRATIONS *IN VITRO* AND *IN VIVO*

During the *ex vivo* culture of cells, we aim to mimic the conditions the cells experience in the tissue *in vivo*. Clearly, *in vitro* culture is still distinctly different from and often only a poor reflection of the *in vivo* situation. Culture media, for example, function as the *in vitro* substitute for the biological fluid present in the normal physiological environment and are designed to support the growth and or differentiation of cells. However, development of these media is rather empirical than scientific and there are striking similarities in the composition of the media used today with those described more than 50 years ago (White 1946). Moreover, the concentration of the nutrients is not constant, but goes up and down with depletion and medium changes, even if only 50% of the medium is replaced per medium change. This could potentially have substantial effects on the behaviour of the cells (Garcia-Montero *et al.* 2001).

Oxygen levels are also distinctly different during culture *in vitro* from those experienced by the cells in *in vivo* tissue. During isolation of cells from *e.g.*, the cartilage or bone marrow, cells will be exposed to considerably higher oxygen levels. Further, standard cultures are performed in medium that is in equilibrium with 21% oxygen, which is far above physiological levels in the human body (Csete 2005).

As a result of the hampered nutrient supply *in vitro*, gradients of nutrients and oxygen will be present in the outer regions of the tissue. These gradients will affect the development of the 3-dimensional neo-tissue. Non-uniform cell differentiation and integration and decreased tissue functionality have, for example, been observed (Malda, Rouwkema *et al.* 2004; Lewis, Macarthur *et al.* 2005; Wendt, Stroebel *et al.* 2006). Mitigation of excessive nutrient gradients by improving nutrient transport is therefore the key to the development of more homogenous engineered constructs with improved extracellular matrix formation and functionality.

STRATEGIES TO IMPROVE NUTRIENT SUPPLIES *IN VITRO*

One of the major obstacles in tissue engineering of thick, complex tissues is to keep the construct viable *in vitro* (during cultivation and formation of the tissue) as well as *in vivo* (upon implantation). Straightforward approaches, for example using seeding techniques which ensure a uniform spatial distribution of cells on a scaffold (Vunjak-Novakovic *et al.* 1998) do not necessarily address the correct underlying mechanisms causing these heterogeneities and have, therefore, been unsuccessful to date. As was discussed in the previous section of this paper, it is the diffusion limitation that will lead to spatial heterogeneities within constructs, which are characterized by significantly higher cell densities in the peripheral layers relative to the interior (Malda, Woodfield *et al.* 2004; Lewis, Macarthur *et al.* 2005) as well as uneven deposition of the tissue matrix (Martin *et al.* 1999). As is clear from equations 1-3, the mass transfer of nutrients to the cells within a tissue can be increased by; instituting convective flow within the tissue, increasing the value for the diffusion coefficient, or decreasing the diffusion distance to reach the cells in the middle of a tissue. Based on these common paradigms, the following sections will discuss some strategies that are currently being explored to decrease the diffusion limitation and, thus, increase the supply of nutrients to the cells within engineered tissues (*Figure 2*).

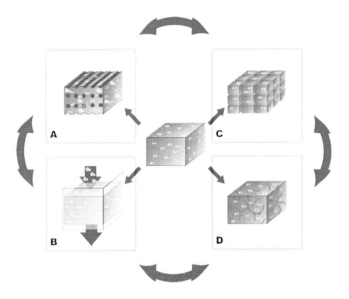

Figure 2. Different strategies have been developed to improve nutrient supply in tissue engineering. These include scaffold-based strategies such as the inclusion of transport channels (increasing the overall diffusion coefficient) or oxygen generating biomaterials (decreasing the diffusion distance) (**A**), medium-based strategies such as perfusion bioreactor systems (increasing convective transport) or oxygen carriers (decreasing the diffusion distance) (**B**), modular tissue engineering approaches to limit the size of engineered tissues (decreasing the diffusion distance) (**C**), and prevascularization strategies to enhance nutrient supply after implantation (increasing convective transport and decreasing the diffusion distance) (**D**). Even though all strategies in itself are capable of enhancing nutrient supply, a combination of multiple strategies is likely to be necessary for the successful production of clinically relevant tissue volumes.

BIOREACTOR CULTURE

Bioreactors can be defined as devices in which biological and/or biochemical processes develop under closely monitored and tightly controlled environmental and operating conditions (*e.g.*, pH, temperature, pressure, nutrient supply, and waste removal) (Martin *et al.* 2004). One of the aspects that initiated bioreactor development in the field of tissue engineering was the improved nutrient and oxygen delivery to the cells inside the tissue, allowing for an increase in the volume of the engineered tissue. This has been achieved with the development of perfusion bioreactor systems. With these systems, culture medium is actively perfused through the engineered tissue instead of around it, meaning that nutrients are no longer only transported to the cells within the tissue by means of diffusion but also by convection. As can be seen in equation 1, the contribution of perfusion to nutrient transport is directly related to the velocity of the medium. In order for the contribution of perfusion to nutrient transport to become significant, the velocity should be high enough so that the Péclet number becomes larger than 1 (see equation 2).

An additional effect of the movement of medium within a bioreactor, either outside the tissue (*e.g.*, spinner flasks and rotating bioreactors) or within the tissue (*e.g.*, perfusion bioreactors), is that shear stresses are imposed on the cells of the tissue. Even though these shear stresses are often beneficial and can result in activation of cell growth, cell differentiation or matrix production, high levels of shear stress can be detrimental for tissue engineering. Therefore, when designing a bioreactor system, care has to be taken that the levels of shear stress do not exceed critical values. Apart from that, the formation of extracellular matrix during the *in vitro* culture of tissues can result in densification of the tissue. Due to this, the perfusion of medium through the tissue can become difficult over time. In such cases, channels can arise within the tissue through which the medium is perfused (Bancroft *et al.* 2002). Depending on the distance between these channels, this again can result in nutrient limitations.

SCAFFOLD ARCHITECTURE

In static cultures, the transport of nutrients within scaffold-based engineered tissues is based on diffusion. Since the diffusion coefficients of nutrients in scaffold materials are generally relatively low, the design characteristics of the scaffold can have a major impact on nutrient limitation within the graft. For example, increasing the pore interconnectivity creates channels through which the nutrients can diffuse. Since the diffusion coefficient in the medium in these channels is higher then the diffusion coefficient in the biomaterial, this reduces the diffusion barrier. In addition, the interconnections between pores facilitate angiogenesis and extracellular matrix infiltration.

When designing scaffold porosity, it should be kept in mind that, apart from influencing the mass transport of nutrients in the scaffold, the porosity will have an effect on tissue development. Apart from that, specific tissue engineering applications may have different porosity and pore size requirements. For instance, a study by Marshall *et al.* indicated that an average interconnecting pore size of 70 μm is optimal for host blood vessel growth in porous collagen type 1 – polyHEMA scaffolds implanted subcutaneously in mice (Marshall *et al.* 2004). On the other hand, a study by Pilliar suggested that scaffold architectures containing interconnecting pores greater than 100-300 μm in diameter are necessary to support spatially uniform cell seeding, osteogenesis and vascular invasion in scaffolds for bone ingrowth (Pilliar 1987). However, even though the scaffold design will need to be optimized for different tissue engineering applications, in general it is preferable that scaffolds have an interconnecting pore volume to maximize the average diffusion coefficient and thus to optimize the mass transport of nutrients (Hutmacher 2000).

Further, recent advances in both computational topology design and solid free-form fabrication, or rapid prototyping, provide the opportunity to create more physiologically realistic grafts with controlled architecture (Sachlos *et al.* 2003; Malda, Woodfield *et al.* 2004; Wilson *et al.* 2004; Hollister 2005). This for example allows the formation of organized networks of channels that can be perfused with culture medium (convection), mimicking vascularized tissue thus reducing the diffusion distance. Such artificial vasculature with organised hierarchical channels can even be lined with endothelial cells, thereby forming an enclosed functional vasculature (Fidkowski *et al.* 2005).

OXYGEN CARRIERS AND OXYGEN GENERATING SCAFFOLDS

As stated previously, oxygen is often regarded as the limiting nutrient in tissue engineering. Therefore, several strategies have been developed that specifically target the distribution of oxygen within engineered tissues. One such strategy, inspired by the role of hemoglobin in oxygen transport in the body, is the inclusion of an oxygen carrier in the culture medium. Two main groups of artificial oxygen carriers have been studied in the field of tissue engineering; those that are based on modified hemoglobin and those that are based on perfluorocarbon (PFC) (Centis *et al.* 2009).

PFC is a compound that is not miscible with water. Therefore, PFC has to be brought into an emulsion with the medium. Since the PFC droplets are immiscible, they serve as rechargeable oxygen reservoirs. As oxygen is consumed by the cells and thus depleted from the aqueous phase of the culture medium, it is replenished by diffusion of dissolved oxygen from PFC particles. This replenishment of oxygen decreases the diffusion distance and as such influences the mass transport of oxygen (see equation 1). As a result, the total oxygen content of the PFC-supplemented medium is higher, supporting higher cell density. However, the oxygen partial pressures measured in the aqueous phase of PFC-supplemented and -unsupplemented medium are the same, and the replenishment of oxygen from the PFC particles does not increase the oxygen concentration in culture medium (Radisic *et al.* 2005; Radisic *et al.* 2006).

Apart from supplying oxygen carriers to the medium, the development of oxygen generating scaffolds in tissue engineering has been reported. Oxygen generating biomaterials were developed by the inclusion of for instance calcium peroxide (Oh *et al.* 2009) or sodium percarbonate (Harrison *et al.* 2007) into polymer scaffolds. Upon contact with water, these salts decompose producing oxygen as well as several biocompatible byproducts. By tuning the degradation properties of the polymer, oxygen can be released over a timeframe ranging from days to weeks thus increasing the oxygen concentration at the cellular level. Although oxygen generating scaffolds could prove to be useful, it should be noted that even though oxygen is generally regarded as the limiting nutrient in tissue engineering, limitations and gradients of compounds like glucose and growth factors are not affected by this strategy. Apart from that, an excess of oxygen in the medium surrounding the cells without an appropriate carrier such as hemoglobin can induce the presence of free radicals, which are cytotoxic (Martin *et al.* 2005).

MODULAR TISSUE ENGINEERING

A recent development to prevent nutrient limitations during *in vitro* culture is bottom-up or modular tissue engineering. The rationale behind this approach is to use a large number of small simple tissue units (micro-tissues) to build larger, more complex tissues. By limiting the size of the micro-tissues, the diffusion distance is minimized thus optimizing the mass transport of nutrients from the medium into the tissue (see equation 3). As such, the micro-tissues can be maintained in standard static cultures without nutrient limitations. By culturing high quantities of micro-tissues, large volumes of tissue can be maintained without the need for extra measures to ensure the sufficient supply of nutrients. Only after the combination of the micro-tissues into the bigger tissue (which could be performed shortly before implantation), nutrition has to be taken into account again

The combination of tissue units into bigger more complex tissues is a subject that has attracted attention over the past few years. While tissue units can be packed into a predefined volume randomly to create bigger tissues (McGuigan *et al.* 2006; McGuigan *et al.* 2007) (Kelm *et al.* 2006), a more controlled method of combining tissue units is often preferable in order to better control tissue architecture. A method that is being explored to achieve this goal is the depositioning or printing of cell units into topologically defined structures (Jakab *et al.* 2004; Jakab *et al.* 2008). This highly versatile technique allows for the precise control of tissue architecture. However, since the number of tissue units that are needed to prepare a tissue of clinically relevant size (in general several mm^3) is very large, the deposition of small single tissue units into their preferred location is often not an option. Therefore, a method that (partly) relies on the spontaneous organization of tissue units can be preferable (Whitesides *et al.* 2002; Du *et al.* 2008).

It is unlikely that nutrient deficiencies will occur in the small tissue units, however deficiencies will inevitably occur when the micro-units are combined into a bigger tissue. Thus, even though modular tissue engineering is a promising strategy to prevent nutrient limitations during an important part of *in vitro* culture, additional measures are necessary to prevent limitations in subsequent steps, for example after implantation of the engineered tissue.

Nutrient transport in implanted engineered tissues

The supply of nutrients to engineered tissues is not only an issue during *in vitro* culture, but also after the implantation of these tissues. Apart from diffusion, the transport of nutrients to the cells within a tissue in the body is facilitated by two processes; convection of blood in the vascular system and convection within the interstitial fluid. The main process that transports nutrients to the close vicinity of most cells in the body is the convection of blood in the vascular system. Three distinct structures can be distinguished in this vascular system. These are the macrovessels (arteries and veins), that branch into microvessels (arterioles and venules) and finally into capillaries. The capillaries facilitate the actual distribution of nutrients to the tissue in the body and as such decrease the diffusion distance from the vascular system to the cells. They distribute the blood over the tissue while lowering the pressure head, allowing blood to diffuse into the tissue. Most tissues in the body have a high density of capillaries to ensure that sufficient nutrients can be transported into close range of all the cells within the tissue. No exact number for the capillary density can be given since this varies between tissues and individuals, but a histological study by Awwad *et al.* shows that in carcinoma of the cervix uteri the average intercapillary distance is approximately 300 microns (Awwad *et al.* 1986). This number, which correlates to a maximum distance of a cell to the nearest capillary of 150 microns, correlates well with the maximum oxygen diffusion range (Carmeliet and Jain 2000).

Even though the convection through blood vessels and the subsequent diffusion of nutrients from the vessels to the cells within the interstitium are the main processes of nutrient transport in natural tissues, convection within the interstitial fluid is another process that can aid in the transport of nutrients. Interstitial flow is closely linked with lymphatic drainage, which returns plasma that has leaked out of the capillaries, owing to hydrostatic and osmotic pressure differences, to the blood circulation (Schmid-Schonbein 1990). As opposed to the flow of blood in the vascular system, interstitial flow is a slow process. Although the exact velocity ranges of this flow are unknown, measurements have suggested that they are in the order of 0,1 to 2 microns per second (Swartz and Fleury 2007). Since the interstitial flow rates are so small, which means that transport due to diffusion is often much faster, the contribution of this process to the transport of nutrients is often ignored. However, the role of convection in the overall distribution of proteins can be significant. Diffusion is inversely related to molecular size and many biologically significant proteins, pharmaceuticals, and delivery vectors are quite large, leading to small diffusion distances over biologically relevant timescales. Apart from that, ionic interactions between solutes and certain matrix components, can further reduce diffusivities. Convection, however, is only weakly tied to molecular size, which means that for larger molecules, convection becomes more important in governing their transport compared to diffusion (Swartz and Fleury 2007).

STRATEGIES TO ENHANCE VASCULARIZATION AFTER IMPLANTATION

Cultured tissues generally do not contain a vascular transport system. Since this results in a lack of convective transport, the cells within the tissue have to rely mainly on diffusion after implantation, which is often insufficient to supply the cells in the inner regions of the tissue due to excessive diffusion distances (see equation 1). After the implantation of engineered tissues, a spontaneous vascularization of the implant is generally seen. This is in part due to an inflammatory wound-healing response, induced by the surgical procedure. Furthermore, the seeded cells often create a hypoxic state in the implant, which stimulates the endogenous release of angiogenic growth factors (Laschke *et al.* 2006). However, as mentioned previously, this induced vessel ingrowth is often too slow to provide adequate nutrient transport to the cells in the middle of the transplanted tissue. Therefore, additional strategies to enhance vascularization are essential to ensure the survival of large tissue engineered grafts.

Several strategies to enhance vascularization of tissue engineered constructs after implantation are currently under investigation. These include scaffold design, the inclusion of angiogenic factors, *in vivo* prevascularization and *in vitro* prevascularization (Rouwkema *et al.* 2008). Although all these strategies can in principle enhance vascularization after implantation, the degree to which these strategies are able to enhance vascularization varies. The first two approaches, scaffold design (Lee *et al.* 2008) and angiogenic factor delivery (Richardson *et al.* 2001; Chen *et al.* 2007), both rely on the ingrowth of host vessels into the entire implanted construct. Therefore, although these strategies are able to increase the rate of vascularization, it would still take multiple days to weeks before the centre of the implant becomes perfused. In contrast, *in vivo* prevascularization can in principle result in the instantaneous perfusion of a construct after implantation at the final site, since the construct is microsurgically connected to the host vasculature (Kneser *et al.* 2006). However, before implantation at the final site, a pre-implantation period is necessary, during which the implant has to rely on a spontaneous angiogenesis from the surrounding vessels into the construct. Therefore, nutrient limitations are likely to occur during this stage. For *in vitro* prevascularization, endothelial cells are added to the tissue construct during culture to form a vascular network before implantation (Black *et al.* 1998; Rouwkema *et al.* 2009). *In vitro* prevascularization does not result in the instantaneous perfusion of a construct, since vessels have to grow from the host into the construct until they reach the vascular network that was formed *in vitro* (Levenberg *et al.* 2005; Tremblay *et al.* 2005). The invading vessels can then connect to the present vasculature, meaning that the entire construct becomes perfused with blood. When compared to scaffold design and angiogenic factor delivery, this method can dramatically decrease the time that is needed to vascularize the implant, since host vessels do not have to grow into the entire construct, but only into its outer regions (that is, until the ingrowing vessels meet the preformed vascular network).

Concluding remarks

It is clear that nutrient supply is a critical and often overlooked issue in the field of tissue engineering. Over the last decades, much research efforts have been directed towards the prevention of nutrient limitations during *in vitro* culture through for instance bioreactor culture, scaffold design and modular tissue engineering. Apart from that, oxygen carriers and oxygen generating scaffolds have been developed to specifically prevent limitations of the component that is commonly regarded as being the limiting factor in the engineering of three-dimensional tissues. Even though these latter strategies can contribute to the prevention of oxygen limitations, an increase in tissue size will inevitably lead to other limiting nutrients. Therefore, we feel that the further development of strategies that target nutrient transport in general, such as modular tissue engineering and perfusion bioreactors, are most promising for *in vitro* engineering of large tissue volumes.

However, engineering a large tissue construct *in vitro* is only useful when the cells in the construct are able to survive the initial period after implantation. This should be considered when designing engineered tissues for implantation purposes. Although a natural vascularization process is initiated after the implantation of a tissue construct, this response is generally insufficient to prevent nutrient limitations in large tissue volumes. Since tissue engineering is rapidly progressing towards the translation from a laboratory situation to clinical applications (Leeuwenburgh *et al.* 2008), we predict that strategies to decrease nutrient limitations *in vivo* will attract increasing attention in the coming years.

Acknowledgements

Dr. Malda is supported by a VENI fellowship from the Dutch Technology Foundation STW, Applied Science Division of NWO and the Technology Program of the Ministry of Economic Affairs.

References

AWWAD, HK, EL NAGGAR, M, MOCKTAR, N and BARSOUM, M (1986) Intercapillary distance measurement as an indicator of hypoxia in carcinoma of the cervix uteri. *International Journal of Radiation Oncology*Biology*Physics* **12**(8), 1329-33.

BANCROFT, GN, SIKAVITSAS, VI, VAN DEN DOLDER, J, SHEFFIELD, TL, AMBROSE, CG, JANSEN, JA and MIKOS, AG (2002) Fluid flow increases mineralized matrix deposition in 3D perfusion culture of marrow stromal osteoblasts in a dose-dependent manner. *Proceedings of the National Academy of Sciences of the United States of America* **99**(20), 12600-5.

BLACK, AF, BERTHOD, F, L'HEUREUX, N, GERMAIN, L and AUGER, FA (1998) In vitro reconstruction of a human capillary-like network in a tissue-engineered skin equivalent. *The FASEB Journal* **12**(13), 1331-40.

CARMELIET, P and JAIN, RK (2000) Angiogenesis in cancer and other diseases. *Nature* **407**(6801), 249-57.

CENTIS, V and VERMETTE, P (2009) Enhancing oxygen solubility using hemoglobin- and perfluorocarbon-based carriers. *Frontiers in Bioscience* **14**, 665-88.

CHEN, RR, SILVA, EA, YUEN, WW and MOONEY, DJ (2007) Spatio-temporal VEGF and PDGF delivery patterns blood vessel formation and maturation. *Pharmaceutical Research* **24**(2), 258-64.

CIOFFI, M, KUFFER, J, STROBEL, S, DUBINI, G, MARTIN, I and WENDT, D (2008) Computational evaluation of oxygen and shear stress distributions in 3D perfusion culture systems, macro-scale and micro-structured models. *Journal of Biomechanics* **41**(14), 2918-25.

CLARK, ERC (1939) Microscopic observations on the growth of blood capillaries in the living mammal. *American Journal of Anatomy* **64**(2), 251-301.

CSETE, M (2005) Oxygen in the cultivation of stem cells. *Annals of the New York Academy of Sciences* **1049**, 1-8.

DORAN, P (2000). *Bioprocess Engineering Principles*. London, Academic press.

DU, Y, LO, E, ALI, S and KHADEMHOSSEINI, A (2008) Directed assembly of cell-laden microgels for fabrication of 3D tissue constructs. *Proceedings of the National Academy of Sciences of the United States of America* **105**(28), 9522-7.

FIDKOWSKI, C, KAAZEMPUR-MOFRAD, MR, BORENSTEIN, J, VACANTI, JP, LANGER, R and WANG, Y (2005) Endothelialized microvasculature based on a biodegradable elastomer. *Tissue Engineering* **11**(1-2), 302-9.

FRIEDMAN, MH (2008). *Principles and models of biological transport*. New York, Springer.

GARCIA-MONTERO, A, VASSEUR, S, MALLO, GV, SOUBEYRAN, P, DAGORN, JC and IOVANNA, JL (2001) Expression of the stress-induced p8 mRNA is transiently activated after culture medium change. *European Journal of Cell Biology* **80**(11), 720-5.

GRIFFITH, LG and SWARTZ, MA (2006) Capturing complex 3D tissue physiology in vitro. *Nature Reviews Molecular Cell Biology* **7**(3), 211-24.

HARRISON, BS, EBERLI, D, LEE, SJ, ATALA, A and YOO, JJ (2007) Oxygen producing biomaterials for tissue regeneration. *Biomaterials* **28**(31), 4628-34.

HOLLISTER, SJ (2005) Porous scaffold design for tissue engineering. *Nature Materials* **4**(7), 518-24.

HUTMACHER, DW (2000) Scaffolds in tissue engineering bone and cartilage. *Biomaterials*

21(24), 2529-43.

JAKAB, K, NEAGU, A, MIRONOV, V, MARKWALD, RR and FORGACS, G (2004) Engineering biological structures of prescribed shape using self-assembling multicellular systems. *Proceedings of the National Academy of Sciences of the United States of America* **101**(9), 2864-9.

JAKAB, K, NOROTTE, C, DAMON, B, MARGA, F, NEAGU, A, BESCH-WILLIFORD, CL, KACHURIN, A, CHURCH, KH, PARK, H, MIRONOV, V, MARKWALD, R, VUNJAK-NOVAKOVIC, G and FORGACS, G (2008) Tissue engineering by self-assembly of cells printed into topologically defined structures. *Tissue Engineering* **14**(3), 413-21.

KELLNER, K, LIEBSCH, G, KLIMANT, I, WOLFBEIS, O, BLUNK, T, SCHULZ, M and GOPFERICH, A (2002) Determination of oxygen gradients in engineered tissue using a fluorescent sensor. *Biotechnology and Bioengineering* **80**(1), 73-83.

KELM, JM, DJONOV, V, ITTNER, LM, FLURI, D, BORN, W, HOERSTRUP, SP and FUSSENEGGER, M (2006) Design of custom-shaped vascularized tissues using microtissue spheroids as minimal building units. *Tissue Engineering* **12**(8), 2151-60.

KNESER, U, POLYKANDRIOTIS, E, OHNOLZ, J, HEIDNER, K, GRABINGER, L, EULER, S, AMANN, KU, HESS, A, BRUNE, K, GREIL, P, STURZL, M and HORCH, RE (2006) Engineering of vascularized transplantable bone tissues, induction of axial vascularization in an osteoconductive matrix using an arteriovenous loop. *Tissue Engineering* **12**(7), 1721-31.

LASCHKE, MW, HARDER, Y, AMON, M, MARTIN, I, FARHADI, J, RING, A, TORIO-PADRON, N, SCHRAMM, R, RUCKER, M, JUNKER, D, HAUFEL, JM, CARVALHO, C, HEBERER, M, GERMANN, G, VOLLMAR, B and MENGER, MD (2006) Angiogenesis in tissue engineering, breathing life into constructed tissue substitutes. *Tissue Engineering* **12**(8), 2093-104.

LEE, J, CUDDIHY, MJ and KOTOV, NA (2008) Three-dimensional cell culture matrices, state of the art. *Tissue Engineering* **14**(1), 61-86.

LEEUWENBURGH, SC, JANSEN, JA, MALDA, J, DHERT, WA, ROUWKEMA, J, VAN BLITTERSWIJK, CA, KIRKPATRICK, CJ and WILLIAMS, DF (2008) Trends in biomaterials research, An analysis of the scientific programme of the World Biomaterials Congress 2008. *Biomaterials* **29**(21), 3047-52.

LEVENBERG, S, ROUWKEMA, J, MACDONALD, M, GARFEIN, ES, KOHANE, DS, DARLAND, DC, MARINI, R, VAN BLITTERSWIJK, CA, MULLIGAN, RC, D'AMORE, PA and LANGER, R (2005) Engineering vascularized skeletal muscle tissue. *Nature Biotechnology* **23**(7), 879-84.

LEWIS, MC, MACARTHUR, BD, MALDA, J, PETTET, G and PLEASE, CP (2005) Heterogeneous proliferation within engineered cartilaginous tissue, the role of oxygen tension. *Biotechnology and Bioengineering* **91**(5), 607-15.

MA, T, YANG, ST and KNISS, DA (2001) Oxygen tension influences proliferation and differentiation in a tissue-engineered model of placental trophoblast-like cells. *Tissue Engineering* **7**(5), 495-506.

MALDA, J, RADISIC, M, LEVENBERG, S, WOODFIELD, TB, OOMENS, CW, BAAIJENS, FP, SVALANDER, P and VUNJAK-NOVAKOVIC, G (2008). Cell nutrition. In *Tissue Eng.* C.A. van Blitterswijk, *et al.* pp 327-362. London, Academic Press.

MALDA, J, ROUWKEMA, J, MARTENS, DE, LE COMTE, EP, KOOY, FK, TRAMPER, J, VAN BLITTERSWIJK, CA and RIESLE, J (2004) Oxygen gradients in tissue-engineered PEGT/PBT cartilaginous constructs, measurement and modeling. *Biotechnology and Bioengineering* **86**(1), 9-18.

Malda, J, Woodfield, TB, Van Der Vloodt, F, Kooy, FK, Martens, DE, Tramper, J, Blitterswijk, CA and Riesle, J (2004) The effect of PEGT/PBT scaffold architecture on oxygen gradients in tissue engineered cartilaginous constructs. *Biomaterials* **25**(26), 5773-80.

Marshall, AJ, Barker, T, Sage, EH, Hauch, KD and Ratner, BD (2004). *Pore size controls angiogenesis in subcutaneously implanted porous matrices*. Proc. 7th World Biomaterials Congress, Sydney.

Martin, I, Obradovic, B, Freed, L and Vunjak-Novakovic, G (1999) Method for quantitative analysis of glycosaminoglycan distribution in cultured natural and engineered cartilage. *Annals of Biomedical Engineering* **27**(5), 656-62.

Martin, I, Wendt, D and Heberer, M (2004) The role of bioreactors in tissue engineering. *Trends in Biotechnology* **22**(2), 80-6.

Martin, Y and Vermette, P (2005) Bioreactors for tissue mass culture, design, characterization, and recent advances. *Biomaterials* **26**(35), 7481-503.

McGuigan, AP and Sefton, MV (2006) Vascularized organoid engineered by modular assembly enables blood perfusion. *Proceedings of the National Academy of Sciences of the United States of America* **103**(31), 11461-6.

McGuigan, AP and Sefton, MV (2007) Design criteria for a modular tissue-engineered construct. *Tissue Engineering* **13**(5), 1079-89.

Obradovic, B, Meldon, J, Freed, L and Vunjak-Novakovic, G (2000) Glycosaminoglycan deposition in engineered cartilage, Experiments and mathematical model. *American Institute of Chemical Engineering Journal* **46**, 1860-1871.

Oh, SH, Ward, CL, Atala, A, Yoo, JJ and Harrison, BS (2009) Oxygen generating scaffolds for enhancing engineered tissue survival. *Biomaterials* **30**(5), 757-62.

Pilliar, R (1987) Porous-surfaced metallic implants for orthopedic applications. *Journal of Biomedical Materials Research* **21**(A1 Suppl), 1-33.

Radisic, M, Deen, W, Langer, R and Vunjak-Novakovic, G (2005) Mathematical model of oxygen distribution in engineered cardiac tissue with parallel channel array perfused with culture medium containing oxygen carriers. *American Journal of Physiology* **288**(3), H1278-89.

Radisic, M, Malda, J, Epping, E, Geng, W, Langer, R and Vunjak-Novakovic, G (2006) Oxygen gradients correlate with cell density and cell viability in engineered cardiac tissue. *Biotechnology and Bioengineering* **93**(2), 332-43.

Radisic, M, Park, H, Chen, F, Salazar-Lazzaro, JE, Wang, Y, Dennis, R, Langer, R, Freed, LE and Vunjak-Novakovic, G (2006) Biomimetic approach to cardiac tissue engineering, oxygen carriers and channeled scaffolds. *Tissue Engineering* **12**(8), 2077-91.

Richardson, TP, Peters, MC, Ennett, AB and Mooney, DJ (2001) Polymeric system for dual growth factor delivery. *Nature Biotechnology* **19**(11), 1029-34.

Rouwkema, J, Rivron, NC and van Blitterswijk, CA (2008) Vascularization in tissue engineering. *Trends in Biotechnology* **26**(8), 434-41.

Rouwkema, J, Westerweel, PE, de Boer, J, Verhaar, MC and van Blitterswijk, CA (2009) The Use of Endothelial Progenitor Cells for Prevascularized Bone Tissue Engineering. *Tissue Engineering part A* **15**(8), 2015-27.

Sachlos, E, Reis, N, Ainsley, C, Derby, B and Czernuszka, JT (2003) Novel collagen scaffolds with predefined internal morphology made by solid freeform fabrication.

Biomaterials **24**(8), 1487-97.

SCHMID-SCHONBEIN, GW (1990) Microlymphatics and lymph flow. *Physiological Reviews* **70**(4), 987-1028.

SWARTZ, MA and FLEURY, ME (2007) Interstitial flow and its effects in soft tissues. *Annual Review of Biomedical Engineering* **9**, 229-56.

TREMBLAY, PL, HUDON, V, BERTHOD, F, GERMAIN, L and AUGER, FA (2005) Inosculation of tissue-engineered capillaries with the host's vasculature in a reconstructed skin transplanted on mice. *American Journal of Transplantation* **5**(5), 1002-10.

VUNJAK-NOVAKOVIC, G, OBRADOVIC, B, MARTIN, I, BURSAC, P, LANGER, R and FREED, L (1998) Dynamic cell seeding of polymer scaffolds for cartilage tissue engineering. *Biotechnology Progress* **14**(2), 193-202.

WENDT, D, STROEBEL, S, JAKOB, M, JOHN, GT and MARTIN, I (2006) Uniform tissues engineered by seeding and culturing cells in 3D scaffolds under perfusion at defined oxygen tensions. *Biorheology* **43**(3-4), 481-8.

WHITE, P (1946) Cultivation of animal tissue in vitro in nutrients of known concentrations. *Growth* **10**, 231-289.

WHITESIDES, GM and GRZYBOWSKI, B (2002) Self-assembly at all scales. *Science* **295**(5564), 2418-21.

WILSON, CE, DE BRUIJN, JD, VAN BLITTERSWIJK, CA, VERBOUT, AJ and DHERT, WJ (2004) Design and fabrication of standardized hydroxyapatite scaffolds with a defined macro-architecture by rapid prototyping for bone-tissue-engineering research. *Journal of Biomedical Materials Research* **68**(1), 123-32.

Biotechnology and Genetic Engineering Reviews - Vol. 26, 179-204 (2009)

Synthesizing Neurophysiology, Genetics, Behaviour and Learning to Produce Whole-Insect Programmable Sensors to Detect Volatile Chemicals

GLEN C. RAINS[1]*, DON KULASIRI[2], ZHONGKUN ZHOU[2], SANDHYA SAMARASINGHE[2], JEFFERY K. TOMBERLIN[3], DAWN M. OLSON[4]

[1]University of Georgia, Tifton Campus, Biological and Agricultural Engineering Dept., Tifton, GA 31793, USA; [2]Lincoln University, Centre for Advanced Computational Solutions (C-facs), ChristChurch, New Zealand; [3]Texas A&M University, Entomology Dept, College Station, TX, 77843, USA; [4]USDA-ARS, Crop Protection and Management Research Unit, Tifton, GA, 31794, USA

Abstract

Insects have extremely sensitive systems of olfaction. These systems have been explored as potential sensors for odourants associated with forensics, medicine, security, and agriculture application. Most sensors based on insect olfaction utilize associative learning to "program" the insects to exhibit some form of behavioural response to a target odourant. To move to the next stage of development with whole-insect programmable sensors, an examination of how odourants are captured, processed and used to create behaviour is necessary. This review article examines how the neurophysiological, molecular, genetic and behavioural system of olfaction works and how an understanding of these systems should lead the way to future developments in whole-insect programmable sensors.

* To whom correspondence may be addressed (grains@uga.edu)

Abbreviations: ORNs, olfactory receptor neurons; AL, antenna lobes; OB, olfactory bulb; PN, projection neuron; US, unconditioned stimulus; CS, conditioned stimulus; CR, conditioned response; PER, proboscis extension response; EAG, electroantennogram; GL, glomeruli; LIDAR, light detection and ranging; GPCR, G protein-coupled receptor; ODE, ordinary differential equation.

Introduction

Most animals have evolved highly sensitive olfactory systems which respond to odours in their environments. Insects in particular must navigate their environment using visual, olfactory, and tactile signals received and processed by their sensory systems to control behavioural responses to food, predators, mates or hosts necessary for individual and collective survival. Insects have very robust and extremely sensitive olfactory systems with extraordinarily high discriminating abilities to identify specific odours with only a small number of molecules (Angioy *et al.*, 2003). There are a number of known examples in which olfaction is known to play a significant role in the behaviour of insects. Some of the more interesting are: The malarial-vector mosquito *Anopheles gambiae* detects their human host by body odours and CO_2 (Takken, 1996); the males of the sphinx moth *Manduca sexta* use pheromones released by female to find mates (Hildebrand, 1995); the honeybee *Apis mellifera* learns different odours to form navigational memories and use them for foraging behaviour (Galizia and Menzel, 2000; Reinhard *et al.*, 2004); and parasitic wasps, such as *Microplitis croceipes*, detect odours produced by plants in response to host feeding (Lewis and Takasu, 1990).

Learning can be defined as a process by which a change in behaviour is exhibited as a result of experience (Thorpe, 1963). The extremely sensitive nature of the insect olfaction system is enhanced by the ability to learn to associate external stimuli with resources, such as food, hosts, and mates. Associative learning in insects has been studied using classical and operant conditioning. The sophisticated ability of honeybees to learn through the notable work of von Frisch (1915, 1950) and Menzel (1985) and more recently the studies of learning in parasitoid wasps and fly species (reviewed in Vet *et al.*, 1995) have been extensively studied. In addition, insects can be trained to associate simultaneously two different odours with food and host rewards (Lewis and Takasu, 1990) or successively to two different odours with the same reward (e.g., Takasu and Lewis, 1996). This training is particularly simple for some parasitic wasps; they can be trained within 5 minutes (Lewis and Takasu, 1990; Takasu and Lewis, 1996). There are three primary factors in associative learning that affect how well insects are conditioned to an single odour source. These are the number of trials the insect is provided an odour [conditioned stimulus(CS)] immediately followed by feeding on sugar water [reward or unconditioned stimulus (US)], the length of time the insect is given food while exposed to the odour and the time interval between each trial. Another significant factor in the level of conditioning received is the physiological state of the insect; Starved parasitic wasps have been conditioned using food as the unconditioned stimulus, while satiated wasps were shown to be more easily conditioned using host frass as the US (Lewis and Takasu, 1990). This and other studies on associative learning in insects have led to the generally accepted conclusion that associative learning is a ubiquitous trait in insects (Dukas, 2008).

There have been a few studies, including those by the authors, which have demonstrated the ability to condition wasps, bees, and moths to odours of human interest and to detect these odours through measurable behavioural responses. The conditioned response (CR), such as the proboscis extension response (PER) and area-restricted searching are used as active feedback indicators of the presence of the odours of interest. Examples include conditioning parasitic wasps, moths and honey bees to chemicals associated with explosives, food toxins and plant odours. In these studies, insects are contained and observed using electronic interfaces that measure behavioural responses, or released and allowed to search and find an odour source.

There are many characteristics of insect olfaction that makes them useful as chemical detectors, some of which are: 1) the ability to detect low levels of chemical odours (Rains *et al.*, 2004), 2) the ability to detect and respond to odours in a background of non-target odours (Rains *et al.*, 2006), 3) a short generation time which allows many life cycles to be produced in a short period of time, 4) the ease of rearing large numbers, 5) the great diversity of insects which allows us to draw upon different species for use in specific habitats or environments (e.g. flying, ground-dwelling or aquatic, nocturnal or diurnal, piercing-sucking or chewing mouthparts, large or small-sized), and 6) the ability to be conditioned to odours in a matter of minutes.

During the last three decades discoveries in anatomy, neurophysiology and molecular biology that underlie insect olfactory systems have allowed us to uncover, for example, the functionalities of the various components of the olfaction system and some of the mechanisms underlying olfactory memory formation. These discoveries have, however, led to new questions such as: What, where, and how is odourant information processed and stored? How reliable is the information stored and for how long a period can the information be stored? How are memories synthesized with acquired information, either through learning or experience? How does odour discrimination affect behavioural control in the brain? Where do cellular memory traces occur within the olfactory nervous system in response to learning? The answer to these questions and others are difficult challenges. From the time a molecule is captured by odourant binding proteins in the antennal sensillum, to the body movements elicited by motor neurons as a behavioural response, we seek a better understanding of how sensory signals cause behaviour and how behaviour controls sensory processing. Through a clearer understanding of how the behavioural, physiological, genetic and molecular levels interact, we believe that whole-insect programmable sensors can be developed for chemical detection, and serve as models for development of better electronic chemical sensing systems as well.

Current whole-insect programmable sensors

Whole-insect programmable sensors were initially investigated under military funding to detect landmines, explosives and toxins. However, additional application areas are numerous and include, but are not limited to such things as detection of leaks, food toxins, plant and animal disease, drugs, cadavers, gravesites, accelerants used for arson, termite infestations, and minerals.

Insect chemical detectors have been developed and reported using honey bees, hawk moths and parasitic wasps. Honey bees are easily reared and studies of their learning and sensing abilities are extensive, providing much background information. The hawk moth is of interest in agriculture as a pest in the larval stage and its large size has made it a useful model for examining the olfactory system. Parasitic wasps have been studied as beneficial insects in agriculture and their tri-trophic interaction with plants and hosts has revealed an intricate foraging strategy for finding food and hosts.

Honey bees learning ability, and memory of learned stimuli have been examined to determine their ability to forage and navigate landscapes using visual and olfactory cues. These studies have proved useful in determining how they may be used as chemical sensors. Recent studies have examined the discriminatory abilities of honey bees to changing concentrations of odourants the bees were trained to detect along with chemicals of similar molecular structures (Wright and Smith, 2004; Wright *et al.*, 2005). Interestingly, honey bees classically conditioned to a low concentration of a target odourant are able to distinguish that odour from molecularly similar odours; but, recognize these odours as the same when conditioning at higher concentrations. It was concluded that odour intensity may be a salient feature of the odourant at low concentration, but not at higher concentrations. Honey bees also appeared to generalize the identity of an odourant mixture using one odour within that mixture, as long as it remained constant during conditioning. Test trials were conducted with all odours in the mixture but one varied during conditioning. Subsequent tests with individual odours found the strongest PER to the odour that was held constant. A related study also found that at lower concentrations, the time the bee is allowed to sample the odour while feeding on sugar water increases their ability to distinguish that odourant (Wright *et al.*, 2009). Other important studies have examined the effects of latency, overshadowing and blocking of conditioned odours within odour mixtures (Hosler and Smith, 2000, Linster and Smith, 1997). These and other studies provide strong background for how to implement conditioning and testing mechanisms in whole-insect programmable sensors.

In most studies of the response to conditioning, honey bees are held in a harness and the PER measured as odours are passed over the antennae. This method is also utilized in a commercial device developed by Inscentinel Ltd. Other methods have examined the use of free-flying bees and light detection and ranging (LIDAR) to track bee location (Hoffman *et al.*, 2007).

M. sexta, has also been studied extensively and has provided some very useful insights into the physiology and biological processes that govern insect olfaction (reviewed later). A device using 10 hawk moths was developed to electrically moni-tor the feeding muscles to determine when cyclohexanone was detected (King *et al.*, 2004). Although this device was too large as built for easy transportation, it would be trivial to miniaturize to proportions that make it portable. Also an electroantennogram (EAG) quadroprobe device utilizing the antennae and whole organism of two moth species, *Helicoverpa zea* and *Trichoplusia ni*, has been developed and tested to detect odour plumes (Myrick *et al.*, 2009, Park *et al.*, 2002). Antennae are either excised or used directly on the insect and EAG response recorded and analyzed, leaving the interpretation of the raw signals to analysis software. Using whole insects, the probe response was conserved for up to 24 hours and was able to classify individual odour-ant plumes in less than 1 second.

M. croceipes has been studied for its ability to forage effectively for hosts and nectar sources. Several studies have examined the wasp's ability to learn, retain, discriminate, and respond to odourants, both within and outside their natural habitat (Takasu and Lewis, 1996). *M. croceipes* has also been tested against the electronic nose, Cyranose 320 (Smiths Detection, Inc.) and was found to be almost 100 times more sensitive to the chemicals, myrcene and 3-octanone (Rains *et al.*, 2004). In the case of this parasitic wasp, a positive response to the target volatile chemical is measured by a movement of the wasp body and antennae around the location of the odour source when the unconditioned stimulus is food. Parasitic wasps were also demonstrated to respond with context-dependent behavioural movements depending on the resource the odourant was linked to (Olson *et al.,* 2003). *M. croceipes* were conditioned to an odourant while feeding on sugar water. After a minimum of 15 minutes, the same wasps were conditioned to a different odourant while stinging their host, *H. zea*. Consequently, when each wasp was presented with one of the conditioning odourants, the wasps responded with either foraging behaviour (food) or a stinging behaviour (host). Further studies have revealed that *M. croceipes* can also discriminate molecularly similar odourants (Meiners *et al.*, 2002). Further studies are needed to understand the affect of odour mixtures and when an odour becomes unrecognizable to the olfactory sensory system.

To utilize the parasitic wasp as a sensor, a device called the Wasp Hound, was developed and has been used to detect the odours associated with Aflatoxin in foods (Rains *et al.*, 2006) and animal carcasses (Tomberlin *et al.*, 2008). Wasps are placed in a small cartridge and sample air passed through. A web camera records images of wasp behaviour and the Wasp Hound connected to a software program called Visual Cortex that analyzes the wasp response and indicates the response as a real-time graph (Utley *et al.*, 2007).

Future developments

The above described examples of whole-insect programmable sensors, with the exception of the EAG quadroprobe, use crudely adapted methods of direct behavioural response to determine when the insect has detected a target odourant. For example, the Wasp Hound measures the crowding of wasps response to the presence of the target odourant around the inlet where the sample air is pumped into the device (Rains *et al.*, 2006). However, the entire repertoire of behaviours also encompasses the initial recognition of the odour, its quality, and possibly its quantity (concentration) as well. As discussed with honey bees, odourant concentration is a salient property, at least at low concentrations. However, the PER response may also be dependent on the perception of whether the odour source is near or far. A PER response by honey bees would necessarily be perceived as a direct contact with the odour source (close proximity to odour source). Consequently, other behavioural responses may be more evident when the odour is perceived to be a distance away. The basic hypothesis is that odour characteristics (quality and quantity), along with what has already been learned and stored in memory control the signals to motor neurons that control movement and behaviour. If behaviours could be observed and analyzed to determine what sensory characteristics caused that behaviour, insect sensors could be used to measure concentrations and track odours to the source (Rains *et al.*, 2008).

To adapt systems that extract this information from a whole-insect programmable sensor, more targeted research is necessary that examines the processes that direct learned behaviour at the molecular and genetic level. Neuronal signals to the mushroom bodies and lateral horn (*Figure 1*) that are derived from learning new experiences, combined with memory of past experiences in the brain centre, results in signals to motor neurons that result in complex behaviours. Determining the how and why a specific behaviour occurs at the molecular level should lead to methods to better utilize the insect olfaction system as a chemical detector. To that end, we will now review the olfactory system of insects and computational models that examine the process of olfaction. As part of that review, we will introduce potential opportunities and obstacles for chemical detection system development.

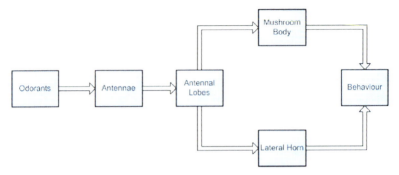

Figure 1. Organization of the insect olfactory system. The first level of odour processing in insects occurs at the antennae. Olfactory receptor neurons (ORNs) inside the insect antennae are compartmentalized into sensory hairs called sensilla (each sensillum contains the dendrites of up to four ORNs). The odourants in the air enter the sensillum through pores in the cuticle of the antennae. Then the ORNs in the sensillum generate odour specific electrical signals (called spikes) in response to the odour. The axons of the ORNs join to the antennal nerve, and project to the antennal lobe. The future processing of the olfactory information occurs in the antennal lobe before it is sent to protocerebrum. The antennal lobe contains globular shaped structures called glomeruli. They are the projection fields of the olfactory receptor neurons onto the second order neurons, which are called projection neurons (PNs). The glomeruli also contain the processes of local interneurons that branch to multiple glomeruli, and transfer the olfactory information between glomeruli. Individual ORNs send axons to only one or a few glomeruli, and individual PNs typically innervate only one single glomerulus. The axons of PNs project to the mushroom body (MB) and lateral horn of the brain. The mushroom bodies are located in the protocerebrum and are the centres of higher order processing in insects. They are a paired structure consisting of thousands of small intrinsic nerve cells (Kenyon cells). These Kenyon cells have their projections within mushroom body structure. They receive sensory information via the dendritic calyx, and send axonal projections to the anterior brain where they bifurcate to form the medial and vertical lobes. The mushroom bodies are also involved in olfactory memory formation. The lateral horn is the PN axon terminal field and is involved in odour recognition.

Features of insect olfactory system

In general, insects use their antennae for olfaction and olfactory receptor neurons (ORNs) in the antennae generate odour-specific electrical signals called spikes (Hallem and Carlson, 2004). These spikes can be recorded using an EAG and have been used to understand how odourants initiate signals in the olfactory system. A signal pathway model given in *Figure 1* has four distinct components: antennae, antennal lobes, mushroom bodies and lateral horn. The antennae have the sensory neurons in the sensilla hairs and their axons (*Figure 2a*) terminating in the antennal lobes (AL) where they synapse with the neurons in the spherical units (*Figure 2b*) called glomeruli (GL). These AL's have two kinds of neurons, excitatory projection neurons (PN) and inhibitory local neurons.

Figure 2a. A parasitic wasp, Microplitis croceipis. Olfactory receptor neurons (ORNs) are compartmentalized into sensory hairs called sensilla on the surface of the antennae.

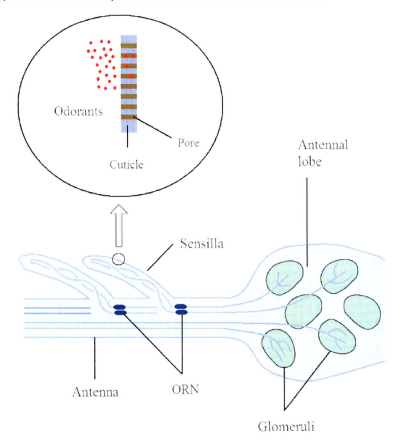

Figure 2b. Insect olfactory sensilla. The ORNs are compartmentalised into sensory hairs to detect odourants which are called sensilla. Each sensillum contains the dendrites of up to four ORNs. The sensillum has three major morphological types: basiconic, coeloconic and trichoid. Odourants in the air enter the sensillum through pores in the cuticle. ORNs project their axons to glomeruli in the antennal lobe.

The projection neurons send their axon terminals to the mushroom body (MB) and lateral horn (LH) both of which are part of the protocerebrum of the insects (*Figures 3 and 4*). The inhibitory local neurons have no axons. Recordings from projection neurons show in some insects strong specialization and discrimination for the odours presented (e.g., the projection neurons of the macroglomeruli, a specialized complex of glomeruli responsible for pheromone detection). How information passed to the mushroom bodies is stored and processed to in-turn send signals via motor neurons to muscles that control complex behaviour is not exactly known though some preliminary studies shed some light on the subject. How this occurs will unlock a key component in understanding the basis for behavioural responses to resource needs, such as food and mates. This in turn is important to understanding and developing sensors that can, from behavioural cues, know what sensory inputs are causing them. Model organisms such as the fruit fly, *Drosophila melanogaster*, help develop general conceptual models of olfactory systems and their operational principles; but each type of insect has its own systemic specificities. For example, *Drosophila* antennae have approximately 1200 ORNs, and the maxillary palps 120 (Shanbhag *et al.*, 1999, 2000; Stocker, 1996).

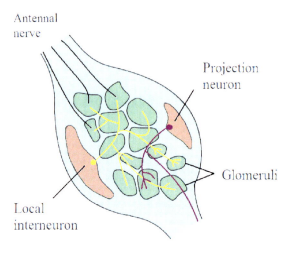

Figure 3. The insect antennal lobe. ORNs send axons from the insect antenna to the antennal lobe via the antennal nerve. ORNs synapse onto second order neurons called projection neurons (PNs) in the AL. The AL can be subdivided into spherical units called glomeruli. Individual ORNs send axons to only one or a few glomeruli, and individual PNs typically innervate only a single glomerulus. The glomeruli also contain the processes of local interneurons that branch in multiple glomeruli, and transfer information between glomeruli. The projection neurons project to higher brain centres such as the mushroom body and lateral horn of the protocerebrum. The local neurons, which are primarily inhibitory, have their neurites restricted to the antennal lobe.

The sensilla hairs in the antenna are categorized into basiconic, coeloconic and trichoid morphological types, the dendrites of up to four ORNs occupy each sensilla. The AL can be subdivided into around 43 GLs, and axons from each ORN can connect to only one or a few glomeruli, and an individual PN typically innervate only a

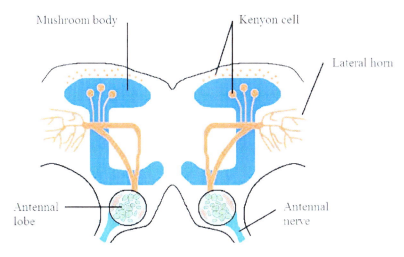

Figure 4. Horizontal section through the brain of the insect showing the main elements of the olfactory pathway. The antennal lobes and their glomeruli are in green, the mushroom bodies are in blue, and the Lateral horns are in orange. Odours are detected by olfactory receptor neurons (ORNs) in the antennae. The ORNs project axons along the antennal nerve to the antennal lobe glomeruli, where they are sorted according to chemo-sensitivity. From there the information is relayed by projection neurons (PNs) to the mushroom body and to the lateral horn. Some PNs bypass the mushroom body calyx and project only to the lateral horn.

single glomerulus (Jefferis *et al.*, 2001; Marin *et al.*, 2002; Wong *et al.*, 2002). The glomeruli also contain the processes of local interneurons that branch to multiple glomeruli (Stocker, 1996; Stocker *et al.*, 1990), providing means for information transfer between glomeruli. The axons of PNs project to the mushroom body (MB) and lateral horn of the brain.

Larvae of *Drosophila* also exhibit a robust olfactory response (Ayyub *et al.*, 1990; Cobb *et al.*, 1992; Monte *et al.*, 1989), which is mediated through the dorsal organ (Heimbeck *et al.*, 1999; Oppliger *et al.*, 2000). Each of the paired dorsal organs contains 21 neurons that project to the antennal lobe of the larval brain (Python and Stocker, 2002).

The ORNs of the antenna and maxillary palp generate action potentials in response to odour stimulation. The odour responses of many of these ORNs have been characterized through extracellular single-unit recordings from individual olfactory sensilla (Clyne *et al.*, 1997; de Bruyne *et al.*, 1999; de Bruyne *et al.*, 2001; Stensmyr *et al.*, 2003). These recordings have revealed that different odourants elicit responses from different subsets of ORNs, and also that ORNs exhibit a remarkable diversity of response properties: responses can be either excitatory or inhibitory and can vary in both intensity and temporal dynamics, depending on the odorant and the ORN (de Bruyne *et al.*, 1999; de Bruyne *et al.*, 2001). Similar ORN response properties have been described in other insects (Heinbockel and Kaissling, 1996; Kaissling *et al.*, 1989; Nikonov and Leal, 2002; Shields and Hildebrand, 2001). Extensive recordings from the antennae and maxillary palps have revealed that ORNs can be categorized into a limited number of functional classes based on their responses to a defined set of chemical odourants. The maxillary palp contains six functional classes of ORNs, which are found in stereotyped pairs within three classes of sensilla (de Bruyne *et al.*, 1999). The antennal basiconic ORNs falls into 18 functional classes that are also found in stereotyped combinations within eight classes of sensilla (de Bruyne *et al.*, 2001; Elmore *et al.*, 2003); the coeloconic and trichoid sensilla on the antenna also contain multiple classes of ORNs (Clyne *et al.*, 1997) but a thorough characterization is not yet available.

The projection neurons project to higher brain centres such as the mushroom body and lateral horn of the protocerebrum. The local neurons, which are primarily inhibitory, have their neurites restricted to the antennal lobe. In *Drosophila*, each olfactory receptor neuron generally expresses a single olfactory receptor gene, and the neurons expressing a given gene all transmit information to one or two spatially invariant glomeruli in the antennal lobe. Moreover, each projection neuron generally receives information from a single glomerulus. The interaction between the olfactory receptor neurons, local neurons and projection neurons reformats the information input from the receptor neurons into a spatio-temporal code before it is sent to higher brain centres.

The mushroom bodies or corpora pedunculata are a pair of structures in the brain of insects and other arthropods. They are usually described as neuropils, i.e. as dense networks of neurons and glia. They get their name from their roughly hemispherical calyx, a protuberance that is joined to the rest of the brain by a central nerve tract or peduncle. Mushroom bodies are known to be involved in learning and memory, particularly for smell. They are largest in the Hymenoptera (bees, wasps, etc.) which are known to have particularly elaborate olfactory control over behaviour. The mushroom bodies have been compared to the cerebral cortex of mammals and are currently the subject of intense research. Because they are small compared to the brain structures of vertebrates, and yet many arthropods are capable of quite complex learning, it is hoped that investigations of the mushroom bodies will allow a clearer view of the neurophysiology of animal perception and cognition. The most recent research is also beginning to use new tools to reveal the genetic control of processes within the mushroom bodies (e.g. Olsen and Wilson, 2008). Also, through classical conditioning, the mushroom bodies incorporate the coded odourant signals from the antennal lobes and make association between that code and the resource with which it was associated. Future studies that link the behavioural response with the neural signals and genetic expression in the mushroom bodies before and after associative learning could be used to better understand and condition insects and potentially other animals to odours. It would also lead to a basic understanding of how animals navigate through their environment.

Most of our current knowledge of the mushroom bodies comes from studies of a few species of insects, especially the cockroach *Periplaneta americana*, the honey bee *Apis mellifera*, the locust, *Schistocerca americana,* and the fruit fly, *Drosophila melanogaster.* Studies of fruit fly mushroom bodies have been particularly important for understanding the genetic basis of their functioning, since the genetics of this species are known in exceptional detail. In the insect brain, the peduncles of the mushroom bodies extend through the midbrain. They are mainly composed of the long, densely packed nerve fibres of the Kenyon cells, the intrinsic neurons of the mushroom bodies. These cells have been found in the mushroom bodies of all species that have been investigated, though their number varies; for example fruit flies have around 2,500 and cockroaches have about 200,000.

Insects and other invertebrates do not have their olfaction system directly connected to respiration. As such, odours are not "sniffed", but are brought into contact with ORN by antennal sensilla and behavioural movements. Insects in particular have a wide array of morphological features that make up the antennae. It is still unclear as to what evolutionary adaptations these features represent (Hannson *et al.*, 1991). However, it would be prudent to recognize the structural design and use of these organs as potential designs for improved detection of odourants using man-made sensing materials. Currently, electronic nose technology has focused on mimicking the combinatorial strategy that is well accepted as the mammalian and insect olfaction method of identifying odours from patterns of responses to different sensors (normally 16 or 32 sensors). However, little effort has examined the method in which odourant molecules are captured or presented to the sensors.

Biology of learning and memory

One of the important aspects of olfactory systems must be the ability to learn different odours and form memories of specific odours to mediate behavioural responses. Understanding the processes of learning and memory of odours could help improve methods of conditioning and measuring behavioural responses of whole-insect programmable sensors. To understand the physiological changes that occur in cellular and neuronal pathways, electrophysiological recordings have been used in mammalian model systems (Liu and Davis, 2006). Such studies are difficult to perform in insects because of their small size; therefore, functional optical imaging techniques have been employed to study physiological differences within the insect olfactory system. Synthetic chemical reporters and fluorescent proteins are used to report the activity of neurons in the system when an odour is introduced to an insect before or after olfactory learning has occurred (Faber *et al.*, 1999). In that work Faber and colleagues studied the honey bee antennal lobe using the calcium sensitive dye, calcium-green-2AM by observing the activity in response to odour stimuli presented before and after associative conditioning with sucrose as a rewarding stimulus. Specific areas of the antennal lobe exhibited calcium signals in response to odour, which increased for more than 30 min after conditioning. Learning changed the activation pattern in the antennal lobe. They discovered that the responses in specific sets of glomeruli were odour –specific, but conserved for each odour across honey bees. Protein-based reporters of neuronal activity were employed in recent studies, which have confirmed that memory traces form in the antennal lobes (Liu and Davis, 2006). However, these memory traces existed only for about five minutes after training corroborating the data from locusts (Bazhenov *et al.*, 2005) and moths (Daly *et al.*, 2004) which supports the hypothesis that the memory formation in antennal lobes are short-term.

The above mentioned studies and other recent studies show that olfactory memory is distributed among diverse types of neurons within the olfactory system, which are classified into first-, second- and third- order neurons (Davis, 2004). Perceptual olfactory learning happens in first-odour neurons which may be mediated by changes in the odorant receptive fields of second and/or third order neurons, and in the coherency of activity among ensembles of second odour neurons (Davis, 2004). The coherent population activity of these neurons increases during operant olfactory conditioning. The odour responsiveness and synaptic activity of second and perhaps third order neurons increase during operant and classical conditioning (Davis, 2004).

Associative and non-associative processes influence odour-driven responses in the insect antennal lobe (AL). Daly *et al.* (2004) studied the changes in AL network activity during learning employing an *in vivo* protocol in *M. sexta* for continuous monitoring of neural ensembles and feeding behaviour over the course of olfactory conditioning. Daly *et al.* (2004) showed that the neural units in the AL responded to Pavlovian conditioning when odour followed food and the response persisted after conditioning. A net loss of neural units responding to odour occurred when odour did not predict food. The experiments showed that odour-specific neural recruitment was positively correlated with changes in the insect's behavioural response to odour. In addition, odour representations in the AL were dynamic and related to olfactory memory formation; learning continually restructures neural network responses spatially in the AL in an odour-specific manner (Daly *et al.*, 2004). The organisation of glomeruli within the AL in animals as diverse as insects (Galizia and Menzel, 2001; Laurent, 1999) and vertebrates (Mori *et al.*, 1999; Xu *et al.*, 2000) is such that a small number of glomeruli are combinatorial conditioned by a large number of odour stimuli (Vosshall *et al.*, 2000), and a distributed system of glomerular PNs is dynamically activated by a compound of multiple odours, and at the same time a specific PN actively represents multiple odour compounds (Christensen *et al.*, 2000; Christensen and White, 2000; Daly *et al.*, 2004; Kay and Laurent, 1999; Laurent, 1999; Laurent *et al.*, 2001; Lei *et al.*, 2002; Mori *et al.*, 1999; Sachse and Galizia, 2002). In vertebrates, the representation of a given odour in the OB is dependent on the experience with an odour (Bhalla and Bower, 1997; Fletcher and Wilson, 2003; Goldberg and Moulton, 1987; Kay and Laurent, 1999; Kendrick *et al.*, 1992; Sullivan *et al.*, 1989; Wilson and Sullivan, 1994; Wilson *et al.*, 1987), and learning-dependent changes in the insect AL have been observed using imaging techniques suggesting that experience-dependent structural changes might be happening in the insect AL (Faber *et al.*, 1999; Yu *et al.*, 2004) as well.

The studies on the experience-dependent plasticity in the AL of *M. sexta*, a model organism of which the anatomy and physiology of the AL are well known (Christensen *et al.*, 2000; Christensen and White, 2000; Daly *et al.*, 2004; Hildebrand and Shepherd, 1997; Kent *et al.*, 1987; Lei *et al.*, 2002; Rospars and Hildebrand, 2000; Tolbert and Hildebrand, 1981), reveal the relationships of behavioural measures of olfactory learning with neuro-physiological measures of odour-evoked ensemble responses in the AL before, during, and after olfactory conditioning (Daly *et al.*, 2001; Daly *et al.*, 2004; Daly and Smith, 2000). As in other insects, an olfactory stimulus followed by food resulted in recruitment of neural units in the AL, and lack of food reinforcement withered the responsive neural units (behaviourally referred to as extinction). Repeated reinforcement always increased the recruitment of neural units. All these findings conclusively show that the insect ALs consist of the synaptic neural circuits which are "plastic" in the sense that structural changes in neural networks in the AL occur dynamically during learning. The output pathways from the AL (PNs) relay dynamic signals which are odour-dependent patterns of inhibition followed by excitation (Daly *et al.*, 2004). The odour-driven responses within the AL are mediated by correlated input from a different sensory modality, as in the case of a gustatory input, through feedback loops; for example, monoaminergic neurons in both *A. mellifera* (Hammer, 1993; Hammer and Menzel, 1995, 1998) and *M. sexta* (Kent *et al.*, 1987) exist in the AL providing signals from other brain areas. In honeybees, for example, a modulatory neuron VUMmx1 releases octopamine (Hammer and Menzel, 1995, 1998) when activated by sucrose. The AL plasticity may have many different roles including magnification of olfactory signals (Linster and Smith, 1997) and formation of the short-term olfactory memory (Faber *et al.*, 1999; Yu *et al.*, 2004). This highlights the need to understand at what point behaviour is altered temporally as sensory input changes. For example, in the Wasp Hound, when parasitic wasps are repeatedly exposed to an odour without food reinforcement, the association between the odour and food is extinguished. Extinction can be reversed with positive reinforcement at appropriate intervals.

Molecular biology of olfactory system

The expression of odourant receptor molecules in ORNs are finely tuned to a specific subset of odour molecules in the environment (Buck and Axel, 1991; Malnic *et al.*, 1999). The olfactory map, a spatial map of the distinct locations for specific odour signals within the AL, is organized according to the type of odourant receptor a particular ORN expresses.

A family of candidate genes, the Or genes, was discovered in *Drosophila* in 1999 (Clyne *et al.*, 1999; Gao and Chess, 1999; Vosshall *et al.*, 1999). These genes control the expression of G protein-coupled receptors (GPCRs) as in the case of mammals (Buck and Axel, 1991) and in the nematode *Caenorhabditis elegans* (Troemel *et al.*, 1995). GPCRs have seven transmembrane domains and are expressed by a diverse range of gene sequences containing 62 members dispersed throughout the genome in small clusters (Clyne *et al.*, 1999; Gao and Chess, 1999; Robertson *et al.*, 2003; Vosshall *et al.*, 1999) which includes two genes that are alternatively spliced. Humans, mice and mosquitoes have approximately 350, 1,000, and 80 functional Or genes (Godfrey *et al.*, 2004; Hill *et al.*, 2002; Malnic *et al.*, 2004; Zhang and Firestein, 2002) respectively, but it is not clear if the number of Or genes are related to odour specificity in olfactory systems. The highly diverse gustatory receptor (Gr) gene family of *Drosophila* consists of 60 genes that encode 68 proteins through alternative splicing (Clyne *et al.*, 2000; Robertson *et al.*, 2003); many of them are up-or-down regulated in gustatory organs (Clyne *et al.*, 2000; Dunipace *et al.*, 2001; Scott *et al.*, 2001); some work as taste receptors (Chyb *et al.* 2003; Dahanukar and Foster, 2001; Ueno *et al.*, 2001); and some are pheromone receptors (Bray and Amrein, 2003). Gr genes are also expressed in the antennae (Scott *et al.*, 2001) and tarsi (Ishimoto and Tanimura, 2004) of *Drosophila* and most probably a majority of insects.

Most of the *Drosophila* antennal odourant receptors, each of which has a unique odour spectrum despite being activated by common ligands, have now been mapped to the ORNs from which they are derived (Hallem *et al.*, 2004), and the complete odour spectrum of an ORN can be understood in terms of the odour spectra of the receptors mapped to that particular ORN. Furthermore, the spontaneous firing rate, response dynamics and signaling mode (excitation or inhibition) of the ORN (Hallem *et al.*, 2004) are also determined by the receptors providing a versatile functionality for each ORN.

As mentioned earlier, approximately 1320 ORNs of the antenna and maxillary palp in *Drosophila* connect with approximately 43 glomeruli in the AL. As in mammals, the axons of ORNs expressing the same odorant receptors connect to only one glomerulus or to a few glomeruli, providing us with a spatial map of ORN projections. Therefore, the pathways of the odour receptors bind the physiological units together functionally in the insect olfactory system although these units are at different locations. In addition, different sets of PNs in the AL are activated by different odours and their responses depend on the odour spectra, signaling mode and response dynamics (Wilson *et al.*, 2004) as in the case of ORNs. The activation of a PN is dependent on its pre-synaptic ORNs (Wang *et al.*, 2003). Wilson *et al.* (2004) suggested that PN output was dependent not only by ORN inputs but also on lateral inputs within the AL (Wilson *et al.*, 2004). Several genes involved in olfactory learning in *Drosophila* encode some of the components of the cAMP signaling pathway: the adenylyl cyclase encoded by the rutabaga (rut) gene; the cAMP phosphodiesterase encoded by the dunce (dnc) gene; the cAMP dependent protein kinase (PKA); the predicted product of the amnesiac (amn) gene; and the transcription factor cAMP-response element binding protein (CREB) (Davis, 2005; McGuire *et al.*, 2005).

Computer modeling of olfactory systems

Computational, mathematical and statistical modeling of various functional aspects of insect olfactory systems has increasingly been reported during the last decade. Getz (1991) developed a preliminary neural network for processing odour stimuli which can learn and identify the quality of an input vector or extract information from a sequence of correlated input vectors. Input vectors can be a sample of time varying olfactory stimuli. A discrete time content-addressable memory (CAM) module was developed to satisfy the Hopfield equations (Getz, 1991) with the addition of a unit time delay feedback, which improved the convergence properties of the network and was used to control a switch which activated the learning or template formation process when the input was "unknown". The network based on CAM had dynamics embedded within a sniff cycle which included a larger time delay that was also controlled the template formation switch. In addition, this time delay modified the input into the CAM module so that the more dominant of two mingling odours or an odour increasing against a background of odours was identified. The network was evaluated using Monte Carlo simulations and it was concluded that a Hopfield type CAM may not be suitable for simulating an olfactory system; however, this pioneering work showed that artificial neural networks, which have multitudes of learning strategies embedded in the network, could be used to model the olfactory systems and their functions.

One of the earliest attempts to model olfactory functions was by Rossokhin and Tsitolovsky (1997) who focused on information processing by neurons. Biochemical reactions that were hypothesised to be controlling the properties of the excitable membranes in the nerve cells were modeled by a set of first order differential equations for chemical reaction kinetics by taking the effect of regulation of the properties of sodium channels into account. The neuron's electrical activity parameters occurring during learning associated with its excitability were simulated, and they showed that the neuronal model exhibited different excitability after the learning procedure relative to the different input signals corresponding to the experimental data.

Pearce *et al.* (2001) modeled the efficiency of odour stimulus encoding within the early stages of an artificial olfactory system using a spiking neuronal model driven by fluorescent microbead chemosensors. The specific objective of the modeling study was to investigate how a rate-coding scheme compared to the direct transmission of graded potentials in terms of the accuracy of the estimate that an ideal observer may make about the stimulus. Their results showed how the charging time-constants of the first stages of neuronal information processing within the OB directly affected the reconstruction of the stimulus.

In an interesting study, Chang *et al.* (1998) developed a general connectionist model for an olfactory system by modeling the dynamical behaviour of each node (neural ensemble) by a second-order ordinary differential equation (ODE) followed by an asymmetric sigmoidal function, with which they modeled the aggregate activity of neurons in terms of system parameters and stimuli from an outside environment. The general connectionist model was used to simulate a mammalian olfactory system having modifiable synaptic connections and spatio-temporal interactions among neural ensembles. They developed a parameter optimization algorithm as an integral component of the model.

Ikeno *et al.* (1999) developed a model of the mushroom body of insects consisting of Kenyon cells based on the ionic currents in the isolated Kenyon cell somata in honey bees as measured by the whole-cell recording method. A rapidly activating and inactivating A-type potassium current, a calcium-activated potassium current and a delayed rectifier-type potassium current, and several types of inward currents were modeled by using Hodgkin/Huxley-type equations. They reconstructed the voltage responses of isolated Kenyon cell based on these mathematical models.

Christensen *et al.* (2001) developed a detailed multi-compartmental model of single local inter-neurons in the AL of the sphinx moth, *Manduca sexta*, using morphometric data from confocal-microscopic images, to study how the complex geometry of local neurons may affect signaling in the AL. Simulations clearly revealed a directionality in the neurons that impeded the propagation of injected currents from the sub-micron-diameter glomerular dendrites toward the much larger-diameter integrating segment in the coarse neuropils. They showed that the background activity typically recorded from LNs *in vivo* could influence synaptic integration and spike transformation in the local neurons. The modeling study supported the experimental findings suggesting that spiking inhibitory local neurons in the AL can operate as multifunctional units under different odour spectra. At low odour intensities, the neurons process mostly intra-glomerular signals; at high odour intensities the same neurons fired overshooting action potentials, resulting in the spread of inhibition globally across the AL. They concluded that the modulation of the passive and active properties of neurons were a deciding factor in defining the multi-glomerular odour representations in the insect brain. Getz *et al.* (1999) also developed a model for the olfactory coding within the in-sect AL using neural networks to investigate how synaptic strengths, feedback circuits and the steepness of neural activation functions influenced the formation of olfactory code in neurons within the AL. They reported that these factors were important in discriminating the dispersed odour spectra. Rospars *et al.* (2001) measured the spike frequency of olfactory receptor cells in response to different odourants experimentally and developed the concentration-response curves which were accounted for by a model of the receptor cell they developed. This model, consisting of three main equations, suggested that most often the variability in sensitivity was due to the variability of odourant receptor binding characteristics.

Gu *et al.* (2007) developed a cross-scale dynamical neural network model to simulate the presentation, amplification and discrimination of host plant odours and sex pheromones to understand the dependence of dynamics of the olfactory maps in the AL on glomerular morphology. They used stochastic dynamical approaches to amplify weak signals and to discriminate odour signals. They used the neural network model to investigate arborizing patterns of the projection neurons (PNs) and timing patterns of the neuronal spiking activity.

There are a significant number of other studies which are rule- or neural network based utilizing similat methodologies of those already discussed (Av-Ron and Rospars, 1995; Chang and Freeman, 1996; Eisenberg *et al.*, 1989; Freeman *et al.*, 1988; French *et al.*, 2006; Getz and Lutz, 1999; Gu and Liljenström, 2007; Ikeno and Usui, 1999; Kaiser *et al.*, 2003; Kanzaki, 1996; Lei *et al.*, 2004; Ma and Krings, 2009; Patterson *et al.*, 2008; Quenet *et al.*, 2002; Rospars *et al.*, 2007; Snopok and Kruglenko, 2002; Webb, 2004); however, to understand insect olfactory systems in the regime of behaviours which are not measurable or observable, we need to develop phenomenological models based on the molecular biology of the olfactory systems so far discovered. Computer models that predict behaviours based on sensory input would need to simulate all the processes from sensory stimuli acquisition to bodily movements of the insect. Models based on actual neurophysiological and genetic processes would not only provide an avenue for understanding mechanisms of odourant conditioning, but also provide insight into how insects and other animals process olfactory signals. Such models could then be used to predict odourant properties based on behavioural observations (Rains *et al.*, 2008) and used as an enhancement to whole-organism programmable sensors.

Genetic breeding

Selective breeding is a traditional approach which allows researchers to sift out desired characteristic combinations in animals or plants without the expense associated with modern molecular techniques. However, depending on the organism of interest, such an approach can be extremely time consuming and expensive. Some organisms require specific environmental parameters while others produce relatively few offspring per generation.

An advantage of using selective breeding with invertebrates, such as insects, is that many species have a short generation cycle allowing for multiple life cycles throughout a year. And, costs associated with colony maintenance are less when compared with managing canines or other vertebrates. For example, in some instances, a cage and a little chicken feed is all that is needed to mass produce an insect (Tomberlin *et al.* 2002). Consequently, phenotypic traits can be isolated in resulting progeny rather quickly in comparison to canines or other vertebrate programs. Selective breeding for certain arthropods, such as the honeybee and silk moth, *Bombyx mori* (Lepidoptera: Bombycidae), proved fruitful resulting in greater efficiency in their management and the production of their associated products. However, repeated inbreeding has resulted in the frequency of recessive traits increasing and producing various debilities. For example, the silk moth has been maintained in captivity for producing silk for well over 5,000 years. However, selective breeding for higher quality silk has resulted in various levels of susceptibility to nucleopolyhedroviruses (Ribeiro *et al.*, 2009).

In contrast, other selective breeding has resulted in greater fitness. Breeding programs for the honeybee have been successful in selecting for greater disease resistance (Evans *et al.*, 2006). Additionally, the simplicity of backcrossing allows for fine-scale genetic mapping and the isolation of behavioral trait-specific genetic regions (Oldroyd and Thompson, 2006). Recent evidence has determined genetic differences in the ability of arthropods to discriminate between target and non-target volatiles (Ferguson *et al.*, 2001). In fact, selection for greater discrimination performance can be achieved through only one generation (Ferguson *et al.*, 2001). Using selective breeding with insects still runs the risk of less desired recessive traits being expressed, and it can be difficult to formulate the right combination and level of expression of physical traits desired in the progeny. In regards to the honeybee, the negative effects of inbreeding are severe due to the genetic load introduced by the sex locus (Oldroyd and Thompson, 2006). Thus, selective breeding is still a gamble with each cross between genetic lines which can be frustrating and time consuming.

Deciphering the genetic mechanisms at work in sensory systems could lead to techniques for manipulating and enhancing the use of arthropods as biological sensors. We provide below examples of manipulations of the olfactory (smell) system in invertebrates which have led to refined sensing in the host organism.

The current insects used as biological sensors are diurnal. Responses to stimuli during non-active hours (i.e. night) result in declined response to external stimuli. Therefore, their use is ultimately limited by day length or light intensity. However, recent efforts have determined that it is possible to develop mutant strains that operate under dark conditions. Cheng and Nash (2008) determined that *D. melanogaster* bearing a mutation in the *inaF* gene, which is responsible for normal trp function (Li *et al.*, 1999) exhibited normal detection ability of the anesthetic halothane under dark conditions; however, once placed in ambient light the sensitivity decreased. Furthermore, Cheng and Nash (2008) determined that mutations in one gene resulted in greater sensitivity under ambient light. Therefore, increased sensitivity through genetic manipulations is a possible avenue in enhancing biosensor sensitivity and utility.

Insects are able to detect and respond to a variety of volatile compounds. Consequently, considerable time and resources are invested in conditioning insects to specific odourants of interest. Another hurdle with whole-insect programmable sensors is the reduction in response to target odourants by the conditioned insect over time (previously discussed extinction). Presently, conditioned arthropods, such as *M. croceipes*, are limited in the time they can be used as their response to target chemicals is reduced after repeated exposure. Behavioural responses can be re-established after re-conditioning (Tertuliano *et al.*, 2004) and can increase the length of time the wasps can be used as a sensor. However, it is possible that these insects could be genetically modified to enhance the behavioural response to select compounds. Therefore, conditioning could result in better responses for longer periods of time, or potentially for a permanent period of time of extinction could also be turned off at the genetic level. Morgan *et al.* (1988) determined that a variety of volatiles detected by a wild strain and three mutant strains of *C. elegans* varied. In some instances the mutant strains were 30% more sensitive to select volatiles than the wild strain. Therefore, it is conceivable that insects used for biosensors could be modified to be more sensitive and specific in regards to what they detect and respond. In addition, the strength of response found in either EAG signals, cardiac response, or changes in observable behaviour, could also be manipulated through genetic engineering to strengthen response to select odourants without associative learning (hard-wired).

References

ANGIOY, A.M., DESOGUS, A., BARBAROSSA, I.T., ANDERSON, P., AND HANSSON, B.S. (2003). Extreme sensitivity in an olfactory system. *Chem. Senses, 28*, 279-284.

AV-RON, E., AND ROSPARS, J.P. (1995). Modeling insect olfactory neuron signaling by a network utilizing disinhibition. *Biosystems, 36*, 101-108.

AYYUB, C., PARANJAPE, J., RODRIGUES, V., AND SIDDIQI, O. (1990). Genetics of olfactory behavior in Drosophila melanogaster. *Journal of Neurogenetics, 6*, 243-262.

BAZHENOV, M., STOPFER, M., SEJNOWSKI, T. J., AND LAURENT, G. (2005). Fast odor learning improves reliability of odor responses in the locust antennal lobe. *Neuron, 46*, 483-492.

BHALLA, U.S., AND BOWER, J.M. (1997). Multiday recordings from olfactory bulb neurons in awake freely moving rats: spatially and temporally organized variability in odorant response properties. *Journal of Computational Neuroscience, 4*, 221-256.

BRAY, S., AND AMREIN, H. (2003). A putative *Drosophila* pheromone receptor expressed in male-specific taste neurons is required for efficient courtship. *Neuron, 39*, 1019-1029.

BUCK, L., AND AXEL, R. (1991). A novel multigene family may encode odorant receptors: a molecular basis for odor recognition. *Cell, 65*, 175-187.

CHANG, H.J., AND FREEMAN, W.J. (1996). Parameter optimization in models of the olfactory neural system. *Neural Networks, 9*, 1-14.

CHANG, H.J., FREEMAN, W.J., AND BURKE, B.C. (1998). Optimization of olfactory model in software to give 1/f power spectra reveals numerical instabilities in solutions governed by aperiodic (chaotic) attractors. *Neural Networks, 11*, 449-466.

CHENG, Y. AND NASH, H.A. (2008). Visual mutations reveal opposing effects of illumination on arousal in *Drosophila*. *Genetics* 178: 2413-2416.

CHRISTENSEN, T.A., D'ALESSANDRO, G., LEGA, J., AND HILDEBRAND, J.G. (2001). Morphometric modeling of olfactory circuits in the insect antennal lobe: I. Simulations of spiking local interneurons. *Biosystems, 61*, 143-153.

CHRISTENSEN, T.A., PAWLOWSKI, V.M., LEI, H., AND HILDEBRAND, J.G. (2000). Multi-unit recordings reveal context-dependent modulation of synchrony in odor-specific neural ensembles. *Nature Neuroscience, 3*, 927-931.

CHRISTENSEN, T.A., AND WHITE, J. (2000). Representation of olfactory information in the brain. *The Neurobiology of Taste and Smell, 2*, 201–232.

CHYB, S., DAHANUKAR, A., WICKENS, A., AND CARLSON, J.R. (2003). *Drosophila* Gr5a encodes a taste receptor tuned to trehalose. *National Acad Sciences, 100*, 14526-14530

CLYNE, P., GRANT, A., O'CONNELL, R., AND CARLSON, J.R. (1997). Odorant response of individual sensilla on the *Drosophila* antenna. *Invertebrate Neuroscience, 3*,

127-135.

CLYNE, P.J., WARR, C.G., AND CARLSON, J.R. (2000). Candidate taste receptors in *Drosophila*. *Science,* **287**, 1830.

CLYNE, P.J., WARR, C.G., FREEMAN, M.R., LESSING, D., KIM, J., AND CARLSON, J.R. (1999). A novel family of divergent seven-transmembrane proteins: candidate odorant receptors in *Drosophila*. *Neuron,* **22**, 327-338.

COBB, M., BRUNEAU, S., AND JALLON, J.M. (1992). Genetic and developmental factors in the olfactory response of *Drosophila melanogaster* larvae to alcohols. *Proceedings: Biological Sciences,* **248**, 103-109.

DAHANUKAR, A., AND FOSTER, K. (2001). A Gr receptor is required for response to the sugar trehalose in taste neurons of *Drosophila*. *Nature Neuroscience,* **4**, 1182-1186.

DALY, K.C., CHANDRA, S., DURTSCHI, M.L., AND SMITH, B.H. (2001). The generalization of an olfactory-based conditioned response reveals unique but overlapping odour representations in the moth *Manduca sexta. Journal of Experimental Biology,* **204**, 3085-3095.

DALY, K.C., CHRISTENSEN, T.A., LEI, H., SMITH, B.H., AND HILDEBRAND, J.G. (2004). Learning modulates the ensemble representations for odors in primary olfactory networks. *Proceedings of the National Academy of Sciences,* **101**, 10476-10481.

DALY, K.C., DURTSCHI, M.L., AND SMITH, B.H. (2001). Olfactory-based discrimination learning in the moth, *Manduca sexta. Journal of Insect Physiology,* **47**, 375-384.

DALY, K.C., AND SMITH, B.H. (2000). Associative olfactory learning in the moth Manduca sexta. *Journal of Experimental Biology,* **203**, 2025-2038

DALY, K.C., WRIGHT, G.A., AND SMITH, B.H. (2004). Molecular features of odorants systematically influence slow temporal responses across clusters of coordinated antennal lobe units in the moth *Manduca sexta. Journal of Neurophysiology,* **92**, 236-254.

DAVIS, R.L. (2004). Olfactory learning. *Neuron,* **44**, 31-48.

DAVIS, R.L. (2005). Olfactory memory formation in *Drosophila*: from molecular to systems neuroscience. *Neuron,* **28**, 275-302.

DE BRUYNE, M., CLYNE, P.J., AND CARLSON, J.R. (1999). Odor coding in a model olfactory organ: the *Drosophila* maxillary palp. *Journal of Neuroscience, 19*(11), 4520-4532.

DE BRUYNE, M., FOSTER, K., AND CARLSON, J.R. (2001). Odor coding in the *Drosophila* antenna. *Neuron,* **30**, 537-552.

DOBRITSA, A.A., VAN DER GOES VAN NATERS, W., WARR, C.G., STEINBRECHT, R.A., AND CARLSON, J.R. (2003). Integrating the molecular and cellular basis of odor coding in the *Drosophila* antenna. *Neuron,* **37**, 827-841.

DUNIPACE, L., MEISTER, S., MCNEALY, C., AND AMREIN, H. (2001). Spatially restricted expression of candidate taste receptors in the *Drosophila* gustatory system. *Current Biology,* **11**, 822-835.

DUKAS, R. (2008). Evolutionary biology of insect learning. *Annual Review of Entomology,* **53**, 145-160.

EISENBERG, J., FREEMAN, W.J., AND BURKE, B. (1989). Hardware architecture of a neural network model simulating pattern recognition by the olfactory bulb. *Neural Networks,* **2**, 315-325.

ELMORE, T., IGNELL, R., CARLSON, J.R., AND SMITH, D.P. (2003). Targeted mutation of a *Drosophila* odor receptor defines receptor requirement in a novel class of sensillum.

Journal of Neuroscience, **23**, 9906-9912.

Evans, J.D., Aronstein, K. Chen, Y.P., Hetru, J., Imler, J.L., Jiang, H., Kanost, M., Thompson, G.J., Zou, Z., and Hultmark, D. (2006). Immune pathways and defense mechanisms in honey bees *Apis mellifera. Insect Molecular Biology.* **15**: 645-656.

Faber, T., Joerges, J., and Menzel, R. (1999). Associative learning modifies neural representations of odors in the insect brain. *Nature Neuroscience,* **2**, 74-78.

Ferguson, H.J., Cobey, S. and Smith, B.H.. (2001). Sensitivity to a change in reward is heritable in the honeybee, *Apis mellifera. Animal Behavior,* **61**: 527-534.

Fletcher, M.L., and Wilson, D.A. (2003). Olfactory bulb mitral-tufted cell plasticity: odorant-specific tuning reflects previous odorant exposure. *Journal of Neuroscience,* **23**, 6946-6955.

Freeman, W.J., Yao, Y., and Burke, B. (1988). Central pattern generating and recognizing in olfactory bulb: A correlation learning rule. *Neural Networks,* **1**, 277-288.

French, D.A., Flannery, R.J., Groetsch, C.W., Krantz, W.B., and Kleene, S.J. (2006). Numerical approximation of solutions of a nonlinear inverse problem arising in olfaction experimentation. *Mathematical and Computer Modelling,* **43**, 945-956.

Frisch Von K. (1950). Bees, Their Vision, Chemical Senses and Language. Ithaca, New York, Great Seal Books, Cornell University Press.

Frisch Von K. (1915). Der Farbensinn und formensinn der biene. *Zool. Abt. Allg. Zool. Physiol.* **35**: 1–182.

Galizia, C.G., and Menzel, R. (2000). Odour perception in honeybees: coding information in glomerular patterns. *Current Opinion in Neurobiology,* **10**, 504-510.

Galizia, C.G., and Menzel, R. (2001). The role of glomeruli in the neural representation of odours: results from optical recording studies. *Journal of Insect Physiology,* **47**, 115-130.

Gao, Q., and Chess, A. (1999). Identification of candidate *Drosophila* olfactory receptors from genomic DNA sequence. *Genomics,* **60**, 31-39.

Getz, W.M. (1991). A neural network for processing olfactory-like stimuli. *Bulletin of Mathematical Biology,* **53**, 805-823.

Getz, W.M., and Lutz, A. (1999). A neural network model of general olfactory coding in the insect antennal lobe. *Chem. Senses,* **24**, 351-372.

Godfrey, P.A., Malnic, B., and Buck, L. B. (2004). The mouse olfactory receptor gene family. *Proceedings of the National Academy of Sciences,* **101**, 2156-2161.

Goldberg, S.J., and Moulton, D.G. (1987). Olfactory bulb responses telemetered during an odor discrimination task in rats. *Experimental Neurology,* **96**, 430.

Gu, Y., and Liljenström, H. (2007). Modelling efficiency in insect olfactory information processing. *Biosystems,* **89**, 236-243.

Hallem, E.A., and Carlson, J.R. (2004). The odor coding system of *Drosophila. Trends in Genetics,* **20**, 453-459.

Hallem, E.A., Ho, M.G., and Carlson, J. R. (2004). The molecular basis of odor coding in the *Drosophila* antenna. *Cell,* **117**, 965-979.

Hammer, M. (1993). An identified neuron mediates the unconditioned stimulus in associative olfactory learning in honeybees. *Nature,* **366**, 59-63.

Hammer, M., and Menzel, R. (1995). Learning and memory in the honeybee. *Journal of Neuroscience,* **15**, 1617-1630.

Hammer, M., and Menzel, R. (1998). Multiple sites of associative odor learning as revealed by local brain microinjections of octopamine in honeybees. *Learning &*

Memory, **5**, 146-156.

HANNSON, B.S. (1991). Insect Olfaction. Springer Verlag, 1ˢᵗ Edition, New York.

HEIMBECK, G., BUGNON, V., GENDRE, N., HABERLIN, C., AND STOCKER, R.F. (1999). Smell and taste perception in *Drosophila melanogaster* larva: toxin expression studies in chemosensory neurons. *Journal of Neuroscience*, **19**, 6599-6609.

HEINBOCKEL, T., AND KAISSLING, K.E. (1996). Variability of olfactory receptor neuron responses of female silkmoths (*Bombyx mori* L.) to benzoic acid and (±)-linalool. *Journal of Insect Physiology*, **42**, 565-578.

HILDEBRAND, J.G. (1995). Analysis of chemical signals by nervous systems. *Proceedings of the National Academy of Sciences*, **92**, 67-74.

HILDEBRAND, J.G., AND SHEPHERD, G.M. (1997). Mechanisms of olfactory discrimination: converging evidence for common principles across phyla. *Annual Review of Neuroscience*, **20**, 595-631.

HILL, C.A., FOX, A.N., PITTS, R.J., KENT, L.B., TAN, P.L., CHRYSTAL, M.A., *ET AL.* (2002). G protein-coupled receptors in *Anopheles gambiae*. *Science*, **298**, 176-178.

HOFFMAN, D.S., NEHRIR, A.R., REPASKY, K.S., SHAW, J.A., AND CARLSTEN, J.L. (2007) Range-resolved optical detection of honeybees by use of wing-beat modulation of scattered light for locating land mines. *Applied Optics* **46**, 3007-3012.

HOSLER, J.S., AND SMITH, B.H. (2000). Blocking and the detection of odor components in blends. *Journal of Experimental Biology*, **203**, 2797-2806.

IKENO H., AND USUI, S. (1999). Mathematical description of ionic currents of the Kenyon cell in the mushroom body of honeybee. *Neurocomputing*, **26**, 177-184.

ISHIMOTO, H AND TANIMURA, T. (2004). Molecular Neurophysiology of Taste in *Drosophila*. *Cellular and Molecular Life Sciences*, **61**: 10-18.

JEFFERIS, G., MARIN, E.C., STOCKER, R.F., AND LUO, L. (2001). Target neuron prespecification in the olfactory map of *Drosophila*. *Nature*, **414**, 204-208.

KAISER, L., PEREZ-MALUF, R., SANDOZ, J.C., AND PHAM-DELEGUE, M.H. (2003). Dynamics of odour learning in *Leptopilina boulardi*, a hymenopterous parasitoid. *Animal Behaviour*, **66**, 1077-1084.

KAISSLING, K.E., MENG, L.Z., AND BESTMANN, H.J. (1989). Responses of *Bombykol* receptor cells to (Z, E)-4, 6-hexadecadiene and linalool. *Journal of Comparative Physiology A: Sensory, Neural, and Behavioral Physiology*, **165**, 147-154.

KANZAKI, R. (1996). Behavioral and neural basis of instinctive behavior in insects: odor-source searching strategies without memory and learning. *Robotics and Autonomous Systems*, **18**, 33-43.

KAY, L.M., AND LAURENT, G. (1999). Odor-and context-dependent modulation of mitral cell activity in behaving rats. N*ature Neuroscience*, **2**, 1003-1009.

KENDRICK, K.M., LEVY, F., AND KEVERNE, E.B. (1992). Changes in the sensory processing of olfactory signals induced by birth in sleep. *Science*, **256**, 833-836.

KENT, K.S., HOSKINS, S.G., AND HILDEBRAND, J.G. (1987). A novel serotonin-immunoreactive neuron in the antennal lobe of the sphinx moth *Manduca sexta* persists throughout postembryonic life. *Journal of neurobiology*, **18**, 451-465.

KING, T.L., HORINE, F.M., DALY, K.C., SMITH, B.H. (2004) Explosives detection with hard-wired moths. *IEEE Transactions on Instrumentation and Measurement* **53**, 1113-1118.

LAURENT, G. (1999). A systems perspective on early olfactory coding. *Science*, **286**, 723.

Laurent, G., Stopfer, M., Friedrich, R.W., Rabinovich, M.I., Volkovskii, A., and Abarbanel, H.D. I. (2001). Odor encoding as an active, dynamical process: experiments, computation and theory theory. . *Annual Review of Neuroscience,* **24,** 263-297.

Lei, H., Christensen, T.A., and Hildebrand, J.G. (2002). Local inhibition modulates odor-evoked synchronization of glomerulus-specific output neurons. *Nature Neuroscience,* **5,** 557-565.

Lei, H., Christensen, T.A., and Hildebrand, J.G. (2004). Spatial and temporal organization of ensemble representations for different odor classes in the moth antennal lobe. *Journal of Neuroscience,* **24,** 11108-11119.

Lewis W.J. and Takasu K. 1990. Use of learned odours by a parasitic wasp in accordance with host and food needs. *Nature* **348,** 635-6.

Li, C., Geng, C. Leung, H.T., Hong, Y.S., Strong, L.L. *et al.,* (1999). INAF, a protein required for transient receptor potential Ca(2+) channel function. *Proceedings of the National Academy of Science USA.* **96:** 13474–13479.

Linster, C., and Smith, B.H. (1997). A computational model of the response of honey bee antennal lobe circuitry to odor mixtures: overshadowing, blocking and unblocking can arise from lateral inhibition. *Behavioural Brain Research,* **87,** 1-14.

Liu, X., and Davis, R.L. (2006). Insect olfactory memory in time and space. *Current Opinion in Neurobiology,* **16,** 679-685.

Ma, Z., and Krings, A.W. (2009). Insect sensory systems inspired computing and communications. *Ad Hoc Networks,* **7,** 742-755.

Malnic, B., Godfrey, P.A., and Buck, L.B. (2004). The human olfactory receptor gene family. *Proceedings of the National Academy of Sciences,* **101,** 2584-2589.

Malnic, B., Hirono, J., Sato, T., and Buck, L. B. (1999). Combinatorial receptor codes for odors. *Cell,* **96,** 713-724.

Marin, E.C., Jefferis, G., Komiyama, T., Zhu, H., and Luo, L. (2002). Representation of the glomerular olfactory map in the *Drosophila* brain. *Cell,* **109,** 243-255.

McGuire, S.E., Deshazer, M., and Davis, R.L. (2005). Thirty years of olfactory learning and memory research in *Drosophila melanogaster. Progress in Neurobiology,* **76,** 328-347.

Meiners, T., Wackers, F., and W.J. Lewis. (2002). The effect of molecular structure on olfactory discrimination by the parasitoid *Microplitis croceipes. Chem. Senses,* 27, 811-816.

Menzel R. (1985). Learning in Honeybees in an Ecological and Behavioral Context. Pages 55–74 in *Experimental Behavioral Ecology* eds B. Hölldobler and M. Lindauer. Gustav Fischer, Stuttgart.

Monte, P., Woodard, C., Ayer, R., Lilly, M., Sun, H., and Carlson, J. (1989). Characterization of the larval olfactory response in *Drosophila* and its genetic basis. *Behavior Genetics,* **19,** 267-283.

Morgan, P.G., Sedensky, M.M., Meneely, P.M. and Cascorbi, H.F. (1988). The effect of two genes on anesthetic response in the nematode *Caenorhabditis elegans. Anesthesiology.* **69**: 246-251.

Mori, K., Nagao, H., and Yoshihara, Y. (1999). The olfactory bulb: coding and processing of odor molecule information. *Science,* **286,** 711-715.

Nikonov, A.A., and Leal, W.S. (2002). Peripheral coding of sex pheromone and a behavioral antagonist in the Japanese beetle, *Popillia japonica. Journal of Chemical*

Ecology, **28**, 1075-1089.

OLDROYD, B.P. AND G.J. THOMPSON. (2006). Behavioural genetics of the honey bee *Apis mellifera. Advanced Insect Physiology.* **33**: 1-49.

OLSEN, S.R., AND WISLON, R.I. (2008). Cracking neural circuits in a tiny brain: new approaches for understanding the neural circuitry of *Drosophila. Trends in Neuroscience,* 31, 512-520.

OLSON, D.M., RAINS, G.C., MEINERS, T., TAKASU, K., TERTULIANO, M., TUMLINSON, J.H, WACKERS, F.L. AND LEWIS, W.J. 2003. Parasitic wasps learn and report diverse chemicals with unique conditionable behaviors. *Chem. Senses* **28**:545-549.

OPPLIGER, F.Y., GUERIN, P.M., AND VLIMANT, M. (2000). Neurophysiological and behavioural evidence for an olfactory function for the dorsal organ and a gustatory one for the terminal organ in *Drosophila melanogaster* larvae. *Journal of Insect Physiology,* **46**, 135-144.

PATTERSON, T.A., THOMAS, L., WILCOX, C., OVASKAINEN, O., AND MATTHIOPOULOS, J. (2008). State-space models of individual animal movement. *Trends in Ecology & Evolution,* **23**, 87-94.

PEARCE, T.C., VERSCHURE, P., WHITE, J., AND KAUER, J.S. (2001). Stimulus encoding during the early stages of olfactory processing: A modeling study using an artificial olfactory system. *Neurocomputing,* **38**, 299-306.

PYTHON, F., AND STOCKER, R.F. (2002). Adult-like complexity of the larval antennal lobe of *D. melanogaster* despite markedly low numbers of odorant receptor neurons. *The Journal of Comparative Neurology,* **445**, 374-387.

QUENET, B., DUBOIS, R., SIRAPIAN, S., DREYFUS, G., AND HORN, D. (2002). Modelling spatiotemporal olfactory data in two steps: from binary to Hodgkin-Huxley neurones. *Biosystems,* **67**, 203-211.

RAINS, G.C., TOMBERLIN, J.K., D'ALESSANDRO, M., AND LEWIS, W.J. (2004). Limits of Volatile Chemical Detection of a Parasitoid Wasp, *Microplitis croceipes,* and an Electronic Nose: A Comparative Study. *Transactions of the ASAE,* **47**: 2145-2152.

RAINS, G.C., UTLEY, S.L., AND LEWIS, W.J. (2006). Behavioral monitoring of trained insects for chemical detection. *Biotechnology Progress,* **22**: 2-8.

RAINS, G.C., TOMBERLIN, J.K., KULASIRI, D. (2008). Using insect sniffing devices for detection, *Trends in Biotechnology,* **26**(6), 288-294.

REINHARD, J., SRINIVASAN, M.V., AND ZHANG, S. (2004). Olfaction: Scent-triggered navigation in honeybees. *Nature,* **427**, 411-412.

RIBEIRO, L.F.C., D.B. ZANATTA, J.P. BRAVO, R.M.C. BRANCALHÃO, AND M.A. FERNANDEZ. (2009). Molecular markers in commercial *Bombyx mori* (Lepidoptera: Bombycidae) hybrids susceptible to multiple nucleopolyhedrovirus. *Genetics and Molecular Research.* **8**: 144-153.

ROBERTSON, H.M., WARR, C.G., AND CARLSON, J.R. (2003). Molecular evolution of the insect chemoreceptor gene superfamily in *Drosophila melanogaster. National Acad Sciences,* **100**, 14537-14542.

ROSPARS, J.P., AND HILDEBRAND, J.G. (2000). Sexually dimorphic and isomorphic glomeruli in the antennal lobes of the sphinx moth *Manduca sexta. Chemical senses,* **25**, 119-129.

ROSPARS, J.P., LANSKY, P., DUCHAMP-VIRET, P., AND DUCHAMP, A. (2001). Characterizing and modeling concentration-response curves olfactory receptor cells. *Neurocomputing,* **38-40**, 319-325.

ROSPARS, J.P., LUCAS, P., AND COPPEY, M. (2007). Modelling the early steps of transduction in insect olfactory receptor neurons. *Biosystems*, **89**, 101-109.

ROSSOKHIN, A.V., AND TSITOLOVSKY, L.E. (1997). A mathematical model of neural information processing at the cellular level. *Biosystems*, **40**, 159-167.

SACHSE, S., AND GALIZIA, C.G. (2002). Role of inhibition for temporal and spatial odor representation in olfactory output neurons: a calcium imaging study. *Journal of neurophysiology*, **87**, 1106-1117.

SCOTT, K., BRADY, R., CRAVCHIK, A., MOROZOV, P., RZHETSKY, A., ZUKER, C., *ET AL.* (2001). A chemosensory gene family encoding candidate gustatory and olfactory receptors in *Drosophila. Cell*, **104**, 661-673.

SHANBHAG, S.R., MÜLLER, B., AND STEINBRECHT, R.A. (1999). Atlas of olfactory organs of *Drosophila melanogaster* 1. Types, external organization, innervation and distribution of olfactory sensilla. *International Journal of Insect Morphology and Embryology*, **28**, 377-397.

SHANBHAG, S.R., MÜLLER, B., AND STEINBRECHT, R.A. (2000). Atlas of olfactory organs of *Drosophila melanogaster* 2. Internal organization and cellular architecture of olfactory sensilla. *Arthropod Structure and Development*, **29**, 211-229.

SHIELDS, V.D.C., AND HILDEBRAND, J.G. (2001). Responses of a population of antennal olfactory receptor cells in the female moth *Manduca sexta* to plant-associated volatile organic compounds. *Journal of Comparative Physiology A: Neuroethology, Sensory, Neural, and Behavioral Physiology*, **186**, 1135-1151.

SNOPOK, B.A., AND KRUGLENKO, I.V. (2002). Multisensor systems for chemical analysis: state-of-the-art in electronic nose technology and new trends in machine olfaction. *Thin Solid Films*, **418**, 21-41.

STENSMYR, M.C., GIORDANO, E., BALLOI, A., ANGIOY, A.M., AND HANSSON, B.S. (2003). Novel natural ligands for *Drosophila* olfactory receptor neurones. *Journal of Experimental Biology*, **206**, 715-724.

STOCKER, R.F. (1996). The organization of the chemosensory system in *Drosophila melanogaster*: a rewiew. *Cell and tissue research*, **275**, 3-26.

STOCKER, R.F., LIENHARD, M.C., BORST, A., AND FISCHBACH, K.F. (1990). Neuronal architecture of the antennal lobe in *Drosophila melanogaster*. *Cell and Tissue Research*, **262**, 9-34.

STÖRTKUHL, K.F., AND KETTLER, R. (2001). Functional analysis of an olfactory receptor in *Drosophila melanogaster*. *Proceedings of the National Academy of Sciences*, **98**, 9381-9385.

STRAUSFELD, N.J., HANSEN, L., LI, Y., GOMEZ, R.S., AND ITO, K. (1998). Evolution, discovery, and interpretations of arthropod mushroom bodies. *Learning & Memory*, **5**, 11-37.

SULLIVAN, R.M., WILSON, D.A., AND LEON, M. (1989). Norepinephrine and learning-induced plasticity in infant rat olfactory system. *Journal of Neuroscience*, **9**, 3998-4006.

TAKASU, K., AND LEWIS, W.J. (1996). The role of learning in adult food location by the larval parasitoid, *Microplitis croceipes* (Hymenoptera: Braconidae). *Journal of Insect Behavior* **9**: 265-281.

TAKKEN, W. (1996). Synthesis and future challenges: the response of mosquitoes to host odours. *Ciba Found Symp.*, **200**, 302-312.

THORPE W.H. (1963) Learning and Instinct in Animals, 2nd edn. Harvard University

Press, Cambridge, MA, USA.

TERTULIANO, M., OLSON, D.M., RAINS, G.C., AND LEWIS, W.J. (2004). "Influence of handling and training protocol on learning and memory of *Microplitis croceipes* (Cresson) (Hymenoptera: Braconidae), *Entomologica Experimentalis et Applicata*, **110**: 165-172.

TOLBERT, L.P., AND HILDEBRAND, J.G. (1981). Organization and synaptic ultrastructure of glomeruli in the antennal lobes of the moth *Manduca sexta*: a study using thin sections and freeze-fracture. *Proceedings of the Royal Society of London. Series B, Biological Sciences*, **213**, 279-301.

TOMBERLIN, J.K., D.C. SHEPPARD, AND J.A. JOYCE. (2002). A comparison of selected life history traits of the black soldier fly (Diptera: Stratiomyidae) when reared on three diets. *Annals of the Entomological Society of America*. **95**: 379-387.

TOMBERLIN, J.K., RAINS, G.C., AND SANFORD, M.R. (2008). Development of *Microplitis croceipes* as a biological sensor, *Entomologia Experimentalis*, **128**: 249-257.

TROEMEL, E.R., CHOU, J.H., DWYER, N.D., COLBERT, H.A., AND BARGMANN, C.I. (1995). Divergent seven transmembrane receptors are candidate chemosensory receptors in C. elegans. *Cell*, **83**, 207.

UENO, K., OHTA, M., MORITA, H., MIKUNI, Y., NAKAJIMA, S., YAMAMOTO, K., *ET AL*. (2001). Trehalose sensitivity in *Drosophila* correlates with mutations in and expression of the gustatory receptor gene Gr5a. *Current Biology*, **11**, 1451-1455.

UTLEY, S.L., RAINS, G.C., AND LEWIS, W.J. (2007). Behavioral monitoring of *Microplitis croceipes*, A parasitoid wasp, for detecting target odorants using a computer vision system. *Transactions of ASABE*, **50**: 1843-1849.

VASSAR, R., CHAO, S.K., SITCHERAN, R., NUNEZ, J.M., VOSSHALL, L.B., AND AXEL, R. (1994). Topographic organization of sensory projections to the olfactory bulb. *Cell*, **79**, 981.

VET L.E.M., LEWIS W.J., AND CARDÉ R.T. (1995). Parasitoid Foraging and Learning. Pages 65–101 in *Chemical Ecology of Insects 2*. eds. Cardé R.T., Bell W.J. New York: Chapman & Hall.

VOSSHALL, L.B., AMREIN, H., MOROZOV, P.S., RZEHTSKY, A., AND AXEL, R. (1999). A spatial map of olfactory receptor expression in the *Drosophila* antenna. *Cell*, **96**, 725-736.

VOSSHALL, L.B., WONG, A.M., AND AXEL, R. (2000). An olfactory sensory map in the fly brain. *Cell*, **102**, 147-159.

WANG, F., NEMES, A., MENDELSOHN, M., AND AXEL, R. (1998). Odorant receptors govern the formation of a precise topographic map. *Cell*, **93**, 47-60.

WANG, J.W., WONG, A.M., FLORES, J., VOSSHALL, L.B., AND AXEL, R. (2003). Two-photon calcium imaging reveals an odor-evoked map of activity in the fly brain. *Cell*, **112**, 271-282.

WEBB, B. (2004). Neural mechanisms for prediction: do insects have forward models? *Trends in Neurosciences*, **27**, 278-282.

WETZEL, C.H., BEHRENDT, H.J., GISSELMANN, G., STÖRTKUHL, K.F., HOVEMANN, B., AND HATT, H. (2001). Functional expression and characterization of a *Drosophila* odorant receptor in a heterologous cell system. *Proceedings of the National Academy of Sciences*, **98**, 9377.

WILSON, D.A., AND SULLIVAN, R.M. (1994). Neurobiology of associative learning in the neonate: early olfactory learning. *Behavioral and Neural Biology*, **61**, 1-18.

WILSON, D.A., SULLIVAN, R.M., AND LEON, M. (1987). Single-unit analysis of postnatal olfactory learning: Modified olfactory bulb output response patterns to learned attractive odors. *Journal of Neuroscience,* **7**, 3154-3162.

WILSON, R.I., TURNER, G.C., AND LAURENT, G. (2004). Transformation of olfactory representations in the *Drosophila* antennal lobe. *American Association for the Advancement of Science*, **303**, 366-370.

WRIGHT, G.A., AND SMITH, B.H. (2004). Variation in complex olfactory stimuli and its influence on odour recognition. *Proceedings of the Royal Society B-Biological Sciences.* **271**, 147-152.

WRIGHT, G.A., THOMSON, M.G., AND SMITH, B.H. (2005). Odour concentration affects odour identity in honey bees. *Proceedings of the Royal Society B-Biological Sciences,* **272**, 2417-2422.

WRIGHT, G.A., CARLTON, M., SMITH, B.H. (2009). A honey bee's ability to learn, recognize and discriminate odours depends upon odour sampling time and concentration. *Behavioral Neuroscience,* **123**(1), 36-43.

WONG, A.M., WANG, J.W., AND AXEL, R. (2002). Spatial representation of the glomerular map in the *Drosophila* protocerebrum. *Cell,* **109**, 229-241.

XU, F., GREER, C.A., AND SHEPHERD, G.M. (2000). Odor maps in the olfactory bulb. *The Journal of Comparative Neurology,* **422**, 489-495.

YU, D., PONOMAREV, A., AND DAVIS, R.L. (2004). Altered Representation of the Spatial Code for Odors after Olfactory Classical Conditioning Memory Trace Formation by Synaptic Recruitment. *Neuron,* **42**, 437-449.

ZHANG, X., AND FIRESTEIN, S. (2002). The olfactory receptor gene superfamily of the mouse. *Nature Neuroscience,* **5**, 124-133.

Biotechnology and Genetic Engineering Reviews - Vol. 26, 205-222 (2009)

Microbial Transglutaminase Production: Understanding the Mechanism

DONGXU ZHANG[1], YANG ZHU[2] AND JIAN CHEN[1]*

[1]Key Laboratory of Industrial Biotechnology of Ministry of Education, School of Biotechnology, Jiangnan University, Wuxi, Jiangsu 214122, China; [2]Department of Biosciences, TNO Quality of Life, Netherlands Organization for Applied Scientific Research, P.O. Box 360, 3700 AJ Zeist, Netherlands

Abstract

Microbial transglutaminase is an important enzyme in food processing for improving protein properties by catalyzing the cross-linking of proteins. Recently, this enzyme has been shown to exhibit wider potential application in tissue engineering, textiles and leather processing, site-specific protein conjugation and wheat gluten allergy reduction. The production of microbial transglutaminase has been significantly improved thanks to advances in bioprocess engineering and genetic engineering during the last three decades. More recently, studies on the biological mechanism of transglutaminase synthesis have further contributed towards the understanding of microbial transglutaminase production by *Streptomyces*. This will further facilitate improving the production of recombinant microbial transglutaminase. In this paper, we will review the progress in bioprocess engineering and genetic engineering in microbial transglutaminase production. We will highlight our understanding of the biological mechanisms of microbial transglutaminase synthesis, including biotechnological approaches used based on these biological mechanisms as a way of improving transglutaminase production. We address in addition the future research needs for microbial transglutaminase production.

*To whom correspondence may be addressed (jchen@jiangnan.edu.cn)

Abbreviations: TGase, transglutaminase; Pro-TGase, pro-transglutaminase; TAP, transglutaminase-activating protease; TAPI, transglutaminase-activating protease inhibitor; CTAB, cetyltrimethyl ammonium bromide.

Introduction

Transglutaminases (EC 2.3.2.13) are a family of enzyme that catalyze the formation of a crosslink between a free amine group and the γ-carboxamide group of protein-bound glutamine (for example the crosslinking of proteins by forming N^{ε}-(γ-glutamyl) lysine bonds, the incorporation of polyamines into protein and the deamidation of protein-bound glutamines) (*Figure 1*) (Zhu and Tramper, 2008). Bonds formed by transglutaminase exhibit a high resistance to proteolytic degradation (Griffin, *et al.*, 2002). *In vitro*, the enzyme is able to catalyze crosslinking of whey proteins, soy proteins, wheat proteins, beef myosin, casein and crude actomyosin, leading to their texturization (Motoki, *et al.*, 1998; Zhu, *et al.*, 1995). This capacity of the enzyme has been used in attempts to improve the functional properties of foods.

(a) R-Glu-CO-NH$_2$ + H$_2$N-Lys-R′ → R-Glu-CO-NH-Lys-R′ + NH$_3$
(b) R-Glu-CO-NH$_2$ + H$_2$N-R′   → R-Glu-CO-NHR′ + NH$_3$
(c) R-Glu-CO-NH$_2$ + H$_2$O   → R-Glu-CO-OH + NH$_3$

Figure 1. Reactions catalyzed by transglutaminase. (a) Transglutaminase catalyses the acyl-transfer reaction between the g-carboxyamide group of a glutamine residue present in one protein and the e-amino group of a lysine residue of a second protein. With this reaction, two proteins are crosslinked by a covalent isopeptide bond to form a new protein. (b) In an alternative reaction, a primary amine is bound to a second protein instead of the e-amino group of a lysine residue. (c) When water replaces amine donor substrates, deamidation of the glutamine occurs and changes it into glutamate. (Adapted from Zhu *et al.* (Zhu, *et al.*, 1995) with kind permission of Springer Science and Business Media.)

Transglutaminase has been found in animals, plants, and microorganisms (Zhu and Tramper, 2008). Until the end of the 1980s, commercial transglutaminase could only be obtained from animal tissues, most commonly from guinea pig liver. The rare source and complicated downstream procedure resulted in an extremely high price for the enzyme, which hampered a wide application in food processing. More recently however, production in microorganisms has received increased interest, and an enzyme of microbial origin is now commonly used for food treatment and has been shown to improve food flavor, appearance and texture. In addition, the enzyme is able to increase the shelf life of certain foods and reduce their allergenicity. Although the main application of transglutaminase remains in the food sector, novel potential applications have emerged during the last decade. These applications cover the areas of biomedical engineering, material science, textiles and leather processing (Zhu and Tramper, 2008). Several excellent reviews on the application of microbial transglutaminase in food and other areas are already available in the literature (Motoki, *et al.*, 1998; Yokoyama, *et al.*, 2004; Zhu, *et al.*, 1995; Zhu, *et al.*, 2008), so we will focus here on the *production* of microbial transglutaminase.

The production of microbial transglutaminase by *Streptoverticillium mobaraense* was first reported by Ando *et al.* in 1989 (Ando, *et al.*, 1989). *Streptoverticillium mobaraense* was later classified as *Streptomcyes mobaraensis* (Date, *et al.*, 2004). Since then, bioengineering and genetic engineering have been used to improve the production of this enzyme. Production strain isolation and fermentation have addressed and increased the production of this enzyme. There have also been several attempts to clone and express this enzyme. Below we will discuss and evaluate the technological development of microbial transglutaminase production and the trends of new technology based on recent advances in understanding the biochemical mechanism of microbial transglutaminase activation and physiological function.

Transglutaminase production by *Streptomyces* spp. and genetically modified strains

(1) TGASE PRODUCTION BY WILD STRAIN

Since Ando *et al.* first reported on microbial transglutaminase production by *Streptoverticillium mobaraense* (Ando, *et al.*, 1989), new strains have been selected continuously for higher enzyme yield and production process has been optimized through various methods. Many studies have addressed and increased the production of microbial transglutaminase. An overview of recent developments of transglutaminase production by *Streptomyces* spp. and other microorganisms is given in *Table 1*.

Table 1. Milestones in microbial transglutaminase production by wild strains. (Adapted from Zhu and Tramper (Zhu and Tramper, 2008) with kind permission of Elsevier.)

Year	Strain	Focus of the development	Yield (unit/ml)[a]	References
1989	*Streptoverticillium mobaraense*	Strain isolation	~2.0	(Ando, *et al.*, 1989)
1996	*Streptoverticillium mobaraense*	Substrate optimization	~1.0	(Zhu, *et al.*, 1996)
1997	*Streptoverticillium cinnamoneum*	Substrate optimization	~0.3	(Junqua, *et al.*, 1997)
1998	*Streptoverticillium mobaraense*	Metabolic optimization	~1.8	(Zhu, *et al.*, 1998)
2000	*Actinomadura sp.*	Strain isolation	n/a	(Kim, *et al.*, 2000)
2001	*Streptoverticillium mobaraense*	Environmental control strategies	3.37	(Zheng, *et al.*, 2001)
2002	*Streptoverticillium mobaraense*	Environmental control strategies	2.94	(Zheng, *et al.*, 2002b)
2002	*Streptoverticillium mobaraense*	Environmental control strategies	3.40	(Zheng, *et al.*, 2002a)
2004	*Streptoverticillium ladakanum*	Strain isolation	0.348	(Téllez-Luis, *et al.*, 2004a)
2004	*Streptoverticillium ladakanum*	Substrate optimization	0.725	(Téllez-Luis, *et al.*, 2004b)
2005	*Streptoverticillium mobaraense*	Environmental control strategies	3.32	(Yan, *et al.*, 2005)
2006	*Bacillus circulans*	Strain isolation and Substrate optimization	0.306	(Souza, *et al.*, 2006)
2007	*Streptomyces sp.*	Strain isolation and Substrate optimization	1.4	(Macedoa, *et al.*, 2007)
2007	*Streptomyces hygroscopicus*	Strain isolation and Environmental control strategies	5.04	(Cheng, *et al.*, 2007)
2008	*Several Streptomyces*	Solid fermentation	n/a	(Nagy, *et al.*, 2008)
2009	*Streptomyces hygroscopicus*	Fermentation strategies	5.79	(Xu, *et al.*, 2009)

[a]n/a = not applicable.

The isolation of high yield production strains has played an important role in improving the production of transglutaminase. Up until now, at least 6 different strains have been studied for production of microbial transglutaminase. These strains have included: *Streptoverticillium mobaraense, Streptoverticillium cinnamoneum, Actinomadura sp., Streptoverticillium ladakanum, Bacillus circulans, Streptomyces sp. and Streptomyces hygroscopicus.* The production of transglutaminase by *Streptomyces* spp. has significantly higher yield than other strains, and therefore, this strain is currently used industrially as a microbial transglutaminase producer.

Additional efforts have been made in substrate optimization, metabolic optimization and environmental control strategies (pH, dissolved oxygen and temperature) to improve the yield of microbial transglutaminase (*Table 1*). For example, Zheng *et al.* developed a two stage environmental control strategy (Zheng, *et al.*, 2002a). With this strategy, the highest transglutaminase activity was improved from 2.0 U/mL to 3.40 U/mL. These studies promoted the production of transglutaminase from various strains isolated from nature. More recently, solid-state fermentation has been studied to produce microbial transglutaminase by *Streptomyces* spp., which enables microbial transglutaminase generation with much reduced production cost (Nagy, *et al.*, 2008).

After two decades of development of microbial transglutaminase production, transglutaminase activity in fermentation broth has been increased from 2.0 to 6.0 U/mL. However, such levels of productivity of microbial transglutaminase is still low. The high cost and price of microbial transglutaminae hamper the wide application of the enzyme in the food industry and other areas. Recombinant technology has however been shown to offer great potential in microbial strain improvement. Therefore, many studies have switched the focus from conventional optimization to genetic engineering as a way of constructing recombinant transglutaminase production strains for high yield.

(2) TGASE PRODUCTION BY GENETICALLY MODIFIED STRAINS

In order to further improve the production of microbial transglutaminase, scientists have attempted to express transglutaminase to high-yields by using genetically modified strains. Due to the well recognized crystal structure of *Streptomyces* transglutaminase (Kashiwagi, *et al.*, 2002), the *Streptomyces* transglutaminase gene became the best origin for exogenous expression. Takehana *et al.* reported in 1994 a reconstructed transglutaminase producing strain. They expressed a chemical synthesized transglutaminase gene from *Streptomyces mobaraensis* in *Escherichia coli* (Takehana, *et al.*, 1994). However, the enzyme activity was very low, though the induced gene product was identical with native TGase transglutaminase in size and in immunological properties. At the same time, Washizu *et al.* expressed pre-pro-transglutaminase gene from *S. mobaraensis* in *Streptomyces lividan*. As a result, active and mature recombinant transglutaminase was obtained (Washizu, *et al.*, 1994). Following these pioneering studies of transglutaminase production by genetically modified strains, a broad variety of hosts and exogenesis expression strategies have subsequently been used to improve transglutaminase production (*Table 2*).

At present, *Escherichi. coli, Corynebacterium glutamicum, Candida boidinii* and *Streptomyces* spp. have been used as hosts for recombinant transglutaminase

Table 2. Milestones in microbial transglutaminase production by genetically modified strains. (Adapted from Zhu and Tramper (Zhu and Tramper, 2008) with kind permission of Elsevier.)

Year	Focus of the development	Yield of pro-transglutaminase (mg/L)[a]	Yield of transglutaminase (unit/ml)[a]	References
1994	Chemical synthesis of the gene for microbial transglutaminase from *Streptoverticillium* and its expression in *Escherichia coli*	n/a	n/a	(Takehana, *et al.*, 1994)
1994	Molecular cloning of the gene for microbial transglutaminase from *Streptoverticillium* and its expression in *Streptomyces lividans*	n/a	n/a	(Washizu, *et al.*, 1994)
1997	High-level expression of the chemically synthesized gene for microbial transglutaminase from *Streptoverticillium* in *Escherichia coli*	n/a	n/a (inclusion bodies)	(Kawai, *et al.*, 1997)
2000	Overproduction of microbial transglutaminase in *Escherichia coli*, in vitro refolding and characterization of the refolded form	n/a	n/a (inclusion bodies)	(Yokoyama, *et al.*, 2000)
2003	Secretion of an active *Streptoverticillium mobaraense* transglutaminase by *Corynebacterium glutamicum*; processing of the pro-transglutaminase by a co-secreted subtilisin-like protease from *Streptomyces albogriseolus*	235	n/a (142 mg/L)	(Kikuchi, *et al.*, 2003)
2003	Production of native-type *Streptoverticillium mobaraense* transglutaminase in *Corynebacterium glutamicum*	n/a	n/a (132 mg/L)	(Date, *et al.*, 2003)
2004	High level expression of *Streptomyces mobaraensis* transglutaminase in *Corynebacterium glutamicum* using a chemical pro-region from *Streptomyces cinnamoneus* transglutaminase	627	n/a	(Date, *et al.*, 2004)
2004	The pro-peptide of *Streptomyces mobaraensis* transglutaminase functions in cis and in trans to mediate efficient secretion of active enzyme from methylotrophic yeasts	n/a	1.83	(Yurimoto, *et al.*, 2004)
2004	Cloning and expression of the transglutaminase gene from *Streptoverticillium ladakanum* in *Streptomyces lividans*	n/a	1.46	(Lin, *et al.*, 2004)

Year	Focus of the development	Yield of pro-transglutaminase (mg/L)[a]	Yield of transglutaminase (unit/ml)[a]	References
2006	Cloning of the transglutaminase gene from *Streptomyces platensis* and its expression in *Streptomyces lividans*	n/a	2.2	(Lin, *et al.*, 2006)
2006	Cloning of transglutaminase gene from *Streptomyces fradiae* and its enhanced expression in the original strain	n/a	3.2	(Liu, *et al.*, 2006)
2008	Overproduction of soluble recombinant transglutaminase from *Streptomyces netropsis* in *Escherichia coli*	4.5	n/a	(Yu, *et al.*, 2008)
2008	Secretion of *Streptomyces mobaraensis* pro-transglutaminase by coryneform bacteria	2500	n/a	(Itaya, *et al.*, 2008)
2008	Characterization and large-scale production of recombinant *Streptoverticillium platensis* transglutaminase	n/a	5.32	(Lin, *et al.*, 2008)

[a]n/a = not applicable.

expression. Recombinant expression of transglutaminase in *E. coli* and *Candida boidinii* has however resulted in little success (Kawai, *et al.*, 1997; Yokoyama, *et al.*, 2000; Yurimoto, *et al.*, 2004). Replacing of the signal peptide or the pro-region of transglutaminase and other approaches have been used to improve the production of recombinant transglutaminase (Date, *et al.*, 2004; Kikuchi, *et al.*, 2003; Yu, *et al.*, 2008). However, improved transglutaminase yield was not observed by using genetically modified strains when compared with the wild-type strain.

An exception however is if *Streptomcyes* is used as the host. Recently, production of recombinant *Streptoverticillium platensis* transglutaminase has reached 5.32 U/mL: for the first time a production of recombinant transglutaminase has been reported comparable with the wild strain (Lin, *et al.*, 2008).

It was learned from microbial transglutaminase exogenous expression that the production of recombinant transglutaminase not only has to incorporate gene clone and expression but also the activation process of this enzyme (Date, *et al.*, 2003; Kikuchi, *et al.*, 2003). Therefore, further understanding of the microbial transglutaminase activation mechanism is crucial to significantly improve recombinant transglutaminase production.

Biological mechanism of transglutaminase synthesis from *Streptomyces*

With an increasing accumulated knowledge on microbial transglutaminase production, scientists have recognized the importance of understanding the biological mechanism of microbial transglutaminase synthesis so as to more accurately control the biosynthesis process. The understanding of biological mechanisms can further improve microbial transglutaminase production by using transglutaminase-producing wild stain or by enabling a better control of the fermentation conditions. In addition, a better understanding of the synthesis mechanism of microbial transglutaminase will help construct recombinant strains more efficiently. *Table 3* gives an overview of recent studies focused on this topic. These advances mainly focus on the enzymatic characterization, gene sequencing, molecular structure, and activation mechanism of microbial transglutaminase.

Table 3. Progress in understanding the biological mechanism of microbial transglutaminase synthesis.

Year	Focus of the development	Refs
1989	Purification and characteristics of a novel transglutaminase derived from microorganisms	(Ando, *et al.*, 1989)
1993	Primary structure of microbial transglutaminase from *Streptoverticillium* sp. strain s-8112	(Kanaji, *et al.*, 1993)
1998	Bacterial pro-transglutaminase from *Streptoverticillium mobaraense* Purification, characterization and sequence of the zymogen	(Pasternack, *et al.*, 1998)
2002	Crystal structure of microbial transglutaminase from *Streptoverticillium mobaraense*	(Kashiwagi, *et al.*, 2002)
2003	Transglutaminase from *Streptomyces mobaraensis* is activated by an endogenous metalloprotease	(Zotzel, *et al.*, 2003a)
2003	Activated transglutaminase from *Streptomyces mobaraensis* is processed by a tripeptidyl aminopeptidase in the final step	(Zotzel, *et al.*, 2003b)
2004	High level expression of *Streptomyces mobaraensis* transglutaminase in *Corynebacterium glutamicum* using a chemical pro-region from *Streptomyces cinnamoneus* transglutaminase	(Date, *et al.*, 2004)
2004	The pro-peptide of *Streptomyces mobaraensis* transglutaminase functions in cis and in trans to mediate efficient secretion of active enzyme from methylotrophic yeasts	(Yurimoto, *et al.*, 2004)
2008	Surfactant protein of the *Streptomyces* subtilisin inhibitor family inhibits transglutaminase activation in *Streptomyces hygroscopicus*	(Zhang, *et al.*, 2008a)
2008	Two different proteases from *Streptomyces hygroscopicus* are involved in transglutaminase activation	(Zhang, *et al.*, 2008b)
2009	Transglutaminase is involved in *Streptomyces hygroscopicus* differentiation	(Zhang, *et al.*, submitted)

The discovery of transglutaminase zymogen (pro-transglutaminase) has revealed the activation mechanism of transglutaminase from *Streptomyces*. In 1998, Pasternack and colleagues found that transglutaminase from *Streptomyces* was secreted as a pro-transglutaminase and could be activated by several exogenous proteases, such as bovine trypsin, intestinal chymotrypsin or dispase from *Bacillus polymyxa* (Pasternack, *et al.*, 1998). Subsequently a metalloprotease was isolated from *Streptomyces mobaraensis* as an endogenous transglutaminase-activating protease (Zotzel, *et al.*, 2003a). Later, Zhang *et al.* found that, in *Streptomyces hygroscopicus*, not only endogenous metalloprotease but also endogenous serine protease is involved in transglutaminase activation (Zhang, *et al.*, 2008b). In actuality, *Streptomyces* pro-transglutaminases appear to have a conserved amino acid sequence preceding the N-terminal of transglutaminase, which contains cleavage sites for both serine protease and metalloprotease, indicating activation of pro-transglutaminase is not a specific process (Zhang, *et al.*, 2008b).

Research on both *S. mobaraensis* and *S. hygroscopicus* has proven that transglutaminase activation process is inhibited by a transglutaminase-activating protease inhibitor. N-terminal amino acid sequencing and homology study of the purified transglutaminase-activating protease inhibitor have revealed that it is a member of the *Streptomyces* subtilisin inhibitor family (Zhang, *et al.*, 2008a; Zotzel, *et al.*, 2003a). This inhibitor could be deposited by cetyltrimethyl ammonium bromide. Furthermore, transglutaminase-activating protease inhibitor possesses surface activity. This surface activity of transglutaminase-activating protease inhibitor makes the activation process of pro-transglutaminase possible. Possessing the surface activity, transglutaminase-activating protease inhibitor molecules are distributed mostly at the air-liquid interface, which allows sufficient free transglutaminase-activating protease molecules to exist in the submerged liquid and perform its function to activate pro-transglutaminase. A sketch map for transglutaminase activation mechanism in *Streptomcyes* is shown in *Figure 2*.

The activation mechanism reveals the physiological function of *Streptomyces* transglutaminase. Metalloprotease, serine protease and *Streptomyces* subtilisin inhibitor protein, the key factors involved in the activation process of transglutaminase (*Figure 2*), are all under regulation by the *A-factor* (Hirano, *et al.*, 2006; Kato, *et al.*, 2005; Kato, *et al.*, 2002; Tomono, *et al.*, 2005). The A-factor is a microbial hormone controlling the differentiation of *Streptomyces* (Kato, *et al.*, 2002). Transglutaminase has been found to be secreted and activated during differentiation rather than during nutrition growth. Furthermore, differentiation of *S. hygroscopicus* on solid culture has been shown to be inhibited by cystamine, a competitive inhibitor of transglutaminase. These studies demonstrate that transglutaminase is involved in differentiation of *Streptomyces* (Zhang, *et al.*, submitted).

The progress in understanding the biological mechanism of transglutaminase synthesis in *Streptomyces* stimulates further improvement in microbial transglutaminase production.

Fermentation strategies based on biological mechanism of transglutaminase synthesis

Based on the activation mechanism of pro-transglutaminase in *Streptomyces*, strategies have been developed to improve the production. Because cetyltrimethyl ammonium bromide could remove the inhibition of transglutaminase-activating protease inhibitor to the activation of pro-transglutaminase, this detergent was added to the fermentation broth to enhance the activation degree. By adding cetyltrimethyl ammonium bromide at later stage of fermentation, transglutaminase activity in fermentation broth increased by 21.8% (Cheng, *et al.*, 2007). Though

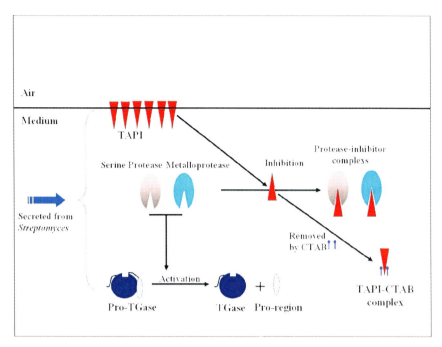

Figure 2. Activation mechanism of transglutaminase in *Streptomcyes.*

cetyltrimethyl ammonium bromide enhances the activation rate, it cannot surpass the degradation, and therefore, it is not ideal to accumulate stable transglutaminase. As shown in *Figure 3*, transglutaminase activity in the medium decreased quickly when transglutaminase-activating protease inhibitor was removed. Therefore, the inhibition by transglutaminase-activating protease inhibitor is a very important factor to control the transglutaminase-activating protease activity.

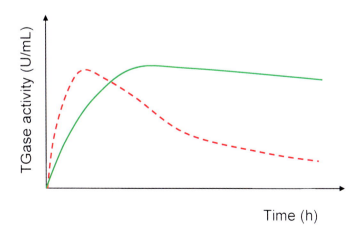

Figure 3. Transglutaminase activation in fermentation with (full line) or without (broken line) transglutaminase-activating protease inhibitor. transglutaminase-activating protease inhibitor was removed by adding of cetyltrimethyl ammonium bromide.

As shown in *Figure 4*, transglutaminase-activating protease can both activate pro-transglutaminase and degrade mature transglutaminase for its nonspecific protease activity. Fortunately, the activation of transglutaminase is more efficient than the degradation. Therefore, a mature transglutaminase can exist with or without the inhibition of transglutaminase-activating protease inhibitor (*Figure 3*). Actually, the activation process benefits from the surfactant of transglutaminase-activating protease inhibitor, which provide a delayed inhibition of transglutaminase-activating protease. This kind of inhibition to transglutaminase-activating protease has maximized the activation of transglutaminase and minimized the degradation of transglutaminase, which is a remarkable control mechanism. Considering such a mechanism, one of the possible ways to enhance the production of transglutaminase is to enlarge the efficiency of transglutaminase activation.

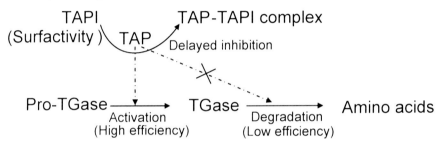

Figure 4. Delayed inhibition of transglutaminase-activating protease mediated by surfactant transglutaminase-activating protease inhibitor.

By addition of protease in the prophase of fermentation, the highest transglutaminase activity increased by 26.6% and the culture time was shortened from 44 h to 36 h (Xu, *et al.*, submitted). This result showed that this method transformed all pro-transglutaminase into mature transglutaminase in fermentation broth. Protease was also inhibited in the end of the fermentation. The large amount of protease (1000 U/mL) added at the prophase were almost inhibited by transglutaminase-activating protease inhibitor to a very low final activity (5 U/mL).

Based on the fact that transglutaminase secretion is associated with differentiation of *Streptomcyes* (Zhang, *et al.*, submitted), a feeding strategy was used to enhance transglutaminase production. As shown in *Figure 5*, a carbon source was fed during the nutritional growth phase, to achieve maximum cell amount (cell factory). Then in the transglutaminase synthesis phase, a nitrogen source was fed to supply the synthesis of transglutaminase. By using this strategy, cell mass and transglutaminase activity increased 80 % and 83 %, respectively (Xu, *et al.*, 2009).

As demonstrated by the examples mentioned above, the understanding of transglutaminase synthesis mechanism of *Streptomyces* provided a rationale control of the fermentation process. Consequently, the production of transglutaminase was significantly improved.

Trends in microbial production of transglutaminase

Our understanding of the biological mechanism of transglutaminase synthesis has also provided great potential in improving the production of recombinant microbial transglutaminase, as we will now consider.

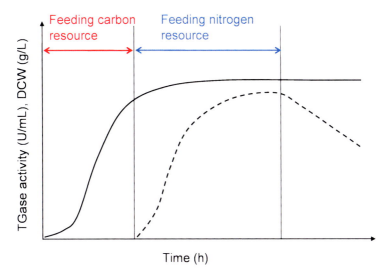

Figure 5. Sketch map of feeding strategy for production of microbial transglutaminase. Transglutaminase activity (broken line) and DCW (real line) were shown with time.

CONSTRUCTION OF GENETIC ENGINEERING STRAINS: CO-EXPRESSION OF PRO-TRANSGLUTAMINASE, TRANSGLUTAMINASE-ACTIVATING PROTEASE AND TRANSGLUTAMINASE-ACTIVATING PROTEASE INHIBITOR

A benefit from understanding the biological mechanism of transglutaminase synthesis is to efficiently construct genetic engineering transglutaminase production strains. Among all the hosts used, mature transglutaminase was only successfully expressed in *Streptomyces* (*Table 2*). This indicates that the protease-inhibitor system is necessary for the bioproduction of recombinant transglutaminase. Because there are no appropriate activation mechanisms (transglutaminase-activating protease-transglutaminase-activating protease inhibitor) existing in *E. coli*, *Corynebacterium glutamicum* or *Candida boidinii*, recombinant pro-transglutaminase is difficult for activation in these hosts. Although it has been attempted to co-express transglutaminase-activating protease in *C. glutamicum*, the activity of transglutaminase-activating protease was not under the control, which would lead to a fast decline of transglutaminase activity (Kikuchi, *et al.*, 2003). It is possible to add exogenous transglutaminase-activating protease and transglutaminase-activating protease inhibitor to activate pro-transglutaminase correctly. However, the problem with this is the high cost for production and purification of transglutaminase-activating protease and transglutaminase-activating protease inhibitor.

Though it is the only successful host for production of recombinant transglutaminase, unfortunately, *Streptomyces* is not an efficient host for the production of recombinant transglutaminase. The yield of *Streptomyces* is only close to that of using the wild strain. Therefore, co-expression of transglutaminase-activating protease and transglutaminase-activating protease inhibitor with pro-transglutaminase in a high efficient host has been considered. As shown in *Table 2*, up until now, *C. glutamicum* has been used as an efficient host, which expresses pro-transglutaminase to 2.5 mg/mL maximally. If calculated as an average specific activity of 13 U/mg, the theoretical highest yield of recombinant transglutaminase in *C. glutamicum* would reach 32.5 U/ mL. Therefore co-expression of transglutaminase, transglutaminase-activating protease and transglutaminase-activating protease inhibitor in *C. glutamicum* is a hopeful way (*Figure 6*). Certainly, it also needs a more detailed examination especially when compared with transglutaminase activation process in *Streptomyces*.

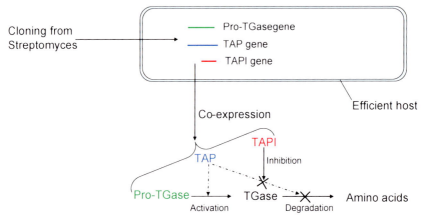

Figure 6. Co-expression of transglutaminase-activating protease, transglutaminase-activating protease inhibitor and pro-transglutaminase in host.

PRO-REGION HELP TGASE FOLDING

Another benefit from understanding the biological mechanism of transglutaminase synthesis is the highly efficient and correct expression of recombinant transgluatminase. A report on endogenous expression of the *Streptomyces* transglutaminase gene in methylotrophic yeasts first revealed that pro-transglutaminase helps the folding of transglutaminase (Yurimoto, *et al.*, 2004). In that study, production of active transglutaminase required a pro-peptide from transglutaminase. When an artificial endopeptidase recognition site was placed between the pro-peptide and mature transglutaminase, secretion and *in vitro* maturation of transglutaminase depended on Kex2-dependent cleavage. Co-expression of unlinked pro-peptide with mature transglutaminase yielded efficient secretion of the active enzyme. These results indicate that the pro-peptide help the formation of mature transglutaminase not only in an intramolecular but also in an intermolecular manner. (Yurimoto, *et al.*, 2004).

Furthermore, a chemical pro-region, combining with *S. mobaraensis* and *S. cinnamoneus* transglutaminase gene, has been used for the production of transglutaminase in *C. glutamicum*. Secretion of transglutaminase by using the chemical pro-region was increased compared to that using the native pro-region (Date, *et al.*, 2004). This result implies that transglutaminase activity could be improved by change the sequence of pro-region.

In 1998, Shinde, *et al.* found a 77-residue pro-peptide of subtilisin acts as an intramolecular chaperone that organizes the correct folding of its own protease domain. The intramolecular chaperone of subtilisin facilitates folding by acting as a template for its protease domain, although it does not form part of that domain (Shinde, *et al.*, 1998). Sequence analysis showed that *Streptomyces* transglutaminase was evolved from protease (Makarova, *et al.*, 1999). Therefore pro-region of transglutaminase may also inherit this character to be an intramolecular chaperone. It is possible to enhance the industrial performance of transglutaminase, such as high activity, high stability or altered specificity, by control of the pro-region (*Figure 7*).

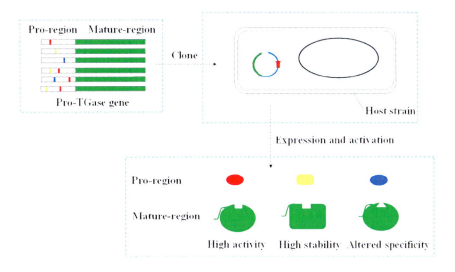

Figure 7. The Pro-region helps folding of transglutaminase.

MOLECULAR MECHANISM OF TRANSGLUTAMINASE INVOLVED IN STREPTOMYCES DIFFERENTIATION

Besides the successful approaches for recombinant transglutaminase production, the molecular mechanism of transglutaminase involved in *Streptomyces* differentiation is worth clarifying. Transglutaminase is involved in *Streptomyces* differentiation (Zhang, *et al.*, submitted), however, the molecular mechanism of this physiological function has not been revealed as yet. Gene knockout of transglutaminase and observation of the phenotype are necessary to know more about the physiological mechanism of transglutaminase in *Streptomyces* differentiation. At the same time, searching the location of transglutaminase and the target protein (or other molecule) of transglutaminase are necessary to reveal the biological mechanism of transglutaminase in *Streptomyces* differentiation. Research on the expression and regulation mechanism of the *Streptomyces* transglutaminase gene is also useful to explain why transglutaminase was involved in *Streptomyces* differentiation. Such a basic understanding will further improve microbial transglutaminase production by using both recombinant and wild strains.

Conclusion

Microbial transglutaminase, as an important enzyme for food processing and poten-tially for other fields where protein modification is concerned, attracts great interest with regards its cost-effective and large-scale production. Efforts have been made during the last two decades on the bioprocess engineering and genetic engineering for the improvement of the production of this enzyme. More recently, studies on understanding the biological mechanism of transglutaminase synthesis have helped with the production of microbial transglutaminase in *Streptomyces*. An understanding of the biological mechanism of transglutaminase synthesis in *Streptomyces* has great potential in further improving the production of microbial transglutaminase either by wild strains or by recombinant strains. We believe that future development in studying microbial production of transglutminase will greatly reduce the cost and improve productivity of this enzyme. As a result, it will accelerate the application of this enzyme.

Acknowledgments

This work was supported by the National Outstanding Youth Foundation of China (No. 20625619), the Major State Basic Research Development Program of China (973 Program) (No. 2007CB714306), the national natural Science Foundation of China (30770055, 20776063), the Key Program of National Natural Science Foundation of China (No. 20836003), the National Natural Science Foundation of Jiangsu Province (No. BK2007018), High-Tech Research Project of Jiangsu Province, China (No.BG2007009) and the grant from the Ph.D. Programs Foundation of Ministry of Education of China (No. 20060295002).

References

Ando, H., Adachi, M. and Umeda, K. (1989) Purification and characteristics of a novel transglutaminase derived from microorganisms. *Agricultural Biology and Chemistry* **53**, 2613-2617.

Cheng, L., Zhang, D.X., Du, G.C. and Chen, J. (2007) Improvement on the activity of microbial transglutaminase with *Streptomyces hygroscopicus* by the addition of surfactant CTAB. *Chinese journal of biotechnology* **23**, 497-501.

Date, M., Yokoyama, K., Umezawa, Y., Matsui, H. and Kikuchi, Y. (2003) Production of native-type *Streptoverticillium mobaraense* transglutaminase in *Corynebacterium glutamicum*. *Applied and Environmental Microbiology* **69**, 3011-3014.

Date, M., Yokoyama, K., Umezawa, Y., Matsui, H. and Kikuchi, Y. (2004) High level expression of Streptomyces mobaraensis transglutaminase in *Corynebacterium glutamicum* using a chimeric pro-region from *Streptomyces cinnamoneus* transglutaminase. *Journal of Biotechnology* **110**, 219-226.

Griffin, M., Casadio, R. and Bergamini, C.M. (2002) Transglutaminases: nature's biological glues. *Biochemical Journal* **368**, 377-396.

Hirano, S., Kato, J.Y., Ohnishi, Y. and Horinouchi, S. (2006) Control of the *Streptomyces* Subtilisin inhibitor gene by AdpA in the A-factor regulatory cascade in *Streptomyces griseus*. *Journal of Bacteriology* **188**, 6207-6216.

Itaya, H. and Kikuchi, Y. (2008) Secretion of *Streptomyces mobaraensis* protransglutaminase by coryneform bacteria. *Applied Microbiology and Biotechnology* **78**, 621-625.

JUNQUA, M., DURAN, R., GANCET, C. AND GOULAS, P. (1997) Optimization of microbial transglutaminase production using experimental designs. *Applied Microbiology and Biotechnology* **48**, 730-734.

KANAJI, T., OZAKI, H., TAKAO, T., KAWAJIRI, H., IDE, H., MOTOKI, M. AND SHIMONISHI, Y. (1993) Primary structure of microbial transglutaminase from *Streptoverticillium* sp. strain s-8112. *Journal of Biological Chemistry* **268**, 11565-11572.

KASHIWAGI, T., YOKOYAMA, K., ISHIKAWA, K., ONO, K., EJIMA, D., MATSUI, H. AND SUZUKI, E. (2002) Crystal structure of microbial transglutaminase from *Streptoverticillium mobaraense. Journal of Biological Chemistry* **277**, 44252-44260.

KATO, J.Y., CHI, W.J., OHNISHI, Y., HONG, S.K. AND HORINOUCHI, S. (2005) Transcriptional control by A-factor of two trypsin genes in *Streptomyces griseus. Journal of Bacteriology* **187**, 286-295.

KATO, J.Y., SUZUKI, A., YAMAZAKI, H., OHNISHI, Y. AND HORINOUCHI, S. (2002) Control by A-factor of a metalloendopeptidase gene involved in aerial mycelium formation in *Streptomyces griseus. Journal of Bacteriology* **184**, 6016-6025.

KAWAI, M., TAKEHANA, S. AND TAKAGI, H. (1997) High-level expression of the chemically synthesized gene for microbial transglutaminase from *Streptoverticillium* in *Escherichia coli. Bioscience, Biotechnology and Biochemistry* **61**, 830-835.

KIKUCHI, Y., DATE, M., YOKOYAMA, K., UMEZAWA, Y. AND MATSUI, H. (2003) Secretion of active-form Streptoverticillium mobaraense transglutaminase by Corynebacterium glutamicum: processing of the pro-transglutaminase by a cosecreted subtilisin-Like protease from Streptomyces albogriseolus. *Applied and Environmental Microbiology* **69**, 358-366.

KIM, H.-S., JUNG, S.-H., LEE, I.-S. AND YU, T.-S. (2000) Production and Characterization of a Novel Microbial Transglutaminase from Actinomadura sp.T-2. *Journal of Microbiology and Biotechnology* **10**, 187-194.

LIN, S.J., HSIEH, Y.F., LAI, L.A., CHAO, M.L. AND CHU, W.S. (2008) Characterization and large-scale production of recombinant Streptoverticillium platensis transglutaminase. *Journal of Industrial Microbiology and Biotechnology* **35**, 981-990.

LIN, Y.-S., CHAO, M.-L., LIU, C.-H. AND CHU, W.-S. (2004) Cloning and expression of the transglutaminase gene from Streptoverticillium ladakanum in Streptomyces lividans. *Process Biochemistry* **39**, 591-598.

LIN, Y.-S., CHAO, M.-L., LIU, C.-H., TSENG, M. AND CHU, W.-S. (2006) Cloning of the gene coding for transglutaminase from *Streptomyces platensis* and its expression in *Streptomyces lividans Process Biochemistry* **41**, 519-524.

LIU, X., YANG, X., XIE, F. AND QIAN, S. (2006) Cloning of transglutaminase gene from *Streptomyces fradiae* and its enhanced expression in the original strain. *Biotechnology Letters* **28**, 1319-1325.

MACEDOA, J., SETTEA, L. AND SATOA, H. (2007) Optimisation studies for the production of microbial transglutaminase from a newly isolated strain of *Streptomyces sp. Journal of Biotechnology* **131**, S213-S214

MAKAROVA, K.S., ARAVIND, L. AND KOONIN, E.V. (1999) A superfamily of archaeal, bacterial, and eukaryotic proteins homologous to animal transglutaminases. *Protein Science* **8**, 1714-1719.

MOTOKI, M. AND SEGURO, K. (1998) Transglutaminase and its use for food processing *Trends in Food Science and Technology* **9**, 204-210.

NAGY, V. AND SZAKACS, G. (2008) Production of transglutaminase by *Streptomyces* isolates in solid-state fermentation. *Letters in Applied Microbiology* **47**, 122-127.

PASTERNACK, R., DORSCH, S., OTTERBACH, J.T., ROBENEK, I.R., WOLF, S. AND FUCHSBAUER, H.-L. (1998) Bacterial pro-transglutaminase from *Streptoverticillium mobaraense* Purification, characterisation and sequence of the zymogen. *European Journal of Biochemistry* **257**, 570-576.

SHINDE, U.P., LIU, J.J. AND INOUYE, M. (1998) Protein memory through altered folding mediated by intramolecular chaperones. *Nature* **392**, 520-522.

SOUZA, C.F.D., FLÔRES, S.H. AND AYUB, M.A.Z. (2006) Optimization of medium composition for the production of transglutaminase by *Bacillus circulans* BL32 using statistical experimental methods. *Process Biochemistry* **41**, 1186-1192.

TÉLLEZ-LUIS, S.J., GONZÁLEZ-CABRIALES, J.J., RAMÍREZ, J.A. AND VÁZQUEZ, M. (2004a) Production of Transglutaminase by Streptoverticillium ladakanum NRRL-3191 Grown on Media Made from Hydrolysates of Sorghum Straw *Food Technology and Biotechnology* **42**, 1-4.

TÉLLEZ-LUIS, S.J., RAMÍREZ, J.A. AND VÁZQUEZ, M. (2004b) Production of transglutaminase by *Streptoverticillium ladakanum* NRRL-3191 using glycerol as carbon source. *Food Technology and Biotechnology* **42**, 75-81.

TAKEHANA, S., WASHIZU, K., ANDO, K., KOIKEDA, S., TAKEUCHI, K., MATSUI, H., MOTOKI, M. AND TAKAGI, H. (1994) Chemical synthesis of the gene for microbial transglutaminase from *Streptoverticillium* and its expression in *Escherichia coli*. *Bioscience, Biotechnology and Biochemistry* **58**, 88-92.

TOMONO, A., TSAI, Y., OHNISHI, Y. AND HORINOUCHI, S. (2005) Three chymotrypsin genes are members of the AdpA regulon in the A-factor regulatory cascade in *Streptomyces griseus*. *Journal of Bacteriology* **187**, 6341-6353.

WASHIZU, K., ANDO, K., KOIKEDA, S., HIROSE, S., MATSUURA, A., TAKAGI, H., MOTOKI, M. AND TAKEUCHI, K. (1994) Molecular cloning of the gene for microbial transglutaminase from *Streptoverticillium* and its expression in *Streptomyces lividans*. *Bioscience, Biotechnology and Biochemistry* **58**, 82-87.

XU, X., CHENG, L., ZHANG, D., REN, Z., LI, J., WANG, M., CHEN, J. AND DU, G. (2009) Improved production of microbial transglutaminase from *Streptomyces hygroscopicus* by the addition of trypsin. (Manuscript submitted).

XU, X., ZHANG, D., WANG, M., LI, J., CHEN, J. AND DU, G. (2009) Improvement of MTG Production by Fed-batch Fermentation. *China Biotechnology*, (in press).

YAN, G., DU, G., LI, Y., CHEN, J. AND ZHONG, J. (2005) Enhancement of microbial transglutaminase production by *Streptoverticillium mobaraense*: application of a two-stage agitation speed control strategy. *Process Biochemistry* **40**, 963-968.

YOKOYAMA, K., NIO, N. AND KIKUCHI, Y. (2004) Properties and applications of microbial transglutaminase. *Applied Microbiology and Biotechnology* **64**, 447-454.

YOKOYAMA, K.I., NAKAMURA, N., SEGURO, K. AND KUBOTA, K. (2000) Overproduction of microbial transglutaminase in *Escherichia coli*, in vitro refolding, and characterization of the refolded form. *Bioscience, Biotechnology and Biochemistry* **64**, 1263-1270.

YU, Y.J., WU, S.C., CHAN, H.H., CHEN, Y.C., CHEN, Z.Y. AND YANG, M.T. (2008) Overproduction of soluble recombinant transglutaminase from Streptomyces netropsis in Escherichia coli. *Applied Microbiology and Biotechnology* **81**, 523-532.

YURIMOTO, H., YAMANE, M., KIKUCHI, Y., MATSUI, H., KATO, N. AND SAKAI, Y. (2004) The pro-peptide of *Streptomyces mobaraensis* transglutaminase functions in cis and in trans to mediate efficient secretion of active enzyme from methylotrophic yeasts. *Bioscience, Biotechnology and Biochemistry* **68**, 2058-2069.

ZHANG, D., JU, X., WANG, M., WU, J., DU, G. AND CHEN, J. (2009) Transglutaminase is involved in *Streptomyces hgroscopicus* differentiation. (Manuscript submitted).

ZHANG, D., WANG, M., DU, G., ZHAO, Q., WU, J. AND CHEN, J. (2008a) Surfactant protein of the *Streptomyces* subtilisin inhibitor family inhibits transglutaminase activation in *Streptomyces hygroscopicus. Journal of Agricultural and Food Chemistry* **56**, 3403-3408.

ZHANG, D., WANG, M., WU, J., CUI, L., DU, G. AND CHEN, J. (2008b) Two different proteases from *Streptomyces hygroscopicus* are involved in transglutaminase activation. *Journal of Agricultural and Food Chemistry* **56**, 10261-10264.

ZHENG, M., DU, G. AND CHEN, J. (2002a) pH control strategy of batch microbial transglutaminase production with *Streptoverticillium mobaraense. Enzyme and Microbial Technology* **31**, 477-481.

ZHENG, M., DU, G., CHEN, J. AND LUN, S. (2002b) Modelling of temperature effects on batch microbial transglutaminase fermentation with *Streptoverticillium mobaraense. World Journal of Microbiology and Biotechnology* **18**, 767-771

ZHENG, M., DU, G., GUO, W. AND CHEN, J. (2001) A temperature-shift strategy in batch microbial transglutaminase fermentation. *Process Biochemistry* **36**, 525-530.

ZHU, Y., RINZEMA, A., TRAMPER, J. AND BOL, J. (1995) Microbial transglutaminase - a review of its production and application in food processing *Applied Microbiology and Biotechnology* **44**, 277-282.

ZHU, Y., RINZEMA, A., TRAMPER, J. AND BOL, J. (1996) Medium design based on stoichiometric analysis of microbial transglutaminase production by *Streptoverticillium mobaraense. Biotechnology and Bioengineering* **50**, 291-298.

ZHU, Y., RINZEMA, A., TRAMPER, J., BRUIN, E.D. AND BOL, J. (1998) Fed-batch fermentation dealing with nitrogen limitation in microbial transglutaminase production by *Streptoverticillium mobaraense Applied Microbiology and Biotechnology* **49**, 251-257.

Zhu, Y. and Tramper, J. (2008) Novel applications for microbial transglutaminase beyond food processing. *Trends in Biotechnology* **26**, 559-565.

Zotzel, J., Keller, P. and Fuchsbauer, H.-L. (2003a) Transglutaminase from *Streptomyces mobaraensis* is activated by an endogenous metalloprotease. *European Journal of Biochemistry* **270**, 3214-3222.

Zotzel, J., Pasternack, R., Pelzer, C., Ziegert, D., Mainusch, M. and Fuchsbauer, H.-L. (2003b) Activated transglutaminase from *Streptomyces mobaraensis* is processed by a tripeptidyl aminopeptidase in the final step *European Journal of Biochemistry* **270**, 4149-4155.

2A to the Fore – Research, Technology and Applications

GARRY A. LUKE[1*], HELENA ESCUIN[2], PABLO DE FELIPE[3] AND MARTIN D. RYAN[1]

[1]Centre for Biomolecular Sciences, School of Biology, Biomolecular Sciences Building, University of St Andrews, North Haugh, St Andrews KY16 9ST, UK; [2]Department of Surgery, University of California at Los Angeles (UCLA), Los Angeles, CA, USA and [3]Spanish Medicines Agency (AEMPS), Parque Empresarial "Las Mercedes", Campezo 1 - Edificio 8, 28022 Madrid, SPAIN

Abstract

The 2A region of the foot-and-mouth disease virus (FMDV) encodes a short sequence that mediates self-processing by a novel translational effect. Translation elongation arrest leads to release of the nascent polypeptide and re-initiation at the next in-frame codon. In this way discrete translation products are derived from a single open read-

*To whom correspondence may be addressed (gal@st-andrews.ac.uk)

Abbreviations: AAV, adeno-associated virus; BADH, betaine aldehyde dehydrogenase; BLI, bioluminescence imaging; BMT, bone marrow transplant; CAT, chloramphenicol acetyltransferase; CFP, cyan fluorescent protein; CMO, choline monooxygenase; CP, coat protein; CPMV, cowpea mosaic virus; CSF, classical swine fever; CTLA-4, cytotoxic T lymphocyte antigen-4; DmAMP1, *Dahlia merckii* antimicrobial peptide 1; DsRED, Discosoma sp red fluorescent protein; ER, endoplasmic reticulum; eRF, eukaryotic release factor; ERT, enzyme replacement therapy; ESC, embryonic stem cell; FLUC, firefly luciferase; FMDV, foot-and-mouth disease virus; GAGs, glycosaminoglycans; GB, glycine betaine; GFP, green fluorescent protein; GLP-1, glucagon-like peptide 1; GM-CSF, granulocyte macrophage colony-stimulating factor; GUS, β-glucuronidase; HOXB4, homeobox B4; HSCs, haematopoietic stem cells; hESC, human embryonic stem cell; IDUA, iduronidase alpha-L; IFN-γ, interferon-γ; IL-12, interleukin-12; iPS, induced pluripotent stem; IRES, internal ribosome entry site; ITAM, immunoreceptor tyrosine-based activation motif; mAb, monoclonal antibody; MGMT, O^6-methylguanine-DNA-methyltransferase; MPS1, mucopolysaccharidosis type 1; NLS, nuclear localizing sequence; ORF, open reading frame; PVX, potato virus X; RFP, red fluorescent protein; Rg, retrogenic; RsAFP2, *Raphanus sativus* antifungal peptide 2; scFv, single-chain variable fragment; SRP, signal recognition particle; TCR, T-cell receptor; Tg, transgenic; Th1, T helper 1; TM, transmembrane domain; TMV, tobacco mosaic virus; UTR, untranslated region; YFP, yellow fluorescent protein.

ing frame. Active 2A-like sequences have been found in (many) other viruses and trypanosome non-LTR retrotransposons. Exponential growth of 2A technology within the last decade has lead to many biotechnological/biomedical applications including the generation of transgenic plants/animals and genetic manipulation of human embryonic stem cells (hESCs).

Introduction

Viruses frequently use unusual regulatory mechanisms to make essential proteins. Commonly these involve translational control or "recoding" mechanisms. Leaky stop codons may be read-through (the stop recoded to a sense codon) to produce either the predicted translation product, or, at a very low level (1-5%) an extended "read-through", protein. In the case of programmed ribosomal frameshifting, the ribosome switches to an alternative frame at a specific site in response to special signals in the messenger RNA. Frameshifting can produce longer or shorter proteins than those resulting from standard decoding as shown in *Figure 1*. Overlapping (*e.g.* –UAAUG-; -UGAUG-; AUGA-), or highly proximal stop/start sequences, may give rise to termination accompanied by a very low level of re-initiation to produce two, discrete, translation products. Another recoding event is ribosomal "skipping", first described in the foot-and-mouth disease virus (FMDV) 2A protein. Here a termination event occurs at a sense codon, followed by release of the nascent polypeptide

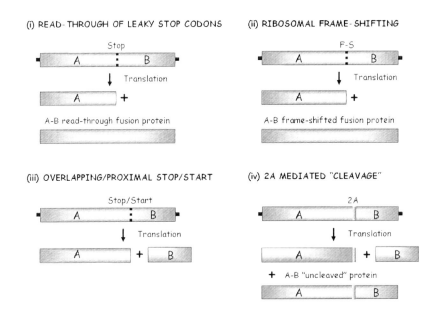

Figure 1. The different outcomes of translating either proximal ORF's (A and B) with a "leaky" stop codon, or a ribosomal frame-shift site (F-S), or overlapping/proximal stop/start codons are compared with a single ORF (comprising domains "A" and "B") with a 2A-like sequence which mediates cleavage at either high or medium levels.

prior to ribosomal translocation to the next in-frame codon (Ryan *et al.*, 1999; Don-nelly *et al.*, 2001a; Doronina *et al.*, 2008a & b). This strategy has the advantage that discrete, translation products can be generated from a single transcription unit. As such it has great potential as a biotechnological tool, reducing the need for multiple vectors. Examples of each recoding event are illustrated schematically in *Figure 1* and are also available in the Database of Translational Recoding, "RECODE", http://recode.genetics.utah.edu/ (Baranov *et al.*, 2003). In this review, historical aspects of the so-called 2A sequence are revised, followed by a more extensive commentary of the current state of knowledge and the progress made in the use of "self-cleaving" 2A peptides for different biotechnological purposes.

2A Ontology

IDENTIFICATION OF THE 2A DOMAIN

The *Picornaviridae* comprise one of the largest and most important families of human and agricultural pathogens. Members of this family include the enteroviruses (*e.g.* coxsackievirus, poliovirus), rhinoviruses (aetiologic agents of the common cold), cardioviruses (*e.g.* encephalomyocarditis virus, Theilovirus), and aphthoviruses (*e.g.* FMDV) (King *et al.*, 2000). The picornavirus genome is a positive-sense RNA molecule with a central open reading frame (ORF) flanked by highly structured 5' and 3' untranslated regions (UTRs). The viral RNA has an oligopeptide (VPg or 3B) covalently attached to the 5' terminus and is 3' terminated with poly(A). The 5'UTR contains an internal ribosome entry site (IRES) which is used for the cap-independent translation of the polyprotein (reviewed in Martínez-Salas, 1999; Martínez-Salas *et al.*, 2008). From a functional point of view, this strategy of protein synthesis allows some viruses - *e.g.* poliovirus, human rhinovirus, foot-and-mouth disease virus - to shut off host-cell protein synthesis and hence usurp the cellular translational ma-chinery for the efficient synthesis of their own proteins. The ORF encodes a large protein precursor (polyprotein) which can be divided into three regions as a result of co-translational cleavages by viral proteins 2A and 3C. These correspond to the N-terminal capsid protein precursor (P1, containing four capsid proteins 1A-1D), the middle of the polyprotein containing three of the nonstructural proteins (P2, the three proteins 2A-2C), and the most C-terminal segment of the polyprotein containing four nonstructural proteins (P3, proteins 3A-3D) (Palmenberg, 1990).

Protein 2A is encoded by all picornaviruses but differs widely in size and function among its different members. In the case of the enteroviruses and rhinoviruses, the 2A polypeptide is a thiol proteinase (designated 2A^pro) whose primary function is to cleave the polyprotein in *cis* between the P1 capsid protein precursor and the replicative domains of the polyprotein (P2/P3) (Sommergruber *et al.*, 1989). The corresponding aphthoviruses and cardioviruses primary cleavage occurs not between P1 and 2A, but at the C-terminus of the 2A region between the capsid protein precursor and 2B. This results in the 2A sequence remaining as a C-terminal extension of the upstream product ([P1-2A] –aphthoviruses; [L-P1-2A] – cardioviruses) until it is cleaved away during secondary processing (Ryan *et al.*, 1991; Ryan and Drew, 1994). While the cardiovirus 2A proteins (between 142 and 157 amino acids [aa]) are of a similar size to the entero- and rhinovirus 2A^pro, no sequence similarity is apparent. The C-terminal region of cardiovirus 2A is, however, highly similar to the much shorter aphthovirus 2A on which this review will focus (Donnelly *et al.*, 1997).

PROCESSING BY THE 2A PROTEIN

Secondary 3C[pro] cleavage of the [1D↓2A] precursor protein between 1D and 2A shows the natural FMDV 2A segment is only 18 aa long (-LLNFDLLKLAGDVES-NPG-) (Belsham, 1993). The core sequence at the C-terminus of 2A, together with the N-terminal proline of 2B (see *Figure 2*) is strongly conserved and contains the canonical motif DxExNPGP (where "x" is any amino acid which is not conserved). The 2A oligopeptide is able to self-cleave at the site corresponding to the 2A/2B junction (-NPG↓P-), the 2A sequence remaining as a C-terminal extension of the upstream product. Following cleavage, the last proline residue forms the N-terminus of the viral 2B protein (Ryan *et al.*, 1991; Ryan and Drew, 1994). Previous work has shown that the 2A region of the FMDV polyprotein is not cleaved through the action of exogenous host proteinases or other virus-encoded proteinases (Ryan *et al.*, 1989; Ryan *et al.*, 1991; Palmenberg *et al.*, 1992).

The "Primary" 2A/2B Polyprotein Cleavage

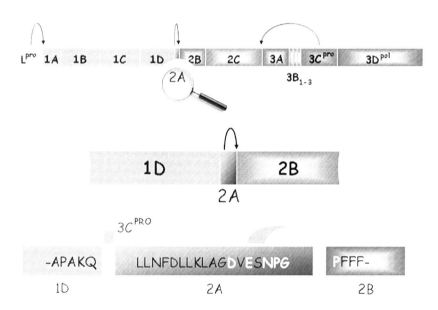

Figure 2. Polyprotein processing in FMDV. The sites of the primary cleavages and the virus proteins responsible are indicated by the curved arrows. The 18aa long 2A sequence is shown along with the conserved DxExNPGP sequence motif.

To verify the ability of FMDV 2A to function independently of all other FMDV se-quences, artificial "self-processing" polyproteins were constructed with the reporter proteins chloramphenicol acetyltransferase (CAT) or green fluorescent protein (GFP) and β-glucuronidase (GUS) flanking the 2A linker peptide – in a single ORF (Ryan and Drew, 1994; Donnelly *et al.*, 1997; Ryan *et al.*, 1999; Donnelly *et al.*, 2001a & b). The 2A sequence within such a context was shown to be highly active in mediating cleavage, and latterly used to study the mechanism of the reaction (see *Figure 3*). When both [CAT2AGUS] and [GFP2AGUS] were analyzed using *in vitro* translation systems, relatively more of the upstream-encoded product, either [CAT2A] or [GFP2A], was generated, suggesting that a nonproteolytic mechanism was involved (Donnelly *et al.*, 1997). Extensive protein degradation studies, examining the effects of non-specific premature termination of transcription/translation, have shown that none of these effects account for this imbalance. This front end loading was due to different rates of synthesis and constitutes a novel type of recoding, reprogrammed genetic decoding (Baranov *et al.*, 2002). Further, frame-shifting the 2A oligopeptide with respect to the reporter proteins demonstrated that activity was mediated by a mechanism reliant on the peptide sequence rather than the nucleotide sequence (Ryan *et al*, 1999). Point mutations introduced into the motif clearly demonstrated that each of the conserved residues contributes to optimal activity (Donnelly *et al.*, 2001b; Ryan *et al.*, 2002).

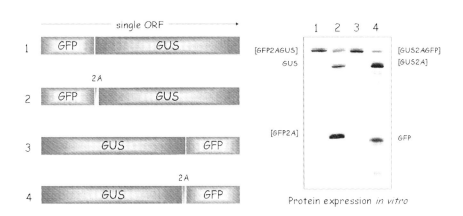

Figure 3. Artificial reporter polyproteins (boxed areas) used to programme *in vitro* translation systems are shown together with translation profiles obtained from rabbit reticulocyte lysates (right). Analysis of the endogenous processing properties shows the polyprotein constructs without the 2A insertion produces a single uncleaved product whereas polyproteins containing 2A self-cleaves with high efficiency to produce three translational products.

Sequences immediately upstream of this motif that were shown to be either critical or very important for activity are not conserved (Donnelly *et al.*, 2001b). Analysis of translation products derived from a series of constructs in which sequences were progressively deleted from the N-terminal region of the FMDV 2A insertion showed that cleavage required a minimum of 13 residues (12N-terminal and 1 C-terminal to the G$^{\downarrow}$P site) (Ryan and Drew, 1994). The N-terminal proline residue of protein 2B is necessary for cleavage. Deletion of the sequences located upstream or downstream of 2A did not abolish the cleavage, although sequences immediately upstream were shown to be either critical, or very important, for activity. Longer versions of 2A with extra sequences (total ~30aa) derived from the capsid protein ("1D") produced higher levels of cleavage (Ryan *et al.*, 1991; Donnelly *et al.*, 2001b; Groot Bramel-Verheije *et al.*, 2000; Klump *et al.*, 2001). Previously described 2A-based multigene vectors have revealed the 2A region functions properly within different contexts, but the cleavage efficiency varies as flanking context changes. By swapping the order of proteins in several artificial polyproteins the stoichiometry is affected by the gene upstream of 2A (Ma and Mitra, 2002; Lengler *et al.*, 2005; Chinnasamy *et al.*, 2006).

OCCURRENCE OF 2A ELEMENTS IN VIRAL AND CELLULAR mRNAs

The C-terminal 18aa of the encephalomyocarditis and Theiler's murine encephalitis virus 2A proteins also mediates a co-translational cleavage at their C-termini (Donnelly *et al.*, 1997). A compilative analysis of genome sequences shows the DxExNPGP-motif to be present in several genera of the *Picornaviridae* (aphtho-, cardio-, erbo-, tescho and certain parechoviruses), single-stranded RNA insect viruses (iflaviruses, dicistroviruses, tetraviruses), double-stranded RNA viruses of the *Reoviridae* (type C and non-ABC rotaviruses, cypoviruses) and penaeid shrimp viruses (*Table 1*). 2A-like sequences are also found in trypanosome repeated sequences (*T.cruzi* – L1Tc: *T.brucei* – TSR1) and cellular sequences, both eukaryotic and prokaryotic (*Tables 2 & 3*).

Cleavage activity of representatives of these 2A-like sequences was studied in our [GFP-GUS] reporter polyprotein system (Luke *et al.*, 2008). At the two "x" positions in the consensus, although there is no conserved amino acid, the choices are limited. The position of the valine in FMDV 2A [-D<u>V</u>ESNPGP-] is represented in all active 2A-like sequences by either valine or isoleucine. At the position of the serine (-DVE<u>S</u>NPGP-), the amino acids can differ between threonine, methionine, leucine, glutamate, glutamine, lysine or proline. Surprisingly, cellular sequences fitting the consensus showed no skipping activity, whereas 2A-like sequences in infectious flacherie virus (IFV; **G**IESNPGP), new adult diarrhoea virus (ADRV-N; **C**IESNPGP) and L1Tc within *T.cruzi* (DIEQ**H**PGP) were active (Donnelly *et al.*, 2001b; Heras *et al.*, 2006; Luke *et al.*, 2008). We therefore favour the idea that this motif alone is not sufficient to confer skipping but requires an appropriate (immediate) upstream element. Previous studies have shown that 2A works efficiently in all eukaryotic cells tested to date: mammalian (Ryan and Drew, 1994), plant (Halpin *et al.*, 1999), insect (Roosien *et al.*, 1990), fungi (Thomas and Maule, 2000) and yeast (de Felipe *et al.*, 2003).

A CO-TRANSLATIONAL MODEL OF 2A-MEDIATED CLEAVAGE

To give an explanation to these findings we propose the nascent 2A-peptide interacts with the exit tunnel of the ribosome to achieve the correct stereo-chemical constraints for the turn motif at the C-terminus of 2A (-ESNPG-) within the peptidyl transferase centre. This conformation of 2A pauses elongation and prevents nucleophilic attack on the carbonyl group of the P site peptidyl-tRNA^{Gly} by the A site prolyl-tRNA^{Pro} amino group. Instead of peptide bond formation, the N-terminal product is somehow released from the ribosome by hydrolysis of the peptidyl(2A)-tRNA ester bond. Here translation either terminates at a sense codon or re-initiates by translocation of the proly-tRNA to the P site. We term this translational effect ribosomal "skipping", highlighting the

Table 1. 2A/2A-like sequences in viruses

Genus	Full name	Abbreviation	Reference
Positive-stranded RNA viruses: *Picornaviridae*			
Aphthovirus	*Foot-and-mouth disease virus*	FMDV	Forss *et al.*, 1984
	Equine rhinitis A virus	ERAV	Li *et al.*, 1996
	Bovine rhinovirus 2	BRV2	E. Rieder, pers. com.
Cardiovirus	*Encephalomyocarditis virus*	EMCV	Duke *et al.*, 1992
	Theiler's murine encephalomyelitis virus	TMEV	Law and Brown, 1990
	Theiler's-like virus of rats	T-LV	Ohsawa *et al.*, 2003
	Saffold virus	SAF-V	Jones *et al.*, 2007
Erbovirus	*Equine rhinitis B virus*	ERBV	Wutz *et al.*, 1996
Teschovirus	*Porcine teschovirus 1*	PTV-1	Doherty *et al.*, 1999
New genus	*Duck hepatitis virus 1*	DHV-1	Kim *et al.*, 2006
	New duck hepatitis virus	N-DHV	Tseng and Tsai, 2007
New genus	*Seneca valley virus*	SVV	acc no: DQ641257
Parechovirus	*Ljungan virus*	LV	Lindberg and Johansson, 2002
Positive-stranded RNA viruses: Iflaviruses (unassigned family)			
Iflavirus	*Infectious flacherie virus*	IFV	Isawa *et al.*, 1998
	Perina nuda picorna-like virus	PnPV	Wu *et al.*, 2002
	Ectropis oblique picorna-like virus	EoPV	Wang *et al.*, 2004
Positive-stranded RNA viruses: *Dicistroviridae*			
Cripavirus	*Cricket paralysis virus*	CrPV	Wilson *et al.*, 2000
	Drosophila C virus	DCV	Johnson and Christian, 1998
	Acute bee paralysis virus	ABPV	Govan *et al.*, 2000
	Kashmir bee virus	KBV	de Miranda *et al.*, 2004
	Israel acute paralysis virus of bees	IAPV	Maori *et al.*, 2007
Positive-stranded RNA viruses: *Tetraviridae*			
Betatetravirus	*Thosea asigna virus*	TaV	Pringle *et al.*, 1999
	Euprosterna elaeasa virus	EeV	Gorbalenya *et al.*, 2002
	Providence virus	PrV	Pringle *et al.*, 2003
Segmented double-stranded RNA viruses: *Reoviridae*			
Rotavirus	*Human rotavirus C*	HuRV-C	Chen *et al.*, 2002
	Bovine rotavirus C	BoRV-C	Jiang *et al.*, 1993
	Porcine rotavirus A	PoRV-C	Bremont *et al.*, 1992
	New adult diarrhoea virus	ADRV-N	Yang *et al.*, 2004
Cypovirus	*Bombyx mori cypovirus 1*	BmCPV-1	Hagiwara *et al.*, 2002
	Dendrolimus punctatus cypovirus 1	DpCPV-1	Zhao *et al.*, 2003
	Lymantria dispar cypovirus 1	LdCPV-1	Rao *et al.*, 2003
	Operophtera brumata cypovirus 18	OpbuCPV-18	Graham *et al.*, 2006
Non-segmented double-stranded RNA viruses: *Totiviridae*			
Unclassified	*Infectious myonecrosis virus*	IMNV	Poulos *et al.*, 2006

Table 2. 2A-like sequences in trypanosomes

Full name	Protein name	Reference
Trypanosoma cruzi	L1Tc non-LTR retrotransposon	Heras *et al.*, 2006 Martin *et al.*, 1995
Trypanosoma brucei	*ingi* non-LTR retrotransposon	Hasan *et al.*, 1984 Murphy *et al.*, 1987

Table 3. 2A-like sequences in eukaryotic and prokaryotic genomes

Full name	Protein name	Reference
Giardia lamblia	Hypothetical protein	acc no: Q7R2B0
Drosophila melanogaster	Mod(mdg4)59.0 protein	Buchner *et al.*, 2000
Mus musculus	Mu opioid receptor variant F, MOR-1F	Pan *et al.*, 2000
Thermatoga maritima	Alpha-glucuronidase	Ruile *et al.*, 1997

crucial aspect of our model - the peptidic bond between the C-terminal glycine of 2A and the N-terminal protein of 2B is not synthesized but skipped (Ryan *et al.*, 1999; Donnelly *et al.*, 2001a; de Felipe *et al.*, 2003). Corroborative results by Atkins *et al.* (2007) has led them to tentatively propose that this "2A" phenomenon be named "StopGo" or some alternative that better reflects its intriguing activity (see also Atkins and Ryan, 2008).

Recent findings suggest that this model needs revision, because release (termination) factors eRF1 and 3 are now known to contribute to 2A translation termination (Doronina *et al.*, 2008a & b). In eukaryotes, these factors associate in a complex which binds to the elongating ribosome when a stop codon enters the A site (Zhouravleva *et al.*, 1995). According to the revised model, the ribosome stalls at the end of 2A where RFs are thought to "recognize" the A site proline codon as a termination signal leading to hydrolysis of the glycine-tRNA ester bond and peptide release. Thus, beyond preventing formation of the peptide bond, the stalled ribosome-2A complex must promote entry of RF without it "reading" the mRNA. RF dissociation from the termination complex facilitates entry of prolyl-tRNA to the A site of the ribosome. The prolyl-tRNA is then translocated from the A to the P site to become the initiating aminoacyl-tRNA of the downstream protein. Whether the ribosome dissociates or continues to translate the downstream context after hydrolysis may depend on the concentration of eukaryotic elongation factors.

Bearing in mind that the two translation products emerge independently from the ribosome, it seems that in all probability targeting information on each one will be processed independently by the cell. Co-translational targeting to the endoplasmic reticulum (ER), Golgi, vacuole, or plasma membrane begins when the signal recognition particle (SRP) binds to a hydrophobic signal sequence of around 15aa present at the N-terminal end of the emerging protein (Zwieb *et al.*, 2005). The nascent chain is subsequently guided, *via* the SRP receptor at the ER, to the protein-conducting channel, or translocon (Keenan *et al.*, 2001; Thomas Rutkowski *et al.*, 2001). This topological arrangement is such that proteins are "insulated" from the cytoplasm as they pass from the peptidyltransferase centre through the ribosomal tunnel to the lumen of the ER. A signal sequence ($D_N\alpha$-factor or Pho8p) fused to the N-terminus of a 2A-based polyprotein expressed in *Saccharomyces cerevisiae* resulted in secretion of the protein upstream of 2A while the protein downstream of 2A (and lacking any signal sequence) remained in the cytoplasm (de Felipe *et al.*, 2003). We suggest these observations are reconciled by our co-translational model in which 2A-mediated cleavage is proteinase independent and occurs within the ribosome (de Felipe *et al.*, 2003; de Felipe and Ryan, 2004).

TARGETED EXPRESSION OF RECOMBINANT PROTEINS

The stability of a protein, its folding and assembly as well as post-translational modification for bioactivity depend on the cellular environment. Consequently, protein subcellular targeting is frequently required to achieve optimal protein expression and is commonly used to improve protein yield. By the inclusion of a range of co-translational signal sequences into our artificial reporter polyprotein systems we have shown that a protein with a signal sequence located at the N-terminus, followed by 2A and then a protein lacking any signal leads to localization of the first protein in the endomembrane system and the second protein in the cytosol (de Felipe *et al.*, 2003; El Amrani *et al.*, 2004). This is the case for plants and yeast, but not for the mammalian translocon system where the second protein is transferred by a "piggyback" effect into the ER (de Felipe and Ryan, 2004). We have shown this effect is due to low cleavage efficiency of the 2A-peptide motif (de Felipe *et al.*, 2009). Interactions between the nascent protein and the translocon may affect the conformation of 2A within the ribosome tunnel and, in consequence, the tight-turn motif at its C-terminus. As a result, a large proportion of the translation products are uncleaved, leading to translocation of the fusion protein into the exocytic pathway. Solutions to this problem include the use of longer 2As and/or, modifying the order of the proteins comprising the polyprotein. For polyproteins encoding a polypeptide destined to remain in the cytoplasm linked to an ER membrane or luminal protein *via* 2A, both polypeptides target to the correct location (de Felipe and Ryan, 2004; El Amrani *et al.*, 2004). More recent work in plant cells has shown if a signal peptide was positioned between the polypeptides, 2A was deemed unnecessary unless both polypeptides were translocated across the ER membrane (Samalova *et al.*, 2006). It appears that separation was dependent on the signal peptide cleavage, presumably by signal peptidase.

In contrast to the co-translational integration of secretory/trans-membrane proteins into the ER membrane, proteins destined for plastids, mitochondria, peroxisomes, and nucleus are synthesized by a common pool of free ribosomes situated in the cytosol. Many of these bear an N-terminal targeting signal, or transit peptide, which transports them through a post-translational targeting pathway to their final destination (Duby and Boutry, 2002; Soll, 2002). Proteins can be targeted post-translationally and imported to nucleus, mitochondria, chloroplast *etc*, when expressed from either anterior or posterior positions within a 2A-polyprotein if they possess the appropriate transit peptide (El Amrani *et al.*, 2004). As a caveat, Samalova *et al.* (2006, 2008) show that 2A, when placed downstream of GFP and RFP (red fluorescent protein) derivatives, does not promote efficient cleavage during translation on plant ribosomes. Since 2A activity can exhibit wider sequence-specific variation in plants than in animals, these findings do not necessarily contradict reports of its successful use in plants (Santa

Cruz *et al.*, 1996; Ma and Mitra, 2002; El Amrani *et al.*, 2004; Lengler *et al.*, 2005). Potential solutions to the problem include: (i) the use of longer 2As containing a favourable upstream context to avoid the potential presence of inhibitory residues at the C-terminus of the heterologous protein upstream of 2A, or (ii) optimization of the short 2A sequence itself to make its activity independent of the context.

2A in yeast biotechnology

Yeasts are attractive hosts for commercial heterologous protein production. These organisms have the significant advantages of a moderately uncomplicated process for generation of an expression strain, rapid growth and relatively low costs for production. The methylotrophic yeast, *Pichia pastoris*, is particularly well suited to foreign protein expression for a number of reasons, including ease of genetic manipulation, high levels of protein expression at the intra- or extracellular level, and the capability of performing many eukaryotic post-translational protein modifications such as glycosylation and proteolytic processing (reviewed in Daly and Hearn, 2005; Macauley-Patrick *et al.*, 2005). It is well established that "environmental" stress situations, and stress reactions, of the host cells can influence the productivity of *Pichia* expression systems (Mattanovich *et al.*, 2004). Particular attention has been paid to the introduction of glycine betaine (GB) synthesis for improving resistance against different stresses (Le Rudulier *et al.*, 1984). GB is generally considered to be the most compatible of the compatible solutes. In higher plants, GB is formed as the result of the two-step oxidation of choline by choline monooxygenase (CMO) and betaine aldehyde dehydrogenase (BADH). The genes encoding the two enzymes of GB synthesis in the halophyte *Suaeda salsa* were cloned and fused with the FMDV 2A in a single open reading frame. The fused genes "*CMO-2A-BADH*" transformed in *P. pastoris* were expressed successfully and the polyprotein was "cleaved" to each functional protein, [CMO-2A] and [BADH]. These GB-synthesizing transgenic lines were more tolerant to salt, methanol, and high temperature stresses (Wang *et al.*, 2007). The successful application of 2A technology in *Saccharomyces cerevisiae* (de Felilpe *et al.*, 2003) and *Pichia pastoris* will be valuable for visually marking transgene-expressing cells, or in any situation where reliable co-expression of multiple transgenes is desired.

2A in plant biotechnology

STABLE EXPRESSION IN PLANTS

Over the past two decades, the plant genome has been engineered with single genes for the production of transgenic crop plants expressing herbicide tolerance (Heck *et al.*, 2005), as well as resistance to fungal (Broglie *et al.*, 1991), viral (Nelson *et al.*, 1988), and bacterial (Mentag *et al.*, 2003) diseases and insect pests (Vaeck *et al.*, 1987). Although this has remained a major area of research, recent years have seen the emergence of plants as "factories" for the commercial production of valuable recombinant proteins. Numerous examples of heterologous proteins produced in a variety of plant hosts have been reported (Smith and Glick, 2000; Howard and Hood, 2005). Genetic engineering is thus moving from the initial phase of introducing single traits to multigenic traits, coding for complete metabolic pathways or biopharmaceuticals that require an assembly of complex multisubunit proteins (for a review see Dafny-Yelin and Tzfira, 2007). Conventional approaches to multigene engineering include sexual crossing, re-transformation, and co-transformation with multiple plasmids, or with single plasmids on which several transgenes are linked (reviewed in François *et al.*, 2002). However, there are drawbacks associated with these strategies such as unpredictable silencing and variability of transgene expression (reviewed by Halpin, 2005). As an alternative, several examples of the polyprotein approach for coordinate and stable expression of multiple proteins have been reported in the literature (François *et al.*, 2002; El Amrani *et al.*, 2004; Liang *et al.*, 2005). These examples differ in the linker sequence used to connect the different proteins and the resulting mode of cleavage of the precursor. In plant biotechnology, 2A has been used successfully as a linker in several examples of metabolome engineering and the introduction of novel product traits (reviewed in Luke *et al.*, 2006; de Felipe *et al.*, 2006).

STRESS RESPONSES IN ENGINEERED PLANTS

Improving the resistance of crops to osmotic stresses was one of the first objectives of plant metabolic engineering (Le Rudulier *et al*; 1984), and remains a major goal today. A common response of ahydrobiotic organisms to drought, salinity, and low temperature stresses is the accumulation of organic compounds of low molecular weight known collectively as compatible solutes. Trehalose is a non-reducing disaccharide of glucose that functions as a compatible solute in the stabilization of biological structures under abiotic stress in bacteria, fungi and invertebrates. With the notable exception of the desiccation-tolerant "resurrection-plants," trehalose is not thought to accumulate to detectable levels in most plants. The engineering of trehalose accumulation in plants has been undertaken not only to improve drought – and salt-tolerance, but also to produce trehalose at low cost for use as a stabilizing agent for pharmaceuticals and other products. Trehalose is biosynthesized by the trehalose-6-phosphate synthase/phosphatase (TPS1/TPS2) enzyme complex in a two-step pathway. Both TPS1 and TPS2 genes of *Zygosaccharomyces rouxii* were introduced simultaneously into potato plants as a ZrTPS2-2A-ZrTPS1 polyprotein in an attempt to generate abiotic stress tolerant plants. Although co-ordinate expression of both proteins was not shown, the resulting transgenic potato plants exhibited increased drought tolerance (Kwon *et al.*, 2004). This one step system also raised the possibility of generating plants resistant to other abiotic stresses.

To defend themselves against fungal pathogens, plants largely depend on the production of a wide array of antifungal molecules, including antimicrobial peptides such as defensins (Lay and Anderson, 2005). The precise mode of action of defensins is not completely understood, although it is generally accepted that they act at the level of the plasma membrane (Thevissen *et al.*, 2003). Further, defensins exhibit synergistically enhanced antifungal activity (François *et al.*, 2002). A polyprotein precursor consisting of two different plant defensins, DmAMP1 (from dahlia seed) and RsAFP2 (from radish seed), with their own signal peptide separated by an LP4-2A hybrid linker peptide, was successfully processed in *Arabidopsis thaliana* (François *et al.*, 2004). The hybrid linker was the first nine amino acids of the fourth linker peptide of the naturally occurring polyprotein precursor originating from seed of *Impatiens balsamina* (**SN↓AADEVAT**) followed by the 20 amino acids of FMDV 2A (Ryan and

Drew, 1994; Tailor *et al.*, 1997). The DmAMP1 was released with 2 additional amino acids at its carboxy-terminus, namely serine and asparagine. The authors speculate that the endogenous proteinase responsible for this cleavage is an asparagine-specific proteinase cleaving at the carboxy-terminus of asparagine. By using a hybrid linker peptide it was possible to produce the individual plant defensins and to target them to mutually exclusive cellular compartments.

IMPROVING THE NUTRITIONAL VALUE OF PLANTS

Making use of the innate protein sorting and targeting mechanism that plant cells normally use for targeting, combinatorial 2A strategies can enhance expression levels of recombinant proteins in plants. In developing maize endosperm cells, the accumulation of zeins (the major storage proteins of maize) involves transport of the newly synthesized proteins into the ER by means of an N-terminal signal peptide, and then assembly into protein bodies. Randall and co-workers (Randall *et al.*, 2004) investigated the potential for producing sulphur-rich β-zein and δ-zein proteins by co-expressing the corresponding genes (each with its signal sequence) as a β-zein-2A-δ-zein construct in transgenic tobacco. Effective processing of the polyprotein was observed, resulting in the stable accumulation of both zein proteins that were targeted to ER-derived protein bodies. When the zeins were fused directly without 2A, the fusion was highly unstable and no separation of β- and δ-zein moieties was observed.

Astaxanthin or its precursors are high-value carotenoids with industrial and biomedical applications. Most astaxanthin available commercially is chemically synthesized. In carotenogenic organisms β-carotene is converted into astaxanthin using several ketocarotenoid intermediates and two enzymes: a β-carotene ketolase encoded by the *crtW* gene, and a β-carotene hydroxylase encoded by the *crtZ* gene. Although the substrate for ketocarotenoid formation, β-carotene, is present in substantial quantities in all photosynthetic plant tissues higher plants do not possess the ability to form ketocarotenoids. Plant-based production offers an attractive cost effective option. The two enzymes crtW and crtZ (from *Paracoccus* sp), carrying N-terminal transit peptide extensions to mediate their post-translational import into chromoplasts, were simultaneously introduced into plants as a crtW-2A-crtZ construct. Subsequent cleavage of the polyprotein, targeting of the two enzymes to the plastid and enzyme activity was shown for both gene products in both tobacco and tomato plants (Ralley *et al.*, 2004).

PLANTS AS BIOREACTORS FOR PHARMACEUTICAL PROTEINS

The use of transgenic plants, seeds and cultured plant cells as bioreactors is now well established. Plant systems offer a safe and cost-effective alternative to microbial or mammalian expression systems for the production of recombinant biopharmaceutical proteins. In particular, genetically modified seeds are very cost-efficient hosts. Yasuda *et al.* (2005) describe a new approach to express the small human peptide hormone, GLP-1 (Glucagon-Like Peptide-1), in rice seed. GLP-1 is a potent blood glucose-lowering hormone that stimulates the secretion of insulin from pancreatic β-cells (Drucker, 2001). The actions of GLP-1 to reduce glycaemia while preventing concomitant weight gain has attracted considerable interest in pharmaceutical approaches to enhancing GLP-1 action for the treatment of type 2 diabetes (Deacon, 2004). The expression of small peptides in genetically modified plant systems is frequently compromised by transgene silencing (Okamoto *et al.*, 1998; François *et al.*, 2002). To avoid silencing, mGLP-1 was fused to GFP with or without the self-processing FMDV 2A sequence, and introduced into rice plants. Both chimeric genes were highly expressed in these transgenic rice seeds, indicating that gene silencing could be avoided by coupling of smaller peptides to a carrier protein (Sugita *et al.*, 2005). Furthermore, the fusion protein containing the 2A sequence was efficiently processed into GFP2A and mGLP-1 peptides. Lack of accumulation of mGLP-1 released by 2A was explained by proteolytic digestion in the cytosolic space of the cells. If the mGLP-1 peptide were targeted into another compartment such as ER or extracellular space, the peptide should accumulate in rice seed cells.

TRANSIENT EXPRESSION IN PLANTS

PLANT VIRUS FUSION PROTEINS

The small size of plant viral genomes, coupled with rapid and convenient engineering, makes viral expression vectors an attractive alternative to production systems based on transgenic plants. Such vectors offer temporary, transient expression systems for the production of valuable proteins, such as vaccines and antibodies in plants. The most widely used viruses are *Potato virus X* (PVX) (Avesani *et al.*, 2003); *Tobacco Mosaic virus* (TMV) (Karasev *et al.*, 2005) and *Cowpea Mosaic virus* (CPMV) (Liu *et al.*, 2005). While the expression and assembly of full-length proteins is often the objective, another important application is the expression of peptides as coat protein (CP) fusions (Cañizares *et al.*, 2005). Such fusions serve to stabilize the expressed protein or peptide, and their purification is straightforward (Porta and Lomonossoff, 1998). For rod-shaped viruses, such as PVX, structural and immunological evidence reveals that the N-terminal part of the CP is presented on the particle surface, enabling the decoration of the particle with a recombinant fusion peptide or protein (Baratova *et al.*, 1992). Inoculation of plants with a recombinant virus genome encoding GFP fused to CP [GFP-CP] have been shown not to produce any particles whereas expression of the fusion protein GFP linked to CP *via* the FMDV 2A [GFP-2A-CP] resulted in the formation of virus decorated with GFP (*Figure 4*). Assembly of "decorated" virions required the presence of free coat protein subunits in addition to the fusion protein subunits. Furthermore, the modified virus retained the ability to move locally and systemically through the plant (Santa Cruz *et al.*, 1996). The crystallographic data of CPMV shows that the N-terminus of the L coat protein is buried inside the virus particles (Lomonossoff and Johnson, 1991). Particularly puzzling then is the finding that some GFP-2A-L fusion proteins appear to be incorporated into the virus particles, resulting in a "green virus" (Gopinath *et al.*, 2000).

Plant viruses have recently been considered as attractive systems for the presentation of foreign epitopes as immunogens in the development of novel vaccination strategies (Streatfield and Howard, 2003). In particular, edible vaccines show great promise as a low-cost delivery mechanism for immunization against various diseases. Classical swine fever (CSF) is an economically important viral infectious disease affecting wild boars and domestic pigs. The envelope glycoprotein (E2) of CSFV was expressed in *Nicotiana benthamiana* plants as an N-terminal fusion *via* 2A-peptide from FMDV with the PVX coat protein [PVX E2-2A-CP]. The resulting chimeric virus particles, purified and used to immunize rabbits, were able to elicit high levels of anti-E2 antibodies. Furthermore, serial infections in *N. benthamiana* demonstrated the epitope coding sequences could still be detected in the third infection cycle. The presence of mixed-type particles assembled from a pool of CP and E2-2A-CP fusion polypeptides contributed to insert stability and maintained virus infectivity (Marconi *et al.*, 2006).

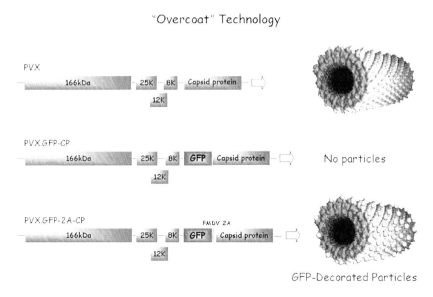

Figure 4. Organization of the wild-type PVX; PVX.GFP-CP, PVX modified to express free GFP and PVX. GFP-2A-CP, PVX modified to express the GFP-2A-CP fusion protein. Boxes represent coding sequences. The predicted Mr values of the four viral proteins common to all constructs are indicated.

ANTIBODY EXPRESSION IN PLANTS (PLANTIBODIES)

An alternative to inducing the immune system to produce antibody is to deliver them directly. Antibodies can be expressed either transiently in plant leaves or stably in transgenic plants, depending on the applications required (Ma *et al.*, 2005). Transient gene expression with modified viral vector is a fast and flexible expression system for high-level production of recombinant antibodies without generation of transgenic plants. Recombinant antibodies include fully assembled whole immunoglobulins (Verch *et al.*, 1998), antigen-binding fragments of immunoglobulins (Sainsbury *et al.*, 2008), and single-chain variable fragment gene fusions (scFv) (Smolenska *et al.*, 1998). To demonstrate the value of this approach a PVX vector was used to express a scFv against the herbicide diuron, as a fusion to the viral coat protein [scFv-2A-CP]. The modified virus accumulated in inoculated *Nicotiana clevelandii* plants and assembled to give virus particles carrying the functional antibody fragment (Smolenska *et al.*, 1998).

The potential advantages which make CPMV an attractive vector system are that the virus grows extremely well in host plants, virus particles are thermostable, and virus particles can easily be separated from infected plant tissue. The genome of CPMV

consists of two positive-stranded RNAs, termed RNA-1 and RNA-2, which are separately encapsidated in identical protein shells (for a review see Goldbach and Wellink,
1996). Both RNAs are translated to give precursor polyproteins, which are processed
into functional polypeptides. RNA-1 encodes proteins involved in RNA replication
and RNA-2, which is dependent on RNA-1 for its replication, encodes the viral movement protein (MP) and the two coat proteins, large (L) and small (S) (Pouwels *et al.*,
2002). When the sequence of GFP was linked to the C-terminus of the S coat protein
via an active FMDV 2A sequence [CPMV/S-2A-GFP], a highly infectious construct
expressing non-fused GFP was obtained (Gopinath *et al.*, 2000). In this context, using
multiple copies of RNA-2 containing different marker genes, it has been demonstrated
that at least two foreign proteins can be produced at high levels and co-directed to the
same location in inoculated tissue (Sainsbury *et al.*, 2008). To this end, it was possible
to use two different RNA-2 molecules containing the heavy and light chains of a murine
antibody to express assembled IgG in *Nicotiana benthamiana*. A notable finding from
this study is that substantially higher levels of assembled antibody could be obtained
using deleted rather than full-length versions of RNA-2. In this system, the region encoding the movement protein and both coat proteins has been removed (Cañizares *et
al.*, 2006). It is clear that deleted versions offer distinct advantages, not only in terms
of expression levels but also in terms of biocontainment.

2A in Animal Biotechnology

IMPROVING THE EFFICACY OF DNA VACCINES

Cytokines are powerful regulators of the immune response. The cytokine interleukin-12
(IL-12) is essential for the induction of interferon-γ (IFN-γ) T helper 1 (Th1)-like cell
responses, and is considered to play a central role in the interaction between the innate
and adaptive arms of immunity (Trinchieri *et al.*, 2003). IL-12 is functionally active
as a heterodimer consisting of p35 and p40 subunits linked by a disulphide bond. The
expression of this cytokine has been complicated by the observation that p40 homodim-
ers exhibit antagonistic activity towards IL-12 (Gillessen *et al.*, 1995). Therefore, it
is desirable to express equal amounts of both chains to generate biologically active
IL-12 (Wolf *et al.*, 1991). Given the stoichiometric production of 2A-peptide linked
proteins, this system is ideally suited for the expression of heterodimeric IL-12. Among
ruminants, recombinant bovine and ovine IL-12 have been produced using the FMDV
2A as a linker between the p40 and p35 subunits (Collins *et al.*, 1998; Chaplin *et al.*,
1999; De Rose *et al.*, 2000; Premraj *et al.*, 2006).
   The ability of plasmid DNA vaccine-elicited immune responses to protect against
viral and bacterial infections, parasites, cancers, and autoimmune diseases has been
well documented in a variety of animal models. One approach to improve the immunogenicity of DNA vaccines is through the co-delivery of cytokine expression
plasmids as genetic adjuvants (Egan and Israel, 2002). Several studies have established
that DNA vaccines encoding mycobacterial antigens induce partial protection against
experimental infection with *Mycobacterium tuberculosis*, *M.leprae* or *M.avium*. Furthermore, co-immunization with IL-12 (p2AIL-12) enhanced the immune response to
these DNA vaccines (Triccas *et al.*, 2002; Palendira *et al.*, 2002; Martin *et al.*, 2003).
IL-12 represents the first heterodimeric protein in an expanding cytokine family, that

includes the two new but so far not well characterized members IL-23 and IL-27 (Trinchieri *et al.*, 2003; Brombacher *et al.*, 2003). Wozniak *et al.* (2006) compared the effects of plasmids expressing both chains of IL-12, IL-23, and IL-27 as adjuvants for DNA immunization against *M. tuberculosis* infection. The genes encoding p19 and p40 chains of IL-23 and Epstein-Barr virus induced gene 3 (EBI3) and p28 chains of IL-27 were cloned on either side of FMDV 2A. Codelivery of plasmids expressing IL-12 and IL-23 (but not IL-27) with the protective antigen DNA85B increased the expansion of antigen-specific IFN-γ secreting T cells and enhanced protective immunity. The ability to produce DNA vaccines expressing multiple epitopes (Yu *et al.*, 2007) means that adjunctive immunotherapy may be tailored for a given pathogen.

THERAPEUTIC MONOCLONAL ANTIBODIES

Engineered antibodies designed as intact molecules or recombinant fragments are currently being developed for the effective treatment of illnesses as varied as cancer, viral diseases and autoimmune diseases. Although *in vivo* transfer of genes encoding specific antibodies could make a significant contribution to medical therapy, the major concern is whether the production will be sufficient to create therapeutically effective plasma levels. In a promising study, Fang *et al.* (2005) describe a recombinant adeno-associated virus (rAAV) gene delivery system that allows sustained production of a full-length antibody at high-concentrations in mice. Monoclonal antibodies (mAbs) as anti-angiogenic agents were expressed from a single ORF by linking the heavy and light chains with a FMDV 2A-peptide. Furthermore, to produce authentic antibody, the 2A sequence at the carboxy-terminus of the heavy chain was removed by adding a furin cleavage site (RAKR) immediately upstream of 2A. Furin is a ubiquitous cellular endoprotease that processes a diverse group of functionally distinct precursors as they pass through the *trans*-Golgi en route to the plasma membrane (Steiner, 1998). This modification resulted in increased mAb serum levels and produced an antibody that closely resembles the fully native protein.

Another concern about gene transfer is the possibility of unwanted and harmful self-directed responses associated with systemic exposure. An approach to uncouple these events with the clinical responses is to administer the antibody locally. Both preclinical murine models and early clinical studies suggest synergy between CTLA-4 antibody blockade and vaccination with irradiated GM-CSF tumour cells (Hurwitz *et al.*, 2000; Hodi *et al.*, 2003; Quezada *et al.*, 2006). Cytotoxic T lymphocyte antigen-4 (CTLA-4) signaling plays an important role in the down-regulation of activated T cells and granulocyte macrophage colony-stimulating factor (GM-CSF) is a cytokine that functions as a white blood cell growth factor. In a combinatorial study, tumour regression as well as decreased levels of autoimmunity associated antibodies have been observed in mice treated with a whole cell immunotherapy for the local secretion of anti-CTLA-4 and GM-CSF (Simmons *et al.*, 2008). The retroviral transfer vector used for expressing the full-length murine anti-CTLA-4 mAb was generated by linking the heavy and light chains using furin and FMDV 2A cleavage sites. Considered together, the longevity of antibody production and the maintenance of tolerance, promising applications of 2A-based therapeutic mAbs might exist in the treatment of chronic diseases such as cancer.

2A-BASED VECTORS IN GENE THERAPY

Gene therapy potentially represents one of the most important developments in modern medicine. The philosophy of gene therapy is simple: "transfer of genetic material to cure a disease or at least to improve the clinical status of a patient" (Pfeifer and Verma, 2001). Gene therapy is being developed for a range of diseases including inherited monogenic disorders (*e.g.* cystic fibrosis), but it is in the treatment of polygenic diseases that this approach has been most evident. For this reason, gene therapy would benefit from vectors capable of expressing multiple genes. Conventional approaches for the production of multicistronic vectors include the use of IRES elements, multiple promoters, and fusion proteins. IRES elements can be large and none of these strategies ensures predictable concordant gene expression. The major advantages of using 2A are its relatively small size and efficient co-expression of transgenes is ensured. To construct multigene vectors, the 2A sequence has been successfully incorporated into adenovirus (Funston *et al.*, 2008), adeno-associated virus (AAV) (Furler *et al.*, 2001; Fang *et al.*, 2005), retrovirus (Schmidt and Rethwilm, 1995; de Felipe *et al.*, 1999; Klump *et al.*, 2001; Szymczak *et al.*, 2004; Milsom *et al.*, 2004), lentivirus (Chinnasamy *et al.*, 2006; Yang *et al.*, 2008; Lee *et al.*, 2008) and plasmid vectors (Osborn *et al.*, 2005; Hasegawa *et al.*, 2007).

OPTICAL IMAGING OF GENE EXPRESSION IN ANIMAL MODELS

Recent advances in molecular imaging technologies should allow the imaging sciences to play a major role in clinical gene therapy (for general reviews on the subject see Massoud and Gambhir, 2003; Jaffer and Weissleder, 2005; Lee *et al.*, 2008). In particular, reporter genes with optical signatures (*e.g* bioluminescence and fluorescence) offer "low"-cost options to "see" the expression of therapeutic genes in targeted tissues. In a recent study, bioluminescence imaging (BLI) with firefly luciferase (*fluc*) of murine haematopoietic stem cells (HSCs) was used for observing the impact of drug selection on the dynamic repopulation process after transplantation (Lee *et al.*, 2008). The chemo-drug resistant gene P140K MGMT and *fluc* gene were linked by the FMDV 2A. O^6-methylguanine-DNA-methyltransferase (MGMT), a DNA repair protein, is known to remove alkyl groups from the O^6 –position of guanine, thus regenerating intact DNA (Daniels and Tainer, 2000). This therapy resulted in the proportional and constant co-expression of both genes over a wide range of transgene expression levels and eliminated the attenuation and tissue-variation problems of the IRES-based approach (Yu *et al.*, 2000).

Combinations of imaging modalities that integrate the strengths of the two modalities offer the prospect of improved therapeutic monitoring and new opportunities for imaging. *In vivo* bioluminescence imaging is a powerful technology based on the use of optically active luciferase reporters. A particular challenge has been to introduce the luciferase gene into cells of interest. The problem has, in part, been circumvented by Cao and colleagues, who generated transgenic mouse strains that constitutively express luciferase (Cao *et al.*, 2005). To extend this approach *fluc* was coupled to GFP *via* FMDV 2A to create a dual function reporter system. Using these transgenic animals, any cell population can be isolated and used as a virtually unlimited source of uniformly labeled cells for stem cell and transplantation studies. Stem cells isolated from this line, were used to successfully demonstrate, in real time, the early events and dynamics of haematopoietic engraftment in living animals.

Mucopolysaccharidosis type 1 (MPS1) is a lysosomal disease due to mutations in the IDUA gene, resulting in deficiency of α-L-iduronidase and accumulation of glycosaminoglycans (GAGs). Individuals presenting with the most severe forms of MPS1 typically suffer from a number of progressively debilitating symptoms (Neufeld and Muenzer, 2001). Bone marrow transplant (BMT), which is a standard therapy for MPS1, is not curative and can lead to morbidity and mortality as complications of the procedure (Peters *et al.*, 1996). Enzyme replacement therapy (ERT) has been developed as a treatment for a number of lysosomal storage diseases. (Kakkis *et al.*, 2001; Wraith *et al.*, 2004). Therapeutic vectors capable of replacing IDUA enzyme levels while at the same time allowing for direct detection would be useful for preclinical and gene therapy studies. The small 2A-like sequences of porcine teschovirus (P2A) and *Thosea asigna* virus (T2A) were used to create a tricistronic vector encoding the luciferase and DsRed2 reporter genes coupled to the therapeutic IDUA gene (IDUA-P2A-luciferase-T2A-DsRED). Efficient cleavage was observed and all three proteins were functional *in vivo* and *in vitro*, allowing for high enzyme levels that could be tracked by non-invasive whole-body luciferase imaging and at the cellular level using DsRed2 (Osborn *et al.*, 2005). The use of dual optical reporter genes in this way, has been shown to enable *in vivo* measurements to facilitate and direct *in vitro* examination.

STEM CELL-BASED GENE THERAPY

Notwithstanding the topical debate on the use of embryonic stem cells (ESCs) versus adult stem cells, both cell types can contribute to the development of regenerative medicine and tissue replacement. Human embryonic stem cells (hESCs) and their derivatives have the advantage of pluripotency and are easily accessible for controlled and specific genetic manipulation. As an initial step in genetic modification, hESCs and hESC-derived cells transfected with plasmid GFP(2A)DsRed showed GFP and DsRed fluorescence (Hasegawa *et al.*, 2007). This is in contrast to IRES-mediated translational initiation where the downstream gene was poorly expressed. Supportive results were also obtained using a tricistronic construct encoding a yellow fluorescent protein with a mitochondrial targeting sequence (YFPmito), cyan fluorescent protein with a nuclear localization signal (CFPnuc) and DsRed. In cells transfected with YFPmito(2A) CFPnuc(2A)DsRed, the proteins have been successfully expressed and no adverse effects on localization detected. This indicates 2A-peptide technology may be invaluable for genetic modification of hESCs and their differentiated progeny *in vitro*, as well as in clinical *ex vivo* gene therapy.

The prototypic example of adult stem cells, the haematopoietic stem cell (HSC), has already proved useful in gene therapy (Aiuti *et al.*, 2002). However, implementation has been hampered by the inability to transduce sufficient cell numbers to exert a phenotypic change. One approach to this has been to protect progenitor cells against the biological effects of alkylating agents by retroviral transduction with an engineered form of MGMT such as P140K that is resistant to inactivation (Jansen *et al.*, 2002; Passagne *et al.*, 2006). An alternative means of achieving *in vivo* selection is to impart a proliferative advantage upon transduced HSCs. The homeobox transcription factor HOXB4 has emerged as an important and potent stimulator of HSC proliferation *in vivo* and *ex vivo* (Antonchuk *et al.*, 2001, 2002). Previous studies have shown that HOXB4 expressed using the 2A strategy retains its ability to support haematopoietic reconstitution by murine HSCs (Klump *et al.*, 2001). Using a murine bone marrow transplantation model, the retroviral delivery of both HOXB4 and mutant MGMT (P140K) *via* the FMDV 2A resulted in enhanced selection of gene-modified cells compared to cells expressing either HOXB4 or MGMT (P140K) alone (Milsom *et al.*, 2004). In addition to efficient generation of cleavage products, it is important that these are transported to the appropriate compartment of the cell. The correct subcellular localization of expressed MGMT and HOXB4 (*tricistronic vectors*: MGMT-2A-HOXB4-IRES-eGFP and HOXB4-2A-MGMT-IRES-eGFP) to the nucleus in multiple cell types shows the addition of 2A sequences did not adversely affect the trafficking of these two proteins (Chinnasamy *et al.*, 2006).

The reprogramming of mouse and human fibroblasts into ESC-like or induced pluripotent stem (iPS) cells (Takahashi and Yamanaka, 2006; Takahashi *et al.*, 2007) has been hailed by many as the holy grail of stem cell research. The "backward step" offers a possible source of tailor-made pluripotent cells and bypasses the ethical controversies that hinder the applications of hESC (see Baertschi and Mauron, 2008). Induction of pluripotent stem cells can be achieved with just four transcription factors, *Oct3/4*, *Sox2*, *Klf4* and *C-Myc* (Takahashi and Yamanaka, 2006; Takahashi *et al.*, 2007; Wernig *et al.*, 2007; Park *et al.*, 2007). However, the possibility of activating oncogenes and generating mutations in the genome through random insertions of the retroviral vectors carrying these genes, may handicap this method for human therapeutic applications. Taking this into account, repeated transient transfections of two expression plasmids, one containing the cDNAs of Oct3/4, Sox2, and Klf4 connected with the FMDV 2A, and the other containing the c-Myc cDNA, into mouse embryonic fibroblasts resulted in iPS cells without integrations. The efficiency of iPS cell generation, however, was significantly reduced and the reprogrammed cell type limited (Okita *et al.*, 2008; see also Stadfeld *et al.*, 2008). A promising approach towards improving efficiency and minimizing the insertion risk by using a single lentiviral vector with a polycistronic cassette has recently been described. The polycistronic cassette comprised the four factors and a combination of two 2A peptides and an IRES element (Sommer *et al.*, 2008) or the four factors and three different 2A peptides (Carey *et al.*, 2009).

ORGANIZATION OF THE T-CELL RECEPTOR (TCR)/CD3 COMPLEX

The T-cell receptor (TCR)/CD3 complex is an elaborate structure designed to recognize antigens on antigen-presenting cells and trigger a number of intracellular signal pathways. Signals, both stimulatory and inhibitory, control T-cell development and the deployment of effector functions during an immune response (Germain and Stefanova, 1999). The variable TCRα and TCRβ chains are expressed on the cell surface with the non-covalently associated CD3γ, CD3δ, CD3ε and ζ-chain transmembrane polypeptides. These six subunits form four dimeric molecules: the ligand-binding TCRαβ heterodimer and the CD3δε, CD3γε and ζζ signalling dimers (reviewed in Call and Wucherpfennig, 2007). Defects in TCR/CD3 expression and function have been reported in a heterogeneous group of diseases, where patients are characterized by their increased susceptibility to infection. Reprogramming the antigen specificity of T cells by TCR gene transfer offers considerable potential for translating basic research into clinical practice (Xue *et al.*, 2005; Engels and Uckert, 2007).

ASSEMBLY OF THE FUNCTIONAL TCR/CD3 COMPLEX IN MICE

TCR transgenic (Tg) mice are a valuable source of antigen-specific T cells for studying T-cell development and responses but are expensive and time-consuming to generate. A novel procedure bypasses the need for transgenesis to achieve this goal, using instead 2A-linked retroviruses to rapidly generate TCR retrogenic (Rg) mice (Arnold *et al.*, 2004; Szymczak *et al.*, 2004; Holst *et al.*, 2006a & b). Using the TCR:CD3 complex as a test system, Szymczak and colleagues reported expression of all four proteins that make up CD3 and the two proteins required to make up TCR using just two retroviral vectors (CD3δγεζ–2A and TCRαβ-2A) (reviewed by Radcliffe and Mitrophanous, 2004). Because retroviruses are known to recombine vector sequences that contain duplications, 2A sequences from FMDV, TaV, and ERAV were used - each cleaving highly efficiently (*Figure 5*). Efficient cell surface expression of TCR/CD3 complexes on bone-marrow derived stem cells was obtained after transduction with the 2A-peptide-linked constructs. The 2A-peptide was also shown to function in derivatives of the transduced stem cells (in the spleen, thymus and blood). Equally impressive results were seen with *in vivo* experiments – all four chains of the CD3 complex were efficiently introduced into CD3-deficient mice, resulting in rescue of T-cell development and immune function.

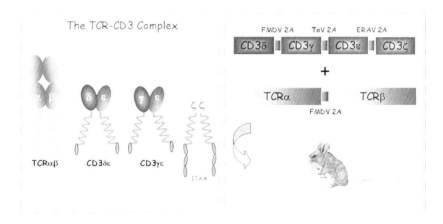

Figure 5. Organization of the TCR-CD3 complex and 2A-linked TCR:CD3 constructs used to restore T-cell development and function in CD3-deficient mice (adapted from Call *et al.*, 2002)

TCR EXPRESSION USING RETROVIRAL VECTORS

Adoptive (or passive) immunotherapy with single retroviral vectors encoding the TCR α and β chains linked by an IRES or 2A-like peptide is an attractive strategy to redirect the antigen specificity of T cells. Both the TCR genes and the vector design can be optimized to improve functional expression at the cell surface of TCR-transferred T cells. Scholten *et al.* (2006) were the first to show that codon modification of α– and β- chain genes, to facilitate TCR mRNA translation, increased the protein expression in TCR transgenic cells (reviewed by Heemskerk, 2006). Besides codon modification, functionality is defined by the transgene cassette. In a comparative study between 2A and IRES-linked TCR genes, the 2A-peptide was superior for expression and function in human and murine recipient cells (Leisegang *et al.*, 2008). Furthermore, IRES-dependent second gene expression was lower than that of the first gene confirming several earlier reports of biased expression of transgenes (Flasshove *et al.*, 2000; Mizuguchi *et al.*, 2000; Hasegawa *et al.*, 2007). Finally, Hart and colleagues combined several strategies to maximize expression of the introduced TCR; the human constant region was replaced with murine sequences (Cohen *et al.*, 2006), the two genes linked using a 2A sequence and the TCR-α-2A-β cassette codon optimized for efficient translation in human cells (Hart *et al.*, 2008).

TCR EXPRESSION USING LENTIVIRAL VECTORS

Lentiviruses are complex retroviruses whose distinguishing feature is their ability to integrate into non-dividing cells without requiring the disassembly of the nuclear membrane. This makes lentiviral expression vectors effective vehicles for gene delivery and expression (Federico, 2003; Heiser, 2004). Recent work by Yang and co-workers examined the feasibility and efficiency of lentiviral TCR gene delivery (Yang *et al.*, 2008). The optimized lentiviral vector combined a furin cleavage site to remove the residual "tag" and a V5 spacer sequence ahead of the 2A-peptide (TCRα-fuV5F2A-TCRβ). How the synthetic V5 peptide correlates with enhanced activities is unknown. In this and other studies, cleavage efficiency of 2A-linked constructs was improved by placing a linker (GSG or SGSG) between the N-terminal protein and the 2A-peptide (Holst *et al.*, 2006b; Provost *et al.*, 2007; Yang *et al.*, 2008; Holst *et al.*, 2008). As alluded to earlier, cleavage efficiency of 2A is affected by the preceding peptide sequence and the upstream protein (Donnelly *et al.*, 1997; Donnelly *et al.*, 2001b). Presumably, the linker sequence adopts an extended conformation to allow for maximal flexibility and enhanced cleavage.

FUNCTIONAL ANALYSIS OF TCR/CD3 ITAMS IN VIVO

Certain receptors expressed by haematopoietic cells have immunoreceptor tyrosine-based activation motifs (ITAMs) in their cytoplasmic domains that initiate cellular activation and differentiation (Lanier, 2006). The TCR/CD3 complex contains a total of 10 ITAMs, one in each of the four CD3 polypeptides and three in each of the two strands of the $\zeta\zeta$ homodimer (see *Figure 5*). It has been suggested that this high ITAM number could contribute to the sensitivity of T cells by amplifying signals generated as a result of encounters with low-density ligands. In a recent study, Rg mice were generated expressing different $CD3^{WT}$ and $CD3^{M}$ combinations using 2A to facilitate the functional analysis of all the CD3 ITAMs *in vivo* (Holst *et al.*, 2008; for brief review, see Malissen, 2008). The signature sequence of a $CD3^{WT}$ (wild-type) ITAM is $YxxL/I-x_{6-8}-YxxL/I$ (where x denotes any amino acid). $CD3^{M}$ represents a Y-F mutant for both ITAM tyrosine residues in that chain. It was demonstrated that a high ITAM number was required for T-cell development and that small differences in ITAM sequence could influence TCR signaling and thymocyte development. Moreover, reduced ITAM numbers, which presumably translate into reduced TCR signal strength, resulted in a lethal autoimmune disease.

Transgenic animals provide a *bona fide* environment for the study of gene function. The practical applications of transgenesis include animals as disease models, animals for the production of therapeutic substances and animals as cell and organ donors. Over the past few years the zebrafish (*Danio rerio*) has emerged as a model system for the study of vertebrate development and disease. Importantly, the transparency and *ex utero* development of the embryo make it possible to monitor transgene effects in real-time and *in vivo*. To test the function of 2A in zebrafish, reporter constructs *eGFP^CAAX-2A-mCherry* and *TMmCherry-2A-NLSeGFP* have been designed to segregate fluorescent proteins to distinct cellular locations (Provost *et al.*, 2007). The CAAX motif, the transmembrane domain (TM) and nuclear localizing sequence (NLS) target proteins to the endomembrane, the plasma membrane and nucleus respectively. In addition, both constructs incorporated a GSG linker on the N-terminal of the PTV-1 2A sequence to improve separation (Holst *et al.*, 2006b). *Tol2*-mediated gene transfer of *eGFP^CAAX-2A-mCherry* into zebrafish embryos produced individual GFP and mCherry proteins indicating efficient cleavage of the 2A-peptide. Furthermore, the expression of *TMmCherry-2A-NLSeGFP* in live embryos resulted in localization of TMmCherry and NLSeGFP to separate and distinct cellular compartments. Lastly, replacing the existing promoter sequence (EF1α) with the insulin promoter (T2K-ins) directed expression to the β cells of the developing zebrafish pancreas. Tissue-specific expression of eGFP and mCherry marker proteins in stably transformed embryos shows this approach could facilitate continuous expression of multiple protein products at various stages of development in zebrafish.

Genetically altered mice offer innovative ways to study gene function as well as model events during developmental processes or human diseases. In this context, Trichas and colleagues report a simple method for generating transgenic mice using a bicistronic reporter construct containing the 2A sequence (Trichas *et al.*, 2008; see also Cao *et al.*, 2005). The reporter comprised a red marker (TdTomato) membrane protein linked *via* the TaV 2A to a green marker (GFP) nuclear protein. Red and green fluorescence was found exclusively in the appropriate subcellular compartments of transfected HeLa cells and infected chick cells, consistent with complete 2A-mediated processing. Predictable expression of reporter sequences such as CAT or GUS in transgenic mice has always been difficult; constructs that work well in cell culture often fail to express in mice (Pardy, 1994; Clark *et al.*, 1997). For the transgenic lines produced in this study targeted expression was apparent in all tissues examined throughout development and into adulthood and remained constant across several generations. This demonstrates the utility of this methodology for generating transgenic mice for both ubiquitous and tissue-specific transgene expression.

So many innovations, so little space

The ability to generate reliable vectors for multigene delivery is at the vanguard of biotechnology and genetic engineering. Conventional approaches for the production of multicistronic vectors using IRESs or multiple promoters are problematic because they cannot ensure that transgenes will be expressed at similar levels. Particularly appealing then, is the stoichiometric expression of 2A-peptide-linked genes and the small size of 2A peptides compared to internal promoters or IRES sequences makes them ideal candidates for use in size-restricted viral and nonviral vectors. Additionally, the diversity of the 2A sequence minimizes the chances for homologous recombination which is an important consideration when using retroviral or lentiviral systems. One outstanding question is the effect of the 2A "tag" attached to the C-terminus of the upstream protein. This may interfere with function, or more importantly may present a new epitope that could be subject to immunological surveillance. However, the attachment of extra amino acids is a routine method for labelling transgene products while leaving their function intact (*e.g.* tags such as the His tag and Myc tag). To our knowledge, the 2A tag does not impair activity and expression - proteins that require authentic termini, or are N-/C- terminally modified, can be introduced as the first or final polyprotein domain, respectively. In any event, strategies have now been devised that allow removal of the 2A linker (see François *et al.*, 2004; Fang *et al.*, 2005). The "unwanted" tag may however stick – antibodies directed against 2A can be used to detect the gene cloned upstream (Ryan and Drew, 1994; de Felipe *et al.*, 2003, 2006). Lastly, the presence of a proline residue at the N-terminus of the downstream protein, as a relict of the 2A self-cleaving process, does not normally interfere with function – it does, however, confer high protein stability (Varshavsky, 1992).

Aware of the factors that influence expression levels it is important to empirically design any co-expression cassette to ensure the polyprotein is the most suitable arrangement in respect to desired function. As a form of control of protein biogenesis, 2A sequences are much more wide-spread than was first suspected. To appease different and opposing sensibilities, 2A variants that are not found in mammalian viruses can be used just as effectively for the production of multiple protein products. Although a relative new-kid-on-the-block in terms of co-expression studies, 2A can safely be considered an "established" player. It is clear that assorted 2A-derived proteins with diverse and distinct localized functions may be stably expressed in several different cell types demonstrating the applicability of this technology in biomedicine and biotechnology. The biotechnological applications of 2A are continually updated on www.st-andrews.ac.uk/ryanlab/Index.htm. We envisage that 2A technology will become one of the predominant strategies for multigene delivery in the coming years.

Acknowledgements

We gratefully acknowledge the long term support of the Wellcome Trust and the Biotechnology and Biological Sciences Research Council. The University of St Andrews is a charity registered in Scotland no. SCO13532.

References

AIUTI, A., SLAVIN, S., AKER, M., FICARA, F., DEOLA, S., MORTELLARO, A., MORECKI, S., ANDOLFI, G., TABUCCHI, A., CARLUCCI, F., MARINELLO, E., CATTANEO, F., VAI, S., SERVIDA, P., MINIERO, R., RONCAROLO, M.G., AND BORDIGNON, C. (2002) Correction of ADA-SCID by stem cell gene therapy combined with nonmyeloablative conditioning. *Science* **296**(5577), 2410-2413.

ANTONCHUK, J., SAUVAGEAU, G., AND HUMPHRIES, R.K. (2001) HOXB4 overexpression mediates very rapid stem cell regeneration and competitive hematopoietic repopulation. *Experimental Hematology* **29**, 1125-1134.

ANTONCHUK, J., SAUVAGEAU, G., AND HUMPHRIES, R.K. (2002) HOXB4 –induced expansion of adult hematopoietic cells *ex vivo*. *Cell* **109**, 39-45.

ARNOLD, P.Y., BURTON, A.R., AND VIGNALI, D.A.A. (2004) Diabetes Incidence Is Unaltered in Glutamate Decarboxylase 65-Specific TCR Retrogenic Nonobese Diabetic Mice: Generation by Retroviral-Mediated Stem Cell Gene Transfer. *The Journal of Immunology* **173**, 3103-3111.

ATKINS, J.F., WILLS, N.M., LOUGHRAN, G., WU, C-Y., PARSAWAR, K., RYAN, M.D., WANG, C-H., AND NELSON, C.C. (2007) A case for "StopGo": Reprogramming translation to augment codon meaning of GGN by promoting unconventional termination (Stop) after addition of glycine and then allowing continued translation (Go). *RNA* **13**, 803-810.

ATKINS, J.F., AND RYAN, M.D. (2008) Foot and Mouth's Achilles' heel? *Nature Biotechnology* **26(12)**, 1335-1336.

AVESANI, L., FALORNI, A., TORNIELLI, G.B., MARUSIC, C., PORCEDDU, A., POLVERARI, A., FALERI, C., CALCINARO, F., AND PEZZOTTI, M. (2003) Improved in planta expression of human islet autoantigen glutamic acid decarboxylase (GAD65). *Transgenic Research* **12**, 203-212.

BAERTSCHI, B., AND MAURON, A. (2008) Moral status revisited: The challenge of reversed potency. *Bioethics* DOI: 10.1111/j.1467-8519.2008.00686.x

BARANOV, P.V., GESTELAND, R.F., AND ATKINS, J.F. (2002) Recoding: translational bifurcations in gene expression. *Gene* **286**, 187-201.

BARANOV, P.V., GURVICH, O.L., HAMMER, A.W., GESTELAND, R.F., AND ATKINS, J.F. (2003). RECODE 2003. *Nucleic Acids Research* **31**, 87-89.

BARATOVA, L.A., GREBENSHCHIKOV, N.I., SHISHKOV, A.V., KASHIRIN, I.A., RADAVSKY, J.L., JARVEKULG, L., AND SAARMA M. (1992) The topography of the surface of potato virus X: tritium planigraphy and immunological analysis. *Journal General Virology* **73**, 229-235.

BELSHAM, G.J. (1993) Distinctive features of the foot-and-mouth disease virus, a member of the picornavirus family; aspects of virus protein synthesis, protein processing and structure. *Progress in Biophysics and Molecular Biology* **60**, 241-260.

BREMONT, M., JUSTE-LESAGE, P., CHABANNE-VAUTHEROT, D., CHARPILIENNE, A., AND COHEN, J. (1992) Sequences of the four larger proteins of a porcine group C rotavirus and comparison with the equivalent group A rotavirus proteins. *Virology* **186**, 684-692.

BROGLIE, K., CHET, I., HOLLIDAY, M., CRESSMAN, R., BIDDLE, P., KNOWLTON, C., MAUVAIS, F., AND BROGLIE, R. (1991) Transgenic plants with enhanced resistance to the fungal pathogen *Rhizoctonia solani*. *Science* **254**, 1194-1197.

BROMBACHER, F., KASTELEIN, R.A., AND ALBER, G. (2003) Novel IL-12 family members shed light on the orchestration of Th1 responses. *Trends in Immunology* **24**, 207-212.

BUCHNER, K., ROTH, P., SCHOTTA, G., KRAUSS, V., SAUMWEBER,H., REUTER, G., AND DOM, R. (2000) Genetic and molecular complexity of the position effect variegation modifier mod(mdg4) in *Drosophila*. *Genetics* **155**, 141-157.

CALL, M.E., PYRDOL, J., WIEDMANN, M., AND WUCHERPFENNIG, K.W. (2002) The Organizing Principle in the Formation of the T-cell Receptor-CD3 Complex. *Cell*

111, 967-979.

CALL, M.E., AND WUCHERPFENNIG, K.W. (2007) Common themes in the assembly and architecture of activating immune receptors. *Nature Reviews Immunology* **7**, 841-850.

CAÑIZARES, M.C., NICHOLSON, L., AND LOMONOSSOFF, G.P. (2005) Use of viral vectors for vaccine production in plants. *Immunology and Cell Biology* **83**, 263-270.

CAÑIZARES, M.C., LIU, L., PERRIN, Y., TSAKIRIS, E., AND LOMONOSSOFF, G.P. (2006) A bipartite system for the constitutive and inducible expression of high levels of foreign proteins in plants. *Plant Biotechnology Journal* **4**, 183-193.

CAO, Y-A., BACHMANN, M.H., BEILHACK, A., YANG, Y., TANAKA, M., SWIJNENBURG, R-J., REEVES, R., TAYLOR-EDWARDS, C., SCHULZ, S., DOYLE, T.C., FATHMAN, C.G., ROBBINS, R.C., HERZENBERG, L.A., NEGRIN, R.S., AND CONTAG, C.H. (2005). Molecular Imaging Using Labeled Donor Tissues Reveals Patterns of Engraftment, Rejection, and Survival in Transplantation. *Transplantation* **80**, 134-139.

CAREY, B.W., MARKOULAKI, S., HANNA, J., SAHA, K., GAO, Q., MITALIPOVA, M., AND JAENISCH, R. (2009) Reprogramming of murine and human somatic cells using a single polycistronic vector. *Proceedings of the National Academy of Sciences of the United States of America* **106**, 157-162

CHAPLIN, P.J., CAMON, E.B., VILLARREAL-RAMOS, B., FLINT, M., RYAN, M.D., AND COLLINS, R.A. (1999) Production of Interleukin-12 as a Self-Processing 2A Polypeptide. *Journal of Interferon and Cytokine Research* **19**, 235-241.

CHEN, Z., LAMBDEN, P.R., LAU, J., CAUL, E.O., AND CLARKE, I.N. (2002) Human group C rotavirus: completion of the genome sequence and gene coding assignments of a non-cultivatable rotavirus. *Virus Research* **83**, 179-187.

CHINNASAMY, D., MILSOM, M.D., SHAFFER, J., NEUENFELDT, J., SHAABAN, A.F., MARGISON, G.P., FAIRBAIRN, L.J., AND CHINNASAMY, N. (2006) Multicistronic lentiviral vectors containing the FMDV 2A cleavage factor demonstrate robust expression of encoded genes at limiting MOI. *Virology Journal* **3**, 14.

CLARK, A.J., HAROLD, G., AND YULL, F.E. (1997) Mammalian cDNA and prokaryotic reporter sequences silence adjacent transgenes in transgenic mice. *Nucleic Acids Research* **25(5)**, 1009-1014.

COHEN, C.J., ZHAO, Y., ZHENG, Z., ROSENBERG, S.A., AND MORGAN, R.A. (2006) Enhanced Antitumor Activity of Murine-Human Hybrid T-Cell Receptor (TCR) in Human Lymphocytes is Associated with Improved Pairing and TCR/CD3 Stability. *Cancer Research* **66(17)**, 8878-8886.

COLLINS, R.A., CAMON, E.B., CHAPLIN, P.J., AND HOWARD, C.J. (1998) Influence of IL-12 on interferon-γ production by bovine leucocyte subsets in response to bovine respiratory syncytial virus. *Veterinary Immunology and Immunopathology* **63**, 69-72.

DAFNY-YELIN, M., AND TZFIRA, T. (2007) Delivery of Multiple Transgenes to Plant Cells. *Plant Physiology* **145**, 1118-1128.

DALY, R., AND HEARN, M.T.W. (2005) Expression of heterologous proteins in *Pichia pastoris*: a useful experimental tool in protein engineering and production. *Journal of Molecular Recognition* **18**, 119-138.

DANIELS, D.S., AND TAINER, J.A. (2000) Conserved structural motifs governing the stoichiometric repair of alkylated DNA by O(6)-alkylguanine-DNA alkyltransferase. *Mutation Research* **460**, 151-163.

DEACON, C.F. (2004) Therapeutic strategies based on glucagons-like peptide 1. *Diabetes*

53, 2181-2189.

DE FELIPE, P., MARTIN, V., CORTÉS, M.L., RYAN, M.D., AND IZQUIERDO, M. (1999) Use of the 2A sequence from foot-and-mouth disease virus in the generation of retroviral vectors for gene therapy. *Gene Therapy* **6**, 198-208.

DE FELIPE, P., HUGHES, L.E., RYAN, M.D., AND BROWN, J.D. (2003) Co-translational, Intraribosomal Cleavage of Polypeptides by the Foot-and-mouth Disease Virus 2A-peptide. *The Journal of Biological Chemistry* **13**, 11441-11448.

DE FELIPE, P., AND RYAN, M.D. (2004) Targeting of Proteins Derived from Self-Processing Polyproteins Containing Multiple Signal Sequences. *Traffic* **5**, 616-626.

DE FELIPE, P., LUKE, G.A., HUGHES, L.E., GANI, D., HALPIN, C., AND RYAN, M.D. (2006) *E unum pluribus*: multiple proteins from a self-processing polyprotein. *Trends in Biotechnology* **24(2)**, 68-75.

DE FELIPE, P., LUKE, G.A., BROWN, J.D., AND RYAN, M.D. (2009) Inhibition of 2A-mediated 'Cleavage' of Certain Artificial Polyproteins Bearing N-terminal Signal Sequences. *Biotechnology Journal* (in press)

DE MIRANDA, J.R., DREBOT, M., TYLER, S., SHEN, M., CAMERON, C.E., STOLTZ, D.B., AND CAMAZINE, S.M. (2004) Complete nucleotide sequence of Kashmir bee virus and comparison with acute bee paralysis virus. *Journal of General Virology* **85**, 2263-2270.

DE ROSE, R., SCHEERLINCK, J-P. Y., CASEY, G., WOOD, P.R., TENNENT, J.M., AND CHAPLIN, P.J. (2000) Ovine Interleukin-12: Analysis of Biologic Function and Species Comparison. *Journal of Interferon and Cytokine Research* **20**, 557-564.

DOHERTY, M., TODD, D., MCFERRAN N., AND HOEY, E.M. (1999) Sequence analysis of a porcine enterovirus serotype 1 isolate: relationships with other picornaviruses. *Journal of General Virology* **80**, 1929-1941.

DONNELLY, M.L.L., GANI, D., FLINT, M., MONAGHAN, S., AND RYAN, M.D. (1997) The cleavage activities of aphthovirus and cardiovirus 2A proteins. *Journal of General Virology* **78**, 13-21.

DONNELLY, M.L.L., LUKE, G.A., MEHROTRA, A., LI, X., HUGHES, L.E., GANI, D., AND RYAN, M.D. (2001a) Analysis of the aphthovirus 2A/2B polyprotein "cleavage" mechanism indicates not a proteolytic reaction, but a novel translational effect: a putative ribosomal "skip". *Journal General Virology* **82**, 1013-1025.

DONNELLY, M.L.L., HUGHES, L.E., LUKE, G.A., MENDOZA, H., TEN DAM, E., GANI, D., AND RYAN, M.D. (2001b) The "cleavage" activities of foot-and-mouth disease virus 2A site-directed mutants and naturally occurring "2A-like" sequences. *Journal General Virology* **82**, 1027-1041.

DORONINA, V.A., WU, C., DE FELIPE, P., SACHS, M.S., RYAN, M.D., AND BROWN, J.D. (2008a) Site-Specific Release of Nascent Chains from Ribosomes at a Sense Codon. *Molecular and Cellular Biology* **28(13)**, 4227-4239.

DORONINA, V.A., DE FELIPE, P., WU, C., SHARMA, P., SACHS, M.S., RYAN, M.D., AND BROWN, J.D. (2008b) Dissection of a co-translational nascent chain separation event. *Biochemical Society Transaction* **36(4)**, 712-716.

DRUCKER, D.J. (2001) Development of glucagons-like peptide-1-based pharmaceuticals as therapeutic agents for the treatment of diabetes. *Current Pharmaceutical Design* **7**, 1399-1412.

DUBY, G., AND BOUTRY, M. (2002) Mitochondrial protein import machinery and targeting information. *Plant Science* **162**, 477-490.

DUKE, G.M., HOFFMAN, M.A., AND PALMENBERG, A.C. (1992) Sequence and structural elements that contribute to efficient encephalomyocarditis virus RNA translation. *Journal of Virology* **66**, 1602-1609.

EGAN, M.A., AND ISRAEL, Z.R. (2002) The use of cytokines and chemokines as genetic adjuvants for plasmid DNA vaccines. *Clinical and Applied Immunology Reviews* **2**, 255-287.

EL-AMRANI, A., BARAKATE, A., ASKARI, B.M., LI, X., ROBERTS, A.G., RYAN, M.D., AND HALPIN, C. (2004) Coordinate Expression and Independent Subcellular Targeting of Multiple Proteins from a Single Transgene. *Plant Physiology* **135**, 16-24.

ENGELS, B., AND UCKERT, W. (2007) Redirecting T lymphocyte specificity by T-cell receptor gene transfer – A new era for immunotherapy. *Molecular Aspects of Medicine* **28**, 115-142.

FANG, J., QIAN, J.J., YI, S., HARDING, T.C., TU, G.H., VANROEY, M., AND JOOSS, K. (2005) Stable antibody expression at therapeutic levels using the 2A-peptide. *Nature Biotechnology* **23(5)**, 584-590.

FEDERICO, M. (2003) Lentivirus gene engineering protocols. In: *Methods in Molecular Biology.* Volume 245 Totowa, NJ: Humana Press.

FLASSHOVE, M., BARDENHEUER, W., SCHNEIDER, A., HIRSCH, G., BACH, P., BURY, C., MORITZ, T., SEEBER, S., AND OPALKA, B. (2000) Type and position of promoter elements in retroviral vectors has substantial effects on the expression level of an enhanced green fluorescent protein reporter gene. *Journal of Cancer Research and Clinical Oncology* **126**, 391-399.

FORSS, S., STREBEL, K., BECK, E., AND SCHALLER, H. (1984) Nucleotide sequence and genome organization of foot-and-mouth disease virus. *Nucleic Acids Research* **12**, 6587-6601.

FRANÇOIS, I.E.J.A., DE BOLLE, M.F.C., DWYER, G., GODERIS, I.J.W.M., VERHAERT, P., PROOST, P., SCHAAPER, W.M.M., CAMMUE, B.P.A., AND BROEKAERT, W.F. (2002) Transgenic expression in Arabidopsis of a polyprotein construct leading to production of two different antimicrobial proteins. *Plant Physiology* **128**, 1346-1358.

FRANÇOIS, I.E.J.A., VAN HEMELRIJCK, W., AERTS, A.M., WOUTERS, P.F.J., PROOST, P., BROEKAERT, W.F., AND CAMMUE, B.P.A. (2004) Processing in *Arabidopsis thaliana* of a heterologous polyprotein resulting in differential targeting of the individual plant defensins. *Plant Science* **166**, 113-121.

FUNSTON, G.M., KALLIOINEN, S.E., DE FELIPE, P., RYAN, M.D., AND IGGO, R.D. (2008) Expression of heterologous genes in oncolytic adenoviruses using picornaviral 2A sequences that trigger ribosome skipping. *Journal of General Virology* **89**, 389-396.

FURLER, S., PATERNA, J.C., WEIBEL, M., AND B ELER, H. (2001) Recombinant AAV vectors containing the foot-and-mouth disease virus 2A sequence confer efficient bicistronic gene expression in cultured cells and rat substantia nigra neurons. *Gene Therapy* **8**, 864-873.

GERMAIN, R.N., AND STEFANOVA, I. (1999) The dynamics of T-cell receptor signaling: complex orchestration and the key roles of tempo and cooperation. *Annual Review Immunology* **17**, 467-522.

GILLESSEN, S., CARVAJAL, D., LING, P., PODLASKI, F.J., STREMLO, D.L., FAMILLETTI, P.C., GUBLER, U., PRESKY, D.H., STERN, A.S., AND GATELY, M.K. (1995) Mouse interleukin-12 (IL-12) p40 homodimer: a potent IL-12 antagonist. *European Journal*

of Immunology **1**, 200-206.

Goldbach, R.W., and Wellink, J. (1996) Comoviruses: molecular biology and replication. In *The Plant Viruses: Polyhedral Virions and Bipartite RNA Genomes* eds. B.D. Harrison and A.F. Murant., pp35-76. New York: Plenum.

Gopinath, K., Wellink, J., Porta, C., Taylor, K.M., Lomonossoff, G.P., and van Kammen, A. (2000) Engineering Cowpea Mosaic Virus RNA-2 into a Vector to Express Heterologous Proteins in Plants. *Virology* **267**, 159-173.

Gorbalenya, A.E., Pringle, F.M., Zeddam, J.L., Luke, B.T., Cameron, C.E., Kalmakoff, J., Hanzlik, T.N., Gordon, K.H.J., and Ward, V.K. (2002) The palm subdomain-based active site is internally permuted in viral RNA-dependent RNA polymerases of an ancient lineage. *Journal of Molecular Biology* **324**, 47-62.

Govan, V.A., Leat, N., Allsopp, M., and Davison, S. (2000) Analysis of the complete genome sequence of acute bee paralysis virus shows that it belongs to the novel group of insect-infecting RNA viruses. *Virology* **277**, 457-463.

Graham, R.I., Rao, S., Possee, R.D., Sait, S.M., Mertens, P.P.C., and Hails, R.S. (2006) Detection and characterization of three novel species of reovirus (*Reoviridae*), isolated from geographically separate populations of the winter moth *Operophtera brumata* (Lepidoptera: *Geometridae*) on Orkney. *Journal of Invertebrate Pathology* **91(2)**, 79-87.

Groot Bramel-Verheije, M.H., Rottier, P.J.M., and Meulenberg, J.J.M. (2000) Expression of a Foreign Epitope by Porcine Reproductive and Respiratory Syndrome Virus. *Virology* **278**, 380-389.

Hagiwara, K., Rao, S., Scott, S.W., and Carner, G.R. (2002) Nucleotide sequences of segments 1, 3 and 4 of the genome of *Bombyx mori* cypovirus 1 encoding putative capsid proteins VP1, VP3 and VP4, respectively. *Journal of General Virology* **83**, 1477-1482.

Halpin, C., Cooke, S.E., Barakate, A., El Amrani, A., and Ryan, M.D. (1999) Self-processing 2A-polyproteins – a system for co-ordinate expression of multiple proteins in transgenic plants. *The Plant Journal* **17(4)**, 453-459.

Halpin, C. (2005) Gene stacking in transgenic plants – the challenge for 21st century plant biotechnology. *Plant Biotechnology Journal* **3(2)**, 141-155.

Hart, D.P., Xue S-A., Thomas, S., Cesco-Gaspere, M., Tranter, A., Willcox, B., Lee, S.P., Steven, N., Morris, E.C., and Stauss, H.J. (2008) Retroviral transfer of a dominant TCR prevents surface expression of a large proportion of the endogenous TCR repertoire in human T cells. *Gene Therapy* **15(8)**, 625-631.

Hasan, G., Turner, M.J., and Cordingley, J.S. (1984) Complete nucleotide sequence of an unusual mobile element from *Trypanosoma brucei*. *Cell* **37**, 333-341.

Hasegawa, K., Cowan, A.B., Nakatsuji, N., and Suemori, H. (2007) Efficient Multicistronic Expression of a Transgene in Human Embryonic Stem Cells. *Stem Cells* **25**, 1707-1712

Heck, G.R., Armstrong, C.L., Astwood, J.D., Behr, C.F., Bookout, J.T., Brown, S.M., Cavato, T.A., DeBoer, D.L., Deng, M.Y., George, C., Hillyard, J.R., Hironaka, C.M., Howe, A.R., Jakse, E.H., Ledesma, B.E., Lee, T.C., Lirette, R.P., Mangano, M.L., Mutz, J.N., Qi, Y., Rodriguez, R.E., Sidhu, S.R., Silvanovich, A., Stoecker, M.A., Yingling, R.A., and You, J. (2005) Genomics, Molecular Genetics and Biotechnology. *Crop Science* **44**, 329-339.

Heemskerk, M.H. (2006) Optimizing TCR gene transfer. *Clinical Immunology* **119**,

121-122.

HEISER, W.C. (2004) Gene delivery to mammalian cells. In: *Methods in Molecular Biology*. Volume 246 Humana Press

HERAS, S.R., THOMAS, M.C., GARCÍA-CANADAS, M., DE FELIPE, P., GARCÍA-PEREZ, J.L., RYAN, M.D., AND LOPEZ, M.C. (2006) L1Tc non-LTR retrotransposons from *Trypanosoma cruzi* contain a functional viral-like self-cleaving 2A sequence in frame with the active proteins they encode. *Cellular and Molecular Life Sciences* **63**, 1449-1460.

HODI, F.S., MIHM, M.C., SOIFFER, R.J., HALUSKA, F.G., BUTLER, M., SEIDEN, M.V., DAVIS, T., HENRY-SPIRES, R., MACRAE, S., WILLMAN, A., PADERA, R., JAKLITSCH, M.T., SHANKAR, S., CHEN, T.C., KORMAN, A., ALLISON, J.P., AND DRANOFF, G. (2003) Biologic activity of cytotoxic T lymphocyte-associated antigen 4 antibody blockade in previously vaccinated metastatic melanoma and ovarian carcinoma patients. *Proceedings of the National Academy of Sciences of the United States of America* **100(8)**, 4712-4717.

HOLST, J., VIGNALI, K.M., BURTON, A.R., AND VIGNALI, D.A.A. (2006a) Rapid analysis of T-cell selection *in vivo* using T-cell-receptor retrogenic mice. *Nature Methods* **3**, 191-197.

HOLST, J., SZYMCZAK-WORKMAN, A.L., VIGNALI, K.M., BURTON, A.R., WORKMAN, C. J., AND VIGNALI DAA. (2006b) Generation of T-cell receptor retrogenic mice. *Nature Protocols* **1(1)**, 406-417.

HOLST, J., WANG, H., EDER, K.D., WORKMAN, C.J., BOYD, K.L., BAQUET, Z., SINGH, H., FORBES, K., CHRUSCINSKI, A., SMEYNE, R., VAN OERS, N.S.C., UTZ, P.J., AND VIGNALI, D.A.A. (2008) Scalable signalling mediated by T-cell antigen receptor-CD3 ITAMs ensures effective negative selection and prevents autoimmunity. *Nature Immunology* **9(6)**, 658-666.

HOWARD, J.A., AND HOOD, E. (2005) Bioindustrial and biopharmaceutical products produced in plants. *Advances in Agronomy* **85**, 91-124.

HURWITZ, A.A., FOSTER, B.A., KWON, E.D., TRUONG, T., CHOI, E.M., GREENBERG, N.M., BURG, M.B., AND ALLISON, J.P. (2000) Combination Immunotherapy of Primary Prostate Cancer in a Transgenic Mouse Using CTLA-4 Blockade. *Cancer Research* **60**, 2444-2448.

ISAWA, H., ASANO, S., SAHARA, K., IIZUKA, T., AND BANDO, H. (1998) Analysis of genetic information of an insect picorna-like virus, infectious flacherie virus of silkworm: evidence for evolutionary relationships among insects, mammalian and plant picorna(-like) viruses. *Archives of Virology* **143**, 127-143.

JAFFER, F.A., AND WEISSLEDER, R. (2005) Molecular imaging in the clinical arena. *The Journal of the American Medical Association* **16**, 855-862.

JANSEN, M., SORG, U.R., RAGG, S., FLASSHOVE, M., SEEBERS, S., WILLIAMS, D.A., AND MORITZ, T. (2002) Hematoprotection and enrichment of transduced cells *in vivo* after gene transfer of MGMT(P140K) into hematopoietic stem cells. *Cancer Gene Therapy* **9**, 737-746.

JIANG, B., TSUNEMITSU, H., GENTSCH, J.R., SAIF, L.J., AND GLASS, R.I. (1993) Nucleotide sequences of genes 6 and 10 of a bovine group C rotavirus. *Nucleic Acids Research* **21**, 2250.

JOHNSON, K.N., AND CHRISTIAN, P.D. (1998) The novel genome organization of the insect picorna-like virus *Drosophila C* virus suggests this virus belongs to a previously

undescribed virus family. *Journal of General Virology* **79**, 191-203.

Jones, M.S., Lukashov, V.V., Ganac, R.D., and Schnurr, D.P. (2007) Discovery of a novel human picornavirus in a stool sample from a pediatric patient presenting with Fever of unknown origin. *Journal of Clinical Microbiology* **45**, 2144-2150.

Kakkis, E.D., Muenzer, J., Tiller, G.E., Waber, L., Belmont, J., Passage, M., Kakkis, E.D., Izykowski, B., Phillips, J., Doroshow, R., Walot, I., Hoft, R., and Neufeld, E.F. (2001) Enzyme-replacement therapy in mucopolysaccharidosis 1. *The New England Journal of Medicine* **344**, 182-188.

Karasev, A.V., Foulke, S., Wellens, C., Rich, A., Shon, K.J., Zwierzynski, I., Hone, D., Koprowski, H., and Reitz, M. (2005) Plant based HIV-1 vaccine candidate: Tat protein produced in spinach. *Vaccine* **23**, 1875-1880.

Keenan, R.J., Freymann, D.M., Stroud, R.M., and Walter, P. (2001) The Signal Recognition Particle. *Annual Review of Biochemistry* **70**, 755-775.

Kim, M.C., Kwon, Y.K., Joh, S.J., Lindberg, A.M., Kwon, J.H., Kim, J.H., and Kim, S.J. (2006) Molecular analysis of duck hepatitis virus type 1 reveals a novel lineage close to the genus *Parechovirus* in the family *Picornaviridae*. *Journal of General Virology* **87**, 3307-3316.

King, A.M.Q., Brown, F., Christian, P., Hovi, T., Hyypia, T., Knowles, J., Lemon, S.M., Minor, P.D., Palmenberg, A.C., Skern, T., Stanway, G. (2000) Picornaviridae. In *Virus Taxonomy. Seventh Report of the International Committee on the Taxonomy of Viruses*. eds. M.H.V. Van Regenmortel, C.M. Fauquet, D.H.L. Bishop, C.H. Calisher, E.B. Carsten, M.K. Estes, S.M. Lemon, J. Maniloff, M.A. Mayo, C.R. Pringle and R.B. Wickner, pp 657-673. New York: Academic Press.

Klump, H., Schiedlmeier, B., Vogt, B., Ryan, M., Ostertag, W., and Baum, C. (2001) Retroviral vector-mediated expression in HoxB4 in hematopoietic cells using a novel expression strategy. *Gene Therapy* **8**, 811-817.

Kwon, S-J., Hwang, E-W., and Kwon, H-B. (2004) Genetic Engineering of Drought Resistant Potato Plants by Co-Introduction of Genes Encoding Trehalose-6-Phosphate Synthase and Trehalose-6-Phosphate Phosphatase of *Zygosaccharomyces rouxii*. *Korean Journal of Genetics* **26(2)**, 199-206.

Lanier, L.L. (2006) Viral immunoreceptor tyrosine-based activation motif (ITAM)-mediated signaling in cell transformation and cancer. *Trends in Cell Biology* **16(8)**, 388-390.

Law, K.M., and Brown, T.D. (1990) The complete nucleotide sequence of the GDVII strain of Theiler's murine encephalomyelitis virus (TMEV). *Nucleic Acids Research* **18**, 6707-6708.

Lay, F.T., and Anderson, M.A. (2005) Defensins – Components of the Innate Immune System in Plants. *Current Protein and Peptide Science* **6**, 85-101.

Lee, Z., Dennis, J.E., and Gerson, S.L. (2008) Imaging Stem Cell Implant for Cellular-Based Therapies. *Experimental Biology and Medicine* **233**, 930-940.

Leisegang, M., Engels, B., Meyerhuber, P., Kieback, E., Sommermeyer, D., Xue, S.A., Reu , S., Stauss, H., and Uckert, W. (2008) Enhanced functionality of T-cell receptor-redirected T cells is defined by the transgene cassette. *Journal of Molecular Medicine* **86**, 573-583.

Lengler, J., Holzmuller, H., Salmons, B., Gunzburg, W.H., and Renner, M. (2005) FMDV-2A sequence and protein arrangement contribute to functionality of CYP2B1-reporter fusion protein. *Analytical Biochemistry* **343**, 116-124.

LE RUDULIER, D., STROM, A.R., DANDEKAR, A.M., SMITH, L.T., AND VALENTINE, R.C. (1984) Molecular biology of osmoregulation. *Science* **224**, 1064-1068.

LI, F., BROWNING, G.F., STUDDERT, M.J., AND CRABB, B.S. (1996) Equine rhinovirus 1 is more closely related to foot-and-mouth disease virus than to other picornaviruses. *Proceedings of the National Academy of Sciences of the United States of America* **93**, 990-995.

LIANG, H., GAO, H., MAYNARD, C.A., AND POWELL, W.A. (2005) Expression of a self-processing, pathogen resistance-enhancing gene construct in Arabidopsis. *Biotechnology Letters* **27**, 435-442.

LINDBERG, A.M., AND JOHANSSON, S. (2002) Phylogenetic analysis of Ljungan virus and A-2 plaque virus, new members of the *Picornaviridae*. *Virus Research* **85**, 61-70.

LIU, L., CAÑIZARES, M.C., MONGER, W., PERRIN, Y., TSAKIRIS, E., PORTA, C., SHARIAT, N., NICHOLSON, L., AND LOMONOSSOFF, G.P. (2005) Cowpea mosaic virus-based systems for the production of antigens and antibodies in plants. *Vaccine* **23**, 1788-1792.

LOMONOSSOFF, G.P., AND JOHNSON, J.E. (1991) The synthesis and structure of comovirus capsids. *Progress in Biophysics and Molecular Biology* **55**, 107-137.

LUKE, G.A., DE FELIPE, P., COWTON, V.M., HUGHES, L.E., HALPIN, C., AND RYAN, M.D. (2006) Self-Processing Polyproteins: A Strategy for Co-expression of Multiple Proteins in Plants. *Biotechnology and Genetic Engineering Reviews* **23**, 239-252.

LUKE, G.A., DE FELIPE, P., LUKASHEV, A., KALLIOINEN, S.E., BRUNO, E.A., AND RYAN, M.D. (2008) The Occurrence, Function, and Evolutionary Origins of "2A-like" Sequences in Virus Genomes. *Journal of General Virology* **89**, 1036-1042.

MA, C., AND MITRA, A. (2002) Expressing multiple genes in a single open reading frame with the 2A region of foot-and-mouth disease virus as a linker. *Molecular Breeding* **9**, 191-199.

MA, JK-C., BARROS, E., BOCK, R., CHRISTOU, P., DALE, P.J., DIX, P.J., FISCHER, R., IRWIN, J., MAHONEY, R., PEZZOTTI, M., SCHILLBERG, S., SPARROW, P., STOGER, E., AND TWYMAN, R.M. (2005) Molecular farming for new drugs and vaccines. *EMBO Reports* **6(7)**, 593-599.

MACAULEY-PATRICK, S., FAZENDA, M.L., MCNEIL, B., AND HARVEY, L.M. (2005) Heterologous protein production using the *Pichia pastoris* expression system. *Yeast* **22**, 249-270.

MALISSEN, B. (2008) CD3 ITAMs count! *Nature Immunology* **9(6)**, 583-584.

MAORI, E., TANNE, E., AND SELA, I. (2007) Reciprocal sequence exchange between non-retro viruses and hosts leading to the appearance of new host phenotypes. *Virology* **328**, 151-157.

MARCONI, G., ALBERTINI, E., BARONE, P., DEMARCHIS, F., LICO, C., MARUSIC, C., RUTILI, D., VERONESI, F., AND PORCEDDU, A. (2006) *In planta* production of two peptides of the Classical Swine Fever Virus (CSFV) E2 glycoprotein fused to the coat protein of potato virus X. *BMC Biotechnology* **6**, 29.

MARTIN, F., MARANON, C., OLIVARES, M., ALONSO, C., AND LOPEZ, M.C. (1995) Characterization of a non-long terminal repeat retrotransposon cDNA (L1Tc) from *Trypanosoma cruzi*: homology of the first ORF with the ape family of DNA repair enzymes. *Journal of Molecular Biology* **247**, 49-59.

MARTIN, E., KAMATH, A.T., BRISCOE, H., AND BRITTON, W.J. (2003) The combination of plasmid interleukin-12 with a single DNA vaccine is more effective than *Mycobacterium bovis* (bacilli Calmette-Guèrin) in protecting against systemic

Mycobacterium avium infection. *Immunology* **109**, 308-314.

Martínez-Salas, E. (1999) Internal ribosome entry site biology and its use in expression vectors. *Current Opinion in Biotechnology* **10,** 458-464.

Martínez-Salas, E., Pacheco, A., Serrano, P., and Fernandez, N. (2008) New insights into internal ribosome entry site elements relevant for viral gene expression. *Journal of General Virology* **89(3)**, 611-626.

Massoud, T.F., and Gambhir, S.S. (2003) Molecular imaging in living subjects: seeing fundamental biological processes in a new light. *Genes and Development* **17**, 545-580.

Mattanovich, D., Gasser, B., Hohenblum, H., and Sauer, M. (2004) Stress in recombinant protein producing yeasts. *Journal of Biotechnology* **113**, 121-135.

Mentag, R., Luckevich, M., Morency, M.J., and Seguin, A. (2003) Bacterial disease resistance of transgenic hybrid poplar expressing the synthetic antimicrobial peptide D4E1. *Tree Physiology* **23**, 405-411.

Milsom, M.D., Woolford, L.B., Margison, G.P., Humphries, R.K., and Fairbairn, L.J. (2004). Enhanced *In Vivo* Selection of Bone Marrow Cells by Retroviral-Mediated Coexpression of Mutant O6-methylguanine-DNA-methyltransferase and HOXB4. *Molecular Therapy* **10(5)**, 862-873.

Mizuguchi, H., Xu, Z., Ishii-Watabe, A., Uchida, E., and Hayakawa, T. (2000) IRES-dependent second gene expression is significantly lower than cap-dependent first gene expression in a bicistronic vector. *Molecular Therapy* **1**, 376-382.

Murphy, N.B., Pays, A., Tebabi, P., Coquelet, H., Guyaux, M., Steinert, M., and Pays, E. (1987) *Trypanosoma brucei* repeated element with unusual structural and transcriptional properties. *Journal of Molecular Biology* **195**, 855-871.

Nelson, R.S., McCormick, S.M., Delannay, X., Dube, P., Layton, J., Anderson, E.J., Kaniewska, M., Proksch, R.K., Horsch, R.B., Rogers, S.G., Fraley, RT., and Beachy, R.N. (1988) Virus tolerance, plant growth, and field performance of transgenic tomato plants expressing coat protein from tobacco mosaic virus. *Biotechnology Journal* **6**, 403-409.

Neufeld, E.F., and Muenzer, J. (2001) The mucopolysaccharidoses. In *The metabolic and molecular bases of inherited disease* eds C.R. Scriver, A.L. Beaudet, W.S. Sly and D. Valle, pp3421-3452. New York: McGraw-Hill.

Ohsawa, K., Wantanabe, Y., Miyata, H., and Sato, H. (2003) Genetic analysis of a Theiler-like virus isolated from rats. *Comparative Medicine* **53**, 191-196.

Osborn, M.J., Panoskaltsis-Mortari, A., McElmurry, R.T., Bell, S.K., Vignali, D.A.A., Ryan, M.D., Wilber, A.C., Scott McIvor, R., Tolar, J., and Blazar, B.R. (2005) A Picornaviral 2A-like Sequence-Based Tricistronic Vector Allowing for High-Level Therapeutic Gene Expression Coupled to a Dual-Reporter System. *Molecular Therapy* **12**, 569-574.

Okamoto, M., Mitsuhara, I., Ohshima, M., Natori, S., and Ohashi, Y. (1998) Enhanced expression of an antimicrobial peptide Sarcotoxin 1A by GUS fusion in transgenic tobacco plants. *Plant & Cell Physiology* **39**, 57-63.

Okita, K., Nakagawa, M., Hyenjong, H., Ichisaka, T., and Yamanaka, S. (2008) Generation of Mouse Induced Pluripotent Stem Cells Without Viral Vectors. *Science* **322**, 949-952.

Palendira, U., Kamath, A.T., Feng, C.G., Martin, E., Chaplin, P.J., Triccas, J.A., and Britton, W.J. (2002) Coexpression of Interleukin-12 Chains by a Self-Splicing Vector

Increases the Protective Cellular Immune Response of DNA and *Mycobacterium bovis* BCG Vaccines against *Mycobacterium tuberculosis*. *Infection and Immunity* **70(4)**, 1949-1956.

PALMENBERG, A.C. (1990) Proteolytic processing of picornaviral polyprotein. *Annual Review of Microbiology* **44**, 603-623.

PALMENBERG, A.C., PARKS, G.D., HALL, D.J., INGRAHAM, R.H., SENG, T.W., PALLAI, P.V. (1992) Proteolytic processing of the cardioviral P2 region: 2A/2B cleavage in clone-derived precursors. *Virology* **190**, 754-762.

PAN, Y-X., XU, J., BOLAN, E., CHANG, A., MAHURTER, L., ROSSI, G., AND PASTERNAK, G.W. (2000) Isolation and expression of a novel alternatively spliced mu opioid receptor isoform MOR-1F. *FEBS Letters* **466**, 337-340.

PARDY, K. (1994) Reporter enzymes for the study of promoter activity. *Molecular Biotechnology* **2**, 23-27.

PARK, I.H., ZHAO, R., WEST, J.A., YABUUCHI, A., HUO, H., INCE, T.A., LEROU, P.H., LENSCH, M.W., AND DALEY, G.Q. (2007) Reprogramming of human somatic cells to pluripotency with defined factors. *Nature* **451(7175)**, 141-146

PASSAGNE, I., EVRARD, A., DEPEILLE, P., CUQ, P., CUPISSOL, D., AND VIAN, L. (2006) O^6-methylguanine DNA-methyltransferase (MGMT) overexpression in melanoma cells induces resistance to nitrosoureas and temozolomide but sensitizes to mitomycin C. *Toxicology and Applied Pharmacology* **211**, 97-105.

PETERS, C., BALTHAZOR, M., SHAPIRO, E.G., KING, R.J., KOLLMAN, C., HEGLAND, J.D., HENSLEE-DOWNEY, J., TRIGG, M.E., COWAN, M.J., SANDERS, J., BUNIN, N., WEINSTEIN, H., LENARSKY, C., FALK, P., HARRIS, R., BOWEN, T., WILLIAMS, T.E., GRAYSON, G.H., WARKENTIN, P., SENDER, L., COOL, V.A., CRITTENDEN, M., PACKMAN, S., KAPLAN, P., LOCKMAN, L.A., ANDERSON, J., KRIVIT, W., DUSENBERY, K., AND WAGNER, J. (1996) Outcome of unrelated donor bone marrow transplantation in 40 children with Hurler syndrome. *Blood* **87**, 4894-4902.

PFEIFER, A. AND VERMA, I.M. (2001) Gene Therapy: Promises and Problems. *Annual Review of Genomics and Human Genetics* **2**, 177-211.

PORTA, C., AND LOMONOSSOFF, G.P. (1998) Scope for using plant viruses to present epitopes from animal pathogens. *Reviews in Medical Virology* **8**, 25-41.

POULOS, B.T., TANG, K.F., PANTOJA, C.R., BONAMI, J.R., AND LIGHTNER, D.V. (2006) Purification and characterization of infectious myonecrosis virus of penaeid shrimp. *Journal of General Virology* **87**, 987-996.

POUWELS, J., VAN DER KROGT, G.N., VAN LENT, J., BISSELING, T., AND WELLINK, J. (2002) The cytoskeleton and the secretory pathway are not involved in targeting the cowpea mosaic virus movement protein to the cell periphery. *Virology* **297**, 48-56.

PREMRAJ, A., SREEKUMAR, E., JAIN, M., AND RASOOL, T.J. (2006) Buffalo (*Bubalus bubalis*) interleukin-12: Analysis of expression profiles and functional cross-reactivity with bovine system. *Molecular Immunology* **43**, 822-829.

PRINGLE, F.M., GORDON, K.H.J., HANZLIK, T.N., KALMAKOFF, J., SCOTTI, P.D., AND WARD, V.K. (1999) A novel capsid expression strategy for *Thosea asigna* virus (*Tetraviridae*). *Journal of General Virology* **80**, 1855-1863.

PRINGLE, F.M., JOHNSON, K.N., GOODMAN, C.L., MCINTOSH, A.H., AND BALL, L.A. (2003) Providence virus: a new member of the *Tetraviridae* that infects cultured insect cells. *Virology* **306**, 359-370.

PROVOST, E., RHEE, J., AND LEACH, S.D. (2007) Viral 2A peptides allow expression of multiple proteins from a single ORF in transgenic zebrafish embryos. *Genesis* **45(10)**, 625-629.

Quezada, S.A., Peggs, K.S., Curran, M.A., and Allison, J.P. (2006) CTLA4 blockade and GM-CSF combination immunotherapy alters the intratumor balance of effector and regulatory T cells. *The Journal of Clinical Investigation* **116(7)**, 1935-1945.

Radcliffe, P.A., and Mitrophanous, K.A. (2004) Multiple gene products from a single vector: "self-cleaving" 2A peptides. *Gene Therapy* **11**, 1673-1674.

Ralley, L., Enfissi, E.M.A., Misawa, N., Schuch, W., Bramley, P.M., and Fraser, P.D. (2004) Metabolic engineering of ketocarotenoid formation in higher plants. *The Plant Journal* **39**, 477-486.

Randall, J., Sutton, D., Ghoshroy, S., Bagga, S., and Kemp, J.D. (2004) Co-ordinate expression of β- and δ- zeins in transgenic tobacco. *Plant Science* **167**, 367-372.

Rao, S., Carner, G.R., Scott, S.W., Omura, T., and Hagiwara, K. (2003) Comparison of the amino acid sequences of RNA-dependent RNA polymerases of cypoviruses in the family *Reoviridae*. *Archives of Virology* **148**, 209-219.

Roosien, J., Belsham, G.J., Ryan, M.D., King, A.M.Q., and Vlak, J.M. (1990) Synthesis of foot-and-mouth disease virus capsid proteins in insect cells using baculovirus expression vectors. *Journal of General Virology* **71**, 1703-1711.

Ruile, P., Winterhalter, C., and Liebl, W. (1997) Isolation and analysis of a gene encoding α-glucuronidase, an enzyme with a novel primary structure involved in the breakdown of xylan. *Molecular Microbiology* **23**, 267-279.

Ryan, M.D., Belsham, G.J., and King, A.M. (1989) Specificity of enzyme-substrate interactions in foot-and-mouth disease virus polyprotein processing. *Virology* **173(1)**, 35-45.

Ryan, M.D., King, A.M., and Thomas, G.P. (1991) Cleavage of foot-and-mouth disease virus polyprotein is mediated by residues located within a 19 amino acid sequence. *Journal of General Virology* **72**, 2727-2732.

Ryan, M.D., and Drew, J. (1994) Foot-and-mouth disease virus 2A oligopeptide mediated cleavage of an artificial polyprotein. *The EMBO Journal* **13**, 928-933.

Ryan, M.D., Donnelly, M.L.L., Lewis, A., Mehrotra, A.P., Wilkie, J., and Gani, D. (1999) A model for Nonstoichiometric, Co-translational Protein Scission in Eukaryotic Ribosomes. *Bioorganic Chemistry* **27**, 55-79.

Ryan, M.D., Luke, G.A., Hughes, L.E., Cowton, V.M., Ten-Dam, E., Xuejun,L., Donnelly, M.L.L., Mehrotra, A., and Gani, D. (2002) The Aphtho- and Cardiovirus "Primary" 2A/2B Polyprotein "Cleavage". In *Molecular Biology of Picornaviruses* eds. B.L. Semler and E. Wimmer, pp 61-70. Washington: ASM Press.

Sainsbury, F., Lavoie, P-O., D'Aoust, M-A., Vézina, L-P., and Lomonossoff, G.P. (2008) Expression of multiple proteins using full-length and deleted versions of cowpea mosaic virus RNA-2. *Plant Biotechnology Journal* **6**, 82-92.

Samalova, M., Fricker, M., and Moore, I. (2006) Ratiometric Fluorescence-Imaging Assays of Plant Membrane Traffic Using Polyproteins. (2006) *Traffic* **7**, 1701-1723.

Samalova, M., Fricker, M., and Moore, I. (2008) Quantitative and Qualitative Analysis of Plant Membrane Traffic Using Fluorescent Proteins. *Methods in Cell Biology* **85**, 353-380.

Santa Cruz, S., Chapman, S., Roberts, A.G., Roberts, I.M., Prior, D.A.M., and Oparka, K.J. (1996) Assembly and movement of a plant virus carrying a green fluorescent protein overcoat. *Proceedings of the National Academy of Sciences of the United States of America* **93**, 6286-6290.

SCHMIDT, M., AND RETHWILM, A. (1995) Replicating Foamy Virus-Based Vectors Directing High Level Expression of Foreign Genes. *Virology* **210**, 167-178.

SCHOLTEN, K.B.J., KRAMER, D., KUETER, E.W.M., GRAF, M., SCHOEDL, T., MEIJER, C.J.L.M., SCHREURS, M.W.J., AND HOOIJBERG, E. (2006) Codon modification of T-cell receptors allows enhanced functional expression in transgenic human T cells. *Clinical Immunology* **119**, 135-145.

SIMMONS, A.D., MOSKALENKO, M., CRESON, J., FANG, J., YI, S., VANROEY, MJ., ALLISON, J.P., AND JOOSS, K. (2008) Local secretion of anti-CTLA-4 enhances the therapeutic efficacy of a cancer immunotherapy with reduced evidence of systemic autoimmunity. *Cancer Immunology, Immunotherapy* **57**, 1263-1270.

SMITH, M.D., AND GLICK, B.R. (2000) The production of antibodies in plants. *Biotechnology Advances* **18**, 85-89.

SMOLENSKA, L., ROBERTS, I.M., LEARMONTH, D., PORTER, A.J., HARRIS, WJ., MICHAEL, T., WILSON, A., AND SANTA CRUZ, S. (1998) Production of a functional single chain antibody attached to the surface of a plant virus. *FEBS Letters* **441**, 379-382.

SOLL, J. (2002) Protein import into chloroplasts. *Current Opinion in Plant Biology* **5**, 529-535.

SOMMER, C.A., STADFELD, M., MURPHY, G.J., HOCHEDLINGER, K., KOTTON, D.N., AND MOSTOSLAVSKY, G. (2008) iPS Cell Generation Using a Single Lentiviral Stem Cell Cassette. DOI: 10.1634/stemcells.2008-1075.

SOMMERGRUBER, W., ZORN, M., BLAAS, D., FESSL, F., VOLKMANN, P., MAURER-FOGY, I., PALLAI, P., MERLUZZI, V., MATTEO, M., SKERN, T., AND KEUCHLER, E. (1989) Polypeptide 2A of human rhinovirus type 2: identification as a protease and characterisation by mutational analysis. *Virology* **169**, 68-77.

STEINER, D.F. (1998) The proprotein convertases. *Current Opinion in Chemical Biology* **2**, 31-39.

STADFELD, M., NAGAYA, M., UTIKAL, J., WEIR, G., AND HOCHEDLINGER K. (2008) Induced Pluripotent Stem Cells Generated Without Viral Integration. *Science* **322**, 945-949.

STREATFIELD, S.J., AND HOWARD, J.A. (2003) Plant production systems for vaccines. *Expert Reviews of Vaccines* **2(6)**, 763-775.

SUGITA, K., ENDO-KASAHARA, S., TADA, Y., LIJUN, Y., YASUDA, H., HAYASHI, Y., JOMORI, T., EBINUMA, H., AND TAKIAWA, F. (2005) Genetically modified rice seeds accumulating GLP-1 analogue stimulate insulin secretion from a mouse pancreatic beta-cell line. *FEBS Letters* **579**, 1085-1088.

SZYMCZAK, A.L., WORKMAN, C.J., WANG, Y., VIGNALI, K.M., DILIOGLOU, S., VANIN, E.F., AND VIGNALI, D.A. (2004) Correction of multi-gene deficiency *in vivo* using a single "self-cleaving" 2A-peptide-based retroviral vector. *Nature biotechnology* **22(5)**, 589-594.

TAILOR, R.H., ACLAND, D.P., ATTENBOROUGH, S., CAMMUE, B.P.A., EVANS, I.J., OSBORN, R.W., RAY, J.A., REES, S.B., AND BROEKAERT, W.F. (1997) A Novel Family of Small Cysteine-rich Antimicrobial Peptides from Seed of *Impatiens balsamina* Is Derived from a Single Precursor Protein. *The Journal of Biological Chemistry* **272(39)**, 24480-24487.

TAKAHASHI, K., AND YAMANAKA, S. (2006) Induction of Pluripotent Stem Cells from

Mouse Embryonic and Adult Fibroblast Cultures by Defined Factors. *Cell* **126**, 663-676.

TAKAHASHI, K., TANABE,K., OHNUKI, M., NARITA, M., ICHISAKA, T., TOMODA, K., AND YAMANAKA, S. (2007) Induction of Pluripotent Stem Cells from Adult Human Fibroblasts by Defined Factors. *Cell* **131**, 861-872.

THEVISSEN, K., FERKET, K.K.A., FRANÇOIS, I.E.J.A., AND CAMMUE, B.P.A. (2003) Interactions of antifungal plant defensins with fungal membrane components. *Peptides* **24**, 1705-1712.

THOMAS, C.L., AND MAULE, A.J. (2000) Limitations on the use of fused green fluorescent protein to investigate structure-function relationships for the cauliflower mosaic virus movement protein. *Journal General Virology* **81**, 1851-1855.

THOMAS RUTKOWSKI, D., LINGAPPA, V.R., AND HEGDE, R.S. (2001) Substrate-specific regulation of the ribosome-translocon junction by N-termianl signal sequences *Proceedings of the National Academy of Sciences of the United States of America* **98(14)**, 7823-7828.

TRICCAS, J.A., SUN, L., PALENDIRA, U., AND BRITTON, W.J. (2002) Comparative affects of plasmid-encoded interleukin-12 and interleukin-18 on the protective efficacy of DNA vaccination against Mycobacterium tuberculosis. *Immunology & Cell Biology* **80**, 358-363.

TRICHAS, G., BEGBIE, J., AND SRINIVAS, S. (2008) Use of the viral 2A-peptide for bicistronic expression in transgenic mice. *BioMed Central BMC Biology* **6**, 40 (15 Sept 2008).

TRINCHIERI, G., PFLANZ, S., AND KASTELEIN, R.A. (2003) The IL-12 family of heterodimeric cytokines: new players in the regulation of T-cell responses. *Immunity* **19(5)**, 641-644.

TSENG, C.H., AND TSAI, H.J. (2007) Molecular characterization of a new serotype of duck hepatitis virus. *Virus Research* **126**, 19-31.

VARSHAVSKY, A. (1992) The N-End Rule. *Cell* **69**, 725-735.

VAECK, M., REYNAERTS, A., HOFTE, H., JANSENS, S., DE BEUCKELEER, M., AND DEAN, C. (1987) Transgenic plants protected from insect attack. *Nature* **328**, 33-37.

VERCH, T., YUSIBOV, V., AND KOPROWSKI, H. (1998) Expression and assembly of a full-length monoclonal antibody in plants using a plant virus vector. *Journal of Immunological Methods* **220**, 69-75.

WANG, X., YHANG, J., LU, J., YI, F., LIU, C., AND HU, Y. (2004) Sequence analysis and genomic organization of a new insect picorna-like virus, Ectropis obliqua picorna-like virus, isolated from *Ectropis obliqua*. *Journal of General Virology* **85**, 1145-1151.

WANG, S., YAO, Q., TAO, J., QIAO, Y., AND ZHANG, Z. (2007) Co-ordinate expression of glycine betaine synthesis genes linked by the FMDV 2A region in a single open reading frame in *Pichia pastoris*. *Applied Microbiology and Biotechnology* **77**, 891-899.

WERNIG, M., MEISSNER, A., FOREMAN, R., BRAMBRINK, T., KU, M., HOCHEDLINGER, K., BERNSTEIN, B.E., AND JAENISCH, R. (2007) *In vitro* reprogramming of fibroblasts into a pluripotent ES-cell-like state. *Nature* **448(7151)**, 318-324.

WILSON,J.E., POWELL, M.J., HOOVER, S.E., AND SARNOW, P. (2000) Naturally occurring dicistronic cricket paralysis virus RNA is regulated by two internal ribosome entry sites. *Molecular and Cellular Biology* **20**, 4990-4999.

WOLF, S.F., TEMPLE, P.A., KOBAYASHI, M., YOUNG, D., DICIG, M., LOWE, L., DZIALO, R.,

FITZ, L., FERENZ, C., AND HEWICK, R.M. (1991) Cloning of cDNA for natural killer cell stimulatory factor, a heterodimeric cytokine with multiple biologic effects on T and natural killer cells. *The Journal of Immunology* **146**, 3074-3080.

WOZNIAK., T.M., RYAN, A.A., AND BRITTON, W.J. (2006) Interleukin-23 Restores Immunity to Mycobacterium tuberculosis Infection in IL-12p40-Deficient Mice and Is Not Required for the Development of IL-17 Secreting T-cell Responses. *The Journal of Immunology* **177**, 8684-8692.

WRAITH, J.E., CLARKE, L.A., BECK, M., KOLODNY, E.H., PASTORES, G.M., MUENZER, J., RAPOPORT, D.M., BERGER, K.I., SWIEDLER, S.J., KAKKIS, E.D., BRAAKMAN, T., CHADBOURNE, E., WALTON-BOWEN, K., AND COX, G.F. (2004) Enzyme replacement therapy in mucopolysaccharidosis 1: a randomized, double-blinded, placebo-controlled, multinational study of recombinant human alpha-L-iduronidase (laronidase) *Journal of Pediatrics* **144**, 581-588.

WU, C., LO, C.F., HUANG, C.J., YU, H.T., AND WANG, C.H. (2002) The complete genome sequence of Perina nuda picorna-like virus, an insect-infecting RNA virus with a genome organization similar to that of the mammalian picornaviruses. *Virology* **294**, 312-323.

WUTZ, G., AUER, H., NOWOTNY, N., GROSSE, B., SKERN, T., AND KUECHLER, E. (1996) Equine rhinovirus serotypes 1 and 2: relationship to each other and to aphthoviruses and cardioviruses. *Journal of General Virology* **77**, 1719-1730.

XUE, S., GILLMORE, R., DOWNS, A., TSALLIOS, A., HOLLER, A., GAO, L., WONG, V., MORRIS, E., AND STAUSS, H.J. (2005) Exploiting T-cell receptor genes for cancer immunotherapy. *Clinical and Experimental Immunology* **139**, 167-172.

YANG, H., MAKEYEV, E.V., KANG, Z., JI, S., BAMFORD, D.H., AND VAN DIJK, A.A. (2004) Cloning and sequence analysis of dsRNA segments 5, 6 and 7 of a novel non-group A, B, C adult rotavirus that caused an outbreak of gastroenteritis in China. *Virus Research* **106**, 15-26.

YANG, S., COHEN, C.J., PENG, P.D., ZHAO, Y., CASSARD, L., YU, Z., ZHENG, Z., JONES, S., RESTIFO, N.P., ROSENBERG, S.A., AND MORGAN, R.A. (2008) Development of optimal bicistronic lentiviral vectors facilitates high-level TCR gene expression and robust tumor cell recognition. *Gene Therapy* **15(21)**, 1411-1423.

YASUDA, H., TADA, Y., HAYASHI, Y., JOMORI, T., AND TAKAIWA, F. (2005) Expression of the small peptide GLP-1 in transgenic plants. *Transgenic Research* **14**, 677-684.

YU, D.H., LI, M., HU, X.D., AND CAI, H. (2007) A combined DNA vaccine enhances protective immunity against *Mycobacterium tuberculosis* and *Brucella abortus* in the presence of an IL-12 expression vector. *Vaccine* **25**, 6744-6754.

YU, Y., ANNALA, A.J., BARRIO, J.R., TOYOKUNI, T., SATYAMURTHY, N., NAMAVARI, M., CHERRY, S.R., PHELPS, M.E., HERSCHMAN, H.R., AND GAMBHIR, S.S. (2000) Quantification of target gene expression by imaging reporter gene expression in living animals. *Nature Medicine* **6**, 933-937.

ZHAO, S. L., LIANG, C.Y., HONG, J.J., AND PENG, H.Y. (2003) Genomic sequence analysis of segments 1 to 6 of *Dendrolimus punctatus* cytoplasmic polyhedrosis virus *Archives of Virology* **148**, 1357-1368.

ZHOURAVLEVA, G., FROLOVA, L., LE GOFF, X., LE GUELLEC, R., INGE-VECHTOMOV, S., KISSELEV, L., AND PHILIPPE, M. (1995) Termination of translation in eukaryotes is governed by two interacting polypeptide chain release factors, eRF1 and eRF3. *The EMBO Journal* **14(16)**, 4065-4072.

Zwieb, C., van Nues, R.W., Rosenblad, M.A., Brown, J.D., and Samuelsson, T. (2005) A nomenclature for all signal recognition particle RNAs. *RNA* **11**, 7-13.

Biotechnology and Genetic Engineering Reviews - Vol. 26, 261-280 (2009)

Advances in mRNA Silencing and Transgene Expression: a Gateway to Functional Genomics in Schistosomes

ELISSAVETA B. TCHOUBRIEVA AND BERND H. KALINNA*

Centre for Animal Biotechnology, Faculty of Veterinary Science, The University of Melbourne, Parkville, 3010 VIC, Australia

Abstract

The completion of the WHO *Schistosoma* Genome Project in 2008, although not fully annotated, provides a golden opportunity to actively pursue fundamental research on the parasites genome. This analysis will aid identification of targets for drugs, vaccines and markers for diagnostic tools as well as for studying the biological basis of drug resistance, infectivity and pathology. For the validation of drug and vaccine targets, the genomic sequence data is only of use if functional analyses can be conducted (in the parasite itself). Until recently, gene manipulation approaches had not been seriously addressed. This situation is now changing and rapid advances have been made in gene silencing and transgenesis of schistosomes.

Introduction

Schistosomes are digenetic blood trematodes that cause schistosomiasis in humans by depositing eggs in blood vessels surrounding the bladder or gut of the infected host. In terms of public health, five species of schistosomes are considered most important. They are divided into intestinal schistosomes (*S. mansoni*, *S. intercalatum*, *S. japonicum*, *S. mekongi*) and urinary schistosomes (*S. haematobium*) due to

---

* To whom correspondence may be addressed (bernd.kalinna@unimelb.edu.au)

Abbreviations: WHO, World Health Organization; DALY, Disability Adjusted Life Years; dsRNA, double stranded RNA; RNAi, RNA interference; RISC, RNA-Induced Silencing Complex; siRNA, small interfering RNA; MMLV, Moloney Murine Leukaemia Virus; VSVG, Vesicular Stomatitis Virus Glycoprotein; HSP70, Heat Shock Protein 70; RT-PCR, Real Time Polymerase Chain Reaction; CMV, Cytomegalovirus; ORF, Open Reading Frame; UTR, Untranslated Region.

the different pathologies they induce, and can be distinguished by the specific snail hosts required for transmission. Schistosomes are spread across Sub-Saharan Africa, South-East Asia and Brazil. The typical transmission site is fresh water ponds (rivers) where the intermediate and the primary hosts come in contact to ensure transmission of the parasite and completion of the life cycle. Schistosomiasis is a major cause of morbidity and mortality and it ranks with malaria and tuberculosis in that regard. It is also amongst the most severe parasitic diseases, threatening millions of people with chronic illness, disfigurement, or death that can result from the parasitic infection and occurs mainly in rural areas of the developing world. The disease is endemic in 76 countries; about 250 million people are globally infected, and almost 800 million are at risk of contracting the infection. The major cause of pathology in schistosomiasis is the schistosome egg, associated with formation of granulomas around trapped eggs in the intestinal wall or in the liver causing fibrosis and hepatosplenomegaly, or eggs can become lodged in the wall of the bladder, resulting in inflammation and blood in the urine. The chronic form of the disease has severe outcomes including ascites (accumulation of fluid in the peritoneal cavity, causing abdominal swelling), liver fibrosis, hepatosplenomegaly, secondary cancers of the bladder and often death. Estimates of the global burden of schistosomiasis range from 1.7 to 4.5 million DALY (disability adjusted life years) (W.H.O., 2002; W.H.O, 2004), with even the higher figure under-estimating the true burden as it does not consider chronic long-term disabilities (Hotez *et al.*, 2006). Revised burden estimates for schistosomiasis are in the scale of 3-70 million DALY (King and Dangerfield-Cha, 2008). The current therapeutic approach, chemotherapy with praziquantel, is efficacious against adult worms. But its prolonged use *en mass* and the fact that praziquantel is ineffective against immature and young adult worms, has already raised concerns about future development of tolerance and drug resistant schistosomes (Utzinger and Keiser, 2004). Vaccines are still not available on the market. It is therefore important to actively pursue fundamental research on parasite genome analysis to identify new targets for drug, vaccine and diagnostics development, and for studying the biological basis of drug resistance, antigenicity, infectivity and pathology. Until very recently, tools for manipulating gene expression in schistosomes have been unavailable, mainly due to the complex developmental life cycle, large genome size and lack of immortalised cell lines.

Schistosomes live as parasites in the liver, gut, lungs or blood vessels of vertebrates and their life cycle involves a mammalian host, where adult worms of the two sexes mate and deposit eggs; a free-living aquatic stage (miracidium), derived from eggs and released into the environment; a molluscan stage; and a second free-living aquatic stage (cercaria). Unlike other trematodes, in the schistosomes the sexes are separate. The male is considerably larger than the female and encloses her within his gynaeco-phoric canal for the entire adult lives of the worms, where they reproduce sexually. *S. mansoni* has a very large and complex genome (haploid genome is 270Mb) organised in 8 pairs of chromosomes (Short, 1983). There are 15-20,000 expressed genes. With the recent completion of the *S. mansoni* sequence genome project (Berriman *et al.*, 2009) and emerging abundance of molecular information, the development of new molecular tools such as RNAi, and the promise of new reliable reagents and techniques for transfection, we have now reached the exciting stage of being able to address important issues in the biology of schistosomes in some detail. This review will focus on the current transgenesis tools available to study the biology and functional genomics of schistosomes and speculate how these can enable us to discover novel genes and biochemical pathways involved in growth, reproduction and survival of the parasite as potential targets for drug and vaccine development.

RNA interference technology (gene silencing)

The discovery of sequence-specific gene silencing in response to double-stranded RNAs (dsRNA) has had an enormous impact on molecular biology by uncovering an unsuspected layer of gene expression regulation. The process, also known as RNA interference (RNAi) or RNA silencing, involves complementary pairing of dsRNAs with their homologous messenger RNA targets, thereby preventing the expression and leading to subsequent degradation of these mRNAs or interfering with protein translation.

The first evidence that dsRNA could lead to gene silencing came from work in *Caenorhabditis elegans* published in 1995 by Guo and Kemphues attempting to use antisense RNA to shut down expression of the *par-1* gene in order to assess its function and although the injection of the antisense RNA successfully disrupted gene expression, injection of the sense-strand control was equally efficient (Guo and Kemphues, 1995). The explanation for this phenomenon was published a few years later when Fire and colleagues injected a mixture of both sense and antisense strand RNA (dsRNA) into *C. elegans*, which completely silenced the homologous gene's expression (Fire *et al.*, 1998).

Since then, RNAi technology has been used as a reverse genetics tool in *C. elegans*, *Drosophila* and a wide variety of other organisms, including zebrafish, plants, human, mouse and mammalian cell culture, to inhibit gene activity on a post-transcriptional level generating loss-of-function mutants to study gene function or identify and validate novel therapeutic targets (reviewed in Siomi and Siomi, 2009).

Silencing was found to have high potency and specificity in *C. elegans* since only a few molecules of dsRNA per cell could trigger gene silencing throughout the treated animal (Grishok *et al.*, 2001; Zamore, 2001). The silencing effect usually lasts only for several days however, and although it does appear to be transferred to daughter cells, it eventually diminishes. In some cases it has been shown to have longer-lasting effects as the RNA interference was detected in the first generation progeny of treated *C. elegans* (Tabara *et al.*, 1998; Grishok and Mello, 2002).

The discovery of RNAi has brought to the scientific community not only a new and exciting biological paradigm, but also an incredibly potent tool for addressing questions of gene function in an easy and high-throughput fashion. Genome-wide RNA interference screens have identified near-complete sets of genes involved in cellular processes. Inducible hairpin RNAi constructs are being applied in studies of complex developmental processes in a tissue-specific manner in *Drosophila* (Mummery-Widmer *et al.*, 2009).

Recently a number of groups have developed expression vectors to continually express dsRNAs in transiently and stably transfected mammalian cells to overcome the short-term effects of traditional RNAi. Some of these vectors have been engineered to express small hairpin RNAs (shRNAs), which are processed *in vivo* into dsRNAs (Yu *et al.*, 2002). These vectors contain the shRNA sequence between a promoter and transcription termination site. The transcript folds into a stem-loop structure and the ends of the shRNAs are processed *in vivo* by an enzyme known as Dicer (ribonuclease III (RNaseIII), converting the shRNAs into ~21 nt dsRNA molecules, which in turn initiate RNAi mediated through the RNA-induced silencing complex (RISC).

In schistosomes, the presence of transcripts encoding dicer and RISC-associated proteins (piwi/argonaute orthologues) was relatively recently described by Verjovski-Almeida and colleagues in their report on the transcriptome of *S. mansoni* (Verjovski-Almeida *et al.*, 2003). This finding indicated that an intact RNAi pathway had evolved in schistosomes. Since then it has been shown that RNAi can be applied in schistosomes and appropriate transformation protocols have been adapted and developed (*Table 1*).

RNAi in schistosomes

The first report for successful RNAi in schistosomes, in which cercariae of *S. mansoni* were soaked in dsRNA for 6 days, was published in 2003 (Skelly *et al.*, 2003). The primary objective of this report was to demonstrate the utility of RNAi in schistosomes using the major gut-associated proteinase, cathepsin B (SmCB1or Sm31) as target. Cathepsins are proteolytic enzymes involved in the digestion of hosts haemoglobin during feeding and a principal source of amino acid nutrients for the parasites. Exposure of larval schistosomes to dsRNA, specific to cathepsin B, was demonstrated to inhibit expression of the enzyme. Subsequently, Boyle and colleagues successfully silenced the glucose transporter (SGTP1) and glyceraldehyde-3-phosphate dehydrogenase (GAPDH) genes in sporocysts of *S. mansoni* (Boyle *et al.*, 2003). Exposure of the parasite to SGTP1 dsRNA reduced larval glucose-uptake capacity by 40%, demonstrating a functional phenotype of reduced glucose transport activity. These two publications established that schistosomes were susceptible to RNAi and silencing was strong, specific and long lasting (up to 28 days) in either miracidia or sporocysts.

Within a short time thereafter several laboratories employed RNAi targeting a variety of genes and using different delivery techniques in schistosomes to study gene function. The proteins attracting the most interest were proteolytic enzymes (metallo-, cysteine, and serine proteases) and genes belonging to signalling pathways implicated in adult worm pairing and/or egg deposition. These two groups of proteins have key functions in schistosomes and therefore are potential targets for novel anti-parasite chemotherapy and immunotherapy. The proteases in schistosomes and other parasitic organisms, have been long implicated in pathologies associated with helminth infections and shown to facilitate host invasion (McKerrow and Doenhoff, 1988; McKerrow *et al.*, 1990), parasite feeding (McKerrow and Doenhoff, 1988) and immune evasion (Tamashiro *et al.*, 1987). Schistosomes feed on ingested host blood and the haemoglobin released from erythrocytes is essential for parasite development, growth and reproduction. Blocking proteolysis of host haemoglobin with broad-spectrum protease inhibitors results in profound anti-schistosomal and anti-pathology effects (Bogitsh *et al.*, 1992; Wasilewski *et al.*, 1996), demonstrating the essential role of this pathway in schistosome metabolism. Studies using RNAi approaches alone or in combination with protease specific inhibitors have now been used methodically to study the network of endopeptidases important for the intestinal protein digestion in

Table 1. Overview of RNAi studies in schistosomes

Species	Delivery method	Life cycle stage	Gene(s) targeted	Phenotype	Reference
S. mansoni	soaking liposomes	schistosomula	cathepsin B	decrease in enzyme activity	(Skelly *et al.*, 2003)
	soaking	miracidia	glucose transporter (SGTP1)	decreased transcript levels of SGTP1 and GAPDH reduced glucose uptake	(Boyle *et al.*, 2003)
	electroporation	schistosomula	cathepsin B1	decreased transcript levels growth retardation	(Correnti *et al.*, 2005)
	soaking	miracidia	class B scavenger receptor (SRB)	reduced SRB transcripts reduction in acetylated LDL binding decrease in sporocyst length	(Dinguirard and Yoshino, 2006)
	particle bombardment	adult worms	Sm TβRII	reduction of Sm TβRII expression reduction of gynaecophoral canal protein expression	(Osman *et al.*, 2006)
	electroporation/ soaking	schistosomula, adult worms	alkaline phosphatase (SmAP)	reduced SmAP RNA and enzyme activity	(Ndegwa *et al.*, 2007)
	soaking	adult worms	Inhibin/Activin (SmInAct)	10 fold decrease in protein levels and aborted egg development at an early stage	(Freitas *et al.*, 2007)
	electroporation	adult worms	thioredoxin glutathione reductase (TGR)	relieved the worm burden	(Kuntz *et al.*, 2007)
	soaking	schistosomula	proteasome subunit SmRPN11/ POH1	developmental regulation of the proteasome	(Nabhan *et al.*, 2007)
	electroporation	schistosomula	Cathepsin D	reduced transcript levels, growth retardation, suppressed aspartic protease activity	(Morales *et al.*, 2008)
	electroporation	adult worms	asparaginyl endopeptidase SmAE (Sm32 or legumain)	In the absence of detectable Sm AE protein, SmCB1 was fully processed and active *in vivo*	(Krautz-Peterson and Skelly, 2008)
	injection in mice	adult worms	hypoxanthine guanine phosphoribosyl transferase (HGPRTase)	reduced number of parasites	(Pereira *et al.*, 2008)
S. japonicum	soaking	schistosomula	gynaecophoral canal protein	reduction of gynaecophoral canal protein expression. inhibition of early pairing and reduced parasite burden	(Cheng *et al.*, 2005)
	electroporation	schistosomula	Mago nashi	reduced transcript and protein levels, pronounced phenotypic changes in the testicular lobes	(Zhao *et al.*, 2008)
	soaking	schistosomula	Peroxiredoxin-1 (Prx-1)	larvae susceptible to hydrogen peroxide	(Kumagai *et al.*, 2009)

Sm: *Schistosoma mansoni*

S. mansoni (Correnti *et al.*, 2005; Delcroix *et al.*, 2006; Morales *et al.*, 2008). It has been shown that initial degradation of host blood proteins is ordered, occasionally redundant, and substrate-specific, while the RNAi mediated silencing effect was strong (up to 80%), target specific and prolonged. The schistosomes treated with dsRNA to SmCB1 were viable, with typical intestinal haematin pigmentation (the result of hae-moglobin digestion) and exhibited a significant growth retardation phenotype (*Figure 1*), indicating the essential nature of this complex protease network which is highly conserved throughout invertebrate evolution (Correnti *et al.*, 2005). In experiments targeting another endopeptidase, cathepsin D, by electroporation with dsRNA it was shown that haematin was apparently not deposited in the gut of schistosomules as it appeared red in colour, indicating the presence of intact rather than digested host haemoglobin (Morales *et al.*, 2008). These schistosomules did not survive to maturity after transfer into mice confirming the essential function this enzyme has in parasite nutrition and its potential as target for novel anti-schistosomal interventions.

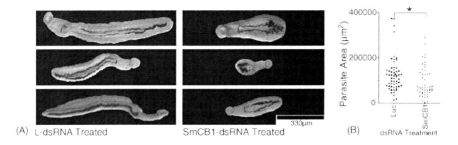

(A) L-dsRNA Treated   SmCB1-dsRNA Treated   (B) dsRNA Treatment

Figure 1. In *Schistosoma mansoni*, RNAi induced by introduction of dsRNA at 3 h after mechanical transformation of schistosomules from cercariae has effects that persist for at least 3 weeks *in vivo*. (A) Examples of schistosomes recovered by portal vein perfusion 3 weeks following electroporation with either luciferase-dsRNA (left) or SmCB1-dsRNA (right) and intramuscular injection into mice. (B) Population quantitation of schistosome size 3 weeks following dsRNA treatment (*$p < 0.005$). Scale bar = 330 μm. (Correnti *et al.*, 2005). Reprinted from *Mol Biochem Parasitol*, 143, Correnti, J. M., Brindley, P. J. and Pearce, E. J., Long-term suppression of cathepsin B levels by RNA interference retards schistosome growth, 209-215, (2005), with permission from Elsevier.

Another schistosome protease - the asparaginyl endopeptidase SmAE (also known as Sm32, or legumain), has been proposed to proteolytically convert the inactive precursor of SmCB1 into its mature catalytic form *in vitro* (Dalton and Brindley, 1996; Caffrey *et al.*, 2004). Although a substantial and specific suppression (>90%) of SmAE transcripts was achieved by RNAi and SmAE enzyme activity could not be detected, it was shown that SmCB1 was fully processed and active. This finding indicated that SmAE was not essential in SmCB1 activation *in vivo* (Krautz-Peterson and Skelly, 2008).

The protocols for effective gene silencing by RNAi targeting cathepsin B in life stages of *S. mansoni* have been summarised by Krautz-Peterson and co-workers (Krautz-Peterson *et al.*, 2007). In brief, the conclusions were that electroporation was more effective in delivering dsRNA in schistosomula compared to soaking and that both small interfering RNAs (siRNAs) (~21 bp) and long dsRNA (>405 bp) demonstrated similar silencing efficiency. Furthermore, neither the persistence of the dsRNA in culture medium, nor transformation of cercariae to schistosomula adversely affected RNAi susceptibility and silencing. Interestingly, complete suppression of the cathepsin B gene was never achieved regardless of the dsRNA dose, possibly due to difficulties to achieve gene silencing uniformly in a mixed population of cells in a living worm.

Research efforts in a number of laboratories are aimed at understanding the role of signal transduction pathways and their role in parasite interaction with its host environment and amongst themselves. For example, the TGF-β signalling pathway is essential for schistosome embryogenesis and has implications for the development of novel therapeutic strategies and prevention of the disease (reviewed in Ting-An and Hong-Xiang, 2009). Schistosomes express a TGF-β type II receptor (SmTβRII) and are exceptional amongst the trematodes in the way that they have evolved separate sexes and the sexual development of the female requires constant contact with the male. This unique phenomenon is associated with pathogenesis since eggs are the triggering factor, not the adult worms themselves (reviewed in Hoffmann, 2004). Thus, blocking components of the parasite TGF-β signalling pathway by RNAi would likely abolish worm pairing and egg production, rendering them potential targets for novel intervention strategies development towards transmission and disease control (Osman *et al.*, 2006; Freitas *et al.*, 2007).

RNAi was also applied in *S. japonicum* to investigate the role of the gynaecophoral canal protein SjGCP during the pairing process (Cheng *et al.*, 2005). The pairing of a male worm with a female worm residing in the gynaecophoral canal of the male plays a critical role in the development of the female parasite. Because the male-specific SjGCP is found in significant quantities in the adult female worm after pairing, it could play an important role in parasite pairing. Dose-dependent inhibition with up to 75% suppression was observed on the SjGCP transcript level in schistosomules 7 days after treatment with siRNA. In further studies, the effect of siRNA duplexes targeting the SjGCP gene was evaluated *in vitro* as well as in mice infected with *S. japonicum in vivo* (Cheng *et al.*, 2009). Two out of three siRNAs employed gave rise to significant suppression of SjGCP transcript levels, reduction of protein amount, and totally abolished the parasite pairing. Evaluation of the pairing inhibitory effect *in vivo*, in mice infected with *S. japonicum* and treated with siRNA, revealed significant inhibition of early parasite pairing and reduced parasite burden, a demonstration of the important role of SjGCP in pairing and subsequent development of *S. japonicum.*

Vector-mediated gene silencing by siRNAs derived from shRNA expressed by mammalian Pol III promoter H1 is also applicable in schistosomes (Zhao *et al.*, 2008). Schistosomula were electroporated with a Mago nashi shRNA expression vector. The shRNA expressed from the mammalian Pol III promoter H1 specifically reduced the levels of Mago nashi mRNA and proteins in *S. japonicum* accompanied by pronounced phenotypic changes in the testicular lobes.

RNAi in schistosomes has provided means for functional analysis of genes such as a CD36-like class B scavenger receptor (SRB) which is potentially involved in some aspect of larval growth and development (Dinguirard and Yoshino, 2006) and an *S. mansoni* alkaline phosphatase (SmAP) (Ndegwa *et al.*, 2007). The use of RNAi technology further suggested that the proteasome may be downregulated during the early stages of schistosomula development and is subsequently upregulated as the parasite matures to the adult stage (Nabhan *et al.*, 2007). Kumagai and colleagues uncovered the function of Peroxiredoxin-1 (Prx-1) in *S. japonicum* as a scavenger against hydrogen peroxide showing its potential as novel target for drug and vaccine development for schistosomiasis (Kumagai *et al.*, 2009). Both genetic and biochemical approaches were employed in the discovery and validation of potential drug target meeting all the major criteria for anti-schistosomal chemotherapy development. Silencing of the thioredoxin glutathione reductase (TGR) by RNAi led to rapid parasite death within 4 days of treatment, proving its essential role in the survival of *S. mansoni*. Use of specific and efficient inhibitors of TGR reduced the worm burden by ~60%, suggesting the importance of TGR for parasite survival and its potential as key target for therapy with those compounds (Kuntz *et al.*, 2007).

The first successful *in vivo* demonstration and evaluation of the therapeutic application of RNAi against schistosomiasis in a chronic infection model has been recently published by Pereira and colleagues. Small interfering RNAs were produced against the hypoxanthine guanine phosphoribosyl transferase (HGPRTase) gene in *S. mansoni* and intravenously injected into infected mice. As a result the total number of parasites was reduced by approximately 27%. RT-PCR analysis showed a significant reduction in parasite target mRNA but not in the hosts homologue. The survival rate of treated mice was not affected by the dose of siRNAs. Further optimization in molecule delivery and siRNA dose could be expected to have a more pronounced effect on the parasite and possibly lead to its complete elimination (Pereira *et al.*, 2008).

Transgenesis of schistosomes

Since completion of the *Schistosoma mansoni* genome sequencing project in 2008, the challenge we now face is how to determine the function of unknown genes and pathways involved in host immune evasion, growth, reproduction and survival, all of which potentially represent novel more effective targets for drug and vaccine development. The development of tools and methods for gene manipulation in schistosomes has been slow due to their complex developmental life cycle, large size genome and absence of immortalised cell lines. Several approaches for the introduction of transgenes (transgenesis) in the form of reporter gene RNA or plasmid based cDNA into schistosomes have been adapted and significant advances have been made (*Table 2*). A number of approaches are theoretically feasible to introduce reporter transgenes into schistosomes. These include microinjection, electroporation, lipofection-type approaches, biolistics (particle bombardment), or the use of infectious vectors such as retrovirus or mobile genetic elements.

In pioneering studies transgenes in the form of mRNA or plasmids were introduced into the parasites by particle bombardment (Davis *et al.*, 1999; Wippersteg *et al.*, 2002a; Heyers *et al.*, 2003). It is now a decade ago that in a landmark paper Davis and colleagues described the delivery of luciferase by mRNA or encoded on a DNA plasmid into adult schistosomes by particle bombardment (Davis *et al.*, 1999). The DNA plasmid contained the *S. mansoni* SL RNA gene fused upstream of the luciferase ORF followed by an *S. mansoni* enolase UTR and polyadenylation signal. With both mRNA and plasmid-encoded luciferase the authors detected reporter protein expression by luminometry. Luciferase was present and expressed 24 hours after particle bombardment.

After this initial paper a number of reports were published in short succession, also employing particle bombardment (Wippersteg *et al.*, 2002a; Wippersteg *et al.*, 2002b; Heyers *et al.*, 2003; Rossi *et al.*, 2003; Wippersteg *et al.*, 2003). Wippersteg *et al.* constructed a plasmid expressing the Green Fluorescent Protein (GFP) reporter gene

Table 2. Overview of transfection studies in schistosomes

Species	Delivery method	Life cycle stage	Promoter	Reporter	Genetic material	Tissue/localisation	Reference
S. mansoni	particle bombardment	adult worms	n/a Sm sl-RNA	luciferase	RNA DNA	no data	(Davis et al., 1999)
	particle bombardment	male adults sporocysts	Sm HSP70	GFP	DNA	tegument surface	(Wippersteg et al., 2002a)
	particle bombardment	sporocysts	Sm ER60	GFP	DNA	excretory/secretory system	(Wippersteg et al., 2002b)
	particle bombardment	male adults sporocysts	Sm Calcineurin A	GFP	DNA	excretory/secretory system	(Rossi et al., 2003)
	particle bombardment	male adults miracidia	Sm HSP70	EGFP	DNA	tegument, transfected miracidia infective for snail	(Heyers et al., 2003)
	particle bombardment	male adults	Sm cathepsin L1 Sm cathepsin B2	GFP	DNA	gut, tegument	(Wippersteg et al., 2005)
	retroviral transduction	Sporocysts schistosomules	retroviral LTR Sm Actin 1.1	neomycin phosphotransferase, luciferase	retroviral genome	tegument	(Kines et al., 2006)
	electroporation	schistosomules	Sm Actin 1.1	luciferase	DNA	not specified	(Correnti et al., 2007)
	electroporation	schistosomules	Sm Actin 1.1 Sm HSP70	luciferase	RNA DNA	not specified	(Morales et al., 2007)
	retroviral transduction	schistosomules	retroviral LTR Sm Actin 1.1	luciferase	retroviral genome	not specified	(Kines et al., 2008)
S. japonicum	electroporation	cultured cells			DNA	cytoplasm	(Yuan et al., 2005)

Sm: *Schistosoma mansoni*

under the control of the *S. mansoni* HSP70 promoter and terminator. This plasmid was introduced into sporocysts and adults and expression of GFP could be shown after heat shock induction by confocal microscopy 3 days after transfection. Fluorescence was mainly visible on the surface of adult worms and inside sporocysts. The authors also employed RT-PCR to detect GFP transcripts and Western Blotting to identify the GFP protein (Wippersteg *et al.*, 2002a).

Expanding on this report Wippersteg and co-workers in the same year published further experiments where the SmHSP70 promoter and terminator elements in the plasmid used for transfection were exchanged for *cis*-acting elements of the *S. mansoni* ER60 (SmER60) gene (Wippersteg *et al.*, 2002b). SmER60 encodes a cysteine protease which in earlier studies had been localised to the endoplasmic reticulum in excretory tissues in adult parasites (Finken-Eigen and Kunz, 1997). After bombardment of sporocysts the observed expression of GFP was tissue-specific, and the localisation of ER60 in the excretory/secretory (ES) system of the larval parasites suggested that ER60 might have a role in penetration and migration of miracidia in the intermediate snail host. In an additional follow-up report the same authors verified this tissue specific expression of the ER60 protease by employing Texas Red-labelled BSA, which accumulates in the ES system, together with biolistic transformation. The ER60-GFP and the Texas Red-BSA co-localised in the same compartments (Wippersteg *et al.*, 2003).

The same approach to co-localise Texas Red-BSA to the ES system was used by Rossi *et al.* to study *S. mansoni* calcineurin A and its expression in the ES system (Rossi *et al.*, 2003). A plasmid containing 5'- and 3'-elements of the calcineurin A gene, driving the expression of GFP was constructed and biolistically transferred into sporocysts and male adult worms. Similar to the results discussed above, fluorescent signals for GFP and Texas Red co-localised in the ES system of the parasite.

In our laboratory we introduced plasmid DNA by particle bombardment into miracidia, sporocysts and adults of *S. mansoni* (Heyers *et al.*, 2003). In these studies the Enhanced Green Fluorescent Protein (EGFP) was used in a plasmid similar to that used be Wippersteg described above. In our studies we particularly focused on the miracidial life cycle stage, because this larval stage offers the unique opportunity to introduce transgenes into the germline and additionally to reintroduce transgenic organisms into the parasite life cycle. Bombarded miracidia were able to infect *B. glabrata* snails and two weeks after infection gold particles could clearly be identified within the developing sporocysts in paraffin-sections of infected snail tissues. Interestingly, these gold particles were located close to the nuclei of germ ball cells (*Figure 2*). Transfection of these totipotent cells could, theoretically, lead to the transformation of the germ line and the derivation of transgenic schistosomes. Reporter gene activity could also be determined at 10 days post-infection by RT-PCR. These findings indicated that it is possible to return transgenic miracidia to the parasite life cycle, a crucial step for the establishment of a transgenesis system for schistosomes.

The most recent publication describing the transfection of schistosomes using biolistic methods reports the tissue-specific expression of GFP driven by the promoters of two of *S. mansoni* protease genes, cathepsin L1 and cathepsin B2 (Wippersteg *et al.*, 2005). As predicted from earlier reports (Bogitsh *et al.*, 2001), the *S. mansoni* cathepsin L1 promoter drove GFP expression throughout the gut whereas transformation with the SmCB2 (Caffrey *et al.*, 2002) construct resulted in GFP fluorescence localised in the tegument.

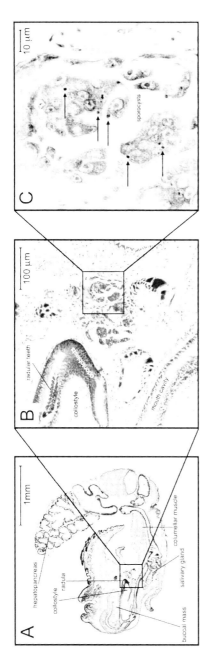

Figure 2. Haematoxylin and eosin-stained paraffin section of *Biomphalaria glabrata* snail two weeks after infection with miracidia transformed by particle bombardment with gold micro-carriers. Panel A: Overview section through the mouth, mid-body, head-body, mid-foot and hepatopancreas/kidney region of the snail. The box outlines the kidney region that is infected with sporocyst. This boxed region is shown at higher magnifications in panels B and C. The locations of other key anatomical features are indicated by text. Panel B: Higher magnification of snail tissues shown in panel A: the regions in and around the box outline developing mother sporocysts. The radula of the snail is prominent in the upper left of the panel. The locations of other key anatomical features are indicated by text. Panel C: Higher power magnification of snail tissues from panel B, with a close-up image of sporocysts. The arrows show the position of gold particles in close proximity to the germ ball cells in the sporocysts. Scale bars are included at the top of each panel. (Heyers *et al.*, 2003).

Reprinted from *Exp Parasitol*, 105, Heyers, O., Walduck, A. K., Brindley, P. J., Bleiss, W., Lucius, R., Dorbic, T., Wittig, B. and Kalinna, B. H., *Schistosoma mansoni* miracidia transformed by particle bombardment infect *Biomphalaria glabrata* snails and develop into transgenic sporocysts, 174-178, (2003), with permission from Elsevier.

More recently the efficacy of electroporation for the introduction of plasmid-based DNA constructs was tested in *S. japonicum* and *S. mansoni* (Yuan *et al.*, 2005; Correnti *et al.*, 2007). Yuan *et al.* using the commercial plasmid pEGFP-C1 (Clontech) were able to show that the CMV promoter could drive EGFP expression in primary cells cultures of *S. japonicum*. Additionally, the plasmids were introduced into schistosomula and adult worms by electroporation and EGFP expression was demonstrated using RT-PCR, Western Blotting and confocal microscopy, which revealed EGFP fluorescence along the tegumental surface of the worms (Yuan *et al.*, 2005). In their report Correnti and colleagues describe the use of electroporation for the delivery of plasmid DNA into larval *S. mansoni* schistosomes, combined with a preliminary analysis of the *S. mansoni* Actin 1.1 (SmAct1.1) promoter sequence (Correnti *et al.*, 2007). Expression of luciferase driven by the SmAct1.1 gene promoter was transient. The authors suggest that the loss of expression over time was probably not due to loss of plasmid, because transfected parasites that were no longer expressing the luciferase remained PCR-positive for luciferase DNA for 8 weeks following electroporation. This finding is similar to that reported by Yuan *et al.* (Yuan *et al.*, 2005). These results also indicated that the electroporation protocol described was either insufficient to deliver the transgene to the germline or that the transgene was not integrated at high frequency to be able to be detected in transgenic F1 parasites.

Although these reports clearly showed advances in the development of molecular tools to characterise schistosome genes, it is unlikely that transfection with plasmid based transgenes will result in integration into the genome. Integration into the genome would represent a key milestone for the study of long-term transgene expression, investigation of schistosome gene function, development of transgenic schistosomes, and insertional mutagenesis approaches. This might be achievable with gene therapy-type approaches utilising retroviruses, lentiviruses, retrotransposons, or transposons which enhance the likelihood of development of heritable, transgenic lines of schistosomes. This is particularly likely if germ-line cells can be targeted for transduction. Additionally, transposons or retroviruses can carry gene cassettes for production of short interfering RNAs and thereby combine a powerful knock-down technology with an efficient delivery system offering the possibility for heritable RNAi targeting specific host cell genes (Brown *et al.*, 2003; Paddison *et al.*, 2004).

Together with colleagues we have recently reported approaches using retroviruses and transposons to transduce schistosomes (Kines *et al.*, 2006; Morales *et al.*, 2007). In Kines *et al.* we produced replication incompetent Moloney Murine Leukaemia Virus (MMLV) virions pseudotyped with Vesicular Stomatitis Virus Glycoprotein (VSVG) and carrying a luciferase reporter gene. After exposure of schistosomes to virions harvested from producer cells cultures, immunofluorescence studies indicated that the VSVG envelope interacted with the schistosome surface and that the retroviral capsid and RNA genome were released within the surface cells (*Figure 3*). We were also able to show the presence of proviral forms of the retrovirus within the schistosome genome by Southern hybridization analysis, while transcription from the transgenes was indicated by the presence of transcripts encoding neomycin phosphotransferase and luciferase. These results strongly suggested integration of the retroviral transgenes into schistosome chromosomes (Kines *et al.*, 2006).

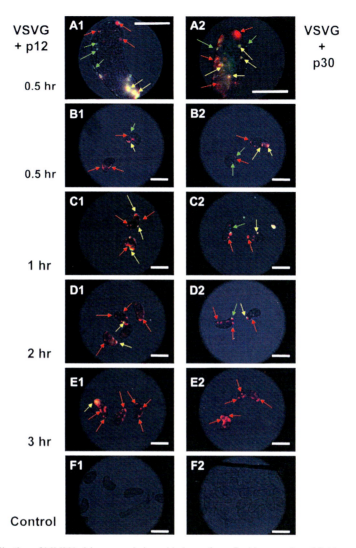

Figure 3. Visualization of MMLV virions associating with the surface of schistosomules of *Schistosoma mansoni*. Schistosomula were exposed to supernatants of MMLV-VSVG (pLNHX-D70-EGFP) virions for 30 min to 3 h, and then fixed and processed for dual immunofluorescence. Immunolabeling protocol involved anti-VSVG antibody with Alexa 488 goat anti-rabbit antibody (green signal, green arrows) and either anti-p12gag (A1–E1) or anti-p30gag (A2–E2) antisera with Alexa 594 donkey anti-goat antibody (red signal, red arrows). F1, F2, control schistosomules not exposed to virions but labeled with the antibodies. Images for green and red fluorescence were acquired separately and then overlaid (colocalization of red and green signals indicated by yellow signal, yellow arrows). Merged images of red, green, and bright field (blue) are presented. Bars: A1 and A2, 100 μm; B1–F1, B2–F2, 50 μm (Kines *et al.*, 2006). Reprinted from *Exp Parasitol*, 112, Kines, K. J., Mann, V. H., Morales, M. E., Shelby, B. D., Kalinna, B. H., Gobert, G. N., Chirgwin, S. R. and Brindley, P. J., Transduction of *Schistosoma mansoni* by vesicular stomatitis virus glycoprotein-pseudotyped Moloney murine leukemia retrovirus, 209-220, (2006), with permission from Elsevier.

We have also used the transposon *piggyBac* to accomplish transformation of *S. mansoni* (Morales *et al.*, 2007). A *piggyBac* donor plasmid containing the luciferase-coding sequence under the control of the SmAct1.1 promoter (Correnti *et al.*, 2007) was constructed and transfected into schistosomes by electroporation of larval stages together with mRNAs encoding the *piggyBac* transposase. The activity of *piggyBac* was determined by plasmid excision assays, and the recovery of excised plasmids from tissues of transformed schistosomules in these assays indicated that *piggyBac* was active in the worm. Southern blot hybridisation analysis of genomic DNAs from populations of schistosomules transformed with donor constructs and helper transposase-mRNA detected numerous variable length luciferase-positive signals. These findings further indicated *piggyBac* transposon insertions into the schistosome chromosomes. The integration sites were analysed by a PCR technique. Numerous *piggyBac* integrations were detected and, after cloning, the fragments sequenced ranged in size from ~1.5 kb to 4 kb. Sequence analysis indicated that integration of *piggyBac* took place at numerous loci in the schistosome genome at target TTAA sites.

Current studies in our laboratory are aimed at vector mediated RNAi of the schistosomal cathepsin B1 protease (SmCatB1), using this viral system described above (Kines *et al.*, 2006). We have designed viral vectors to express targeted dsRNA, thus coupling a powerful delivery vehicle with a potent mechanism to specifically silence a gene of interest. We have targeted SmCatB1 because RNAi knockdown of this protein delivers a visible phenotype (Correnti *et al.*, 2005). Preliminary studies have shown that the newly developed viral vector for the production of virions carrying a dsRNA hairpin loop specific for SmCatB1 resulted in 80% reduction in transcript levels of this protease 72 h after exposure to the virus and an effective silencing of the cathepsin B enzyme (Tchoubrieva *et al.*, unpublished).

Concluding remarks and outlook

The genome of *S. mansoni* (Berriman *et al.*, 2009; Zerlotini *et al.*, 2009) marks the first step toward a better understanding of this destructive, blood-feeding parasite of humans. Genomic sequence data provide a rich resource for gene expression profiling using microchip gene array analysis, and for gene deletion studies and/or gene silencing. Thus, the availability of the complete genome sequence of *S. mansoni* and emerging technologies now provide a golden opportunity to explore the function of schistosome genes and gene products. Driven by this rapidly growing information on genome sequences the progress in schistosome RNAi and transgenesis over the last few years has been remarkable and existing technologies have been rapidly adapted for use with these parasites.

An exciting approach which shows promise for conducting effective functional analyses in schistosomes is the development of promoter trap or polyA trap vectors. Gene trapping is a high-throughput approach used to introduce insertional mutations across the genome. Gene traps are able to simultaneously inactivate and report the expression of the trapped gene at the insertion site, and provide a DNA tag for the rapid identification of the disrupted gene (Zambrowicz *et al.*, 1998; Stanford *et al.*, 2001). Gene trapping has been used in mouse, zebrafish and *Drosophila* genomics and has proven to be a powerful method to identify functional genes (Wurst *et al.*, 1995; Hicks *et al.*, 1997; Lukacsovich and Yamamoto, 2001; Jao *et al.*, 2008). We are now pursuing the gene trapping approach as the next step towards functional genomics of schistosomes to ultimately establish a correlation between the physical and genetic maps of the schistosome genome.

Acknowledgements

We thank Anna Walduck for helpful discussions and critical review of the manuscript. Support from the National Health & Medical Research Council of Australia (454422) and the US National Institutes of Health (RO1AI072773) is gratefully acknowledged.

References

BERRIMAN, M., HAAS, B. J., LOVERDE, P. T., WILSON, R. A., DILLON, G. P., CERQUEIRA, G. C., MASHIYAMA, S. T., AL-LAZIKANI, B., ANDRADE, L. F., ASHTON, P. D., ASLETT, M. A., BARTHOLOMEU, D. C., BLANDIN, G., CAFFREY, C. R., COGHLAN, A., COULSON, R., DAY, T. A., DELCHER, A., DEMARCO, R., DJIKENG, A., EYRE, T., GAMBLE, J. A., GHEDIN, E., GU, Y., HERTZ-FOWLER, C., HIRAI, H., HIRAI, Y., HOUSTON, R., IVENS, A., JOHNSTON, D. A., LACERDA, D., MACEDO, C. D., MCVEIGH, P., NING, Z., OLIVEIRA, G., OVERINGTON, J. P., PARKHILL, J., PERTEA, M., PIERCE, R. J., PROTASIO, A. V., QUAIL, M. A., RAJANDREAM, M. A., ROGERS, J., SAJID, M., SALZBERG, S. L., STANKE, M., TIVEY, A. R., WHITE, O., WILLIAMS, D. L., WORTMAN, J., WU, W., ZAMANIAN, M., ZERLOTINI, A., FRASER-LIGGETT, C. M., BARRELL, B. G. AND EL-SAYED, N. M. (2009). The genome of the blood fluke *Schistosoma mansoni. Nature* **460**, 352-358.

BOGITSH, B. J., DALTON, J. P., BRADY, C. P. AND BRINDLEY, P. J. (2001). Gut-associated immunolocalization of the *Schistosoma mansoni* cysteine proteases, SmCL1 AND SmCL2. *J Parasitol* **87**, 237-41.

BOGITSH, B. J., KIRSCHNER, K. F. AND ROTMANS, J. P. (1992). *Schistosoma japonicum*: immunoinhibitory studies on hemoglobin digestion using heterologous antiserum to bovine cathepsin D. *J Parasitol* **78**, 454-9.

BOYLE, J. P., WU, X. J., SHOEMAKER, C. B. AND YOSHINO, T. P. (2003). Using RNA interference to manipulate endogenous gene expression in *Schistosoma mansoni* sporocysts. *Mol Biochem Parasitol* **128**, 205-15.

BROWN, A. E., BUGEON, L., CRISANTI, A. AND CATTERUCCIA, F. (2003). Stable and heritable gene silencing in the malaria vector *Anopheles stephensi. Nucleic Acids Res* **31**, e85.

CAFFREY, C. R., MCKERROW, J. H., SALTER, J. P. AND SAJID, M. (2004). Blood 'n' guts: an update on schistosome digestive peptidases. *Trends Parasitol* **20**, 241-8.

CAFFREY, C. R., SALTER, J. P., LUCAS, K. D., KHIEM, D., HSIEH, I., LIM, K. C., RUPPEL, A., MCKERROW, J. H. AND SAJID, M. (2002). SmCB2, a novel tegumental cathepsin B from adult *Schistosoma mansoni. Mol Biochem Parasitol* **121**, 49-61.

CHENG, G., FU, Z., LIN, J., SHI, Y., ZHOU, Y., JIN, Y. AND CAI, Y. (2009). In vitro and in vivo evaluation of small interference RNA-mediated gynaecophoral canal protein silencing in *Schistosoma japonicum. J Gene Med* **11**, 412-21.

CHENG, G. F., LIN, J. J., SHI, Y., JIN, Y. X., FU, Z. Q., JIN, Y. M., ZHOU, Y. C. AND CAI, Y. M. (2005). Dose-dependent inhibition of gynecophoral canal protein gene expression in vitro in the schistosome (*Schistosoma japonicum*) by RNA interference. *Acta*

Biochim Biophys Sin (Shanghai) **37**, 386-90.

CORRENTI, J. M., BRINDLEY, P. J. AND PEARCE, E. J. (2005). Long-term suppression of cathepsin B levels by RNA interference retards schistosome growth. *Mol Biochem Parasitol* **143**, 209-15.

CORRENTI, J. M., JUNG, E., FREITAS, T. C. AND PEARCE, E. J. (2007). Transfection of *Schistosoma mansoni* by electroporation and the description of a new promoter sequence for transgene expression. *Int J Parasitol* **37**, 1107-15.

DALTON, J. P. AND BRINDLEY, P. J. (1996). Schistosome asparaginyl endopeptidase SM32 in hemoglobin digestion. *Parasitol Today* **12**, 125.

DAVIS, R. E., PARRA, A., LOVERDE, P. T., RIBEIRO, E., GLORIOSO, G. AND HODGSON, S. (1999). Transient expression of DNA and RNA in parasitic helminths by using particle bombardment. *Proc Natl Acad Sci U S A* **96**, 8687-8692.

DELCROIX, M., SAJID, M., CAFFREY, C. R., LIM, K. C., DVORAK, J., HSIEH, I., BAHGAT, M., DISSOUS, C. AND MCKERROW, J. H. (2006). A multienzyme network functions in intestinal protein digestion by a platyhelminth parasite. *J Biol Chem* **281**, 39316-29.

DINGUIRARD, N. AND YOSHINO, T. P. (2006). Potential role of a CD36-like class B scavenger receptor in the binding of modified low-density lipoprotein (acLDL) to the tegumental surface of *Schistosoma mansoni* sporocysts. *Mol Biochem Parasitol* **146**, 219-30.

FINKEN-EIGEN, M. AND KUNZ, W. (1997). *Schistosoma mansoni*: gene structure and localization of a homologue to cysteine protease ER 60. *Exp Parasitol* **86**, 1-7.

FIRE, A., XU, S., MONTGOMERY, M. K., KOSTAS, S. A., DRIVER, S. E. AND MELLO, C. C. (1998). Potent and specific genetic interference by double-stranded RNA in *Caenorhabditis elegans*. *Nature* **391**, 806-11.

FREITAS, T. C., JUNG, E. AND PEARCE, E. J. (2007). TGF-beta signaling controls embryo development in the parasitic flatworm *Schistosoma mansoni*. *PLoS Pathog* **3**, e52.

GRISHOK, A. AND MELLO, C. C. (2002). RNAi (Nematodes: *Caenorhabditis elegans*). *Adv Genet* **46**, 339-60.

GRISHOK, A., PASQUINELLI, A. E., CONTE, D., LI, N., PARRISH, S., HA, I., BAILLIE, D. L., FIRE, A., RUVKUN, G. AND MELLO, C. C. (2001). Genes and mechanisms related to RNA interference regulate expression of the small temporal RNAs that control *C. elegans* developmental timing. *Cell* **106**, 23-34.

GUO, S. AND KEMPHUES, K. J. (1995). par-1, a gene required for establishing polarity in *C. elegans* embryos, encodes a putative Ser/Thr kinase that is asymmetrically distributed. *Cell* **81**, 611-20.

HEYERS, O., WALDUCK, A. K., BRINDLEY, P. J., BLEISS, W., LUCIUS, R., DORBIC, T., WITTIG, B. AND KALINNA, B. H. (2003). *Schistosoma mansoni* miracidia transformed by particle bombardment infect Biomphalaria glabrata snails and develop into transgenic sporocysts. *Exp Parasitol* **105**, 174-8.

HICKS, G. G., SHI, E. G., LI, X. M., LI, C. H., PAWLAK, M. AND RULEY, H. E. (1997). Functional genomics in mice by tagged sequence mutagenesis. *Nat Genet* **16**, 338-44.

HOFFMANN, K. F. (2004). An historical and genomic view of schistosome conjugal biology with emphasis on sex-specific gene expression. *Parasitology* **128 Suppl 1**, S11-22.

HOTEZ, P. J., MOLYNEUX, D. H., FENWICK, A., OTTESEN, E., EHRLICH SACHS, S. AND SACHS, J. D. (2006). Incorporating a rapid-impact package for neglected tropical diseases with programs for HIV/AIDS, tuberculosis, and malaria. *PLoS Med* **3**, e102.

JAO, L. E., MADDISON, L., CHEN, W. AND BURGESS, S. M. (2008). Using retroviruses as a mutagenesis tool to explore the zebrafish genome. *Brief Funct Genomic Proteomic* **7**, 427-43.

KINES, K. J., MANN, V. H., MORALES, M. E., SHELBY, B. D., KALINNA, B. H., GOBERT, G. N., CHIRGWIN, S. R. AND BRINDLEY, P. J. (2006). Transduction of *Schistosoma mansoni* by vesicular stomatitis virus glycoprotein-pseudotyped Moloney murine leukemia retrovirus. *Exp Parasitol* **112**, 209-20.

KINES, K. J., MORALES, M. E., MANN, V. H., GOBERT, G. N. AND BRINDLEY, P. J. (2008). Integration of reporter transgenes into *Schistosoma mansoni* chromosomes mediated by pseudotyped murine leukemia virus. *FASEB J* **22**, 2936-48.

KING, C. H. AND DANGERFIELD-CHA, M. (2008). The unacknowledged impact of chronic schistosomiasis. *Chronic Illn* **4**, 65-79.

KRAUTZ-PETERSON, G., RADWANSKA, M., NDEGWA, D., SHOEMAKER, C. B. AND SKELLY, P. J. (2007). Optimizing gene suppression in schistosomes using RNA interference. *Mol Biochem Parasitol* **153**, 194-202.

KRAUTZ-PETERSON, G. AND SKELLY, P. J. (2008). Schistosome asparaginyl endopeptidase (legumain) is not essential for cathepsin B1 activation in vivo. *Mol Biochem Parasitol* **159**, 54-8.

KUMAGAI, T., OSADA, Y., OHTA, N. AND KANAZAWA, T. (2009). Peroxiredoxin-1 from *Schistosoma japonicum* functions as a scavenger against hydrogen peroxide but not nitric oxide. *Mol Biochem Parasitol* **164**, 26-31.

KUNTZ, A. N., DAVIOUD-CHARVET, E., SAYED, A. A., CALIFF, L. L., DESSOLIN, J., ARNER, E. S. AND WILLIAMS, D. L. (2007). Thioredoxin glutathione reductase from *Schistosoma mansoni*: an essential parasite enzyme and a key drug target. *PLoS Med* **4**, e206.

LUKACSOVICH, T. AND YAMAMOTO, D. (2001). Trap a gene and find out its function: toward functional genomics in Drosophila. *J Neurogenet* **15**, 147-68.

MCKERROW, J. H., BRINDLEY, P., BROWN, M., GAM, A. A., STAUNTON, C. AND NEVA, F. A. (1990). *Strongyloides stercoralis*: identification of a protease that facilitates penetration of skin by the infective larvae. *Exp Parasitol* **70**, 134-43.

MCKERROW, J. H. AND DOENHOFF, M. J. (1988). Schistosome proteases. *Parasitol Today* **4**, 334-340.

MORALES, M. E., MANN, V. H., KINES, K. J., GOBERT, G. N., FRASER, M. J., JR., KALINNA, B. H., CORRENTI, J. M., PEARCE, E. J. AND BRINDLEY, P. J. (2007). piggyBac transposon mediated transgenesis of the human blood fluke, *Schistosoma mansoni*. *FASEB J* **21**, 3479-89.

MORALES, M. E., RINALDI, G., GOBERT, G. N., KINES, K. J., TORT, J. F. AND BRINDLEY, P. J. (2008). RNA interference of *Schistosoma mansoni* cathepsin D, the apical enzyme of the hemoglobin proteolysis cascade. *Mol Biochem Parasitol* **157**, 160-8.

MUMMERY-WIDMER, J. L., YAMAZAKI, M., STOEGER, T., NOVATCHKOVA, M., BHALERAO, S., CHEN, D., DIETZL, G., DICKSON, B. J. AND KNOBLICH, J. A. (2009). Genome-wide analysis of Notch signalling in Drosophila by transgenic RNAi. *Nature* **458**, 987-92.

NABHAN, J. F., EL-SHEHABI, F., PATOCKA, N. AND RIBEIRO, P. (2007). The 26S proteasome in *Schistosoma mansoni*: bioinformatics analysis, developmental expression, and RNA interference (RNAi) studies. *Exp Parasitol* **117**, 337-47.

NDEGWA, D., KRAUTZ-PETERSON, G. AND SKELLY, P. J. (2007). Protocols for gene silencing in schistosomes. *Exp Parasitol* **117**, 284-91.

Osman, A., Niles, E. G., Verjovski-Almeida, S. and LoVerde, P. T. (2006). *Schistosoma mansoni* TGF-beta receptor II: role in host ligand-induced regulation of a schistosome target gene. *PLoS Pathog* **2**, e54.

Paddison, P. J., Caudy, A. A., Sachidanandam, R. and Hannon, G. J. (2004). Short hairpin activated gene silencing in mammalian cells. *Methods Mol Biol* **265**, 85-100.

Pereira, T. C., Pascoal, V. D., Marchesini, R. B., Maia, I. G., Magalhaes, L. A., Zanotti-Magalhaes, E. M. and Lopes-Cendes, I. (2008). *Schistosoma mansoni*: evaluation of an RNAi-based treatment targeting HGPRTase gene. *Exp Parasitol* **118**, 619-23.

Rossi, A., Wippersteg, V., Klinkert, M. Q. and Grevelding, C. G. (2003). Cloning of 5' and 3' flanking regions of the *Schistosoma mansoni* calcineurin A gene and their characterization in transiently transformed parasites. *Mol Biochem Parasitol* **130**, 133-8.

Short, R. B. (1983). Presidential address: Sex and the single schistosome. *J Parasitol* **69**, 3-22.

Siomi, H. and Siomi, M. C. (2009). On the road to reading the RNA-interference code. *Nature* **457**, 396-404.

Skelly, P. J., Da'dara, A. and Harn, D. A. (2003). Suppression of cathepsin B expression in *Schistosoma mansoni* by RNA interference. *Int J Parasitol* **33**, 363-9.

Stanford, W. L., Cohn, J. B. and Cordes, S. P. (2001). Gene-trap mutagenesis: past, present and beyond. *Nat Rev Genet* **2**, 756-68.

Tabara, H., Grishok, A. and Mello, C. C. (1998). RNAi in *C. elegans*: soaking in the genome sequence. *Science* **282**, 430-1.

Tamashiro, W. K., Rao, M. and Scott, A. L. (1987). Proteolytic cleavage of IgG and other protein substrates by *Dirofilaria immitis* microfilarial enzymes. *J Parasitol* **73**, 149-54.

Ting-An, W. and Hong-Xiang, Z. (2009). PTK-pathways and TGF-beta signaling pathways in schistosomes. *J Basic Microbiol* **49**, 25-31.

Utzinger, J. and Keiser, J. (2004). Schistosomiasis and soil-transmitted helminthiasis: common drugs for treatment and control. *Expert Opin Pharmacother* **5**, 263-85.

Verjovski-Almeida, S., DeMarco, R., Martins, E. A., Guimaraes, P. E., Ojopi, E. P., Paquola, A. C., Piazza, J. P., Nishiyama, M. Y., Jr., Kitajima, J. P., Adamson, R. E., Ashton, P. D., Bonaldo, M. F., Coulson, P. S., Dillon, G. P., Farias, L. P., Gregorio, S. P., Ho, P. L., Leite, R. A., Malaquias, L. C., Marques, R. C., Miyasato, P. A., Nascimento, A. L., Ohlweiler, F. P., Reis, E. M., Ribeiro, M. A., Sa, R. G., Stukart, G. C., Soares, M. B., Gargioni, C., Kawano, T., Rodrigues, V., Madeira, A. M., Wilson, R. A., Menck, C. F., Setubal, J. C., Leite, L. C. and Dias-Neto, E. (2003). Transcriptome analysis of the acoelomate human parasite *Schistosoma mansoni*. *Nat Genet* **35**, 148-57.

W.H.O (2004). The global burden of disease: 2004 update.

W.H.O. (2002). Prevention and control of schistosomiasis and soil-transmitted helminthiasis: report of a WHO expert committee, Geneva 2002. *WHO Technical Report Series* **912**, 1-57.

Wasilewski, M. M., Lim, K. C., Phillips, J. and McKerrow, J. H. (1996). Cysteine protease inhibitors block schistosome hemoglobin degradation in vitro and decrease worm burden and egg production in vivo. *Mol Biochem Parasitol* **81**, 179-189.

Wippersteg, V., Kapp, K., Kunz, W. and Grevelding, C. G. (2002b). Characterisation of the cysteine protease ER60 in transgenic *Schistosoma mansoni* larvae. *Int J Parasitol* **32**, 1219-24.

WIPPERSTEG, V., KAPP, K., KUNZ, W., JACKSTADT, W. P., ZAHNER, H. AND GREVELDING, C. G. (2002a). HSP70-controlled GFP expression in transiently transformed schistosomes. *Mol Biochem Parasitol.* **120**, 141-50.

WIPPERSTEG, V., RIBEIRO, F., LIEDTKE, S., KUSEL, J. R. AND GREVELDING, C. G. (2003). The uptake of Texas Red-BSA in the excretory system of schistosomes and its colocalisation with ER60 promoter-induced GFP in transiently transformed adult males. *Int J Parasitol* **33**, 1139-43.

WIPPERSTEG, V., SAJID, M., WALSHE, D., KHIEM, D., SALTER, J. P., MCKERROW, J. H., GREVELDING, C. G. AND CAFFREY, C. R. (2005). Biolistic transformation of *Schistosoma mansoni* with 5' flanking regions of two peptidase genes promotes tissue-specific expression. *Int J Parasitol* **35**, 583-9.

WURST, W., ROSSANT, J., PRIDEAUX, V., KOWNACKA, M., JOYNER, A., HILL, D. P., GUILLEMOT, F., GASCA, S., CADO, D., AUERBACH, A. AND ET AL. (1995). A large-scale gene-trap screen for insertional mutations in developmentally regulated genes in mice. *Genetics* **139**, 889-99.

YU, J. Y., DERUITER, S. L. AND TURNER, D. L. (2002). RNA interference by expression of short-interfering RNAs and hairpin RNAs in mammalian cells. *Proc Natl Acad Sci U S A* **99**, 6047-52.

YUAN, X. S., SHEN, J. L., WANG, X. L., WU, X. S., LIU, D. P., DONG, H. F. AND JIANG, M. S. (2005). *Schistosoma japonicum*: a method for transformation by electroporation. *Exp Parasitol* **111**, 244-9.

ZAMBROWICZ, B. P., FRIEDRICH, G. A., BUXTON, E. C., LILLEBERG, S. L., PERSON, C. AND SANDS, A. T. (1998). Disruption and sequence identification of 2,000 genes in mouse embryonic stem cells. *Nature* **392**, 608-11.

ZAMORE, P. D. (2001). RNA interference: listening to the sound of silence. *Nat Struct Biol* **8**, 746-50.

ZERLOTINI, A., HEIGES, M., WANG, H., MORAES, R. L., DOMINITINI, A. J., RUIZ, J. C., KISSINGER, J. C. AND OLIVEIRA, G. (2009). SchistoDB: a *Schistosoma mansoni* genome resource. *Nucleic Acids Res* **37**, D579-82.

ZHAO, Z. R., LEI, L., LIU, M., ZHU, S. C., REN, C. P., WANG, X. N. AND SHEN, J. J. (2008). *Schistosoma japonicum*: inhibition of Mago nashi gene expression by shRNA-mediated RNA interference. *Exp Parasitol* **119**, 379-84.

Biotechnology and Genetic Engineering Reviews - Vol. 26, 281-296 (2009)

Metabolic Network-Based Interpretation of Gene Expression Data Elucidates Human Cellular Metabolism

TOMER SHLOMI

Department of Computer Science, Technion – Israel Institute of Technology, Haifa 32000, Israel

Abstract

Research into human metabolism is expanding rapidly due to the emergence of metabolism as a key factor in common diseases. Mathematical modeling of human cellular metabolism has traditionally been performed via kinetic approaches whose applicability for large-scale systems is limited by lack of kinetic constants data. An alternative computational approach bypassing this hurdle called constraint-based modeling (CBM) serves to analyze the function of large-scale metabolic networks by solely relying on simple physical-chemical constraints. However, while extensive research has been performed on constraint-based modeling of microbial metabolism, large-scale modeling of human metabolism is still in its infancy. Utilizing constraint-based modeling to model human cellular metabolism is significantly more complicated than modeling microbial metabolism as in multi-cellular organisms the metabolic behavior varies across cell-types and tissues. It is further complicated due to lack of data on cell type- and tissue-specific metabolite uptake from the surrounding micro-environments and tissue-specific metabolic objective functions. To overcome these problems, several studies suggested CBM methods that integrate metabolic networks with gene expression data that is easily measurable under various conditions. This

* To whom correspondence may be addressed (tomersh@cs.technion.ac.il)

Abbreviations: CBM, Constraint-Based modeling; FBA, Flux Balance Analysis; LP, Linear Programming; MILP, Mixed-Integer Linear Programming paper, reviews three CBM methods for analyzing and predicting metabolic states based on gene expression data. These methods lay the foundation for studying normal and abnormal human cellular metabolism in tissue-specific manner.

Introduction

Metabolism is known to play a significant role in human physiology. Research into human metabolism and its regulation has expanded rapidly due to the emergence of metabolic diseases such as diabetes, obesity and cardiovascular as major sources of morbidity and mortality (Lanpher, 2006; Muoio, 2006). Cancers are known to involve abnormal metabolic phenotypes, and metabolic targets have long been used in cancer chemotherapy (Galmarini, 2008; Serkova, 2007). Recently, metabolism was shown to be involved in various brain disorders, from schizophrenia to neurodegenerative disorders (Holmes, 2006; Huang, 2007). Advances in systems biology research of cellular metabolism can hence be expected to have a profound effect on our understanding of human disease.

Mathematical modeling of cellular metabolism has traditionally been performed via kinetic modeling techniques that require detailed information on kinetic constants and on enzyme and metabolite concentrations (Garfinkel, 1964). However, lack of accurate cellular information of that kind limits the current applicability of such methods to small-scale systems. Metabolic Control Analysis (MCA) is a different approach that is commonly used to model the dependencies of metabolic fluxes and metabolite concentrations on network parameters (Heinrich, 1974; Kacser, 1973). The applicability of this approach is also limited by difficulties to systematically measure required control coefficient parameters that underlie MCA. Additionally, various mathematical formulations based on approximated kinetics, such as linear, Lin-Log, and power-law have been proposed which render kinetic, metabolic differential equations systems easier to handle analytically and require less experimental parameters (Heijnen, 2005). However, while such methods indeed enable the analysis of larger networks, lack of parameters still prevent their application on genome-scale metabolic networks.

An alternative approach bypassing these hurdles called constraint-based modeling (CBM) serves to analyze the function of large-scale metabolic networks by solely relying on simple physical-chemical constraints. Applications of CBM for large-scale microbial networks have proven to be highly successful in studies involving: metabolic engineering, biological discovery, assessment of phenotypic behavior (e.g. growth rates, uptake rates, by-product secretion, and knockout lethality), biological network analysis, and studies of metabolic evolution (see (Price, 2004) and (Feist, 2007) for reviews). Specifically, CBM was shown to be a promising tool for identifying potential targets for metabolic engineering of various microbes, enabling the design of genetic alterations that lead the production of target metabolites of interest. By now, genome-scale constraints-based models have been reconstructed for over 30 microbes (Duarte, 2004; Feist, 2008).

While extensive research has been performed on constraint-based modeling of microbial metabolism, large-scale modeling of human metabolism is still in its infancy. Partial reconstructions of human metabolic networks had until recently been performed only for specific cell types and organelles (Chatziioannou, 2003; Vo, 2004; Wiback, 2002). A fundamental step forward has been presented in recent reconstructions of the first global human metabolic network, in studies by Duarte *et al.* (Duarte, 2007) and by Ma *et al.* (Ma, 2007). These network reconstructions enable a shift in the focus of metabolic modeling from the realm of microbes to human (Mo, 2007; Mo, 2009).

Utilizing reconstructed networks to model human cellular metabolism is significantly more complicated than modeling microbial metabolism as in multi-cellular organisms the metabolic behavior varies across cell-types and tissues. In microbes, CBM methods rely on available data regarding the metabolic composition of the growth media (the input metabolites) and regarding the organism's biomass composition (the output metabolites), predicting a steady-state flow of metabolic mass from input to output metabolites (*Figure 1*). In human, however, various tissues and cell-types have different metabolic micro-environments, for which comprehensive data on tissue-specific metabolite uptake and secretion is currently unavailable. Another major difficulty relates to the fact that different tissues have different metabolic objectives that are not well characterized and are largely unknown. This stands in contrast to the modeling of microbial metabolism, where a simple objective function (such as maximizing the biomass production rate) can be used together with the flux balance analysis (FBA) method to predict biologically plausible flux distributions. Another difficulty in modeling human metabolism is the lack of data on regulatory mechanisms that shape cell-type and tissue-specific metabolism. In microbes, such as *E. coli* and *S. cerevisiae*, on the other hand, extensive knowledge on transcriptional regulation exists and its integration with CBM was shown to improve metabolic flux predictions (Covert, 2004; Herrgard, 2006).

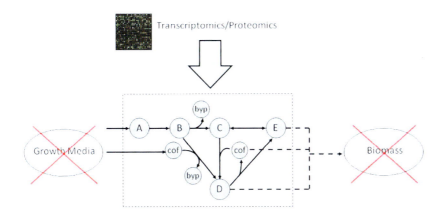

Figure 1: A schematic representation of the data sources available for predicting cell type- and tissue-specific human cellular metabolism. In contrast to microbes grown in laboratory conditions, where the exact composition of the growth media (i.e. the network input) is known, the metabolic environment of specific cell-types and tissues in a multi-cellular organism is largely unknown. Furthermore, while in microbes, a simple assumption regarding a cellular objective function (i.e. the network output), such as that of biomass production maximization, can be used to prediction cellular metabolic states, tissue and cell type-specific objective functions are difficult to characterize. However, while both the input and output of the network are unknown for multi-cellular organisms, data on gene expression is readily available to facilitate predictions of tissue and cell type-specific metabolism. This paper reviews several methods that utilize gene expression data for that purpose.

To account for missing regulatory data and constraints in the modeling of cellular metabolism, several studies suggested CBM methods that integrate metabolic networks with gene expression data that is easily measurable under various conditions (Akesson, 2004; Becker, 2008; Shlomi, 2008). Gene expression is known to play a major role in controlling metabolism, with significant changes in gene expression patterns observed between different growth conditions in microbes and between different tissues in human (Levine, 2006; Son, 2005; Yanai, 2005). Previous studies have found a strong qualitative correspondence between gene expression and measured (Daran-Lapujade, 2004; Fong, 2004) as well as predicted (Akesson, 2004; Bilu, 2006; Famili, 2003; Schuster, 2002) metabolic fluxes in microbes (see (Ovacik, 2008) for review). However, the correlation between expression and metabolic flux is far from maximal and in some cases significant transcriptional changes do not reflect changes in flux (Banta, 2007), and vice-versa, significant changes in measured flux may not reflect transcriptional changes (Tummala, 2003; Yang, 2002). These discrepancies may result from post-transcriptional regulation of protein synthesis and degradation rates, and post-translational modifications that represent additional regulatory mechanisms which affect the potential activity rate of metabolic enzymes (Park, 2005). Furthermore, discrepancies between gene expression and metabolic flux may result from the presence of assembly factors and chaperones that play an important role in regulating the level of various enzyme complexes (Ackerman, 1990; Dibrov, 1998). These kind of regulatory mechanisms are commonly referred to as *hierarchical regulation*. An additional level of flux regulation not reflected in gene expression, called *metabolic regulation*, denotes the effect of metabolite concentrations on the actual enzyme activity through allosteric and mass action effects (Rossell, 2006).

In this paper, we review three CBM methods for analyzing and predicting metabolic states based on gene expression data (Akesson, 2004; Becker, 2008; Shlomi, 2008). The first method, by Akesson *et al* (Akesson, 2004), searches for metabolic states in which the flux through absent, non-expressed genes is constrained to zero. This approach was originally applied for microbes, but theoretically can also be applied to study cell type- and tissue-specific human cellular metabolism under various conditions. The second method, by Becker *et al* (Becker, 2008), identifies lowly expressed genes whose coded enzymes are likely to be absent, and searches for metabolic states in which the flux through these enzymes is minimized. This method was applied to study human skeletal-muscle metabolism under various physiological conditions. Finally, we describe our method (Shlomi, 2008) that considers data on both lowly and highly expressed genes in a certain context and employs network integration to accumulate this data into a global, consistent prediction of metabolic behavior. This method is applied to predict tissue-specific metabolic activity for 10 human tissues, showing that the predicted flux distributions contain valuable biological insights that are not directly apparent in the original expression data. In the next section we provide an overview on constraint-based modeling. In following sections we describe the methods of Akesson and colleagues, Becker and Palsson and Shlomi and colleagues in chronological order of publication.

Overview of constraint-based modeling (CBM)

Constraint-based modeling of metabolic networks is based on the fact that cells are subject to certain constraints that limit their possible behaviors (Price, 2004). By imposing such constraints on the space of possible steady-state metabolic behaviors, one can determine which metabolic states are possible and which are not in a large-scale model. The predicted metabolic state represents a feasible flux distribution through all reactions in the network (i.e., a vector of steady-state flux rates), denoted here as $v \in R^m$ (where m denotes the number of reactions in the network). The constraints imposed on this flux distribution are as follows:

Mass balance - At steady-state there is no accumulation or depletion of metabolites within the metabolic network, hence the rate of each metabolite production should be equal to its consumption rate. This balance of fluxes can be formulated mathematically based on a stoichiometric matrix, which represents both the topology of the metabolic network and the stoichiometry of the biochemical reactions (proportions of substances involved in the reactions). Each row in this matrix represents a metabolite and each column a reaction. Hence, denoting the matrix by $S \in M_{n,m}$, $S_{i,j}$ represents the stoichiometric coefficient of metabolite i in reaction j. Mass balance constraint is enforced by the equation: $S \cdot v = 0$. The set of flux distributions satisfying this constraint spans a sub-space of R^m.

Reaction directionality (thermodynamic constraints) - The directionality of many biochemical reactions is limited based on thermodynamic considerations. For these reactions, the sign of the flux is constrained to be positive.

Enzymatic rates - For some enzymes the maximal flux rate can be determined based on physiological data. These constraints are imposed by setting an upper bound on the rate of specific reactions. Specifically, these constraints as well as the above directionality constraints are enforced by the equation: $v_{min} \leq v \leq v_{max}$, where v_{min} and v_{max} represent lower and upper bound vectors on flux rates, respectively.

Nutrient availability - To predict metabolic flux distributions under various metabolic growth environments, additional constraints are required to reflect the availability of different nutrients. This is achieved by constraining the flux through exchange reactions, which represent the production and consumption of extra-cellular metabolites within the growth medium.

The set of constraints described above can be represented as linear equations and hence form a convex solution space. Previous studies have incorporated several additional non-linear constraints to account, for example, for metabolite concentrations (Henry, 2007) or to eliminate thermodynamically infeasible cycles (Beard, 2002). However, although such constraints further reduce the size of the solution space, the resulting space becomes non-convex and hence it is difficult to analyze it. The analysis of a convex solution space obtained with the linear constraints described above is commonly done via the following computational approaches:

Optimization methods - looking for an optimal flux distribution under different optimization criteria. These methods limit the solution space by assuming that cellular metabolism optimizes a certain objective. The most commonly used objective function for microorganisms, employed by the Flux Balance Analysis (FBA) method, is that of biomass maximization (Fell, 1986; Kauffman, 2003). Other optimization criteria may involve the maximization of ATP production rate or minimization of nutrients uptake. The biomass production maximization is implemented by defining an additional pseudo-reaction, v_{growth}, representing the production of essential biomass compounds. The stoichiometric coefficients of this reaction are based on experimentally derived proportions, c_i, of the metabolite precursors X_i which contribute to biomass production: $\sum c_i X_i \rightarrow Biomass$. The search for a feasible flux distribution, v that maximizes v_{growth} is done using Linear Programming (LP). Notably, in many cases LP optimizations do not provide a single optimal solution but rather provide a convex space of optimal solutions. This space can be further explored using topological methods as described below.

Topological methods - exploring the solution space and investigating flux dependencies. The solution space can be characterized by means of *extreme pathways*, which are the edges of a convex space (Papin, 2003). Any point inside the solution space can be represented as a non-negative linear combination of extreme pathways. Alternatively it can be characterized as a collection of *elementary modes*, which are minimal sets of enzymes that can operate at steady state (Schuster, 2000). Dependencies between pairs of fluxes in the solution space can be identified via *flux coupling analysis* (Burgard, 2004), while dependencies between groups of fluxes can be identified by means of *correlated sets* of reactions (Reed, 2004). Alternatively, the solution space can be investigated by randomly sampling different feasible solution from within it (Almaas, 2004). Another method that is commonly used to characterize the solution space is *flux variability analysis* (FVA) which determines the feasible range of each reaction separately within the solution space (Mahadevan, 2003).

In addition to LP, other optimization methods such as Quadratic Programming (QP) (Segre, 2002) and Mixed Integer Linear Programming (MILP) (Burgard, 2003; Shlomi, 2005; Shlomi, 2007) were previously used for CBM analysis. Specifically, MILP was used to predict gene knockout sets that lead to metabolite over-production and to predict metabolic adaptation following gene knockouts. MILP enables efficient solving of linear optimization problems that include integer variables, which makes it very useful in formulating various metabolic optimization problems of interest as demonstrated below.

CBM Analysis of Expression Data

CONSTRAINING THE FLUX THROUGH ABSENT ENZYMES TO ZERO (AKESSON *ET AL* (AKESSON, 2004)):

The study of Akesson and colleagues describes an extended FBA method which incorporates additional constraints on metabolic fluxes based on gene expression data. Specifically, gene expression data is used to identify *absent enzymes* whose coding genes are non-expressed and their flux is hence constrained to zero. Enzymes whose activity depends on essential factors (such as subunits, assembly factors, translational activator, etc) that are non-expressed are also considered to be absent. As in FBA, the method requires a definition of a metabolic growth medium, as well as a definition of a cellular objective function. The resulting linear optimization problem is formulated as following:

max v_{growth}

s.t.

$S \bullet v = 0$, $\qquad\qquad\qquad\qquad$ (1)

$$v_{min} \leq v \leq v_{max},\tag{2}$$

$$v_i = 0, i \in C,\tag{3}$$

where equation (1) enforces mass-balance (where v_{growth} denotes a pseudo growth reaction that represents the production of essential biomass precursors (as described above)), equation (2) enforces reaction directionality and enzyme capacity constraints, and equation (3) constrains the flux through absent enzymes to zero (where C denote the set of absent reactions based on the expression data).

This method was applied to predict metabolic fluxes for the genome-scale metabolic network model of the yeast *S. cerevisiae* (Forster, 2003), based on gene expression data from chemostat and batch cultivations. The predicted metabolic fluxes were shown to be more accurate than those obtained via FBA, by comparison to quantitative measurements of exchange fluxes (biomass yield, ethanol, glycerol, etc), as well as qualitative estimations of changes in intracellular fluxes.

The main limitation of this approach is that expression-derived data on absent enzymes is integrated with the model via *hard constraints* that force the flux through the associated reactions to zero. With this method, even a single gene that is falsely considered to be absent, although it is essential for the production of the organism's biomass, would result in false prediction of no feasible solutions in the model. For example, applying this method to predict metabolic state under a minimal medium that consists of a single carbon nutrient, if the sole membrane transporter of this nutrient is falsely considered to be absent, the method will falsely predict that there are no feasible metabolic states. Hence, an important issue with this approach is to accurately determine the set of genes that are truly absent under a given context. Towards this goal, Akesson and colleagues used oligonucleotide arrays that for each gene have perfect match and mismatch probes that can be used to ascertain presence of transcripts. A gene is considered absent when not detected in any of the replicates used. However, as indicated by the authors, in some cases, "false absents" may occur, in which a low abundant transcript is present although not detected on the array. The paper actually describes a gene (PET122) that is falsely determined to be absent (as of a technical problem with a specific probe in the microarray), leading to the prediction of no feasible metabolic states. In addition to genes that are falsely considered to be absent, local errors in the model can also lead to false prediction of no feasible metabolic states. For example, under aerobic batch cultivations, a certain gene (ARO9) was correctly determined to be absent based on expression data, though when this enzyme's flux was constrained to zero, no feasible metabolic states were found as of a missing isozyme (AAT2) in the model.

While in the paper of Akesson and colleagues this method was applied only to study yeast metabolism, a similar approach can be used to model human metabolism based on gene expression data. However, similarly to FBA, this method depends on an explicit definition of an objective function that is difficult to define for different cell-types and tissues in a multi-cellular organism. Notably though, as demonstrated in the study of Becker *et al* (see, Becker, 2008) discussed below, a simplified objective function, such as that of ATP production, can still be used in some cases to obtain valuable predictions.

MINIMIZING THE USAGE OF POTENTIALLY ABSENT GENES – THE BECKER & PALSSON METHOD

Becker and Palsson proposed a novel CBM method, Gene Inactivity Moderated by Metabolism and Expression (GIMME), for integrating a metabolic network with gene expression data. This method attempts to overcome the problem outlined above of having to accurately determine the set of absent genes, considering the noisy nature of expression data. Instead, genes with expression level below a pre-defined threshold, denoted x_{cutoff}, are considered as *likely to be absent*, and a feasible flux distribution that minimizes the flux through these reactions is predicted. Similarly to the method of Akesson and coworkers, GIMME also requires a definition of a cellular objective function. The method works via a two-step procedure:

1. Applying FBA to find the maximal possible value of the cellular objective function (denoted by the reaction v_{obj}), while allowing all reactions to be active (i.e. regardless of the expression data).
2. Finding a feasible flux distribution that satisfies the given cellular objective function above some level (i.e. a percentage of the maximum found in (1), denoted *objective_threshold*), while minimizing the activity of reactions that are likely to be absent. This is done via the following linear programming formulation:

$$\min \sum_i w_i |v_i|$$

s.t.

$$S \bullet v = 0,$$
$$v_{min} \leq v \leq v_{max},$$

$$v_{obj} \geq objective\_threshold,$$

The optimization function minimizes the flux through reactions that are likely to be absent, by weighting the minimization pressure based on the expression level of their corresponding genes: $w_i = max\{x_{cutoff} - x_i, 0\}$. A simple transformation converts the above problem into LP, by duplicating each reversible reaction to two irreversible reactions (that carry only positive flux) and removing the absolute value operator. The minimal weighted sum of flux found by the optimization problem was used to define a *consistency score* between the expression data and the cellular objective function. An optimally high consistency score (achieved when the weighted sum of fluxes is zero) means that the expression levels are optimally tuned towards the achievement of the objective function.

The functionality of GIMME was demonstrated for both microbes and human. In microbes, the method was applied to predict metabolic flux distributions for *E. coli* strains that undergo adaptive evolution to improve their growth rate. Specifically, GIMME was used to compute a consistency score between the cellular objective of maximizing growth rate and expression data measured before and after the adaptive evolution process. The analysis correctly predicted a higher consistency score for the expression data measured after the adaptive evolution which maximizes *E. coli's* growth rate. A similar result was shown regarding a high consistency score between expression and the objective function of maximizing lactate production, in an engineered *E. coli* strain that was evolved to maximize the production rate of lactate. In the context of human metabolism, GIMME was applied on a human network model (Duarte, 2007) based on various datasets of gene expression data for skeletal muscle cells (covering patients before and after gastric bypass, or glucose/insulin infusion, and various groups of patients with obesity). The cellular objective function used in this case was to produce ATP at no less than half the optimal rate possible in the model (without the usage of the expression data). This analysis showed two interesting results: (i) The metabolic states of a patient before and after gastric bypass or glucose/insulin infusion are more similar to each other than to metabolic states of other patients. (ii) The consistency score between gene expression and the cellular objective of maximizing ATP production increases in expression samples measured after a patient is given a substantial dose of glucose or insulin in the bloodstream, as can be expected.

The lack of expression data for a substantial number of human metabolic reactions led Becker and Palsson to examine the robustness of their method to missing data. They found that GIMME is not robust even when reducing the number of reactions with expression data by 5%, leading to significantly different results. This lack of robustness may be attributed to the following factors, among others: (i) Computing the similarity in metabolic flux distributions derived for different expression samples, only a single optimal flux distribution was considered, although a space of possible optimal solutions commonly exist (Mahadevan, 2003). The method discussed below addresses this issue by comprehensively exploring the space of optimal possible flux distributions when integrating the network with expression data. (ii) For genes that are considered as likely to be absent, the minimization of the weighted sum of flux rates implicitly assumes some quantitative correlation between expression level and flux rates. However, this stands in marked contrast to experimental evidence that show only a qualitative correspondence between expression and flux rates. Another problematic aspect of GIMME is that, similarly to the method of Akesson and colleagues, it requires a specification of a cellular objective function. Specifically, the objective function of maximizing ATP production ignores other, potentially important, metabolic aspects of muscle metabolism.

INTEGRATED ANALYSIS OF BOTH HIGHLY AND LOW EXPRESSED - Shlomi and coworkers

The methods of Akesson and colleagues and Becker and Palsson use gene expression data only to identify genes that are absent or likely to absent and hence their coded enzymes are expected not to carry metabolic flux. However, both methods do not account for a complementary observation that enzyme-coding genes with high expression levels are likely to carry metabolic flux, due to the strong qualitative correspondence between gene expression and measured fluxes. Recently, we suggested a new method that considers data on both lowly and highly expressed genes as cues for the likelihood that their associated reactions carry metabolic flux and employ network integration to accumulate these cues into a global, consistent prediction of metabolic behavior. Utilizing expression data of highly expressed genes as cues for enzyme activity eliminates the requirement for a metabolic objective function (as shown below), which is difficult to define for multi-cellular organisms.

The method of Shlomi and coworkers employs a discrete representation of significantly high or low gene expression levels in a certain context, following previous such representation of expression data in metabolic modeling for other aims (Covert, 2004; Shlomi, 2007). The discretization of the data can be done by defining two thresholds, such that genes whose expression level is below (above) the lower (upper) thresholds are considered to be lowly (highly) expressed (as performed in the methods described above to detect absent genes). In the study of Shlomi *et al* on human tissue-specific metabolism, tissue-specific gene expression data was extracted from a database that already performed the discretization via a different approach (Shmueli, 2003; Yanai, 2005). Given the discrete gene expression data under a certain context, a subset of the reactions in the model (denoted R_H) is defined as highly-expressed, and another subset (denoted R_L) is defined as lowly-expressed. Then, a CBM optimization problem is formulated to find a feasible flux distribution that maximizes the number of enzymes whose flux activity is consistent with their measured expression level – i.e. obtain a flux distribution that maximizes (minimizes) the number of reactions that are associated with highly (lowly) expressed enzymes. The resulting optimization problem is formulated via the following Mixed Integer Linear Programming (MILP) problem:

$$\max_{v,y^+,y^-} \sum_{i \in R_H} (y_i^+ + y_i^-) + \sum_{i \in R_L} y_i^+)$$

$s.t$

$$S \bullet v = 0 \tag{1}$$

$$v_{min} \leq v \leq v_{max} \tag{2}$$

$$v_i + y_i^+ (v_{min,i} - \varepsilon) \geq v_{min,i'} \quad , i \in R_H \tag{3}$$

$$v_i + y_i^- (v_{max,i} + \varepsilon) \leq v_{max,i'} \quad , i \in R_H \tag{4}$$

$$v_{min,i} (1-y_i^+) \leq v_i \leq v_{max,i} (1-y_i^+) \quad , i \in R_L \tag{5}$$

$$v \in Rm$$

$$y_i^+, y_i^- \in [0,1]$$

For each highly expressed reaction, the Boolean variables y^+ and y^- represent whether it is active (i.e. carry metabolic flux) in the forward and backward directions, respectively. A highly-expressed reaction is considered to be active if it carries a significant positive flux that is greater than a positive threshold ε (equation (3)) or a significant negative flux lower than $-\varepsilon$ (equation (4) for reversible reactions). For each lowly-expressed reaction, the Boolean variable y^+ represents whether the reaction is inactive (equation 5). The optimization function maximizes the number of reactions that are highly-expressed (R_H) and active, and the number of lowly-expressed reactions (R_L) that are inactive.

In many cases, the resulting flux distribution obtained from the above MILP is not unique, as multiple optimal flux distributions that obtain the same similarity with the expression data exist, due to isozymes and alternative pathways in the network (Bilu, 2006; Mahadevan, 2003; Shlomi, 2007). To account for these alternative solutions, a variant of Flux Variability Analysis (Mahadevan, 2003) was employed to explore the space of optimal flux distributions. This analysis was used to assign genes with *flux activity states*, reflecting the presence/absence of non-zero flux through their enzymatic reactions. Specifically, a gene is predicted to be active (inactive) if its associated reactions carry (do not carry) metabolite flux in all optimal, feasible flux distributions. Because expression levels are not enforced as exclusive determiners of metabolic flux, the flux activity states of genes may deviate from their expression states. Genes are considered to be post-transcriptionally up- or down-regulated based on a difference between their measured expression level and their predicted flux activity state in a given tissue.

An illustrative example for the application of this method is shown in *Figure 2*. In this example, three reactions are considered to be highly expressed and two reactions are considered lowly expressed based on the expression of their associated genes. The above MILP problem predicts a single optimal flux distribution that is consistent with the expression state of 4 out of the 5 significantly highly or lowly expressed reactions. Based on this flux distribution, two enzymes are predicted to be post-transcriptionally regulated: (i) The enzyme *E7* is predicted to have no metabolic flux although it is highly expressed, and is hence considered to be *down-regulated*. (ii) The enzyme *E4* is predicted to have metabolic flux although it is not significantly highly expressed in this context, and is hence considered to be *up-regulated*.

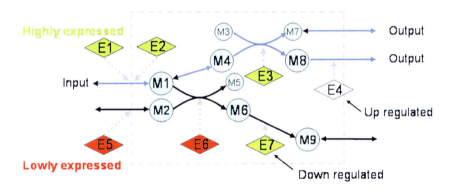

Figure 2: An illustrative example of prediction of flux activity states of genes based on a metabolic network model and gene expression measurements. Circular nodes represent metabolites, whereas diamond nodes represent enzymes. Red, green, and white colors represent significantly low, significantly high, and normal expression of the enzyme-coding genes, respectively. Solid edges represent metabolic reactions. Dashed edges associates enzymes with the reactions they catalyze. The predicted steady-state flux distribution involves the activation of reactions is colored in purple. Enzymes E4 and E7 are predicted to be post-transcriptionally regulated (up and down respectively). The figure was taken from (Shlomi, 2008).

As a basic validation of this method, it was applied to predict metabolic flux distributions for the yeast *S. cerevisiae* based on gene expression data measured under various growth media. The predicted flux distributions were shown to significantly correlate with measured metabolic fluxes in the central carbon metabolism of *S.*

cerevisiae. As a further larger-scale validation, the predicted flux distributions were found to significantly correlate with FBA predictions (computed based on explicit yeast biomass maximization). Specifically, the predicted flux activity patterns were also shown to correctly capture the directionality of metabolite exchange with the growth media, and the production of essential biomass precursors, in agreement with FBA predictions.

Next, the method was employed to compute tissue-specific metabolic behavior based on the metabolic network model of Duarte *et al.*(Duarte, 2007) and expression data from 10 tissues: brain, heart, kidney, liver, lung, pancreas, prostate, spleen, skeletal muscle and thymus. Many of the genes that are predicted to be active (inactive) in a certain tissue are not highly (lowly) expressed there, showing the considerable amount of additional information obtained by integrating gene expression data with the metabolic network to infer metabolic gene activity. Specifically, this approach predicts an average ~20% genes that are post-transcriptionally regulated, with expression data not reflecting their metabolic activity state. To systematically validate these predictions it was shown that the predicted tissue-specificity of genes, reactions, and metabolites significantly correlates with their known tissue-specificity pattern extracted from various databases. Interestingly, focusing on tissue specificity findings of genes, reactions and metabolites that are inferred by post-transcriptional up (down) regulation (i.e., tissue-specificity findings that cannot be inferred by using the original expression data without the metabolic network) provides a significant high (low) number of matches with the known tissue-specific associations in all cases. Specifically, one of the tissue-specificity datasets used as validation was based on tissue-associations of genes that cause in-born errors of metabolism, obtained by mining the OMIM database. The results showed that this method correctly predicts the tissue-specificity of a large fraction of these genes, providing a significant added-value over the original expression data.

Discussion

This paper reviewed three methods for predicting cellular metabolism based on integrating a metabolic network with gene expression data. As described above, these methods clearly show the added value of utilizing expression data to predict biologically plausibly metabolic states in both microbes and human. The methods of Akesson *et al* and Becker and Palsson use gene expression data to identify genes that are absent or likely to be absent in certain contexts and search for metabolic states which prevent (or minimize) the flux through the associated metabolic reactions. The method of Shlomi *et al* considers data on both lowly and highly expressed genes in a given context as cues for the likelihood that their associated reactions carry metabolic flux and employs CBM to accumulate these cues into a global, consistent prediction of the metabolic state. This eliminates the requirement for a definition of metabolic objective function. The lack of dependency on an objective function is a marked advantage as the latter is difficult to define for multi-cellular organisms. Notably though, as shown by Becker and Palsson, a simple definition of an objective function, such as that of producing ATP in a high rate, may in some cases suffice to provide interesting insight into human cellular metabolism. More detailed definitions of tissue and cell type-specific objective functions are expected to improve the prediction accuracy of the presented methods. Such objective functions may be derived based on computational learning that would give optimal correspondence between predicted and measured metabolic phenotypes (Burgard, 2003).

The CBM methods presented here open the way for future computational investigations of metabolic disorders given the relevant expression data. A first attempt to visualize and interpret changes in gene expression data measured following gastric bypass surgery via a genome-scale metabolic network was done by Duarte *et al* (Duarte, 2007). Another potential application would be the prediction of diagnostic biomarkers for metabolic diseases that could be identified via biofluid metabolomics (Kell, 2007). Towards this goal, we have recently developed a CBM method for predicting metabolic biomarkers for in-born errors of metabolism by searching for changes in metabolite uptake and secretion rate due to genetic alterations (Shlomi, 2009). Incorporating cell type- and tissue-specific gene expression data within this framework can potentially improve the identification of diagnostic biomarkers. Overall, the methods presented here lay the foundation for studying normal and abnormal human cellular metabolism in tissue-specific manner based on commonly measured gene expression data.

References

ACKERMAN, S.H. AND TZAGOLOFF, A. (1990) ATP10, a yeast nuclear gene required for the assembly of the mitochondrial F1-F0 complex. *J Biol Chem.* **265**(17), 9952-9.

AKESSON, M., FORSTER, J., AND NIELSEN, J. (2004) Integration of gene expression data into genome-scale metabolic models. *Metab Eng.* **6**(4), 285-93.

ALMAAS, E., *ET AL.* (2004) Global organization of metabolic fluxes in the bacterium Escherichia coli. *Nature.* **427**(6977), 839-43.

BANTA, S., *ET AL.* (2007) Contribution of gene expression to metabolic fluxes in hypermetabolic livers induced through burn injury and cecal ligation and puncture in rats. *Biotechnol Bioeng.* **97**(1), 118-37.

BEARD, D.A., LIANG, S.D., AND QIAN, H. (2002) Energy balance for analysis of complex metabolic networks. *Biophys J.* **83**(1), 79-86.

BECKER, S.A. AND PALSSON, B.O. (2008) Context-specific metabolic networks are consistent with experiments. *PLoS Comput Biol.* **4**(5), e1000082.

BILU, Y., *ET AL.* (2006) Conservation of expression and sequence of metabolic genes is reflected by activity across metabolic states. *PLoS Comput Biol.* **2**(8), e106.

BURGARD, A.P. AND MARANAS, C.D. (2003) Optimization-based framework for inferring and testing hypothesized metabolic objective functions. *Biotechnol Bioeng.* **82**(6), 670-7.

BURGARD, A.P., *ET AL.* (2004) Flux coupling analysis of genome-scale metabolic network reconstructions. *Genome Res.* **14**(2), 301-12.

BURGARD, A.P., PHARKYA, P., AND MARANAS, C.D. (2003) Optknock: a bilevel programming framework for identifying gene knockout strategies for microbial strain optimization. *Biotechnol Bioeng.* **84**(6), 647-57.

CHATZIIOANNOU, A., PALAIOLOGOS, G., AND KOLISIS, F.N. (2003) Metabolic flux analysis as a tool for the elucidation of the metabolism of neurotransmitter glutamate. *Metab*

Eng. **5**(3), 201-10.

COVERT, M.W., *ET AL.* (2004) Integrating high-throughput and computational data elucidates bacterial networks. *Nature.* **429**(6987), 92-6.

DARAN-LAPUJADE, P., *ET AL.* (2004) Role of transcriptional regulation in controlling fluxes in central carbon metabolism of Saccharomyces cerevisiae. A chemostat culture study. *J Biol Chem.* **279**(10), 9125-38.

DIBROV, E., FU, S., AND LEMIRE, B.D. (1998) The Saccharomyces cerevisiae TCM62 gene encodes a chaperone necessary for the assembly of the mitochondrial succinate dehydrogenase (complex II) *J Biol Chem.* **273**(48), 32042-8.

DUARTE, N.C., *ET AL.* (2007) Global reconstruction of the human metabolic network based on genomic and bibliomic data. *Proc Natl Acad Sci U S A.* **104**(6), 1777-82.

DUARTE, N.C., HERRGARD, M.J., AND PALSSON, B.O. (2004) Reconstruction and validation of Saccharomyces cerevisiae iND750, a fully compartmentalized genome-scale metabolic model. *Genome Res.* **14**(7), 1298-309.

FAMILI, I., *ET AL.* (2003) Saccharomyces cerevisiae phenotypes can be predicted by using constraint-based analysis of a genome-scale reconstructed metabolic network. *Proc Natl Acad Sci U S A.* **100**(23), 13134-9.

FEIST, A.M., *ET AL.* (2007) A genome-scale metabolic reconstruction for Escherichia coli K-12 MG1655 that accounts for 1260 ORFs and thermodynamic information. *Mol Syst Biol.* **3**, 121.

FEIST, A.M. AND PALSSON, B.O. (2008) The growing scope of applications of genome-scale metabolic reconstructions using Escherichia coli. *Nat Biotechnol.* **26**(6), 659-67.

FELL, D.A. AND SMALL, J.R. (1986) Fat synthesis in adipose tissue. An examination of stoichiometric constraints. *Biochem J.* **238**(3), 781-6.

FONG, S.S. AND PALSSON, B.O. (2004) Metabolic gene-deletion strains of Escherichia coli evolve to computationally predicted growth phenotypes. *Nat Genet.* **36**(10), 1056-8.

FORSTER, J., *ET AL.* (2003) Genome-scale reconstruction of the Saccharomyces cerevisiae metabolic network. *Genome Res.* **13**(2), 244-53.

GALMARINI, C.M., POPOWYCZ, F., AND JOSEPH, B. (2008) Cytotoxic nucleoside analogues: different strategies to improve their clinical efficacy. *Curr Med Chem.* **15**(11), 1072-82.

GARFINKEL, D. AND HESS, B. (1964) Metabolic Control Mechanisms. Vii.A Detailed Computer Model of the Glycolytic Pathway in Ascites Cells. *J Biol Chem.* **239**, 971-83.

HEIJNEN, J.J. (2005) Approximative kinetic formats used in metabolic network modeling. *Biotechnol Bioeng.* **91**(5), 534-45.

HEINRICH, R. AND RAPOPORT, T.A. (1974) A linear steady-state treatment of enzymatic chains. Critique of the crossover theorem and a general procedure to identify interaction sites with an effector. *Eur J Biochem.* **42**(1), 97-105.

HENRY, C.S., BROADBELT, L.J., AND HATZIMANIKATIS, V. (2007) Thermodynamics-based metabolic flux analysis. *Biophys J.* **92**(5), 1792-805.

HERRGARD, M.J., *ET AL.* (2006) Integrated analysis of regulatory and metabolic networks reveals novel regulatory mechanisms in Saccharomyces cerevisiae. *Genome Res.* **16**(5), 627-35.

HOLMES, E., *ET AL.* (2006) Metabolic profiling of CSF: evidence that early intervention may impact on disease progression and outcome in schizophrenia. *PLoS Med.* **3**(8), e327.

HUANG, C., *ET AL.* (2007) Changes in network activity with the progression of Parkinson's disease. *Brain.* **130**(Pt 7), 1834-46.

KACSER, H. AND BURNS, J.A. (1973) The control of flux. *Symp Soc Exp Biol.* **27**, 65-104.

KAUFFMAN, K.J., PRAKASH, P., AND EDWARDS, J.S. (2003) Advances in flux balance analysis. *Curr Opin Biotechnol.* **14**(5), 491-6.

KELL, D.B. (2007) Metabolomic biomarkers: search, discovery and validation. *Expert Rev Mol Diagn.* **7**(4), 329-33.

LANPHER, B., BRUNETTI-PIERRI, N., AND LEE, B. (2006) Inborn errors of metabolism: the flux from Mendelian to complex diseases. *Nat Rev Genet.* **7**(6), 449-60.

LEVINE, D.M., *ET AL.* (2006) Pathway and gene-set activation measurement from mRNA expression data: the tissue distribution of human pathways. *Genome Biol.* **7**(10), R93.

MA, H., *ET AL.* (2007) The Edinburgh human metabolic network reconstruction and its functional analysis. *Mol Syst Biol.* **3**, 135.

MAHADEVAN, R. AND SCHILLING, C.H. (2003) The effects of alternate optimal solutions in constraint-based genome-scale metabolic models. *Metab Eng.* **5**(4), 264-76.

MO, M.L., JAMSHIDI, N., AND PALSSON, B.O. (2007) A genome-scale, constraint-based approach to systems biology of human metabolism. *Mol Biosyst.* **3**(9), 598-603.

MO, M.L. AND PALSSON, B.O. (2009) Understanding human metabolic physiology: a genome-to-systems approach. *Trends Biotechnol.* **27**(1), 37-44.

MUOIO, D.M. AND NEWGARD, C.B. (2006) Obesity-related derangements in metabolic regulation. *Annu Rev Biochem.* **75**, 367-401.

OVACIK, M.A. AND ANDROULAKIS, I.P. (2008) On the Potential for Integrating Gene Expression and Metabolic Flux Data. *Current Bioinformatics.* **3**, 142-148.

PAPIN, J.A., *ET AL.* (2003) Metabolic pathways in the post-genome era. *Trends Biochem Sci.* **28**(5), 250-8.

PARK, S.J., *ET AL.* (2005) Global physiological understanding and metabolic engineering of microorganisms based on omics studies. *Appl Microbiol Biotechnol.* **68**(5), 567-79.

PRICE, N.D., REED, J.L., AND PALSSON, B.O. (2004) Genome-scale models of microbial cells: evaluating the consequences of constraints. *Nat Rev Microbiol.* **2**(11), 886-97.

REED, J.L. AND PALSSON, B.O. (2004) Genome-scale in silico models of E. coli have multiple equivalent phenotypic states: assessment of correlated reaction subsets that comprise network states. *Genome Res.* **14**(9), 1797-805.

ROSSELL, S., *ET AL.* (2006) Unraveling the complexity of flux regulation: a new method demonstrated for nutrient starvation in Saccharomyces cerevisiae. *Proc Natl Acad Sci U S A.* **103**(7), 2166-71.

SCHUSTER, S., FELL, D.A., AND DANDEKAR, T. (2000) A general definition of metabolic pathways useful for systematic organization and analysis of complex metabolic networks. *Nat Biotechnol.* **18**(3), 326-32.

SCHUSTER, S., *ET AL.* (2002) Use of network analysis of metabolic systems in bioengineering. *Bioprocess and Biosystems Engineering* **24**(6), 363 - 372

SEGRE, D., VITKUP, D., AND CHURCH, G.M. (2002) Analysis of optimality in natural and perturbed metabolic networks. *Proc Natl Acad Sci U S A.* **99**(23), 15112-7.

Serkova, N.J., Spratlin, J.L., and Eckhardt, S.G. (2007) NMR-based metabolomics: translational application and treatment of cancer. *Curr Opin Mol Ther.* **9**(6), 572-85.

Shlomi, T., Berkman, O., and Ruppin, E. (2005) Regulatory on/off minimization of metabolic flux changes after genetic perturbations. *Proc Natl Acad Sci U S A.* **102**(21), 7695-700.

Shlomi, T., *et al.* (2008) Network-based prediction of human tissue-specific metabolism. *Nat Biotechnol.* **26**(9), 1003-10.

Shlomi, T., Cabili, M.N., and Ruppin, E. (2009) Predicting Metabolic Biomarkers of Human Inborn Errors of Metabolism. *Mol Syst Biol.* **5**, 263.

Shlomi, T., *et al.* (2007) A genome-scale computational study of the interplay between transcriptional regulation and metabolism. *Mol Syst Biol.* **3**, 101.

Shmueli, O., *et al.* (2003) GeneNote: whole genome expression profiles in normal human tissues. *C R Biol.* **326**(10-11), 1067-72.

Son, C.G., *et al.* (2005) Database of mRNA gene expression profiles of multiple human organs. *Genome Res.* **15**(3), 443-450.

Tummala, S.B., *et al.* (2003) Transcriptional analysis of product-concentration driven changes in cellular programs of recombinant Clostridium acetobutylicumstrains. *Biotechnol Bioeng.* **84**(7), 842-54.

Vo, T.D., Greenberg, H.J., and Palsson, B.O. (2004) Reconstruction and functional characterization of the human mitochondrial metabolic network based on proteomic and biochemical data. *J Biol Chem.* **279**(38), 39532-40.

Wiback, S.J. and Palsson, B.O. (2002) Extreme pathway analysis of human red blood cell metabolism. *Biophys J.* **83**(2), 808-18.

Yanai, I., *et al.* (2005) Genome-wide midrange transcription profiles reveal expression level relationships in human tissue specification. *Bioinformatics.* **21**(5), 650-9.

Yang, C., Hua, Q., and Shimizu, K. (2002) Integration of the information from gene expression and metabolic fluxes for the analysis of the regulatory mechanisms in Synechocystis. *Appl Microbiol Biotechnol.* **58**(6), 813-22.

Biotechnology and Genetic Engineering Reviews - Vol. 26, 297-334 (2009)

Stem cell biology and cell transplantation therapy in the retina

FUMITAKA OSAKADA[1,2,*], YASUHIKO HIRAMI[1,3], MASAYO TAKAHASHI[1,3]

[1]Laboratory for Retinal Regeneration, Center for Developmental Biology, RIKEN, 2-2-3 Minatojima-minamimachi, Chuo-ku, Kobe 650-0047, Japan; [2]Systems Neurobiology Laboratory, The Salk Institute for Biological Studies, 10010 North Torrey Pines Road, La Jolla, California 92037, USA; [3]Department of Ophthalmology, Institute of Biomedical Research and Innovation, 2-2 Minatojima-minamimachi, Chuo-ku, Kobe 650-0047, Japan

Abstract

Embryonic stem (ES) cells, which are derived from the inner cell mass of mammalian blastocyst stage embryos, have the ability to differentiate into any cell type in the body and to grow indefinitely while maintaining pluripotency. During development, cells undergo progressive and irreversible differentiation into specialized adult cell types. Remarkably, in spite of this restriction in potential, adult somatic cells can be reprogrammed and returned to the naive state of pluripotency found in the early embryo simply by forcing expression of a defined set of transcription factors. These induced pluripotent stem (iPS) cells are molecularly and functionally equivalent to ES cells and provide powerful *in vitro* models for development, disease, and drug screening, as well as material for cell replacement therapy. Since functional impairment results from cell loss in most central nervous system (CNS) diseases, recovery of lost cells

* To whom correspondence may be addressed (fosakada@salk.edu)

Abbreviations: ES cells, embryonic stem cells; iPS cells, induced pluripotent stem cells; CNS, central nervous system; SVZ, subventricular zone; SGZ, subgranular zone; RP, retinitis pigmentosa; AMD, age-related macular degeneration; RPE, retinal pigment epithelium; ONL, outer nuclear layer; INL, inner nuclear layer; GCL, ganglion cell layer; IPL, inner plexiform layer; OPL, outer plexiform layer; BRB, blood-retinal barrier; FGF, fibroblast growth factors; CNTF, ciliary neurotrophic factor; VEGF, vascular endothelial growth factor; PEDF, pigment epithelium derived factor; TGF, transforming growth factor; LGN, lateral geniculate nucleus; rAAV, recombinant adeno-associated viral vectors; ERG, electroretinogram; CNV, choroidal neovascularization; SDIA, stromal cell-derived inducing activity; SFEB, serum-free floating culture of embryoid body-like aggregates.

is an important treatment strategy. Although adult neurogenesis occurs in restricted regions, the CNS has poor potential for regeneration to compensate for cell loss. Thus, cell transplantation into damaged or diseased CNS tissues is a promising approach to treating various neurodegenerative disorders. Transplantation of photoreceptors or retinal pigment epithelium cells derived from human ES cells can restore some visual function. Patient-specific iPS cells may lead to customized cell therapy. However, regeneration of retinal function will require a detailed understanding of eye development, visual system circuitry, and retinal degeneration pathology. Here, we review the current progress in retinal regeneration, focusing on the therapeutic potential of pluripotent stem cells.

Introduction

For many decades, it was believed that neurons in the adult mammalian central nervous system (CNS) could not regenerate after injury, as postulated by Ramón y Cajal in 1913. However, recent evidence has overturned this long-held dogma. Neurogenesis takes place in the adult mammalian CNS, with neural stem cells residing in the subventricular zone (SVZ) of the lateral ventricle and subgranular zone (SGZ) of the hippocampal dentate gyrus (Lledo *et al.*, 2006; Zhao *et al.*, 2008). Nonetheless, the CNS has poor potential for regeneration to compensate for cell loss in diseased or injured tissues.

Visual impairment is usually caused by specific loss of different cell populations within the retina (*Figure 1*). For instance, glaucoma is a retinal degenerative disease in which the retinal ganglion cells forming the optic nerve are selectively lost. In retinitis pigmentosa (RP), photoreceptors are lost due to genetic mutations (Hartong *et al.*, 2006). In age-related macular degeneration (AMD), degeneration of the retinal pigment epithelium (RPE) is followed by loss of photoreceptors (Rattner and Nathans, 2006). Since first order neurons are selectively affected in retinitis pigmentosa and AMD, the neural circuitry mediating higher order visual processing is maintained in the early phase of degeneration (Humayun *et al.*, 2003). Thus, transplantation of photoreceptors or RPE cells permits effective recovery of visual function (Haruta *et al.*, 2004; Lund *et al.*, 2006; MacLaren *et al.*, 2006; Lamba *et al.*, 2009), a process that may be facilitated by the proximity of synaptic target cells to the first order neurons. In contrast, in the case of glaucoma, optic nerve regeneration requires replacement of retinal ganglion cells and reconstruction of distant synaptic connections to the brain.

The strategies for regeneration can be classified into two approaches: (i) activation of endogenous neural stem cells and (ii) transplantation of lost cell types. In the adult retina, Müller glia serve as endogenous retinal progenitors in response to injury (Ooto *et al.*, 2004; Osakada *et al.*, 2007; Karl *et al.*, 2008). Thus, drug therapies that target Müller glia are a promising approach for retinal regeneration (Osakada and Takahashi, 2009). However, retinal transplantation has been investigated extensively over the last decade, particularly the possible use of stem cells, which are attractive due to their self-proliferation and differentiation potential (Haruta et al 2004; Klassen et al 2004; Takahashi et al 1998). In this review, we focus on transplantation therapy in the retina, with an emphasis on pluripotent stem cells.

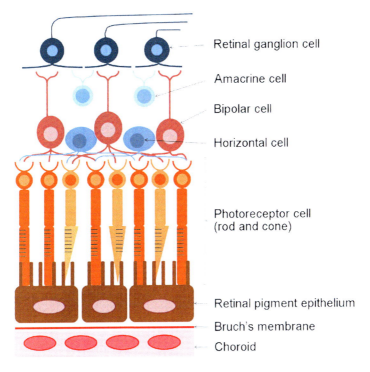

Retinal ganglion cell

Amacrine cell

Bipolar cell

Horizontal cell

Photoreceptor cell
(rod and cone)

Retinal pigment epithelium

Bruch's membrane

Choroid

Figure 1. Cell types and layers in the adult retina.

Visual system

RETINAL DEVELOPMENT

The eye is derived from three types of tissue during embryogenesis: the neural ectoderm gives rise to the retina and the overlying retinal pigment epithelium (RPE), the mesoderm produces the cornea and sclera, and the surface ectoderm generates the lens. The eye develops as a result of interactions between the surface ectoderm and the optic vesicle, an evagination of the diencephalon (forebrain) (*Figure 2A*). The optic vesicle is connected to the developing CNS by a stalk that later becomes the optic nerve. Upon contacting the surface ectoderm, the optic vesicle epithelium forms a lens placode (*Figure 2B*), which subsequently invaginates, pinches off, and eventually becomes the lens. During these events, the optic vesicle folds inward to form a bilayered optic cup (*Figure 2C, D*). The outer layer of the optic cup differentiates into the RPE, whereas the inner layer differentiates into the neural retina (*Figure 2E*). The iris and ciliary body develop from the peripheral edges of the retina, and the sclera is derived from mesenchymal cells of neural crest origin that migrate to form the cornea and trabecular meshwork of the anterior eye chamber (*Figure 2F, G*).

During development, retinal progenitors change their differentiation competence under the control of intrinsic regulators (*Table 1*) as well as extrinsic regulators (*Table 2*). Within the retina, seven types of cells differentiate from common progenitors in the following temporal sequence: retinal ganglion cells, cone photoreceptors, amacrine cells, and horizontal cells, followed by rod photoreceptors, bipolar cells, and Müller glia (*Figure* 3). These cells comprise three cell layers: rod and cone photoreceptors in the outer nuclear layer (ONL), horizontal, bipolar, and amacrine cells as well as Müller glia in the inner nuclear layer (INL), and ganglion and displaced amacrine cells in the ganglion cell layer (GCL). All retinal cells are born at the outer surface of the retina, adjacent to the pigmented epithelium. Postmitotic cells migrate a distance to occupy their final positions within the retina, and then establish synaptic connections to other neurons.

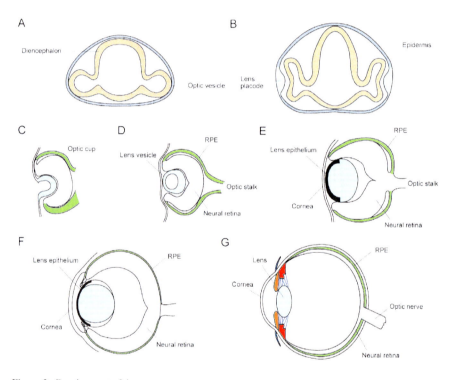

Figure 2. Development of the eye.
(A-F) Mouse embryos at E9.5 (A), E10.0 (B), E10.5 (C), E11.5 (D), E13.5 (E) and E18.5 (F). (G) Adult eyes. (Figures from Experimental Medicine (Osakada and Takahashi, 2006)).

Table 1. Intrinsic factors regulating retinal cell differentiation.

Cell type	Homeobox genes	bHLH genes
Photoreceptor cells	Crx / Otx2	NeuroD / Mash1
Horizontal cells	Pax6 / Six3 / Prox1	Math3
Biopolar cells	Chx10	Mash1 / Math3
Amacrine cells	Pax6 / Six3	NeuroD / Math3
Ganglion cells	Pax6	Math5
Müller glia	Rx	Hes1 / HEs5

Table 2. Extrinsic factors regulating retinal cell differentiation.

Cell type	Soluble factor
Photoreceptor cells	(+) Retinoic acid
	(+) Taurine
	(+) Thyroid hormone
	(+) Shh
	(+) FGF
	(+) CNTF
Horizontal cells	
Biopoloar cells	(+) CNTF
Amacrine cells	
Ganglion cells	(-) Shh
Müller glia	(-) Retinoic acid
	(-) FGF

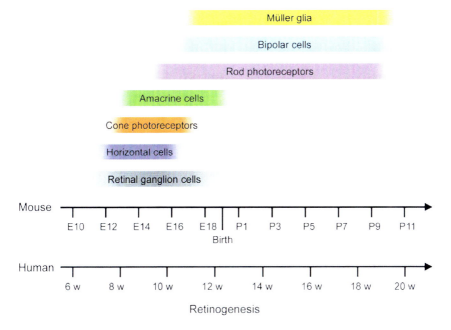

Figure 3. Genesis of seven types of retinal cells during development.
Retinal ganglion cells and horizontal cells differentiate first, followed by cone photoreceptors, amacrine cells, rod photoreceptors, bipolar cells, and finally Müller glia, with overlap in the appearance of these different cell types. The sequence of cell genesis in the vertebrate retina is highly conserved among many species.

In the mouse retina, synapses between retinal ganglion cells and amacrine cells appear in the inner plexiform layer (IPL) during the first postnatal week. Then, rods and cones establish synapses with horizontal cells in the outer plexiform layer (OPL). Finally, during the second postnatal week, bipolar cell dendrites contact photoreceptors and horizontal cells in the OPL, and their axons contact amacrine cells and retinal ganglion cells in the IPL. Retinal ganglion cells first form conventional synapses with amacrine cells between P3 and P20, then develop ribbon synapses with bipolar cells starting at P11, around the time of eye opening, reaching a maximum density around P20. In conjunction with this temporal sequence of synapse formation, waves of electrical and calcium activity propagate across the retina. Retinal ganglion cells exhibit spontaneous bursts of action potentials during the propagation of retinal waves. Retinal waves between P1 and P10 are driven by cholinergic transmission, while subsequent waves are glutamate-dependent, correlating with synapse formation between retinal ganglion cells and bipolar cells. These retinal waves are implicated in the refinement of ganglion cell projection patterns.

ANATOMY AND PHYSIOLOGY OF THE ADULT RETINA

The outermost retinal layer is the RPE, a monolayer of cells whose apical surface faces the photoreceptor outer segments and whose basal surface rests on the extracellular matrix structure known as Bruch's membrane (*Figure* 1). Beyond Bruch's membrane, the fenestrated capillaries of choroidal vessels (choriocapillaris) supply nutrients and oxygen to the outer retina (Marmor, 1998). RPE cells have tight junctions that form the outer blood-retinal barrier (BRB) and pump water and ions from the apical to basal side of the cells, while nutrients are transported via the choroidal circulation to photoreceptors. The RPE also participates in metabolic functions, such as phagocytosis of photoreceptor outer segments, and in the retinoid cycle, in which it isomerizes *all-trans* retinol to *11-cis* retinal. Another function of the RPE is secretion of various trophic factors, including fibroblast growth factors (FGFs) (Schweigerer *et al.*, 1987; Malecaze *et al.*, 1991; Bost *et al.*, 1992), ciliary neurotrophic factor (CNTF) (Wen *et al.*, 1995), vascular endothelial growth factor (VEGF) (Adamis *et al.*, 1993), insulin-like growth factor (IGF) (Waldbillig *et al.*, 1991; Waldbillig *et al.*, 1992), pigment epithelium derived factor (PEDF), transforming growth factor (TGF) β family members (Tanihara *et al.*, 1993), and tissue inhibitors of metalloproteinases (Alexander *et al.*, 1990; Della *et al.*, 1996).

The first steps in vision begin in the retina, where a dense array of photoreceptors converts the incoming pattern of light into electrochemical signals. In the mammalian retina, 95% of photoreceptors are rods, and the remaining 5% are cones. Rod photoreceptors are more sensitive, and thus function under dark-adapted conditions, and their responses are saturated at high light levels. Cone photoreceptors, on the other hand, are less sensitive and function under light-adapted conditions. Rod and cone photoreceptor signaling differs from that of almost every other neuron in the CNS. In the absence of light, photoreceptors continuously release the excitatory neurotransmitter glutamate. A light stimulus activates the phototransduction cascade in the outer segment of the photoreceptor, causing a decrease in the level of intracellular cGMP and closure of cGMP-gated cation channels. This leads to hyperpolarization of the photoreceptor, resulting in decreased glutamate release from the photoreceptor terminal in the OPL. Both rod and cone photoreceptors employ this general cascade, although the molecules that function in the G protein cascades differ. Importantly, photoreceptors do not produce action potentials, but instead have graded potentials that are modulated around a mean level. In most neurons in the CNS, the release of glutamate is dependent on voltage-dependent calcium channels. However, the release of glutamate from photoreceptors is graded, resulting in graded potentials in bipolar and horizontal cells.

Photoreceptors contact the dendrites of both bipolar cells and horizontal cells (Masland, 2001). The synaptic terminals of rods (rod spherules) and cones (cone pedicules) contain synaptic ribbons that are associated with high rates of neurotransmitter release. Bipolar cells are excitatory interneurons that use glutamate as a neurotransmitter and are specialized for sustained transmitter release. Hyperpolarizing bipolar cells hyperpolarize in response to reduced glutamate release from photoreceptors, while depolarizing bipolar cells instead depolarize in response to glutamate. Other cells in the retina are horizontal cells, which are laterally extensive interneurons in the outer row of the INL, and amacrine cells, which form a morphologically and physiologically diverse group of mostly inhibitory interneurons located in the inner row of the INL. Horizontal cells are postsynaptic to photoreceptors, and feed signals back onto photoreceptor synaptic terminals and forward onto the dendrites of bipolar cells. Most amacrine cells use GABA, glycine, or both acetylcholine and GABA as neurotransmitters and provide feedback inhibition to bipolar cell terminals and feed-forward inhibition to ganglion cells. Ganglion cells are physiologically and morphologically diverse, with an estimated 15-20 different types. They can be classified by many physiological and anatomical criteria, including size, response, receptive field, color, ON or OFF, conduction velocity, morphology, branch pattern, stratification, coupling, and coverage (Field and Chichilnisky, 2007).

VISUAL PROCESSING

Vision is initiated in rod and cone photoreceptors when light is converted into an electrical signal. Visual information is processed in parallel pathways known as the midget pathway, the parasol pathway, the blue-ON/yellow-OFF pathway, and the rod pathway (Dacey, 2000; Wassle, 2004). Cone photoreceptors make synapses to ON or OFF cone bipolar cells, which connect to midget retinal ganglion cells, parasol retinal ganglion cells, or bistratified retinal ganglion cells (*Figure* 4). Interestingly, the neural circuits in the rod pathway differ from those in the cone pathway. Rod photoreceptors connect to ON rod bipolar cells, which subsequently send the signal to AII amacrine cells. The AII amacrine cells send the information to ON or OFF cone bipolar cells, which synapse to ON or OFF ganglion cells, respectively (Masland, 2001; Wassle, 2004).

In addition, lateral inhibitory pathways modulate the excitatory signaling in the vertical pathway. First, lateral inhibition occurs in the OPL via feedback from horizontal cells at the photoreceptor-bipolar cell synapse. In the IPL, amacrine cells mediate lateral inhibition via feedback, feed-forward, and serial inhibition. These excitatory and inhibitory pathways are the origin of the fundamental center-surround receptive field characteristic of most neurons in the visual pathway.

Parallel processing is a common strategy within sensory systems of the mammalian brain (Nassi and Callaway, 2009). In the primate visual system, at least three parallel pathways originate in the retina and convey distinct visual signals to primary visual cortex (V1) via the lateral geniculate nucleus (LGN) of the thalamus. The LGN is a six-layered structure with the four most dorsal layers referred to as parvocellular (P) and the two most ventral layers as magnocellular (M), based on the presence of small and large sized cells, respectively. The intercalating zones between each M and P layer and the

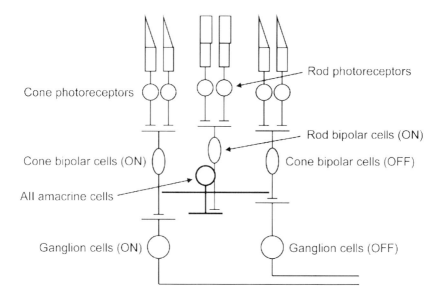

Figure 4. Cone and rod pathway in visual transduction.
Cone photoreceptors contact ON and OFF bipolar cells, which in turn contact ON and OFF ganglion cells, respectively. In contrast, rod photoreceptors send inputs to ON rod bipolar cells, which in turn synapse with AII amacrine cells. ON signals enter the cone pathways via gap junctions between AII amacrine cells and ON cone bipolar cells. OFF signals are produced by a glycinergic synapse between AII amacrine cells and OFF cone bipolar cells.

extremely small cells there are recognized as a separate population and referred to as koniocellular (K). The magnocellular (M) pathway provides signals specialized for the detection of low contrast, quickly moving stimuli, whereas the pavocellular (P) pathway provides signals specialized for form and color processing. The koniocellular (K) pathway is more heterogeneous, and our understanding of the signals it provides remains incomplete. Of a dozen different retinal ganglion cell types projecting to the LGN, three are particularly well characterized and linked to parallel pathways. Midget retinal ganglion cells comprise about 80% of these neurons and terminate in the P layers, parasol retinal ganglion cells make up about 10% and project to the M layers, and bistratified retinal ganglion cells, which form synapses in both the upper and lower sublaminae of the IPL, comprise the remaining 10% and project to the K layers. Each class of LGN neurons then projects to a specific subdivision of primary visual cortex. 4Cα layers in the visual cortex receive inputs from magnocellular neurons, and 4Cβ layers are innervated by parvocellular neurons. Koniocellular neurons project to layers 2 and 3, specifically to regions known as "blobs" that stain densely for the enzyme cytochrome oxidase.

Once the condensed and parallel signals from the retina arrive in the visual cortex, the original components of the visual scene must be extracted, elaborated on, and integrated into a unified percept. Beyond V1, visual information is processed in dorsal and ventral streams, which constitute separate but interconnected pathways that are processed through the occipital, parietal, and temporal extrastriate visual cortex. The dorsal stream is thought to specialize in the analysis of spatial relationships and motion, whereas the ventral stream is thought to specialize in the analysis of shapes and object recognition. However, the relationship between the M, P, and K pathways of the retina and LGN and the dorsal and ventral streams of the visual cortex is not well understood. Recent studies have provided evidence for extensive mixing and convergence of M, P, and K pathway inputs, suggesting that V1 outputs bear little or no systematic relationship to its parallel inputs. Different strategies might be used in V1 to transfer parallel input signals into multiple output streams.

Retinal degenerative diseases

RETINITIS PIGMENTOSA AND ALLIED HEREDITARY DISORDERS

Retinitis pigmentosa (RP) is a group of inherited retinal disorders that are characterized by photoreceptor degeneration and subsequent damage to the inner retinal layers (Hartong *et al.*, 2006). The onset of visual symptoms initially occurs as night blindness and loss of the peripheral visual field in a wide range of ages from childhood to late adulthood, and progressive impairment of visual function in severe cases eventually results in loss of central vision. RP can be inherited as autosomal dominant, autosomal recessive, or X-linked forms. Patients without a family history of the disease are also reported as simplex cases. RP is usually diagnosed with typical visual symptoms and typical ophthalmoscopic fundus abnormalities, including attenuation of retinal vessels, mottling of the RPE, bone-spicule like retinal pigmentation and optic nerve head pallor (*Figure* 5). Perimetry shows typical visual field defects, such as relative scotomas in the mid-periphery in early stages of the disease, while the scotomas enlarge toward the far periphery and constrict the central visual field as the disease progresses. The electroretinogram (ERG) is used for disease diagnosis and follow-up of disease progression. The ERG is an electrical potential generated by the retina in response to a flash of light generated by radial currents arising either directly from retinal neurons or from retinal glia in response to changes in extracellular potassium concentration caused by retinal neuronal activity. In most RP cases, the ERG amplitudes are reduced, and are undetectable in advanced stages.

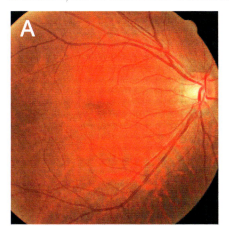 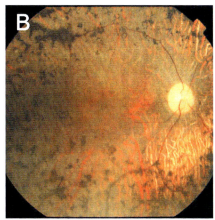

Figure 5. The fundus of a normal subject and a patient with retinitis pigmentosa.
Compared to the fundus of a normal subject (A), the fundus of a patient with retinitis pigmentosa (B) shows attenuation of retinal vessels, mottling of the RPE, bone-spicule like retinal pigmentation, and optic nerve head pallor.

Histopathologic studies have shown that the initial change in the degenerating RP retina is a shortening of the rod outer segments (Li *et al.*, 1995; Milam *et al.*, 1996). As the disease progresses, photoreceptors are gradually lost through apoptosis, likely due to mutant proteins concentrated in the cells. Great advances in understanding the molecular genetics of RP were made in the 1990s. Mutations in a wide variety of photoreceptor-specific genes are responsible for RP (Farrar *et al.*, 2002), with 41 genes and 10 loci identified to date as disease-causing (http://www.sph.uth.tmc.edu/Retnet/). These likely play important roles in photoreceptor-specific functions, such as phototransduction and maintenance of cytoskeletal structure, intracellular signaling, and cell metabolism.

Clinical treatments for RP have not yet been developed, but a number of studies investigating possible therapies are under way in animal models of photoreceptor degeneration and offer hope for the future. Intraocular injections of growth factors such as FGFs, CNTF, PEDF, and brain-derived neurotrophic factor (BDNF) have been shown to protect against photoreceptor degeneration (Faktorovich *et al.*, 1990; LaVail *et al.*, 1992; LaVail *et al.*, 1998; Cao *et al.*, 2001). However, these proteins usually have transient effects, and cells that are genetically modified to express specific factors are used to obtain sustained release (Cayouette and Gravel, 1997; Akimoto *et al.*, 1999; Miyazaki *et al.*, 2003; Imai *et al.*, 2005). A clinical trial is under way in which human RPE cells are transfected with the CNTF gene, encapsulated in a semipermeable membrane, and implanted into human eyes (Sieving *et al.*, 2006). However, although growth factor therapy may be effective for photoreceptor survival, it cannot repair lost cells in the degenerating retina.

The potential of gene replacement therapy has been investigated in animal models of retinal degeneration (Bennett *et al.*, 1996; Takahashi *et al.*, 1999; Ali *et al.*, 2000). Most studies of wild-type gene transfer into eyes used adenoviral or recombinant adeno-associated viral vectors (rAAV). Some drawbacks of adenoviral vector injection include immune rejection with inflammatory changes and reduced duration of transgene expression. These adverse effects can be avoided using selected serotypes of rAAV vectors (Hoffman *et al.*, 1997). rAAV-mediated gene therapy in canine models (Acland *et al.*, 2001), and clinical trials on Leber's congenital amaurosis, a congenital form of RP (Bainbridge *et al.*, 2008; Maguire *et al.*, 2008). This type of disease is caused by mutations in RPE65, which associated with retinoid metabolism in the RPE. Although this biochemical defect results in severe visual impairment, photoreceptor degeneration is delayed. Like growth factor therapy, gene replacement therapy must be performed in the early stages of disease before the progression of photoreceptor cell loss.

A number of studies in human retinal transplantation for advanced RP patients have been conducted using photoreceptor sheets from cadaver eyes (Berger *et al.*, 2003), cell suspensions of fetal retina (Humayun *et al.*, 2000) and sheets of fetal neural retina with the RPE (Radtke *et al.*, 2002; Radtke *et al.*, 2004). Transplantation of adult retinal sheets and fetal retinal cells did not cause any changes in visual acuity or ERG recordings, and in one eye transplantated with a sheet of fetal neural retina with RPE, visual improvement was observed one year after the treatment. During the follow-up period, there was no evidence of immune rejection of grafted tissues in any of the reported cases.

AGE-RELATED MACULAR DEGENERATION

Age-related macular degeneration (AMD) is the leading cause of blindness in patients over 50 years of age in industrialized countries. In AMD, progressive degeneration of the RPE and disruption of Bruch's membrane leads to overlying photoreceptor impairment and loss of central vision. An early characteristic feature in AMD is drusen, which are protein and lipid deposits in Bruch's membrane or between Bruch's membrane and the RPE. Components of these deposits are oxidized products that are presumed to arise from the overlying RPE cells. A major lipid component of the outer segment, docosahexanoate and its oxidized products, as well as carboxyethyl pyrole are more frequently found in the sub-RPE deposits of AMD patients than in controls (Crabb *et al.*, 2002). Other components of drusen include serum proteins, such as immunoglobulins, activated complement components (C3a and C5a), and complement regulators (vitronectin, clusterin and complement factor H (CFH)) (Mullins *et al.*, 2000; Johnson *et al.*, 2001; Nozaki *et al.*, 2006). These immunoglobulins or complement proteins suggest the presence of chronic inflammation in the sub-RPE area associated with progressive RPE degeneration or the pathological growth of choroidal vessels under the retina (choroidal neovascularization; CNV). Recently, it was reported that a variation in the gene encoding CFH is strongly associated with the development of AMD by haplotype mapping and single nucleotide polymorphism analysis (Edwards *et al.*, 2005; Hageman *et al.*, 2005; Haines *et al.*, 2005; Klein *et al.*, 2005).

Classically, laser photocoagulation of the CNV was the only established treatment for AMD. However, laser photocoagulation induces scar formation of targeted abnormal vessels with damage to surrounding normal retinal tissue that results in visual impairment. Recently, several new therapeutic approaches have been used to treat AMD. Photodynamic therapy is conducted with intravenous injections of a photosensitizing agent delivered selectively to the CNV and excitation of the agent by a specific wavelength using a laser. Free radicals arising from excited photosensitizers cause occlusion of the CNV with minimal influence on surrounding normal retinal tissues. An alternative approach is intraocular injections of anti-VEGF agents to prevent the growth of the CNV. Three types of anti-VEGF agents have been used clinically and have been shown to maintain or improve vision for at least one or two years after treatment: anti-VEGF aptamer (pegaptanib), a humanized anti-VEGF monoclonal antibody (bevacizumab), and a high-affinity anti-VEGF Fab (ranibizumab) (Avery *et al.*, 2006; Heier *et al.*, 2006).

Surgical treatment for AMD was also developed in the 1990s. de Juan first reported removal of the CNV in AMD through a small incision site in the retina by micro-tipped forceps inserted into the vitreous cavity (de Juan and Machemer, 1988). However, the outcome of the follow-up reports on this treatment was not satisfactory; visual improvement was observed in about 33% of the treated eyes and deterioration occurred in 27% (Falkner *et al.*, 2007). In addition, excision of the CNV along with the removal of surrounding the RPE is followed by a progressive increase in RPE atrophy (Nasir *et al.*, 1997; Castellarin *et al.*, 1998).

Macular translocation by 360 degree incision and rotation of the retina was first reported by Machemer (Machemer and Steinhorst, 1993). With this surgical technique, the central area of the retina is detached and reattached to the normal RPE area remote from the affected RPE area. The long-term follow-up study showed that visual improvement was observed in 17% of treated eyes and deterioration occurred in 44% at final examination, with progressive atrophy of the RPE being the limiting factor for visual outcome (Aisenbrey *et al.*, 2007).

In addition, Peyman reported RPE transplantation in two patients with end-stage AMD (Peyman *et al.*, 1991). The technique involves the preparation of a large retinal flap encompassing the macula and replacement of the RPE and Bruch's membrane. The first patient with an autologous graft had visual improvement and the patient fixated over the transplanted RPE. However, the second patient with a homologous graft showed no improvement. Since homologous RPE transplantation without systemic immunosuppression resulted in the failure of grafts due to a delayed immune reaction (Algvere *et al.*, 1994; Algvere *et al.*, 1997; Algvere *et al.*, 1999), systemic immunosuppression is necessary to avoid graft rejection, but this technique did not lead to visual improvement (Tezel *et al.*, 2007). Autologous RPE transplantation by using peripheral RPE with choroid grafts has been described by van Meurs and Van Den Biesen, and follow-up reports indicate improved visual function but frequent surgical complications (van Meurs and Van Den Biesen, 2003; Joussen *et al.*, 2006; MacLaren *et al.*, 2007).

Potential donor cells for retinal cell transplantation

RETINAL TISSUE

There are several candidate donor sources for photoreceptor transplantation (*Table 3*). Transplantation of whole sheets of embryonic or neonatal neural retina into the sub-retinal space has been reported to result in the survival and differentiation of the grafted tissue (Seiler *et al.*, 1990; Zhang *et al.*, 2003). However, integration of the transplanted cells is quite limited, and they only rarely made connections with the host tissue. One of the problems is that the inner part of the transplanted tissue contained retinal ganglion cells and displaced amacrine cells, which did not re-form functional synapses with host bipolar cells or outer nuclear cells of the surviving host retina. Thus, a sheet arrayed with orientated photoreceptors might be ideal for photoreceptor replacement in cases in which the host ONL degenerates.

Table 3. Potential of various donor cells for photoreceptor transplantation therapy.

Cell type	Photoreceptor	Quantity	Integration	Immune rejection	Ethical problems	Tumor formation
Retinal tissue	◎	×	△	△	×	○
Developing retinal cells (P5)	◎	×	◎	△	×	○
Brain-derived neural progenitor	×	△	△	△	△	○
Retinal progenitor	◎	△	◎	△	×	○
Ciliary body	○	△	△	△	◎	○
Iris	○	△	△	◎	◎	○
Müller glia	○	△	△	△	◎	○
ES cell	◎	◎	◎	○	×	△
iPS cell	◎	◎	◎	◎	◎	×

◎: good, ○: acceptable, △: marginal, ×: bad

DEVELOPING RETINAL CELLS

Cells from the developing retina can be used as a donor source for transplantation. MacLaren *et al.* have reported that transplantation of P3-6 post-mitotic rod precursors improved visual function (MacLaren *et al.*, 2006). The transplanted cells migrate to the ONL, integrate into the host neural retina, and restore some visual function. Notably, transplanted P3-6 post-mitotic rod precursors are capable of integrating into the normal adult or degenerating retina, whereas proliferating progenitors or stem cells are not. This indicates that the ontogenic stage of transplanted photoreceptors determines the ability of these cells to integrate into the host retina. However, because post-mitotic embryonic or neonatal retinal cells cannot proliferate, a large quantity of retinal tissue is required to harvest enough P3-6 rod precursors. The supply of available fetal retinal cells is limited, and use of human fetal tissue presents ethical problems. Thus, *in vitro* expansion of stem cells is an ideal alternative source of donor retinal cells.

NEURAL STEM CELLS IN THE BRAIN

Adult neural stem cells are present in the SVZ of the lateral ventricle and the SGZ of the hippocampal dentate gyrus (Lledo *et al.*, 2006; Zhao *et al.*, 2008). Neural stem cells in the SVZ generate young neurons that migrate tangentially to the olfactory bulb where they replace multiple types of interneurons. Neural stem cells in the SGZ differentiate into granule neurons of the dentate gyrus, sending axonal projections to area CA3 and dendritic arbors into the molecular layer. These adult-born neurons are integrated functionally into pre-existing neural circuits. Neural stem cells also exist in humans (Eriksson *et al.*, 1998; Sanai *et al.*, 2004), suggesting that they might be harnessed for transplantation therapy. Thus, we examined whether stem cells in the adult rat hippocampus could differentiate into retinal neurons after transplantation into the developing eye (Takahashi *et al.*, 1998). Within four weeks of transplantation, the adult stem cells were integrated into the retina and exhibited morphologies and positions characteristic of Müller, amacrine, bipolar, horizontal, and photoreceptor cells. However, none acquired end-stage markers unique to retinal neurons. We therefore conclude that adult brain-derived stem cells cannot adopt retinal fates even when exposed to the cues present during retinal development. Although the brain and the retina are both generated from the ectodermally derived neural tube, neural progenitors in different CNS regions differ in their competence to generate specific types of mature neurons.

NEURAL STEM CELLS IN THE DEVELOPING RETINA

We also examined the potential of neural stem cells in the embryonic retina. Neural progenitors from the rat fetal retina can be expanded *in vitro* and can differentiate into various types of retinal neurons, including photoreceptors (Akagi *et al.*, 2003). Klassen *et al.* have demonstrated that retinal progenitors differentiate into cells of photoreceptor morphology and express both rhodopsin and recoverin following transplantation to the subretinal space. Notably, transplantation of retinal progenitors improve light-mediated behavior of rhodopsin knockout (Rho$^{-/-}$) mice, a model of photoreceptor dystrophy (Klassen *et al.*, 2004). However, retinal progenitors lose their ability to differentiate into photoreceptors following massive expansion *in vitro* (Akagi *et al.*, 2003). Improving the methods for propagating stem cells and producing large numbers of photoreceptor cells is possible, but use of human fetal tissue presents ethical problems.

PROGENITORS IN THE CILIARY MARGIN AND IRIS

The somatic progenitors in adult eye tissue are another potential source of donor cells. The ciliary marginal zone has been reported to contain stem cells even in adults (Ahmad *et al.*, 2000; Tropepe *et al.*, 2000). When cultured *in vitro*, these cells give rise to retinal neurons, including photoreceptors. Iris-derived cells have also been reported to generate retinal neurons (Haruta *et al.*, 2001; Akagi *et al.*, 2004; Akagi *et al.*, 2005; Asami *et al.*, 2007). Adult tissues have the advantage of being suitable for autografts, which do not cause immune rejection. Autologous iris tissue can be feasibly obtained with peripheral iridectomy. Unlike the hippocampus, both the ciliary margin and iris derive from the optic vesicle and optic cup, suggesting that they may be more competent than brain stem cells to generate retinal neurons. Cells differentiated from adult somatic progenitors in the eye express several photoreceptor marker proteins, but not all the genes responsible for photoreceptor function (our unpublished data). Thus, it is likely that the generation of functional photoreceptors requires a recapitulation of the normal process of retinal development.

MÜLLER GLIA IN THE ADULT RETINA

Müller glia are a specialized form of radial glia that span the entire depth of the retina. Müller glial processes surround neuronal cell bodies in the nuclear layers and contact synapses in the plexiform layers. The distal processes of the Müller glia form the external limiting membrane of the retina, and their endfeet form the inner limiting membrane. Under physiological conditions, Müller glia play an important role in regulating extracellular K+ and pH, in the uptake of the neurotransmitter glutamate, and in the synthesis of glutamine.

In contrast to other endogenous adult cell types, Müller glia have the potential to generate retinal neurons after injury *in vivo* (Ooto *et al.*, 2004; Karl *et al.*, 2008; Wan *et al.*, 2008). Interestingly, glial cells in the SVZ of the lateral ventricle and the SGZ of the hippocampal dentate gyrus also act as neural stem cells (Osakada and Takahashi, 2007; Lledo *et al.*, 2008; Chojnacki *et al.*, 2009). The possibility that Müller glia are an intrinsic source of regeneration was first raised by Braisted *et al.* through experiments in the goldfish showing that laser damage elicited proliferation of Müller glia and the concomitant replacement of the damaged cone photoreceptors (Braisted *et al.*, 1994). In addition, Müller glia in the retinas of post-hatch chicks have been reported to possess regenerative capacity (Fischer and Reh, 2001). Several lines of evidence support a close relationship between Müller glia and retinal progenitors. Recent gene expression profiling studies have demonstrated a large degree of overlap in the genes expressed in the Müller glia and late retinal progenitors (Blackshaw *et al.*, 2003). Moreover, the proliferation and differentiation of Müller glia-derived progenitors can be regulated by both intrinsic (homeobox genes and basic helix-loop-helix genes) and extrinsic factors (Wnt, Notch, Shh, FGF and EGF) (Ooto *et al.*, 2004; Das *et al.*, 2006; Osakada *et al.*, 2007; Wan *et al.*, 2007), similar to what has been observed with retinal progenitors during eye development.

The neural stem cell properties of Müller glia have been also verified *in vitro* (Das *et al.*, 2006). Like neural stem/progenitor cells(Reynolds and Weiss, 1992), dissociated Müller glia form neurospheres *in vitro*. These neurospheres can differentiate into neurons and glia, demonstrating the multipotency of Müller glia. In addition, Müller glia-derived progenitors can be identified and purified as a distinct population by FACS analysis using Hoechst labeling, another characteristic of progenitors (Das *et al.*, 2006).

These neurogenic properties of Müller glia raise the possibility that they may be used as a donor source for transplantation. Indeed, Müller glia-derived cells can differentiate into retinal neurons after transplantation into the retina (Das *et al.*, 2006). Müller glia can be obtained from human retinal tissue at eye banks, but it is impossible to isolate Müller glia from the patient's retina to perform autologous cell transplantation. Because Müller glia serve as endogenous progenitors *in vivo*, drug therapy targeting Müller glia may more suitable than transplantation therapy (Osakada *et al.*, 2007; Osakada and Takahashi, 2009). To repair the retina effectively, we need to identify at least two drugs promoting proliferation of Müller glia-derived progenitors and differentiation into required cell types.

ES CELLS

ES cells are pluripotent cells derived from the inner cell mass of blastocyst stage embryos. They can maintain an undifferentiated state indefinitely *in vitro* and differentiate into derivatives of all three germ layers: the ectoderm, endoderm and mesoderm. These characteristics make ES cells an attractive potential donor source for degenerative diseases such as Parkinson's disease, spinal cord injury, and diabetes, as well as retinal degeneration (McDonald *et al.*, 1999; Kim *et al.*, 2002; Keirstead *et al.*, 2005; Takagi *et al.*, 2005). For this purpose, controlled differentiation and defined culture methods to obtain safe, differentiated cells are indispensable. Based on our knowledge of embryonic development, we have developed methods of inducing retinal cells from ES cells *in vitro* (Ikeda *et al.*, 2005; Osakada *et al.*, 2008; Osakada *et al.*, 2009). Our defined culture method for generating large numbers of photoreceptors and RPE from monkey and human ES cells will greatly contribute to retinal regeneration therapy. As discussed above, because embryonic retina, brain neural stem cells, somatic progenitors derived from the adult ciliary body or iris are limited in differentiation potential and/or proliferation capacity, pluripotent stem cells represent the most promising donor source for transplantation. One of the problems to be resolved is removal of undifferentiated cells for donor cell preparation (Fukuda *et al.*, 2006; Choo *et al.*, 2008).

iPS CELLS

Recent biotechnology has enabled the artificial generation of pluripotent stem cells from somatic cells (Takahashi *et al.*, 2007). Four transcription factors, Oct3/4, Sox2, Klf4, and c-Myc, reprogram mouse embryonic or adult fibroblasts to generate induced pluripotent stem (iPS) cells, which express undifferentiated ES cell markers, have similar gene expression profiles to ES cells, form teratomas, and contribute to all cell types in chimeric animals, including the germ line (Maherali *et al.*, 2007; Okita *et al.*, 2007; Wernig *et al.*, 2007). iPS cells have been generated from fibroblasts, bone marrow cells, hepatocytes, gastric epithelial cells, pancreatic cells, neural stem cells, B lymphocytes, keratinocytes, and blood progenitors(Aasen *et al.*, 2008; Aoi *et al.*, 2008; Hanna *et al.*, 2008; Kim *et al.*, 2008; Stadtfeld *et al.*, 2008a; Loh *et al.*, 2009). The clinical application of human ES cell therapy faces ethical difficulties concerning the use of human embryos, as well as tissue rejection following implantation (Takahashi *et al.*, 2007; Yu *et al.*, 2007). However, iPS cells can resolve these issues, because they can be generated easily from the patient's own somatic cells.

On the other hand, iPS cells present new challenges not faced with ES cells. The establishment of iPS cells has required gene transfer of reprogramming factors. Early reports used retroviral, lentiviral, and adenoviral vectors (Takahashi and Yamanaka, 2006; Brambrink *et al.*, 2008; Mali *et al.*, 2008; Stadtfeld *et al.*, 2008b), but plasmids were also subsequently shown to be effective (Okita *et al.*, 2008). Although transgenes are largely silenced in iPS cells, their reactivation, particularly that of c-Myc, can lead to tumorigenesis. However, c-Myc but is not required for generation of iPS cells, even though it enhances the establishment efficiency (Nakagawa *et al.*, 2008). Recently, recombinant reprogramming factor proteins have been used to establish mouse and human iPS cells (Kim *et al.*, 2009; Zhou *et al.*, 2009), and identification of small molecules that can reprogram somatic cells may enable establishment of gene insertion-free, animal product-free iPS cells. It should be noted that partial or aberrant reprogramming results in impaired ability to differentiate into the required cell type, so selection and validation of iPS cells are critical.

Generation of RPE and photoreceptors from pluripotent stem cells

IN VITRO ES CELL DIFFERENTIATION SYSTEM

Since the establishment of mouse ES cells in 1981 (Evans and Kaufman, 1981) and human ES cells in 1998 (Thomson *et al.*, 1998), much progress has been made in ES cell propagation and differentiation techniques. Over the last decade, several methods have been reported to control differentiation of ES cells into neural cells (Bain *et al.*, 1995; Kawasaki *et al.*, 2000; Lee *et al.*, 2000a; Tropepe *et al.*, 2001; Wichterle *et al.*, 2002; Barberi *et al.*, 2003; Ying *et al.*, 2003; Watanabe *et al.*, 2005; Ueno *et al.*, 2006; Eiraku *et al.*, 2008). Each method has its own advantages and disadvantages, depending on the type of neural cells desired, and can induce differentiation of neural tissues with distinct regional identities within the CNS.

By using a co-culture system, we have established an efficient method of inducing selective neural differentiation of ES cells under serum-free, retinoic acid-free conditions: the stromal cell-derived inducing activity (SDIA) method (Kawasaki *et al.*, 2000). During vertebrate embryogenesis, nervous tissue arises from uncommitted ectoderm during gastrulation. Subsequently, the CNS anlage is patterned to acquire regional specification along the rostral-caudal and dorsal-ventral axes. Thus, we examined the positional identity of ES cell-derived neural cells (Mizuseki *et al.*, 2003;

Irioka *et al.*, 2005; Ueno *et al.*, 2006). RT-PCR analysis with rostral-caudal CNS markers showed that SDIA-treated ES cells express the forebrain marker *Otx2*, the midbrain-hindbrain border marker *En2*, and the rostral hindbrain marker *Gbx2*. In contrast, little expression is detected for the spinal cord markers *Hoxb4*, *Hoxb9*, and *HB9*. The rostral-caudal specification of SDIA-induced neural cells can be modified by adding the caudalizing factor retinoic acid. Treatment with retinoic acid promotes the expression of caudal CNS markers, such as *Gbx2*, *Hoxb4*, *Hoxb9*, and *HB9*, whereas the forebrain marker *Otx2* is suppressed. In addition, we examined positional markers along the dorsal-ventral axis. SDIA treatment induces both dorsal (*Pax7* and *Dbx1*) and ventral (*Irx3* and *HNF3β*) neural tube markers. The ventral-most neural tube markers *Nkx6.1*, *Nkx2.2*, and *HNF3β* are also expressed in SDIA-treated cells. Thus, SDIA-induced neural precursors differentiate into a wide range of CNS cell types that correlate with their positions along the dorsal-ventral and rostral-caudal axes.

The neural crest arises from the juncture of the dorsal CNS and non-neural ectoderm (Knecht and Bronner-Fraser, 2002), where a number of BMP family members are expressed. Although BMP signals inhibit neural induction at the early gastrula stage, the same signals promote neural crest formation when applied at later developmental stages (Jessell, 2000; Knecht and Bronner-Fraser, 2002). Consistent with these *in vivo* events, late BMP4 exposure after the fourth day of SDIA treatment causes differentiation of neural crest cells and dorsal-most CNS cells, with autonomic nervous tissue preferentially induced by high BMP4 concentrations and sensory lineages by low BMP4 concentrations. Moreover, early exposure of SDIA-treated ES cells to BMP4 suppresses neural differentiation and promotes formation of epidermal cells. In contrast, sonic hedgehog (Shh) suppresses the development of dorsal tissues and promotes the differentiation of ventral CNS tissue *in vivo* (Jessell, 2000). Consistent with this activity, Shh suppresses differentiation of ES cells into dorsal cells, including AP2+/NCAM+ neural crest cells, and increases the number of ventral cells, including Nkx2.2+ cells, HNF3β+ floor plate cells, and motor neurons.

We have also established a serum-free, feeder-free culture system that induces efficient neural differentiation from ES cells: the serum-free floating culture of embryoid body-like aggregates (SFEB) method (Watanabe *et al.*, 2005) (*Figure 6*). In the presence of a Wnt antagonist (Dkk-1), SFEB efficiently induces the formation of Bf1+ telencephalic precursors (*Figure 1*). Subregional specification of the telencephalon can be reproduced *in vitro* using embryologically relevant patterning molecules. Wnt inhibits neural differentiation and forebrain development at earlier stages (Aubert *et al.*, 2002; Nordstrom *et al.*, 2002), but positively regulates pallial telencephalic specification at later developmental stages (Galceran *et al.*, 2000; Lee *et al.*, 2000b; Gunhaga *et al.*, 2003). In contrast, Shh has been implicated in the ventral specification of the forebrain (Ericson *et al.*, 1995; Chiang *et al.*, 1996; Rallu *et al.*, 2002). Consistent with these activities, treatment with Wnt3a or Shh during late SFEB culture increases differentiation into the pallial (Pax6+, Bf1+) or basal (Nkx2.1+, Islet1/2+, Bf1+) telencephalic population, respectively (*Figure 1*). Moreover, caudal CNS tissues such as Math1+ cerebellar neurons are induced from ES cells by SFEB culture followed by BMP4/Wnt3a treatment (Su *et al.*, 2006). The induced Math1+ cells are mitotically active and express markers characteristic of granule cell precursors (Pax6, Zic1, and Zipro1). L7+/Calbindin-D28K+ Purkinje cells are also induced under similar culture conditions. These observations indicate that SDIA- or SFEB-treated

ES cells generate naive neural progenitors that are competent to differentiate into the rostral-caudal and dorsal-ventral ranges of neuroectodermal derivatives in response to patterning signals.

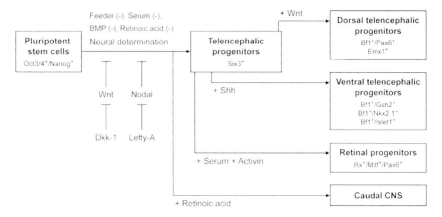

Figure 6. Directed differentiation of pluripotent stem cells into various neural progenitors in response to patterning signals.

Pluripotent stem cells differentiate into the rostral-caudal and dorsal-ventral range of neuroectodermal derivatives in response to patterning signals. SFEB and soluble factor treatment induce telencephalic, retinal, cerebellar, and caudal neural progenitors.

DIRECTED DIFFERENTIATION OF ES CELLS AND iPS CELLS INTO RETINAL CELLS

In vitro differentiation of ES cells mimics at least in part the patterning and differentiation events that occur during embryogenesis. ES cells differentiate into a variety of neural cells with specific spatiotemporal identities. In response to exogenous patterning signals such as Wnt, Shh, BMP4, and retinoic acid, ES cell-derived neural progenitors differentiate into a wide range of neural cells that correlate with their positions along the dorsal-ventral and rostral-caudal axes.

During early embryogenesis, the retinal primordium forms in the rostralmost region of the diencephalon expressing Six3. The transcription factor Rx, an early marker of the eye field, plays an essential role in the specification of the retinal primordium within the Six3+ rostral CNS. During early embryogenesis (E10.5), progenitors in the optic cup express Rx in the inner layer that will give rise to the neural retina, Mitf in the outer layer that will give rise to the RPE, and Pax6 in both layers (Furukawa *et al.*, 1997; Mathers *et al.*, 1997; Bora *et al.*, 1998; Nguyen and Arnheiter, 2000; Baumer *et al.*, 2003) (*Figure 7*). Thus, the neural retinal lineage during early development is characterized by Rx/Pax6 coexpression. We first attempted to induce Six3+ rostralmost CNS progenitors and Rx+/Pax6+ retinal progenitors by applying exogenous patterning signals. Because the extracellular patterning signals that induce the retinal primordia have not yet been identified, we used a candidate approach to identify soluble factors that induce Rx/Pax6 expression (Ikeda *et al.*, 2005). In serum-free and feeder-free aggregate culture (SFEB culture), strong expression of Six3 is found on culture day 5, but not in cells cultured with the caudalizing factor RA. The strongest induction of Rx+/Pax+ cells is observed when ES cells were treated with Dkk-1, Lefty-A, FCS, and Activin (referred to as SFEB/DLFA cells).

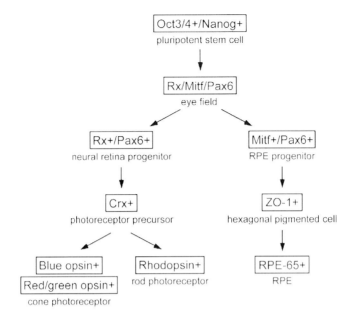

Figure 7. Multi-step commitment in the development of retinal cells.
Markers for the differentiation steps are boxed. ES cells and iPS cells differentiate into photoreceptors and RPE in a stepwise manner along the developmental time course.

The RPE expresses Mitf, Pax6, and Otx2, but not Rx, during retinal development (*Figure 7*). We therefore tested whether RPE differentiation was observed in SFEB/DLFA culture by examining expression of the early RPE marker Mitf. Mitf+ cell aggregates are observed in SFEB/DLFA cultures, whereas SFEB-treated cells rarely express Mitf. Consistent with the *in vivo* expression profile of RPE markers, most Mitf+ cells in the SFEB/DLFA culture are Pax6+ and Otx2+. Thus, we conclude that SFEB/DLFA treatment preferentially induces differentiation of retinal progenitors from mouse ES cells.

Next, we examined photoreceptor differentiation of SFEB/DLFA-treated retinal progenitors (Osakada *et al.*, 2008). Inhibition of Notch-1 has been shown to promote photoreceptor differentiation *in vivo* (Jadhav *et al.*, 2006). In mouse ES cells, purified Rx+ cells express Notch (Notch1-4) and its downstream mediators Hes1, Hes5, and Hey1. Inhibition of Notch signaling in these cells by the γ-secretase inhibitor DAPT increases the frequency of Crx+ photoreceptor precursors generated from Rx-GFP+ cells. We then examined whether Crx+ photoreceptor precursors that were efficiently generated using DAPT could further differentiate into rod and cone photoreceptors. On day 28, 20-25% of ES-derived cells express red/green opsin or blue opsin, cone-specific pigment proteins that are indispensable for color vision. In contrast, fewer cells express rhodopsin, the rod-type visual pigment. To optimize conditions for rod differentiation, we next tested several soluble factors that are reported to promote rod differentiation from embryonic or neonatal retinal progenitors (Levine *et al.*, 2000). When aFGF, bFGF, taurine, Shh, and retinoic acid are added, a large number of cells (15-20%) express the rod photoreceptor marker Rhodopsin.

We have also established a defined culture method for monkey and human ES cells that does not require serum (Osakada *et al.*, 2008). Serum-free, feeder-free suspension culture combined with application of Dkk-1 and Lefty-A induces retinal progenitors positive for RX, PAX6, MITF, and CHX10 on culture days 30-40. Subsequently, the progenitors generate hexagonal pigment epithelium polarized with apical microvilli and basal membranes. The pigment cells expressed RPE-65 and CRALBP, formed ZO-1+ tight junctions, and exhibited phagocytic function on day 100 (*Figure 8A, B*). However, photoreceptor differentiation occurs only infrequently under these conditions. Additional treatment with retinoic acid and taurine significantly promotes differentiation of ES cell-derived progenitors into photoreceptors positive for CRX, NRL, RECOVERIN, RHODOPSIN, BLUE OPSIN, or RED/GREEN OPISN on day 140. The induced photoreceptors express genes responsible for phototransduction (*Figure 8C, D*).

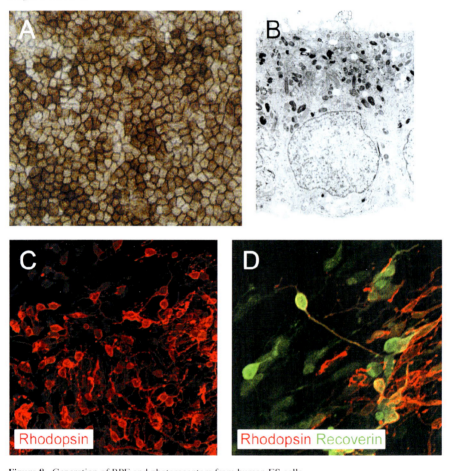

Figure 8. Generation of RPE and photoreceptors from human ES cells.
(A) Hexagonal pigmented cells derived from human ES cells. (B) Electron micrograph of human ES cell-derived RPE. (C) Rhodopsin+ photoreceptors derived from human ES cells. (D) Human ES cell-derived Rhodopsin+ cells co-express Recoverin. (Figures from *Nature Biotechnology* (Osakada *et al.*, 2008)).

These differentiation methods using defined factors for ES cells are applicable to iPS cells (Hirami *et al.*, 2009). It should be noted that the selection and validation of iPS cells are critical for generation of iPS cells and their further differentiation. We have found that mouse iPS cells selected by Nanog expression (Nanog-iPS cells) can differentiate into photoreceptors and RPE, while iPS cells selected by Fbx15 expression (Fbx15-iPS cells) can not. Nanog-iPS cells are fully reprogrammed, and contribute to the germline of adult chimeric mice can to live-late term embryos when injected into tetraploid blastocysts. In contrast, Fbx15-iPS cells are similar to ES cells in morphology, proliferation, and teratoma formation, but show clear differences from ES cells in their global gene expression profile and differential potential.

It is also important to note that because xenogenic factors may cause immune rejection following transplantation, the generation of differentiated cells from human ES cells or iPS cells without contamination from animal-derived substances is essential for the clinical application of transplantation strategies (Martin *et al.*, 2005). For retinal differentiation, the above-mentioned method requires the addition of recombinant Dkk-1 and Lefty-A proteins, which are produced in animal cells or *E. coli*, raising the possibility of infection or immune rejection due to cross-species contamination. By contrast, using chemical compounds to induce differentiation offers several advantages compared with using recombinant proteins. Not only are they non-biological products, but they show stable activity, have small differences between production lots, and are low cost. Thus, establishment of chemical compound-based culture systems will be necessary for human pluripotent cell-based transplantation therapies.

We have succeeded in establishing chemical compound-based culture systems for generating retinal cells from human ES cells and iPS cells (Osakada *et al.*, 2009). This represents another significant step towards clinical application of retinal transplantation therapy. Application of the casein kinase I inhibitor CKI-7, the ALK-4 inhibitor SB-431542, and the Rho-associated kinase inhibitor Y-27632 in serum-free and feeder-free floating aggregate culture induces retinal progenitors positive for RX, MITF, PAX6, and CHX10. The treatment induces hexagonal pigmented cells that express RPE-65 and CRALBP, form ZO-1-positive tight junctions, and exhibit phagocytic functions. Subsequent treatment with retinoic acid and taurine induces photoreceptors that express CRX, NRL, RECOVERIN, RHODOPSIN, and genes involved in phototransduction.

Retinal cell transplantation

PHOTORECEPTOR TRANSPLANTATION

At early stages of retinal degeneration, when second order neurons still have nearly normal functional synapses and donor cells can establish functional synapses on appropriate second order neurons (Humayun *et al.*, 2003), photoreceptor replacement is likely to be an effective way to recover normal retinal function. Thus, an important criterion for the donor cell source in this type of therapy is the ability to generate a sufficient number of functional photoreceptors. In this respect, pluripotent stem cells have an advantage, since they can provide unlimited numbers of donor cells (*Table 3*).

Recently, Lamba *et al.* have reported that human ES cell-derived retinal cells can migrate into the mouse retina, settle into the appropriate layers and express markers for differentiated cells, including both rod and cone photoreceptors (Lamba *et al.*, 2009). They showed that when transplanted into the vitreous humor of newborn wild-type mice, ES cell-derived retinal cells integrated into all the retinal layers and expressed markers appropriate for the lamina in which they settled. When transplanted into the subretinal space of adult mice, the cells integrated into the ONL expressed photoreceptor markers, and differentiated morphologically identifiable outer segments. In a study using post-mitotic photoreceptor precursors from the postnatal mouse retina as donor cells (MacLaren *et al.*, 2006), the optimal donor age for integration into the host newborn mouse retina was P3-P6. Thus, cell type-specific and stage-specific selection of ES cell-derived differentiated cells will raise the efficiency of integration into the host retina (Osakada and Takahashi, 2009).

The function of transplanted photoreceptors was assessed in animal retinal degeneration models by the pupillary light reflex, ERGs, and recording receptive fields across the surface of the superior colliculus. The pupil responses of Rho[-/-] mice transplanted with postnatal Rho[+/+] retinal cells in one eye and a sham injection of Rho[-/-] retinal cells in the other eye were examined by MacLaren *et al.* (MacLaren *et al.*, 2006). In Rho[-/-] mice, rod photoreceptor degeneration gradually progresses by ~12 weeks of age, while these mice retain cone function capable of detecting high intensity stimuli (Toda *et al.*, 1999). The pupil reflex in the treated eyes could be elicited by lower intensity stimuli than in the sham-injected eyes, which showed a pupil reflex similar to that of untreated Rho[-/-] mice. They concluded that transplanted rod photoreceptor precursors are light-responsive and functionally connected to downstream neural circuits.

Although the ERG response is the sum of all retinal cells and consists of overlapping positive and negative component potentials that originate from different stages of retinal processing, it is a useful tool for studying retinal function because it can be recorded noninvasively under physiological conditions. The ERG response consists of light-driven photoreceptor activity (A-wave) and post-synaptic retinal circuitry (B-wave and oscillatory potentials). Crx knockout mice (Crx[-/-] mice) completely lack of any rod or cone ERG response, even though the photoreceptor layer is intact (Furukawa *et al.*, 1999). When human ES cell-derived retinal cells are transplanted into the subretinal space of 4- to 6-week-old Crx[-/-] mice, extensive migration of transplanted cells into the ONL was observed. Moreover, the ERG B-wave response was restored, with the number of integrated rod photoreceptors correlated with the amplitude of the B-wave (Lamba *et al.*, 2009).

Electrophysiological recordings from the superior colliculus provide information about the direct retinal input that is topographically organized according to the area of the retina. Physiological changes take place in the visual centers of the brain parallel to the spatial and temporal progression of the photoreceptor degeneration in the retina. This method is suitable for assessing a confined area of the transplanted retinal tissue. Retinal degeneration model rats that received fetal retinal sheet transplants recovered visual evoked potentials in the superior colliculus (Woch *et al.*, 2001; Thomas *et al.*, 2004). However, visual function mediated by cortical activity is assessed by behavioral tests, such as the visual optokinetic system or visual water task, described by Prusky *et al.* (Prusky *et al.*, 2000; Prusky *et al.*, 2004; McGill *et al.*, 2007).

Finally, the host environment is also critical for photoreceptor transplantation. Retinal degeneration is characterized by the formation of glial scars and microglial activation, which may impede integration and survival of transplanted cells. Robust integration of transplanted retinal cells into the retina of host mice deficient for both vimentin and glial fibrillary acidic protein has been reported (Kinouchi *et al.*, 2003). Moreover, chondroitinases and matrix metalloproteases-2 that degrade the extracellular matrix in the diseased retina aid in the integration of transplanted photoreceptors (Suzuki *et al.*, 2006; Suzuki *et al.*, 2007). Disruption of the outer limiting membrane also increases photoreceptor integration following transplantation (West *et al.*, 2008). Therefore, in addition to immunosuppression, the host retinal environment must be modulated for successful transplantation.

RPE TRANSPLANTATION

Transplantation of the RPE as a therapeutic approach for AMD aims at photoreceptor rescue by restoring phagocytosis of photoreceptor outer segments, secretion of trophic factors, and reconstruction of BRB. The possibility of photoreceptor rescue by transplantation of the RPE was demonstrated in reports showing that transplantation of normal rat RPE into the subretinal space of the Royal College of Surgeons (RCS) rat resulted in phagocytosis of host outer segments and prevention of photoreceptor degeneration (Li and Turner, 1988; Lopez *et al.*, 1989). The RCS rat is a widely used animal model of inherited retinal dystrophy in which the disability of RPE phagocytosis of shed outer segments by a mutation in the gene encoding the receptor tyrosine kinase Mertk leads to progressive loss of photoreceptors (D'Cruz *et al.*, 2000; Feng *et al.*, 2002). The restoration of visual function in RPE cell transplanted RCS rats was confirmed using ERG, pupillary reflex, and recording receptive fields across the surface of the superior colliculus (Yamamoto *et al.*, 1993; Whiteley *et al.*, 1996; Sauve *et al.*, 1998; Coffey *et al.*, 2002).

Transplantation of RPE cells from human fetal and adult eyes into the subretinal space of RCS rats prevented photoreceptor cell death and improved visual function (Little *et al.*, 1996; Castillo *et al.*, 1997; Little *et al.*, 1998). Recently, ES cell-derived RPE cells were used as donor cells for RPE transplantation in the RCS rat model (Haruta *et al.*, 2004; Lund *et al.*, 2006). In these studies, host animals were immunosuppressed postoperatively as they were injected with the xenogenic tissues.

The subretinal space has been shown to have the properties of an immune privileged site, similar to the anterior chamber of the eye (Wenkel and Streilein, 1998). The mechanism involves active suppression of immune responses through FasL (CD95L) expression in the RPE, which induces apoptosis of CD95+ inflammatory cells (Wenkel and Streilein, 2000). Other mechanisms involve suppression of interferon-γ and upregulation of IL-10 by soluble factors secreted by the RPE, such as TGF-β, thrombospondin, somatostatin, and PEDF (Zamiri *et al.*, 2005; Zamiri *et al.*, 2006). However, since the outer BRB is broken down in the RCS rat, chronic rejection occurs and the donor RPE cells that are normally MHC class II-negative express MHC class II mRNA in the subretinal space after transplantation (Zhang and Bok, 1998). The BRB also appears to break down in AMD patients, and the immune privileged status of their subretinal space seems imperfect. Since immune rejection is a serious problem for successful RPE transplantation, it is better to use autologous tissue. The technique of preparing autologous RPE cells or sheets with choroid tissue requires relatively large amounts of ocular tissue and leads to a high risk of surgical complications. The advantage of using differentiated RPE from pluripotent stem cells is that these cells can provide an unlimited number of donor cells (Haruta *et al.*, 2004; Klimanskaya *et al.*, 2004; Lund *et al.*, 2006; Osakada *et al.*, 2008). Moreover, iPS cells offer the possibility of avoiding immune rejection, since RPE cells could be generated from the patient's own somatic cells (Takahashi and Yamanaka, 2006; Hirami *et al.*, 2009).

Finally, improved techniques for effective transplantation need to be explored. For instance, transplantation of RPE sheets rather than cell suspensions may improve functional recovery. In addition, development of specialized surgical tools will be required for human RPE transplantation.

Concluding remarks

iPS cell technology represents a great advance in the field of regenerative medicine (Hanna *et al.*, 2007; Dimos *et al.*, 2008; Park *et al.*, 2008; Ebert *et al.*, 2009; Ye *et al.*, 2009). Many improvements in the methods for establishing iPS cells have been reported; for instance, c-Myc is not essential for the generation of iPS cells (Nakagawa *et al.*, 2008), and reprogramming factor proteins can induce iPS cells without viral vectors or plasmids (Kim *et al.*, 2009; Zhou *et al.*, 2009). The next challenges will be identification of chemical compounds that induce reprogramming, establishment of feeder-free and serum-free culture conditions for establishment of iPS cells, and enhancement of establishment efficiency, as well as elucidation of reprogramming mechanisms (Feng *et al.*, 2009; Hochedlinger and Plath, 2009). For clinical applications, safe reprogramming methods, the choice of target cells, and validation of genuine iPS cells must be carefully optimized. For cell transplantation therapy, safety and effectiveness must be evaluated using *in vivo* animal models. For this purpose, larger animals that are more closely related to humans, such as rabbits, pigs, and monkeys, are useful for preclinical studies (Liu *et al.*, 2008; Esteban *et al.*, 2009). Nevertheless, ES cells are still important for regenerative medicine, since patients with genetic disorders including RP may require transplantation of normal ES cell-derived cells.

At present, clinical applications of RPE and photoreceptor transplantation are at different stages. RPE transplantation may closer to therapeutic use, since transplantation of human ES cell-derived RPE cells has been submitted to the US Food and Drug Administration (FDA) for approval. Autograft and allograft studies using monkey iPS cell-derived cells will greatly help our understanding of the safety of such transplantations, including rejection and tumorigenesis. Thus, the next step in RPE transplantation research will focus on evaluating safety and effectiveness in monkeys and humans, as well as the development of surgical methods.

Transplantation of retinal cells from human ES cells has been reported to restore some visual function in a photoreceptor degeneration model, namely Crx-deficient mice (Lamba *et al.*, 2009), but this study was performed without selection of photoreceptors. Purification of photoreceptors at a specific stage is critical for enhancement of transplantation efficacy (MacLaren *et al.*, 2006; Osakada and Takahashi, 2009). For this purpose, identification of surface antigens marking P3-6 rod photoreceptors is crucial. To integrate transplanted cells into host tissue and to restore retinal function, the normal mechanisms of circuit assembly in the developing retina will likely need to be reactivated. Thus, questions that require further investigation include: How do axons and dendrites make the appropriate connections? How are complex neural circuits assembled? How is visual information encoded and decoded? If we can control these steps artificially, effective retinal regeneration therapy can be achieved.

In addition to donor cell or tissue preparation, manipulation of the host conditions is essential for successful transplantation. Diseased host tissue may exhibit pathological conditions that differ from normal physiological conditions. For instance, inflammation affects the survival, synapse formation, and neural activity of transplanted cells, and except in cases of autotransplantation, immune rejection occurs following cell transplantation. The optimal time windows for photoreceptor transplantation should be determined for every type of retinal degeneration. A better understanding of the pathological host conditions, including immune responses, will be essential for regeneration of retinal function. Optimization of pharmacological drugs and transplantation materials are therefore crucial for successful transplantation.

Acknowledgments

We thank H. Suemori and N. Nakatsuji (Kyoto University, Kyoto, Japan) for the human ES cell line, J. Takahashi (Kyoto University) for the monkey ES cell line, K. Takahashi and S. Yamanaka (Kyoto University) for the human iPS cell line, Y. Sasai (RIKEN, Kobe, Japan) for the SDIA and SFEB methods, and members of the Takahashi laboratory, the Sasai laboratory and the Akaike laboratory for helpful discussions. This work was supported by Grants-in-Aid from the Ministry of Education, Culture, Sports, Science and Technology (F.O. and Y.H.), the Leading Project (M.T.), and the Mochida Memorial Foundation for Medical and Pharmaceutical Research (F.O.).

References

AASEN T, RAYA A, BARRERO MJ, GARRETA E, CONSIGLIO A, GONZALEZ F, VASSENA R, BILIC J, PEKARIK V, TISCORNIA G, EDEL M, BOUE S, BELMONTE JC (2008) Efficient and rapid generation of induced pluripotent stem cells from human keratinocytes. *Nat Biotechnol* **26**:1276-1284.

ACLAND GM, AGUIRRE GD, RAY J, ZHANG Q, ALEMAN TS, CIDECIYAN AV, PEARCE-KELLING SE, ANAND V, ZENG Y, MAGUIRE AM, JACOBSON SG, HAUSWIRTH WW, BENNETT J (2001) Gene therapy restores vision in a canine model of childhood blindness. *Nat Genet* **28**:92-95.

ADAMIS AP, SHIMA DT, YEO KT, YEO TK, BROWN LF, BERSE B, D'AMORE PA, FOLKMAN J (1993) Synthesis and secretion of vascular permeability factor/vascular endothelial growth factor by human retinal pigment epithelial cells. *Biochem Biophys Res Commun* **193**:631-638.

AHMAD I, TANG L, PHAM H (2000) Identification of neural progenitors in the adult mammalian eye. *Biochem Biophys Res Commun* **270**:517-521.

AISENBREY S, BARTZ-SCHMIDT KU, WALTER P, HILGERS RD, AYERTEY H, SZURMAN P, THUMANN G (2007) Long-term follow-up of macular translocation with 360 degrees retinotomy for exudative age-related macular degeneration. *Arch Ophthalmol* **125**:1367-1372.

AKAGI T, HARUTA M, AKITA J, NISHIDA A, HONDA Y, TAKAHASHI M (2003) Different characteristics of rat retinal progenitor cells from different culture periods. *Neurosci Lett* **341**:213-216.

AKAGI T, MANDAI M, OOTO S, HIRAMI Y, OSAKADA F, KAGEYAMA R, YOSHIMURA N,

Takahashi M (2004) Otx2 homeobox gene induces photoreceptor-specific phenotypes in cells derived from adult iris and ciliary tissue. *Invest Ophthalmol Vis Sci* **45**:4570-4575.

Akagi T, Akita J, Haruta M, Suzuki T, Honda Y, Inoue T, Yoshiura S, Kageyama R, Yatsu T, Yamada M, Takahashi M (2005) Iris-derived cells from adult rodents and primates adopt photoreceptor-specific phenotypes. *Invest Ophthalmol Vis Sci* **46**:3411-3419.

Akimoto M, Miyatake S, Kogishi J, Hangai M, Okazaki K, Takahashi JC, Saiki M, Iwaki M, Honda Y (1999) Adenovirally expressed basic fibroblast growth factor rescues photoreceptor cells in RCS rats. *Invest Ophthalmol Vis Sci* **40**:273-279.

Alexander JP, Bradley JM, Gabourel JD, Acott TS (1990) Expression of matrix metalloproteinases and inhibitor by human retinal pigment epithelium. *Invest Ophthalmol Vis Sci* **31**:2520-2528.

Algvere PV, Gouras P, Dafgard Kopp E (1999) Long-term outcome of RPE allografts in non-immunosuppressed patients with AMD. *Eur J Ophthalmol* **9**:217-230.

Algvere PV, Berglin L, Gouras P, Sheng Y (1994) Transplantation of fetal retinal pigment epithelium in age-related macular degeneration with subfoveal neovascularization. *Graefes Arch Clin Exp Ophthalmol* **232**:707-716.

Algvere PV, Berglin L, Gouras P, Sheng Y, Kopp ED (1997) Transplantation of RPE in age-related macular degeneration: observations in disciform lesions and dry RPE atrophy. *Graefes Arch Clin Exp Ophthalmol* **235**:149-158.

Ali RR, Sarra GM, Stephens C, Alwis MD, Bainbridge JW, Munro PM, Fauser S, Reichel MB, Kinnon C, Hunt DM, Bhattacharya SS, Thrasher AJ (2000) Restoration of photoreceptor ultrastructure and function in retinal degeneration slow mice by gene therapy. *Nat Genet* **25**:306-310.

Aoi T, Yae K, Nakagawa M, Ichisaka T, Okita K, Takahashi K, Chiba T, Yamanaka S (2008) Generation of pluripotent stem cells from adult mouse liver and stomach cells. *Science* **321**:699-702.

Asami M, Sun G, Yamaguchi M, Kosaka M (2007) Multipotent cells from mammalian iris pigment epithelium. *Dev Biol* **304**:433-446.

Aubert J, Dunstan H, Chambers I, Smith A (2002) Functional gene screening in embryonic stem cells implicates Wnt antagonism in neural differentiation. *Nat Biotechnol* **20**:1240-1245.

Avery RL, Pieramici DJ, Rabena MD, Castellarin AA, Nasir MA, Giust MJ (2006) Intravitreal bevacizumab (Avastin) for neovascular age-related macular degeneration. *Ophthalmology* **113**:363-372 e365.

Bain G, Kitchens D, Yao M, Huettner JE, Gottlieb DI (1995) Embryonic stem cells express neuronal properties in vitro. *Dev Biol* **168**:342-357.

Bainbridge JW, Smith AJ, Barker SS, Robbie S, Henderson R, Balaggan K, Viswanathan A, Holder GE, Stockman A, Tyler N, Petersen-Jones S, Bhattacharya SS, Thrasher AJ, Fitzke FW, Carter BJ, Rubin GS, Moore AT, Ali RR (2008) Effect of gene therapy on visual function in Leber's congenital amaurosis. *N Engl J Med* **358**:2231-2239.

Barberi T, Klivenyi P, Calingasan NY, Lee H, Kawamata H, Loonam K, Perrier AL, Bruses J, Rubio ME, Topf N, Tabar V, Harrison NL, Beal MF, Moore MA, Studer L (2003) Neural subtype specification of fertilization and nuclear transfer embryonic stem cells and application in parkinsonian mice. *Nat Biotechnol* **21**:1200-1207.

BAUMER N, MARQUARDT T, STOYKOVA A, SPIELER D, TREICHEL D, ASHERY-PADAN R, GRUSS P (2003) Retinal pigmented epithelium determination requires the redundant activities of Pax2 and Pax6. *Development* **130**:2903-2915.

BENNETT J, TANABE T, SUN D, ZENG Y, KJELDBYE H, GOURAS P, MAGUIRE AM (1996) Photoreceptor cell rescue in retinal degeneration (rd) mice by in vivo gene therapy. *Nat Med* **2**:649-654.

BERGER AS, TEZEL TH, DEL PRIORE LV, KAPLAN HJ (2003) Photoreceptor transplantation in retinitis pigmentosa: short-term follow-up. *Ophthalmology* **110**:383-391.

BLACKSHAW S, KUO WP, PARK PJ, TSUJIKAWA M, GUNNERSEN JM, SCOTT HS, BOON WM, TAN SS, CEPKO CL (2003) MicroSAGE is highly representative and reproducible but reveals major differences in gene expression among samples obtained from similar tissues. *Genome Biol* **4**:R17.

BORA N, CONWAY SJ, LIANG H, SMITH SB (1998) Transient overexpression of the Microphthalmia gene in the eyes of Microphthalmia vitiligo mutant mice. *Dev Dyn* **213**:283-292.

BOST LM, AOTAKI-KEEN AE, HJELMELAND LM (1992) Coexpression of FGF-5 and bFGF by the retinal pigment epithelium in vitro. *Exp Eye Res* **55**:727-734.

BRAISTED JE, ESSMAN TF, RAYMOND PA (1994) Selective regeneration of photoreceptors in goldfish retina. *Development* **120**:2409-2419.

BRAMBRINK T, FOREMAN R, WELSTEAD GG, LENGNER CJ, WERNIG M, SUH H, JAENISCH R (2008) Sequential expression of pluripotency markers during direct reprogramming of mouse somatic cells. *Cell Stem Cell* **2**:151-159.

CAO W, TOMBRAN-TINK J, ELIAS R, SEZATE S, MRAZEK D, MCGINNIS JF (2001) In vivo protection of photoreceptors from light damage by pigment epithelium-derived factor. *Invest Ophthalmol Vis Sci* **42**:1646-1652.

CASTELLARIN AA, NASIR M, SUGINO IK, ZARBIN MA (1998) Progressive presumed choriocapillaris atrophy after surgery for age-related macular degeneration. *Retina* **18**:143-149.

CASTILLO BV, JR., DEL CERRO M, WHITE RM, COX C, WYATT J, NADIGA G, DEL CERRO C (1997) Efficacy of nonfetal human RPE for photoreceptor rescue: a study in dystrophic RCS rats. *Exp Neurol* **146**:1-9.

CAYOUETTE M, GRAVEL C (1997) Adenovirus-mediated gene transfer of ciliary neurotrophic factor can prevent photoreceptor degeneration in the retinal degeneration (rd) mouse. *Hum Gene Ther* **8**:423-430.

CHIANG C, LITINGTUNG Y, LEE E, YOUNG KE, CORDEN JL, WESTPHAL H, BEACHY PA (1996) Cyclopia and defective axial patterning in mice lacking Sonic hedgehog gene function. *Nature* **383**:407-413.

CHOJNACKI AK, MAK GK, WEISS S (2009) Identity crisis for adult periventricular neural stem cells: subventricular zone astrocytes, ependymal cells or both? *Nat Rev Neurosci* **10**:153-163.

CHOO AB, TAN HL, ANG SN, FONG WJ, CHIN A, LO J, ZHENG L, HENTZE H, PHILP RJ, OH SK, YAP M (2008) Selection against undifferentiated human embryonic stem cells by a cytotoxic antibody recognizing podocalyxin-like protein-1. *Stem Cells* **26**:1454-1463.

COFFEY PJ, GIRMAN S, WANG SM, HETHERINGTON L, KEEGAN DJ, ADAMSON P, GREENWOOD J, LUND RD (2002) Long-term preservation of cortically dependent visual function in RCS rats by transplantation. *Nat Neurosci* **5**:53-56.

CRABB JW, MIYAGI M, GU X, SHADRACH K, WEST KA, SAKAGUCHI H, KAMEI M, HASAN A, YAN L, RAYBORN ME, SALOMON RG, HOLLYFIELD JG (2002) Drusen proteome analysis: an approach to the etiology of age-related macular degeneration. *Proc Natl Acad Sci U S A* **99**:14682-14687.

D'CRUZ PM, Yasumura D, Weir J, Matthes MT, Abderrahim H, LaVail MM, Vollrath D (2000) Mutation of the receptor tyrosine kinase gene Mertk in the retinal dystrophic RCS rat. *Hum Mol Genet* **9**:645-651.

DACEY DM (2000) Parallel pathways for spectral coding in primate retina. *Annu Rev Neurosci* **23**:743-775.

DAS AV, MALLYA KB, ZHAO X, AHMAD F, BHATTACHARYA S, THORESON WB, HEGDE GV, AHMAD I (2006) Neural stem cell properties of Muller glia in the mammalian retina: regulation by Notch and Wnt signaling. *Dev Biol* **299**:283-302.

DE JUAN E, JR., MACHEMER R (1988) Vitreous surgery for hemorrhagic and fibrous complications of age-related macular degeneration. *Am J Ophthalmol* **105**:25-29.

DELLA NG, CAMPOCHIARO PA, ZACK DJ (1996) Localization of TIMP-3 mRNA expression to the retinal pigment epithelium. *Invest Ophthalmol Vis Sci* **37**:1921-1924.

DIMOS JT, RODOLFA KT, NIAKAN KK, WEISENTHAL LM, MITSUMOTO H, CHUNG W, CROFT GF, SAPHIER G, LEIBEL R, GOLAND R, WICHTERLE H, HENDERSON CE, EGGAN K (2008) Induced pluripotent stem cells generated from patients with ALS can be differentiated into motor neurons. *Science* **321**:1218-1221.

EBERT AD, YU J, ROSE FF, JR., MATTIS VB, LORSON CL, THOMSON JA, SVENDSEN CN (2009) Induced pluripotent stem cells from a spinal muscular atrophy patient. *Nature* **457**:277-280.

EDWARDS AO, RITTER R, 3RD, ABEL KJ, MANNING A, PANHUYSEN C, FARRER LA (2005) Complement factor H polymorphism and age-related macular degeneration. *Science* **308**:421-424.

EIRAKU M, WATANABE K, MATSUO-TAKASAKI M, KAWADA M, YONEMURA S, MATSUMURA M, WATAYA T, NISHIYAMA A, MUGURUMA K, SASAI Y (2008) Self-organized formation of polarized cortical tissues from ESCs and its active manipulation by extrinsic signals. *Cell Stem Cell* **3**:519-532.

ERICSON J, MUHR J, PLACZEK M, LINTS T, JESSELL TM, EDLUND T (1995) Sonic hedgehog induces the differentiation of ventral forebrain neurons: a common signal for ventral patterning within the neural tube. *Cell* **81**:747-756.

ERIKSSON PS, PERFILIEVA E, BJORK-ERIKSSON T, ALBORN AM, NORDBORG C, PETERSON DA, GAGE FH (1998) Neurogenesis in the adult human hippocampus. *Nat Med* **4**:1313-1317.

ESTEBAN MA, XU J, YANG J, PENG M, QIN D, LI W, JIANG Z, CHEN J, DENG K, ZHONG M, CAI J, LAI L, PEI D (2009) Generation of induced pluripotent stem cell lines from tibetan miniature pig. *J Biol Chem.* **284**:17634-17640.

EVANS MJ, KAUFMAN MH (1981) Establishment in culture of pluripotential cells from mouse embryos. *Nature* **292**:154-156.

FAKTOROVICH EG, STEINBERG RH, YASUMURA D, MATTHES MT, LAVAIL MM (1990) Photoreceptor degeneration in inherited retinal dystrophy delayed by basic fibroblast growth factor. *Nature* **347**:83-86.

FALKNER CI, LEITICH H, FROMMLET F, BAUER P, BINDER S (2007) The end of submacular surgery for age-related macular degeneration? A meta-analysis. *Graefes Arch Clin Exp Ophthalmol* **245**:490-501.

Farrar GJ, Kenna PF, Humphries P (2002) On the genetics of retinitis pigmentosa and on mutation-independent approaches to therapeutic intervention. *EMBO J* **21**:857-864.

Feng B, Ng JH, Heng JC, Ng HH (2009) Molecules that promote or enhance reprogramming of somatic cells to induced pluripotent stem cells. *Cell Stem Cell* **4**:301-312.

Feng W, Yasumura D, Matthes MT, LaVail MM, Vollrath D (2002) Mertk triggers uptake of photoreceptor outer segments during phagocytosis by cultured retinal pigment epithelial cells. *J Biol Chem* **277**:17016-17022.

Field GD, Chichilnisky EJ (2007) Information processing in the primate retina: circuitry and coding. *Annu Rev Neurosci* **30**:1-30.

Fischer AJ, Reh TA (2001) Muller glia are a potential source of neural regeneration in the postnatal chicken retina. *Nat Neurosci* **4**:247-252.

Fukuda H, Takahashi J, Watanabe K, Hayashi H, Morizane A, Koyanagi M, Sasai Y, Hashimoto N (2006) Fluorescence-activated cell sorting-based purification of embryonic stem cell-derived neural precursors averts tumor formation after transplantation. *Stem Cells* **24**:763-771.

Furukawa T, Kozak CA, Cepko CL (1997) rax, a novel paired-type homeobox gene, shows expression in the anterior neural fold and developing retina. *Proc Natl Acad Sci USA* **94**:3088-3093.

Furukawa T, Morrow EM, Li T, Davis FC, Cepko CL (1999) Retinopathy and attenuated circadian entrainment in Crx-deficient mice. *Nat Genet* **23**:466-470.

Galceran J, Miyashita-Lin EM, Devaney E, Rubenstein JL, Grosschedl R (2000) Hippocampus development and generation of dentate gyrus granule cells is regulated by LEF1. *Development* **127**:469-482.

Gunhaga L, Marklund M, Sjodal M, Hsieh JC, Jessell TM, Edlund T (2003) Specification of dorsal telencephalic character by sequential Wnt and FGF signaling. *Nat Neurosci* **6**:701-707.

Hageman GS, Anderson DH, Johnson LV, Hancox LS, Taiber AJ, Hardisty LI, Hageman JL, Stockman HA, Borchardt JD, Gehrs KM, Smith RJ, Silvestri G, Russell SR, Klaver CC, Barbazetto I, Chang S, Yannuzzi LA, Barile GR, Merriam JC, Smith RT, Olsh AK, Bergeron J, Zernant J, Merriam JE, Gold B, Dean M, and Allikmets R. (2005) A common haplotype in the complement regulatory gene factor H (HF1/CFH) predisposes individuals to age-related macular degeneration. *Proc Natl Acad Sci U S A* **102**:7227-7232.

Haines JL, Hauser MA, Schmidt S, Scott WK, Olson LM, Gallins P, Spencer KL, Kwan SY, Noureddine M, Gilbert JR, Schnetz-Boutaud N, Agarwal A, Postel EA, Pericak-Vance MA (2005) Complement factor H variant increases the risk of age-related macular degeneration. *Science* **308**:419-421.

Hanna J, Wernig M, Markoulaki S, Sun CW, Meissner A, Cassady JP, Beard C, Brambrink T, Wu LC, Townes TM, Jaenisch R (2007) Treatment of sickle cell anemia mouse model with iPS cells generated from autologous skin. *Science* **318**:1920-1923.

Hanna J, Markoulaki S, Schorderet P, Carey BW, Beard C, Wernig M, Creyghton MP, Steine EJ, Cassady JP, Foreman R, Lengner CJ, Dausman JA, Jaenisch R (2008) Direct reprogramming of terminally differentiated mature B lymphocytes to pluripotency. *Cell* **133**:250-264.

326   F. Osakada *et al.*

Hartong DT, Berson EL, Dryja TP (2006) Retinitis pigmentosa. *Lancet* **368**:1795-1809.

Haruta M, Kosaka M, Kanegae Y, Saito I, Inoue T, Kageyama R, Nishida A, Honda Y, Takahashi M (2001) Induction of photoreceptor-specific phenotypes in adult mammalian iris tissue. *Nat Neurosci* **4**:1163-1164.

Haruta M, Sasai Y, Kawasaki H, Amemiya K, Ooto S, Kitada M, Suemori H, Nakatsuji N, Ide C, Honda Y, Takahashi M (2004) In vitro and in vivo characterization of pigment epithelial cells differentiated from primate embryonic stem cells. *Invest Ophthalmol Vis Sci* **45**:1020-1025.

Heier JS, Antoszyk AN, Pavan PR, Leff SR, Rosenfeld PJ, Ciulla TA, Dreyer RF, Gentile RC, Sy JP, Hantsbarger G, Shams N (2006) Ranibizumab for treatment of neovascular age-related macular degeneration: a phase I/II multicenter, controlled, multidose study. *Ophthalmology* **113**:633 e631-634.

Hirami Y, Osakada F, Takahashi K, Okita K, Yamanaka S, Ikeda H, Yoshimura N, Takahashi M (2009) Generation of retinal cells from mouse and human induced pluripotent stem cells. *Neurosci Lett* **458**:126-131.

Hochedlinger K, Plath K (2009) Epigenetic reprogramming and induced pluripotency. *Development* **136**:509-523.

Hoffman LM, Maguire AM, Bennett J (1997) Cell-mediated immune response and stability of intraocular transgene expression after adenovirus-mediated delivery. *Invest Ophthalmol Vis Sci* **38**:2224-2233.

Humayun MS, de Juan E, Jr., del Cerro M, Dagnelie G, Radner W, Sadda SR, del Cerro C (2000) Human neural retinal transplantation. *Invest Ophthalmol Vis Sci* **41**:3100-3106.

Humayun MS, Weiland JD, Fujii GY, Greenberg R, Williamson R, Little J, Mech B, Cimmarusti V, Van Boemel G, Dagnelie G, de Juan E (2003) Visual perception in a blind subject with a chronic microelectronic retinal prosthesis. *Vision Res* **43**:2573-2581.

Ikeda H, Osakada F, Watanabe K, Mizuseki K, Haraguchi T, Miyoshi H, Kamiya D, Honda Y, Sasai N, Yoshimura N, Takahashi M, Sasai Y (2005) Generation of Rx+/Pax6+ neural retinal precursors from embryonic stem cells. *Proc Natl Acad Sci USA* **102**:11331-11336.

Imai D, Yoneya S, Gehlbach PL, Wei LL, Mori K (2005) Intraocular gene transfer of pigment epithelium-derived factor rescues photoreceptors from light-induced cell death. *J Cell Physiol* **202**:570-578.

Irioka T, Watanabe K, Mizusawa H, Mizuseki K, Sasai Y (2005) Distinct effects of caudalizing factors on regional specification of embryonic stem cell-derived neural precursors. *Brain Research Developmental Brain Research* **154**:63-70.

Jadhav AP, Mason HA, Cepko CL (2006) Notch 1 inhibits photoreceptor production in the developing mammalian retina. *Development* **133**:913-923.

Jessell TM (2000) Neuronal specification in the spinal cord: inductive signals and transcriptional codes. *Nat Rev Genet* **1**:20-29.

Johnson LV, Leitner WP, Staples MK, Anderson DH (2001) Complement activation and inflammatory processes in Drusen formation and age related macular degeneration. *Exp Eye Res* **73**:887-896.

Joussen AM, Heussen FM, Joeres S, Llacer H, Prinz B, Rohrschneider K, Maaijwee KJ, van Meurs J, Kirchhof B (2006) Autologous translocation of the choroid and retinal pigment epithelium in age-related macular degeneration. *Am J Ophthalmol* **142**:17-30.

KARL MO, HAYES S, NELSON BR, TAN K, BUCKINGHAM B, REH TA (2008) Stimulation of neural regeneration in the mouse retina. *Proc Natl Acad Sci U S A* **105**:19508-19513.

KAWASAKI H, MIZUSEKI K, NISHIKAWA S, KANEKO S, KUWANA Y, NAKANISHI S, NISHIKAWA SI, SASAI Y (2000) Induction of midbrain dopaminergic neurons from ES cells by stromal cell-derived inducing activity. *Neuron* **28**:31-40.

KEIRSTEAD HS, NISTOR G, BERNAL G, TOTOIU M, CLOUTIER F, SHARP K, STEWARD O (2005) Human embryonic stem cell-derived oligodendrocyte progenitor cell transplants remyelinate and restore locomotion after spinal cord injury. *J Neurosci* **25**:4694-4705.

KIM D, KIM CH, MOON JI, CHUNG YG, CHANG MY, HAN BS, KO S, YANG E, CHA KY, LANZA R, KIM KS (2009) Generation of Human Induced Pluripotent Stem Cells by Direct Delivery of Reprogramming Proteins. *Cell Stem Cell* **4**:472-476.

KIM JB, ZAEHRES H, WU G, GENTILE L, KO K, SEBASTIANO V, ARAUZO-BRAVO MJ, RUAU D, HAN DW, ZENKE M, SCHOLER HR (2008) Pluripotent stem cells induced from adult neural stem cells by reprogramming with two factors. *Nature* **454**:646-650.

KIM JH, AUERBACH JM, RODRIGUEZ-GOMEZ JA, VELASCO I, GAVIN D, LUMELSKY N, LEE SH, NGUYEN J, SANCHEZ-PERNAUTE R, BANKIEWICZ K, McKAY R (2002) Dopamine neurons derived from embryonic stem cells function in an animal model of Parkinson's disease. *Nature* **418**:50-56.

KINOUCHI R, TAKEDA M, YANG L, WILHELMSSON U, LUNDKVIST A, PEKNY M, CHEN DF (2003) Robust neural integration from retinal transplants in mice deficient in GFAP and vimentin. *Nat Neurosci* **6**:863-868.

KLASSEN HJ, NG TF, KURIMOTO Y, KIROV I, SHATOS M, COFFEY P, YOUNG MJ (2004) Multipotent retinal progenitors express developmental markers, differentiate into retinal neurons, and preserve light-mediated behavior. *Invest Ophthalmol Vis Sci* **45**:4167-4173.

KLEIN RJ, ZEISS C, CHEW EY, TSAI JY, SACKLER RS, HAYNES C, HENNING AK, SANGIOVANNI JP, MANE SM, MAYNE ST, BRACKEN MB, FERRIS FL, OTT J, BARNSTABLE C, HOH J (2005) Complement factor H polymorphism in age-related macular degeneration. *Science* **308**:385-389.

KLIMANSKAYA I, HIPP J, REZAI KA, WEST M, ATALA A, LANZA R (2004) Derivation and comparative assessment of retinal pigment epithelium from human embryonic stem cells using transcriptomics. *Cloning Stem Cells* **6**:217-245.

KNECHT AK, BRONNER-FRASER M (2002) Induction of the neural crest: a multigene process. *Nat Rev Genet* **3**:453-461.

LAMBA DA, GUST J, REH TA (2009) Transplantation of human embryonic stem cell-derived photoreceptors restores some visual function in crx-deficient mice. *Cell Stem Cell* **4**:73-79.

LAVAIL MM, UNOKI K, YASUMURA D, MATTHES MT, YANCOPOULOS GD, STEINBERG RH (1992) Multiple growth factors, cytokines, and neurotrophins rescue photoreceptors from the damaging effects of constant light. *Proc Natl Acad Sci U S A* **89**:11249-11253.

LAVAIL MM, YASUMURA D, MATTHES MT, LAU-VILLACORTA C, UNOKI K, SUNG CH, STEINBERG RH (1998) Protection of mouse photoreceptors by survival factors in

328   F. Osakada *et al.*

retinal degenerations. *Invest Ophthalmol Vis Sci* **39**:592-602.

Lee SH, Lumelsky N, Studer L, Auerbach JM, McKay RD (2000a) Efficient generation of midbrain and hindbrain neurons from mouse embryonic stem cells. *Nat Biotechnol* **18**:675-679.

Lee SM, Tole S, Grove E, McMahon AP (2000b) A local Wnt-3a signal is required for development of the mammalian hippocampus. *Development* **127**:457-467.

Levine EM, Fuhrmann S, Reh TA (2000) Soluble factors and the development of rod photoreceptors. *Cell Mol Life Sci* **57**:224-234.

Li LX, Turner JE (1988) Inherited retinal dystrophy in the RCS rat: prevention of photoreceptor degeneration by pigment epithelial cell transplantation. *Exp Eye Res* **47**:911-917.

Li ZY, Kljavin IJ, Milam AH (1995) Rod photoreceptor neurite sprouting in retinitis pigmentosa. *J Neurosci* **15**:5429-5438.

Little CW, Cox C, Wyatt J, del Cerro C, del Cerro M (1998) Correlates of photoreceptor rescue by transplantation of human fetal RPE in the RCS rat. *Exp Neurol* **149**:151-160.

Little CW, Castillo B, DiLoreto DA, Cox C, Wyatt J, del Cerro C, del Cerro M (1996) Transplantation of human fetal retinal pigment epithelium rescues photoreceptor cells from degeneration in the Royal College of Surgeons rat retina. *Invest Ophthalmol Vis Sci* **37**:204-211.

Liu H, Zhu F, Yong J, Zhang P, Hou P, Li H, Jiang W, Cai J, Liu M, Cui K, Qu X, Xiang T, Lu D, Chi X, Gao G, Ji W, Ding M, Deng H (2008) Generation of induced pluripotent stem cells from adult rhesus monkey fibroblasts. *Cell Stem Cell* **3**:587-590.

Lledo PM, Alonso M, Grubb MS (2006) Adult neurogenesis and functional plasticity in neuronal circuits. *Nat Rev Neurosci* **7**:179-193.

Lledo PM, Merkle FT, Alvarez-Buylla A (2008) Origin and function of olfactory bulb interneuron diversity. *Trends Neurosci* **31**:392-400.

Loh YH, Agarwal S, Park IH, Urbach A, Huo H, Heffner GC, Kim K, Miller JD, Ng K, Daley GQ (2009) Generation of induced pluripotent stem cells from human blood. *Blood* **113**:5476-5479.

Lopez R, Gouras P, Kjeldbye H, Sullivan B, Reppucci V, Brittis M, Wapner F, Goluboff E (1989) Transplanted retinal pigment epithelium modifies the retinal degeneration in the RCS rat. *Invest Ophthalmol Vis Sci* **30**:586-588.

Lund RD, Wang S, Klimanskaya I, Holmes T, Ramos-Kelsey R, Lu B, Girman S, Bischoff N, Sauve Y, Lanza R (2006) Human embryonic stem cell-derived cells rescue visual function in dystrophic RCS rats. *Cloning Stem Cells* **8**:189-199.

Machemer R, Steinhorst UH (1993) Retinal separation, retinotomy, and macular relocation: II. A surgical approach for age-related macular degeneration? *Graefes Arch Clin Exp Ophthalmol* **231**:635-641.

MacLaren RE, Pearson RA, MacNeil A, Douglas RH, Salt TE, Akimoto M, Swaroop A, Sowden JC, Ali RR (2006) Retinal repair by transplantation of photoreceptor precursors. *Nature* **444**:203-207.

MacLaren RE, Uppal GS, Balaggan KS, Tufail A, Munro PM, Milliken AB, Ali RR, Rubin GS, Aylward GW, da Cruz L (2007) Autologous transplantation of the retinal pigment epithelium and choroid in the treatment of neovascular age-related macular degeneration. *Ophthalmology* **114**:561-570.

Maguire AM, Simonelli F, Pierce EA, Pugh EN Jr, Mingozzi F, Bennicelli J, Banfi S,

MARSHALL KA, TESTA F, SURACE EM, ROSSI S, LYUBARSKY A, ARRUDA VR, KONKLE B, STONE E, SUN J, JACOBS J, DELL'OSSO L, HERTLE R, MA JX, REDMOND TM, ZHU X, HAUCK B, ZELENAIA O, SHINDLER KS, MAGUIRE MG, WRIGHT JF, VOLPE NJ, McDONNELL JW, AURICCHIO A, HIGH KA, BENNETT J (2008) Safety and efficacy of gene transfer for Leber's congenital amaurosis. *N Engl J Med* **358**:2240-2248.

MAHERALI N, SRIDHARAN R, XIE W, UTIKAL J, EMINLI S, ARNOLD K, STADTFELD M, YACHECHKO R, TCHIEU J, JAENISCH R, PLATH K, HOCHEDLINGER K (2007) Directly reprogrammed fibroblasts show global epigenetic remodeling and widespread tissue contribution. *Cell Stem Cell* **1**:55-70.

MALECAZE F, MATHIS A, ARNE JL, RAULAIS D, COURTOIS Y, HICKS D (1991) Localization of acidic fibroblast growth factor in proliferative vitreoretinopathy membranes. *Curr Eye Res* **10**:719-729.

MALI P, YE Z, HOMMOND HH, YU X, LIN J, CHEN G, ZOU J, CHENG L (2008) Improved efficiency and pace of generating induced pluripotent stem cells from human adult and fetal fibroblasts. *Stem Cells* **26**:1998-2005.

MARTIN MJ, MUOTRI A, GAGE F, VARKI A (2005) Human embryonic stem cells express an immunogenic nonhuman sialic acid. *Nat Med* **11**:228-232.

MASLAND RH (2001) The fundamental plan of the retina. *Nat Neurosci* **4**:877-886.

MATHERS PH, GRINBERG A, MAHON KA, JAMRICH M (1997) The Rx homeobox gene is essential for vertebrate eye development. *Nature* **387**:603-607.

McDONALD JW, LIU XZ, QU Y, LIU S, MICKEY SK, TURETSKY D, GOTTLIEB DI, CHOI DW (1999) Transplanted embryonic stem cells survive, differentiate and promote recovery in injured rat spinal cord. *Nat Med* **5**:1410-1412.

McGILL TJ, LUND RD, DOUGLAS RM, WANG S, LU B, SILVER BD, SECRETAN MR, ARTHUR JN, PRUSKY GT (2007) Syngeneic Schwann cell transplantation preserves vision in RCS rat without immunosuppression. *Invest Ophthalmol Vis Sci* **48**:1906-1912.

MILAM AH, LI ZY, CIDECIYAN AV, JACOBSON SG (1996) Clinicopathologic effects of the Q64ter rhodopsin mutation in retinitis pigmentosa. *Invest Ophthalmol Vis Sci* **37**:753-765.

MIYAZAKI M, IKEDA Y, YONEMITSU Y, GOTO Y, SAKAMOTO T, TABATA T, UEDA Y, HASEGAWA M, TOBIMATSU S, ISHIBASHI T, SUEISHI K (2003) Simian lentiviral vector-mediated retinal gene transfer of pigment epithelium-derived factor protects retinal degeneration and electrical defect in Royal College of Surgeons rats. *Gene Ther* **10**:1503-1511.

MIZUSEKI K, SAKAMOTO T, WATANABE K, MUGURUMA K, IKEYA M, NISHIYAMA A, ARAKAWA A, SUEMORI H, NAKATSUJI N, KAWASAKI H, MURAKAMI F, SASAI Y (2003) Generation of neural crest-derived peripheral neurons and floor plate cells from mouse and primate embryonic stem cells. *Proc Natl Acad Sci U S A* **100**:5828-5833.

MULLINS RF, RUSSELL SR, ANDERSON DH, HAGEMAN GS (2000) Drusen associated with aging and age-related macular degeneration contain proteins common to extracellular deposits associated with atherosclerosis, elastosis, amyloidosis, and dense deposit disease. *FASEB J* **14**:835-846.

NAKAGAWA M, KOYANAGI M, TANABE K, TAKAHASHI K, ICHISAKA T, AOI T, OKITA K, MOCHIDUKI Y, TAKIZAWA N, YAMANAKA S (2008) Generation of induced pluripotent stem cells without Myc from mouse and human fibroblasts. *Nat Biotechnol* **26**:101-106.

NASIR MA, SUGINO I, ZARBIN MA (1997) Decreased choriocapillaris perfusion following surgical excision of choroidal neovascular membranes in age-related macular degeneration. *Br J Ophthalmol* **81**:481-489.

Nassi JJ, Callaway EM (2009) Parallel processing strategies of the primate visual system. *Nat Rev Neurosci* **10**:360-372.

Nguyen M, Arnheiter H (2000) Signaling and transcriptional regulation in early mammalian eye development: a link between FGF and MITF. *Development* **127**:3581-3591.

Nordstrom U, Jessell TM, Edlund T (2002) Progressive induction of caudal neural character by graded Wnt signaling. *Nat Neurosci* **5**:525-532.

Nozaki M, Raisler BJ, Sakurai E, Sarma JV, Barnum SR, Lambris JD, Chen Y, Zhang K, Ambati BK, Baffi JZ, Ambati J (2006) Drusen complement components C3a and C5a promote choroidal neovascularization. *Proc Natl Acad Sci U S A* **103**:2328-2333.

Okita K, Ichisaka T, Yamanaka S (2007) Generation of germline-competent induced pluripotent stem cells. *Nature* **448**:313-317.

Okita K, Nakagawa M, Hyenjong H, Ichisaka T, Yamanaka S (2008) Generation of mouse induced pluripotent stem cells without viral vectors. *Science* **322**:949-953.

Ooto S, Akagi T, Kageyama R, Akita J, Mandai M, Honda Y, Takahashi M (2004) Potential for neural regeneration after neurotoxic injury in the adult mammalian retina. *Proc Natl Acad Sci U S A* **101**:13654-13659.

Osakada F, Takahashi M (2006) Retinal regeneration by somatic stem cells. *Experimental Medicine* **24**:256-262.

Osakada F, Takahashi M (2007) Neurogenic potential of Müller glia in the adult mammalian retina. *Inflamm Regen* **27**:499-505.

Osakada F, Takahashi M (2009) Drug development targeting the glycogen synthase kinase-3beta (GSK-3beta)-mediated signal transduction pathway: targeting the Wnt pathway and transplantation therapy as strategies for retinal repair. *J Pharmacol Sci* **109**:168-173.

Osakada F, Ikeda H, Sasai Y, Takahashi M (2009) Stepwise differentiation of pluripotent stem cells into retinal cells. *Nat Protoc* **4**:811-824.

Osakada F, Ooto S, Akagi T, Mandai M, Akaike A, Takahashi M (2007) Wnt signaling promotes regeneration in the retina of adult mammals. *J Neurosci* **27**:4210-4219.

Osakada F, Jin ZB, Hirami Y, Ikeda H, Danjyo T, Watanabe K, Sasai Y, Takahashi M (2009) *In vitro* differentiation of retinal cells from human pluripotent stem cells by small-molecule induction. *J Cell Sci* **122**:3169-3179.

Osakada F, Ikeda H, Mandai M, Wataya T, Watanabe K, Yoshimura N, Akaike A, Sasai Y, Takahashi M (2008) Toward the generation of rod and cone photoreceptors from mouse, monkey and human embryonic stem cells. *Nat Biotechnol* **26**:215-224.

Park IH, Arora N, Huo H, Maherali N, Ahfeldt T, Shimamura A, Lensch MW, Cowan C, Hochedlinger K, Daley GQ (2008) Disease-specific induced pluripotent stem cells. *Cell* **134**:877-886.

Peyman GA, Blinder KJ, Paris CL, Alturki W, Nelson NC, Jr., Desai U (1991) A technique for retinal pigment epithelium transplantation for age-related macular degeneration secondary to extensive subfoveal scarring. *Ophthalmic Surg* **22**:102-108.

Prusky GT, West PW, Douglas RM (2000) Behavioral assessment of visual acuity in mice and rats. *Vision Res* **40**:2201-2209.

Prusky GT, Alam NM, Beekman S, Douglas RM (2004) Rapid quantification of adult and developing mouse spatial vision using a virtual optomotor system. *Invest*

Ophthalmol Vis Sci **45**:4611-4616.

RADTKE ND, SEILER MJ, ARAMANT RB, PETRY HM, PIDWELL DJ (2002) Transplantation of intact sheets of fetal neural retina with its retinal pigment epithelium in retinitis pigmentosa patients. *Am J Ophthalmol* **133**:544-550.

RADTKE ND, ARAMANT RB, SEILER MJ, PETRY HM, PIDWELL D (2004) Vision change after sheet transplant of fetal retina with retinal pigment epithelium to a patient with retinitis pigmentosa. *Arch Ophthalmol* **122**:1159-1165.

RALLU M, MACHOLD R, GAIANO N, CORBIN JG, MCMAHON AP, FISHELL G (2002) Dorsoventral patterning is established in the telencephalon of mutants lacking both Gli3 and Hedgehog signaling. *Development* **129**:4963-4974.

RATTNER A, NATHANS J (2006) Macular degeneration: recent advances and therapeutic opportunities. *Nat Rev Neurosci* **7**:860-872.

REYNOLDS BA, WEISS S (1992) Generation of neurons and astrocytes from isolated cells of the adult mammalian central nervous system. *Science* **255**:1707-1710.

SANAI N, TRAMONTIN AD, QUINONES-HINOJOSA A, BARBARO NM, GUPTA N, KUNWAR S, LAWTON MT, MCDERMOTT MW, PARSA AT, MANUEL-GARCIA VERDUGO J, BERGER MS, ALVAREZ-BUYLLA A (2004) Unique astrocyte ribbon in adult human brain contains neural stem cells but lacks chain migration. *Nature* **427**:740-744.

SAUVE Y, KLASSEN H, WHITELEY SJ, LUND RD (1998) Visual field loss in RCS rats and the effect of RPE cell transplantation. *Exp Neurol* **152:**243-250.

SCHWEIGERER L, MALERSTEIN B, NEUFELD G, GOSPODAROWICZ D (1987) Basic fibroblast growth factor is synthesized in cultured retinal pigment epithelial cells. *Biochem Biophys Res Commun* **143**:934-940.

SEILER M, ARAMANT RB, EHINGER B, ADOLPH AR (1990) Transplantation of embryonic retina to adult retina in rabbits. *Exp Eye Res* **51**:225-228.

SIEVING PA, CARUSO RC, TAO W, COLEMAN HR, THOMPSON DJ, FULLMER KR, BUSH RA (2006) Ciliary neurotrophic factor (CNTF) for human retinal degeneration: phase I trial of CNTF delivered by encapsulated cell intraocular implants. *Proc Natl Acad Sci U S A* **103**:3896-3901.

STADTFELD M, BRENNAND K, HOCHEDLINGER K (2008a) Reprogramming of pancreatic beta cells into induced pluripotent stem cells. *Curr Biol* **18**:890-894.

STADTFELD M, NAGAYA M, UTIKAL J, WEIR G, HOCHEDLINGER K (2008b) Induced pluripotent stem cells generated without viral integration. *Science* **322**:945-949.

SU HL, MUGURUMA K, MATSUO-TAKASAKI M, KENGAKU M, WATANABE K, SASAI Y (2006) Generation of cerebellar neuron precursors from embryonic stem cells. *Developmental Biology* **290**:287-296.

SUZUKI T, MANDAI M, AKIMOTO M, YOSHIMURA N, TAKAHASHI M (2006) The simultaneous treatment of MMP-2 stimulants in retinal transplantation enhances grafted cell migration into the host retina. *Stem Cells* **24**:2406-2411.

SUZUKI T, AKIMOTO M, IMAI H, UEDA Y, MANDAI M, YOSHIMURA N, SWAROOP A, TAKAHASHI M (2007) Chondroitinase ABC treatment enhances synaptogenesis between transplant and host neurons in model of retinal degeneration. *Cell Transplant* **16**:493-503.

TAKAGI Y, TAKAHASHI J, SAIKI H, MORIZANE A, HAYASHI T, KISHI Y, FUKUDA H, OKAMOTO Y, KOYANAGI M, IDEGUCHI M, HAYASHI H, IMAZATO T, KAWASAKI H, SUEMORI H, OMACHI S, IIDA H, ITOH N, NAKATSUJI N, SASAI Y, HASHIMOTO N (2005) Dopaminergic neurons generated from monkey embryonic stem cells function in a Parkinson primate model. *J Clin Invest* **115**:102-109.

Takahashi K, Yamanaka S (2006) Induction of pluripotent stem cells from mouse embryonic and adult fibroblast cultures by defined factors. *Cell* **126**:663-676.

Takahashi K, Tanabe K, Ohnuki M, Narita M, Ichisaka T, Tomoda K, Yamanaka S (2007) Induction of pluripotent stem cells from adult human fibroblasts by defined factors. *Cell* **131**:861-872.

Takahashi M, Palmer TD, Takahashi J, Gage FH (1998) Widespread integration and survival of adult-derived neural progenitor cells in the developing optic retina. *Mol Cell Neurosci* **12**:340-348.

Takahashi M, Miyoshi H, Verma IM, Gage FH (1999) Rescue from photoreceptor degeneration in the rd mouse by human immunodeficiency virus vector-mediated gene transfer. *J Virol* **73**:7812-7816.

Tanihara H, Yoshida M, Matsumoto M, Yoshimura N (1993) Identification of transforming growth factor-beta expressed in cultured human retinal pigment epithelial cells. *Invest Ophthalmol Vis Sci* **34**:413-419.

Tezel TH, Del Priore LV, Berger AS, Kaplan HJ (2007) Adult retinal pigment epithelial transplantation in exudative age-related macular degeneration. *Am J Ophthalmol* **143**:584-595.

Thomas BB, Seiler MJ, Sadda SR, Aramant RB (2004) Superior colliculus responses to light - preserved by transplantation in a slow degeneration rat model. *Exp Eye Res* **79**:29-39.

Thomson JA, Itskovitz-Eldor J, Shapiro SS, Waknitz MA, Swiergiel JJ, Marshall VS, Jones JM (1998) Embryonic stem cell lines derived from human blastocysts. *Science* **282**:1145-1147.

Toda K, Bush RA, Humphries P, Sieving PA (1999) The electroretinogram of the rhodopsin knockout mouse. *Vis Neurosci* **16**:391-398.

Tropepe V, Hitoshi S, Sirard C, Mak TW, Rossant J, van der Kooy D (2001) Direct neural fate specification from embryonic stem cells: a primitive mammalian neural stem cell stage acquired through a default mechanism. *Neuron* **30**:65-78.

Tropepe V, Coles BL, Chiasson BJ, Horsford DJ, Elia AJ, McInnes RR, van der Kooy D (2000) Retinal stem cells in the adult mammalian eye. *Science* **287**:2032-2036.

Ueno M, Matsumura M, Watanabe K, Nakamura T, Osakada F, Takahashi M, Kawasaki H, Kinoshita S, Sasai Y (2006) Neural conversion of ES cells by an inductive activity on human amniotic membrane matrix. *Proc Natl Acad Sci USA* **103**:9554-9559.

van Meurs JC, Van Den Biesen PR (2003) Autologous retinal pigment epithelium and choroid translocation in patients with exudative age-related macular degeneration: short-term follow-up. *Am J Ophthalmol* **136**:688-695.

Waldbillig RJ, Schoen TJ, Chader GJ, Pfeffer BA (1992) Monkey retinal pigment epithelial cells in vitro synthesize, secrete, and degrade insulin-like growth factor binding proteins. *J Cell Physiol* **150**:76-83.

Waldbillig RJ, Pfeffer BA, Schoen TJ, Adler AA, Shen-Orr Z, Scavo L, LeRoith D, Chader GJ (1991) Evidence for an insulin-like growth factor autocrine-paracrine system in the retinal photoreceptor-pigment epithelial cell complex. *J Neurochem* **57**:1522-1533.

Wan J, Zheng H, Xiao HL, She ZJ, Zhou GM (2007) Sonic hedgehog promotes stem-cell potential of Muller glia in the mammalian retina. *Biochem Biophys Res Commun* **363**:347-354.

WAN J, ZHENG H, CHEN ZL, XIAO HL, SHEN ZJ, ZHOU GM (2008) Preferential regeneration of photoreceptor from Muller glia after retinal degeneration in adult rat. *Vision Res* **48**:223-234.

WASSLE H (2004) Parallel processing in the mammalian retina. *Nat Rev Neurosci* **5**:747-757.

WATANABE K, KAMIYA D, NISHIYAMA A, KATAYAMA T, NOZAKI S, KAWASAKI H, WATANABE Y, MIZUSEKI K, SASAI Y (2005) Directed differentiation of telencephalic precursors from embryonic stem cells. *Nat Neurosci* **8**:288-296.

WEN R, SONG Y, CHENG T, MATTHES MT, YASUMURA D, LAVAIL MM, STEINBERG RH (1995) Injury-induced upregulation of bFGF and CNTF mRNAS in the rat retina. *J Neurosci* **15**:7377-7385.

WENKEL H, STREILEIN JW (1998) Analysis of immune deviation elicited by antigens injected into the subretinal space. *Invest Ophthalmol Vis Sci* **39**:1823-1834.

WENKEL H, STREILEIN JW (2000) Evidence that retinal pigment epithelium functions as an immune-privileged tissue. *Invest Ophthalmol Vis Sci* **41**:3467-3473.

WERNIG M, MEISSNER A, FOREMAN R, BRAMBRINK T, KU M, HOCHEDLINGER K, BERNSTEIN BE, JAENISCH R (2007) In vitro reprogramming of fibroblasts into a pluripotent ES-cell-like state. *Nature* **448**:318-324.

WEST EL, PEARSON RA, TSCHERNUTTER M, SOWDEN JC, MACLAREN RE, ALI RR (2008) Pharmacological disruption of the outer limiting membrane leads to increased retinal integration of transplanted photoreceptor precursors. *Exp Eye Res* **86**:601-611.

WHITELEY SJ, LITCHFIELD TM, COFFEY PJ, LUND RD (1996) Improvement of the pupillary light reflex of Royal College of Surgeons rats following RPE cell grafts. *Exp Neurol* **140**:100-104.

WICHTERLE H, LIEBERAM I, PORTER JA, JESSELL TM (2002) Directed differentiation of embryonic stem cells into motor neurons. *Cell* **110**:385-397.

WOCH G, ARAMANT RB, SEILER MJ, SAGDULLAEV BT, MCCALL MA (2001) Retinal transplants restore visually evoked responses in rats with photoreceptor degeneration. *Invest Ophthalmol Vis Sci* **42**:1669-1676.

YAMAMOTO S, DU J, GOURAS P, KJELDBYE H (1993) Retinal pigment epithelial transplants and retinal function in RCS rats. *Invest Ophthalmol Vis Sci* **34**:3068-3075.

YE L, CHANG JC, LIN C, SUN X, YU J, KAN YW (2009) Induced pluripotent stem cells offer new approach to therapy in thalassemia and sickle cell anemia and option in prenatal diagnosis in genetic diseases. *Proc Natl Acad Sci U S A* **106**:9826-9830.

YING QL, STAVRIDIS M, GRIFFITHS D, LI M, SMITH A (2003) Conversion of embryonic stem cells into neuroectodermal precursors in adherent monoculture. *Nat Biotechnol* **21**:183-186.

YU J, VODYANIK MA, SMUGA-OTTO K, ANTOSIEWICZ-BOURGET J, FRANE JL, TIAN S, NIE J, JONSDOTTIR GA, RUOTTI V, STEWART R, SLUKVIN, II, THOMSON JA (2007) Induced pluripotent stem cell lines derived from human somatic cells. *Science* **318**:1917-1920.

ZAMIRI P, MASLI S, STREILEIN JW, TAYLOR AW (2006) Pigment epithelial growth factor suppresses inflammation by modulating macrophage activation. *Invest Ophthalmol Vis Sci* **47**:3912-3918.

ZAMIRI P, MASLI S, KITAICHI N, TAYLOR AW, STREILEIN JW (2005) *Thrombospondin* plays a vital role in the immune privilege of the eye. *Invest Ophthalmol Vis Sci* **46**:908-919.

Zhang X, Bok D (1998) Transplantation of retinal pigment epithelial cells and immune response in the subretinal space. *Invest Ophthalmol Vis Sci* **39**:1021-1027.

Zhang Y, Caffe AR, Azadi S, van Veen T, Ehinger B, Perez MT (2003) Neuronal integration in an abutting-retinas culture system. *Invest Ophthalmol Vis Sci* **44**:4936-4946.

Zhao C, Deng W, Gage FH (2008) Mechanisms and functional implications of adult neurogenesis. *Cell* **132**:645-660.

Zhou H, Wu S, Joo JY, Zhu S, Han DW, Lin T, Trauger S, Bien G, Yao S, Zhu Y, Siuzdak G, Scholer HR, Duan L, Ding S (2009) Generation of induced pluripotent stem cells using recombinant proteins. *Cell Stem Cell* **4**:381-384.

Biotechnology and Genetic Engineering Reviews - Vol. 26, 335-352 (2009)

Functional Metagenomics: Recent Advances and Future Challenges

LUDMILA CHISTOSERDOVA*

Department of Chemical Engineering, University of Washington, Seattle WA 98195, USA

Abstract

Metagenomics is a relatively new but fast growing field within environmental biology directed at obtaining knowledge on genomes of environmental microbes as well as of entire microbial communities. With the sequencing technologies improving steadily, generating large amounts of sequence is becoming routine. However, it remains difficult to connect specific microbial phyla to specific functions in the environment. A number of 'functional metagenomics' approaches have been implemented in the recent years that allow high-resolution genomic analysis of uncultivated microbes, connecting them to specific functions in the environment. These include analysis of niche-specialized low complexity communities, reactor enrichments, and the use labeling technologies. Metatranscriptomics and metaproteomics are the newest sub-disciplines within the metagenomics field that provide further levels of resolution for functional analysis of uncultivated microbes and communities. The recent emergence of new (next generation) sequencing technologies, resulting in higher sequence output and dramatic drop in the price of sequencing, will be defining a new era in metagenomics. At this time the sequencing effort will be taken to a new level to allow addressing new, previously unattainable biological questions as well as accelerating genome-based discovery for medical and biotechnological applications.

*To whom correspondence may be addressed (milachis@u.washington.edu)

Abbreviations: WGS, Whole-Genome Shotgun; bp, base pairs; Mb, 1,000,000 base pairs; Gb, 1,000,000,000 base pairs.

Introduction

Metagenomics is a fast growing and diverse field within environmental biology directed at obtaining knowledge on genomes of environmental microbes, without prior
cultivation, as well as of entire microbial communities. Other terms are also used to
describe this methodology: environmental genomics, ecogenimics, community genomics, megagenomis. For the purposes of this review, all these terms are interchangeable.
The term 'functional metagenomics', in a broad sense, is meant to reflect a connection
between the identity of a microbe, or a community, uncovered via metagenomics
and their respective function(s) in the environment. The power of metagenomics is
in allowing one to tap into the vast metabolic potential of uncultivated microbes that
represent the majority of microbes on Earth, including entirely novel microbes and
novel metabolic pathways. The two main and principally distinct outcomes of the
metagenomic approach are the emerging new outlook at the complexity of microbial
communities, in terms of both species diversity and community dynamics, and identification of genetic determinants for production of biologically active molecules and
processes that carry a potential for medical and biotechnologcial applications. Recent
advances in next-generation (ultra-high throughput) sequencing technologies, resulting
in a dramatic drop in the price of DNA sequencing, bring about the promise of a new
era in metagenomics, when the sequencing effort will be taken to a new level, to allow
addressing new, previously unattainable biological questions, as well as accelerating
genome-based discovery for biotechnological applications.

Despite its short history, metagenomics has recently become a mainstream approach
in every field under the umbrella of biological sciences, and the term itself became
a household name. As a reflection of the community interest, the already profound
impact, and the future potential of metagenomics, the field is a subject of intense
discussion, resulting in a steady stream of reviews covering every topic in the field,
from strategies and methodologies of metagenomics (Handelsman 2004; Schloss
and Handelsman, 2005; Snyder et al., 2009), to industrial and medical applications
(Warnecke and Hess, 2009; Li et al., 2009; Preidis and Versalovic, 2009), to the bioinformatics issues (Markovitz et al., 2008; Kunin et al., 2008; Lapidus 2009). By no
means does this review intend to comprehensively cover all the work conducted in
the past that involved metagenomic or functional metagenomic approaches. Instead, it
will pursue two major goals: highlighting the recent advances specifically addressing a
connection between the function and phylogeny in the environment, via high resolution
metagenomics, metatranscriptomics and metaproteomics, and offering an outlook at
the future of metagenomics that will rely on the newest sequencing technologies as
well as on considerably evolved data management infrastructures.

Brief history of metagenomics

As genomics, starting in mid-nineties, has revolutionized the entire range of biological
sciences, so did metagenomics a decade or so later. However, the history of metagenomics should probably be traced back to the work of Staley and Konopka (1985), first
reporting on 'great plate count anomaly', and the works of the Woese group, identifying
the 16S rRNA gene as a marker molecule for assessing microbial diversity (Woese

1987). The practical use of 16S rRNA analysis as a tool for phylogenetic profiling of microbial communities has been pioneered by the Pace group in early nineties (Schmidt et al., 1991), constituting the onset of metagenomics as a sub-field of microbial ecology, albeit without an official name. The term 'metagenomics' was coined almost a decade later by the Handelsman group (Handelsman et al., 1998) and was instantly embraced by the scientific community. In this latter case, the term referred to the functional analysis of mixed environmental DNA captured as large size inserts after a screen for a specific activity, thus also providing one meaning to 'functional metagenomics'. The two seminal works that defined the most widely accepted meaning of metagenomics, the random whole-genome shotgun (WGS) sequencing-based analysis of microbial populations, were published in quick succession in 2004. One describes analysis of an artificially simple community, of a biofilm growing on the surface of an acid mine drainage (Tyson et al., 2004), and the other describes a much more complex community of the Sargasso Sea (Venter at al., 2004). The significance of these two early studies is two-fold. On the one hand, they ultimately defined the path for future metagenomic projects. On the other, they provided important insights into the scale of the sequencing effort that would be required for analyzing communal DNA using this method, spanning a range of scenarios from very simple to very complex. Accordingly, the outcomes of these projects regarding the knowledge on specific members of the communities interrogated were dramatically different. The former (with only 76 Mb sequencing effort) resulted in assembly and analysis of almost complete genomes of the dominant species, including accurate metabolic reconstruction and detection of strain-specific genomic variants. The latter, with a much larger sequencing effort (almost 2 Gb) resulted in very fragmented assemblies even for the most abundant species, with most of the dataset being represented by singleton sequencing reads. The stage has been set for a flood of WGS-sequencing-based projects to follow. At the moment of writing, 167 projects are listed in the GOLD database (Liolios et al., 2008), and results from 57 of these have been already published.

Over the same time, breakthroughs in developing alternative sequencing technologies occurred, promising a significantly higher throughput and significantly reduced cost of sequencing, and these new (known as next-generation) sequencing technologies have been immediately tested in metagenomic applications (Edwards et al., 2006). These new sequencing technologies have also enabled a new subfield of metagenomics, termed metatranscriptomics, (i.e. shotgun characterization of environmental transcripts) that is developing incredibly quickly. Another subfield, metaproteomics (i.e. analysis of community protein pools) is emerging, signaling the arrival of a whole new era of metagenomics.

Current state of metagenomics (and how it can be improved)

It has been just over five years since the onset of WGS-sequencing-based metageomics. How has the field progressed towards its maturity and how has it changed since the seminal works published in 2004? About a decade ago, while discussing the fate of microbial genomics, Dr. Woese has lamented the fact that microbiologists have never developed an appropriate, overarching concept for the field, the need for which he thought critical (Woese 1998). The same can be said today about metagenomics.

For the past five years, the field has mostly operated in a 'Wild West' fashion, with no concerted effort, little broad-scale coordination, and in the absence of established 'gold standards'. Individual scientists have been spearheading individual metagenomic projects, mostly on a small scale, as dictated by the economics of sequencing, often times without prior knowledge on the structure or the complexity of the community in question. However, the complexity of natural communities as a challenge to metagenomics has been recognized from the very start. Venter and colleagues (2004) have modeled a sequence coverage level that would be required to identify most of the genomes in the Sargasso Sea sample, concluding that at least an order of magnitude larger sequencing effort was necessary. Another early metagenomic project that interrogated an even more complex community, the one inhabiting soil, resulted in a similar conclusion, admitting significant under-sampling (Tringe et al., 2005). However, the effort toward significantly deeper sampling was deemed not feasible and thus not necessary. As a result, most of the metagenomic projects carried out so far have been following the path of under-sampling, the disregard for community complexity validated by the broad use of a method called gene-centric analysis (Tringe et al., 2005). This method (a poor man's approach to metagenomics) treats a community (mostly represented by singleton reads, frequently of poor quality) as an aggregate, ignoring the context of individual species. Each read is automatically assigned to a functional category, and this way functional profiles of communities can be created. Communities then can be compared to each other in terms of functional profiles (Tringe et al., 2005). This approach performs rather well with singleton sequencing reads generated by the Sanger technology, with approximately 90% of genes being found to encode at least one and sometimes two putative polypeptides. However, the resolution of this approach drops further when it comes to functional gene annotation. This task relies on the content of the current gene and protein databases, which are heavily biased toward model organisms that do not fully represent the diversity of the organisms in the environment (Woese 1998). The problem of annotation of environmental genes persists beyond the lack of close homologs for the genes represented in metagenomic databases. In databases most frequently used to aid in annotation of metagenomic sequences (such as the non-redundant NCBI database, the SEED database), many of the specialized biochemical pathways are poorly annotated. Thus, even if close homologs are present, their most likely functions may be called incorrectly. Even if the functions of genes can be predicted with precision based on a homolog match, placing them into the context of specific metabolic pathways is not always possible with the gene-centric approach and out of the context of a metabiolic make up of an individual organism. For example, the citric acid cycle (whose main role is in energy generation) and the methylcitric acid cycle (whose main role is in propionate utilization) share a number of genes and enzymes in common. The precise differentiation between the two pathways can only be achieved having the knowledge on an entire or almost entire gene complement, and in the context of an individual genome.

The problems described above are even more severe when it comes to the analysis and annotation of the shorter reads currently produced by the 454 sequencing technology (Wommack et al., 2008). In the works published so far, mostly based on the earlier versions of the 454 technology that produced 100 to 200 bp reads, up to 70 per cent of reads could not be classified (Edwards et al., 2006; Dinsdale et al., 2008; Brulc et al., 2009; Thurber et al., 2009). Ironically, despite the hype about the significantly higher throughput and lower cost of 454 sequencing (compared to the Sanger-based sequencing), the metagenomic projects employing this technology tend to produce datasets of relatively small size, deeming the functional profiling and comparative genomics tasks almost useless exercises when it comes to communities of high complexity.

How can this situation be improved? One obvious way is to commeasure the sequencing effort with the complexity of the community. Predictions can be easily made for how much raw sequence is required to obtain good coverage for dominant species (Kunin et al., 2008). If the goal of the study is to obtain insights into the genomes of the minor members of the community, creative approaches such as specific enrichment strategies can be applied. Such approaches, guided by a specific goal, should enable functional insights into the community as a whole or into specific members of the community, resulting in a more complete and meaningful interpretation of the sequence data. This in turn will enable a high-resolution biological knowledge that constitutes functional genomics in a broader sense.

Approaches to functional metagenomics

One of the simple questions typically asked via metagenomics is "Who is there?". Phylogenetic profiling via metagenomics is straightforward as the 16S (or 18S) rRNA genes are easily recognizable and the growing databases of these genes allow for rather precise phylogenetic assignments (Tringe and Hugenholtz, 2008). The more difficult question asked via metagenomics is "What are they doing?". This question can be approached via the gene-centric analysis as a number of functional genes that are hallmarks for major biogeochemical processes are well recognizable (Tringe et al., 2005; Kunin et al., 2008). The most difficult question asked via metagenomics is "Who is doing what?" as this requires establishing a connection between the function and an organism. The resolution of such knowledge can range from as low as a phylum level to as high as strain level. Below, a number of exemplary projects that succeeded in this goal are highlighted. Examples of single species genomes assembled from metagenomic data are presented in Table 1.

METAGENOMICS OF LOW COMPLEXITY COMMUNITIES

The acid main drainage (AMD) community remains the poster child of metagenomics (Tyson et al., 2004). A modest sequencing effort of 76 Mb has been sufficient for high coverage of the genomes of the dominant species of this low complexity community. Two main lessons learned from assembling separate genomes from a communal sequence pool have been the use of relaxed stringency criteria for sequence alignment, in anticipation of polymorphisms, and the necessity of binning, i.e. assignment of scaffolds to organism types (in this case by the G+C content as well as read depth). As a result, nearly complete genomes (at 10X coverage) were assembled of a *Leptospirillum* group II bacterium and a *Ferroplasma* group II archaeon. Two other genomes (*Leptospirillum* group III and *Ferroplasma* group I) have been covered at 3X. Reconstruction of the metabolism of these species has provided important insights into their ecological roles and the function of the community. Both the *Leptospirillum*

Table 1. Examples of single species genomes extracted from metagenomic sequences

Organism/ environment	Sequencing effort (Mb)	Number of contigs/ scaffolds	Largest contig/ scaffold (Mb)	Sequence coverage (x)	Genome size (Mb)	Reference
C. Cloacamonas acidiminovorans/ reactor	1,120	1	2,246	10.9	2.2	Pelletier et al., 2008
Cenarchaeum symbiosum/ sponge	50	1	2,045	>8	2.0	Hallam et al. 2006
Kuenenia stuttgartensis/ reactor	150	5	2,200	22	4.2	Strous et al., 2006
Leptospirillum sp. group II/ AMD	76	70	137	10	2.2	Tyson et al., 2004
Ferroplasma sp. group II/ AMD	76	59	138	10	1.8	Tyson et al., 2004
Gammaproteobacterium 3/ worm	204	22	1,908	5.2	NA	Woyke et al. 2006
Gammaproteobacterium 1/ worm	204	91	333	3.0	NA	Woyke et al. 2006
Deltaproteobacterium 4/ worm	204	172	252	3.3	NA	Woyke et al. 2006
Deltaproteobacterium 1/ worm	204	226	407	8.4	NA	Woyke et al. 2006
C. Accumulibacter phosphatis/ US sludge	98	33	3,027	8	5.6	García Martín et al., 2006
C. Accumulibacter phosphatis/ OZ sludge	78	254	69	5	5.6	García Martín et al., 2006
Methylotenera mobilis/ lake	60	4078	16	4.4	2.5	Kalyuzhnaya et al., 2008

NA, information not available

and the *Ferroplasma* species appear to possess multiple pathways for carbon fixation, while the *Ferroplasma* species are also equipped for a heterotrophic life style. However, none of the major species are predicted to be able to fix nitrogen, suggesting that a less abundant species, *Leptospirillum* group III is responsible for this activity. A detailed reconstruction of putative electron transfer chains in the *Leptospirillum* and the *Ferroplasma* species has revealed markedly different strategies for harnessing energy from iron oxidation.

Another example of a low complexity, highly specialized community metagenomics is the analysis of symbionts of a marine oligochaete *Olavius algarvensis* (Woyke et al., 2006). In the course of evolution, this worm has lost a mouth, a gut and nephridia, their functions completely taken over by the four symbiotic bacteria, two gammaproteobacterial types and two deltaproteobacterial types. Nearly complete genomes of these four symbionts have been recovered in scaffolds as large as 1.6 Mb, via a sequencing effort of slightly over 200 Mb, allowing for metabolic reconstruction of each species as well as predicting the specific roles for each bacterium in the syntrophic elemental cycling providing metabolic energy and biomass that feed the host. Two species (the gammaproteobacterial strains) were determined to specialize in oxidation of reduced sulphur compounds, while two others (the deltaproteobacterial srains) were predicted to be sulphate reducers. All four species are equipped for autotrophic CO_2 fixation, two former possessing genes for the Calvin-Benson-Bassham cycle and the two latter possessing genes for the reductive acetyl-CoA and the reductive tricarboxylic acid cycles. A suite of metabolic pathways that would enable recycling of the host waste has been identified, including ammonium and urea uptake and metabolism systems. In addition, systems were identified for degradation of osmolytes such as taurine, glycine betaine and trimethylamine N-oxide. The nearly complete genome recovery for the symbionts has also allowed to question whether signatures of symbiont dependence on the host were present, such as genome size reduction, loss of essential metabolic pathways etc. None of these have been found, suggesting that the symbionts of *O. algarvensis* should be able to survive as free-living bacteria. Finally, a scenario has been envisioned of how the symbionts switch between different pathways for energy generation while the host migrates between oxic and unoxic zones in its natural habitat.

In another study, the WGS sequencing approach has been combined with fosmid walking to obtain a complete genome sequence for an uncultivated marine crenarchaeote, a sponge symbiont (Hallam et al., 2006). Based on the initial 2, 779 fosmid end reads, fosmids have been selected containing DNA of the dominant ribotype of *Cenarchaeum symbiosum*, and eventually 155 of the fosmid inserts have been shotgun sequenced, to produce a circular genome representing a composite of closely related but potentially non-identical strains. Metabolic reconstruction revealed important insights into the metabolic pathways of the symbiont that make it useful for the host, such as nitrogen metabolism and CO_2 assimilation, and also revealed metabolic dependence of the symbiont on the supply of essential amino acids and vitamins. While the strategy chosen in this project, of fosmid walking, required a smaller sequencing effort compared to the projects relying on the assembly of shotgun reads, it was a laborious and time consuming effort, taking over four years.

A termite gut is a habitat for a highly specialized microbial community devoted to lignocellulose degradation. Warnecke et al. (2007) targeted a community from a hindgut of a wood-feeding 'higher' termite in order to obtain insights into the diversity of genes enabling cellulose and xylan hydrolysis by the bacterial symbionts. While only a modest sequencing effort was committed (71 Mb) considering the relatively species-rich community, a gene-centric approach resulted in identification of more than 700 glycoside hydrolase catalytic domains, representing 45 different carbohydrate-active enzymes. Compositional binning of the assembled sequences allowed assignment of glycoside hydrolases and other carbohydrate-binding module enzymes to the specific phylogenetic groups, most prominently to *Treponema* and to *Fibrobacter* species. The knowledge gained from metagenomic sequencing was augmented by the proteomic analysis that detected some of the most highly expressed hydrolytic enzymes, as well as by *in vitro* activity tests.

ENRICHMENT-BASED METAGENOMICS

Some organisms not available in pure cultures can be enriched in laboratory conditions, or in industrial bioreactors, to reach a relative proportion in mixed population large enough to warrant high coverage for the respective genome(s) when using the traditional WGS sequencing approach. One study applied this strategy to sequence, assemble and annotate the genome of *Kuenenia stuttgartiensis*, a novel bacterium representing a functional guild involved in anaerobic ammonium oxidation (anammox), from a complex bioreactor community. The annamox bacteria were discovered just over a decade ago and were demonstrated to possess specialized biochemical pathways enabling anaerobic oxidation of ammonium. While these organisms are known for very slow growth and none are available in pure culture, their role has been established as key participants of the global nitrogen cycle, contributing up to 50% to the removal of fixed nitrogen from oceans. Strous and colleagues (2006) enriched *K. stuttgartiensis* in a laboratory reactor in which its population comprised approximately 73% of total cell counts. A rather massive sequencing effort was applied (192,713 reads, approximately 154 Mb) allowing not only for assembly, but also for closing most of the gaps in the *K. stuttgartiensis* genome, resulting in only five contigs, with more than 98% of the genome captured. Metabolic reconstruction revealed unexpectedly high metabolic versatility of the organism and high degree of functional redundancy. Over 200 genes were identified involved in catabolism and respiration. Metabolic pathways for ammonium oxidation, electron transfer, energy conservation and carbon metabolism were reconstructed with great precision, and some were confirmed experimentally. Most significantly, candidate gene clusters were identified as being responsible for biological hydrazine metabolism and for ladderane biosynthesis, functions uniquely connected to annamox. To identify these functions in a novel organism, the availability of a complete or nearly complete genome was a prerequisite, as these genes could only be predicted in the context of a specific genome.

A similar approach was used by García Martín and colleagues (2006) to reconstruct nearly complete genomes of two strains of *Candidatus Accumulibacter phosphatis*, a polyphosphate-accumulating bacterium harnessed for inorganic phosphorus removal in wastewater treatment plants. Sludge samples from two laboratory scale reactors containing 60 to 80% of A. phosphatis as a proportion of total cell counts were subjected to WGS sequencing, producing 5x and 8x sequence coverage, respectively, for the two A. phosphatis genomes. Sequence reads were assembled using two alternative assembler tools (to compare their performance, not reviewed here), resulting in contigs as large as 170 kb and scaffolds as large as 3 Mb, covering at least 97% of one of the genomes. The high quality of this genome allowed for obtaining important insights into the biochemical details of polyphosphate biosynthesis as well the as details of other key metabolic pathways. One of the important problems in biologically enhanced phosphate removal has been the lack of understanding of the sources of reducing power for polyhydroxyalkanoate synthesis in the anaerobic phase. Based on the genomic analysis, a novel cytochrome has been proposed to function as a quinol-NAD(P)H reductase, allowing for anaerobic functioning of the tricarboxylic acid cycle. Despite previous evidence for denitrifying capability of A. phosphatis, no traditional respiratory nitrate reductase was encoded, suggesting that a different enzyme may be carrying out the function. Reconstruction of the complete nitrogen fixation pathway was also a surprise. Again, only from complete or nearly complete genomic sequence could the metabolic blueprint be reconstructed with such precision, representing a turning point in the understanding of the genetics of the process of enhanced biological phosphate removal. This knowledge opens ways for predicting the efficiency of bioreactors for phosphorus removal and suggests optimal conditions for their operation.

Another study has demonstrated that, with an adequate sequencing effort, genomes of minor or at least non-dominant members of the communities could also be sequenced to completion (Pelletier et al., 2008). The genome of *Candidatus* Cloacamonas acidaminovorans that is part of a complex community of industrial anaerobic digesters has been targeted in order to obtain insights into the physiology and metabolism of this representative of a candidate division WWE1, with no cultivated members. The strategy implemented was a massive sequencing effort (1.7 million reads, 1.12 Gb) to end-sequence 1 million fosmids, followed by an iterative assembly approach (similar to the one used by Hallam et al., 2006) to reconstruct the entire genome. Indeed the genomic insights uncovered were worth the effort. Analysis of the genome revealed low gene density (81%) unusual for bacteria, 40% of genes being unique. The best matches for the genes in *Candidatus* C. acidiaminovorans were distributed among distantly related taxa, including Proteobacteria, Firmicutes and Planctomycetes. Some of the proteins encoded in the genome were more related to their eukaryotic than their prokaryotic counterparts. The organism was also predicted to use pyrophosphate-dependent enzymes, a relatively rare feature. A deficiency in biosynthesis of 12 amino acids and several vitamins and cofactors were predicted from the genome, offering one explanation for this organism not existing in pure culture. No typical respiratory chains could be predicted from the genome, suggesting that the fermentation metabolism must be responsible for energy production. Overall, the genome of this novel syntrophic bacterium is an important contribution to the gene pool that determines the quality of annotation for the newly emerging genomes and metagenomes.

More recently, a community of a production-scale biogas reactor was analyzed using the 454 pyrosequencing (FLX System) technology, with en average read length of 230 bases, at a 142 Mb sequencing effort (Schlüter et al., 2008). Attempts of assembly resulted in a large number of contigs over 500 bp, some contigs acceding 10 kb in size and the largest contig being 31.5 Kb in size. The contig sequences were matched to the related genomes available in the non-redundant database. This study demonstrated that de novo assembly from shorter 454 sequences is in principle possible for metagenomic sequences, at least in cases of relatively low community complexity.

BROMODEOXYURIDINE LABELING AS A MEANS FOR FUNCTIONAL METAGENOMICS

The principle of this approach is in targeting species actively replicating their DNA in response to the addition of a specific compound, by labeling the newly synthesized DNA with bromodeoxyuridine (an analogue of thymidine). So far this method was employed on a large scale only in one project, as an attempt to identify the species active in utilization of dissolved organic carbon (DOC) in the coastal ocean (Mou at al, 2008). Dimethylsulphoniopropionate (DMSP) and vanillate were used as model DOC compounds in microcosm incubations supplemented with bromodeoxyuridine, followed by pyrosequencing of the DNA captured by immunoprecipitation. However, in this case, no enrichment was observed for known species involved in degradation of the target compounds and no key genes involved in this process were found to be over-represented. From this attempt only, the potential of this method for characterizing natural populations remains uncertain, and further exploration of this approach is required. Some reasons for this failure are obvious, such as the insufficient sampling (only 4 to 10 Mb per sample), while others are less clear. It is possible that the efficiency of the label incorporation is not uniform among different taxa. It is also possible that label incorporation takes place independently of substrate stimulation, thus the results obtained represent random sampling rather than selection for functional types.

TARGETING FUNCTIONAL TYPES VIA STABLE ISOTOPE PROBING

One way to directly link a function in the environment to a specific guild performing this function is to feed the population a substrate of interest, labeled by a heavy isotope, followed by characterization of the heavy fraction of communal DNA that is enriched in DNA of microbes that actively metabolized the labeled substrate. This technique is known as Stable Isotope Probing and it has been effective in identifying microbes involved in specific biogeochemical transformations such as methylotrophy, phenol degradation, glucose metabolism etc. (Friedrich 2006). Typically, small amounts of DNA are isolated from these experiments, and these are used for phylogenetic profiling and detection of key functional genes, after PCR amplification. So far, there is only one example of scaling this method up to obtain amounts of DNA enabling the WGS sequencing approach, applied to communities of a freshwater lake sediment involved in utilization of C1 compounds (methylotrophs; Kalyuzhnaya et al., 2008). The goal of this targeted metagenomic approach has been two-fold: to reduce the complexity of the community that has been estimated at approximately 5000 species and to directly link specific substrate repertoires to functional guilds. Five different labeled substrates have been employed, methane, methanol, methylamine, formaldehyde and formate, resulting in five 'functional' metagenomes (26 to 58 Mb in size). Community complexity in each microcosm was found to be dramatically reduced compared to the complexity of non-enriched community. From the present 16S rRNA genes, the communities shifted toward specific functional guilds that included bona fide methylotroph species as well as organisms distantly related to cultivated species, implicating them in methylotrophy. The methylamine microcosm metagenome (37 Mb) was found least complex and it was dominated by a single species, *Methylotenera mobilis* represented by a number of closely related strains, while in the non-enriched community *Methylotenera* species comprised less than 0.5% of the population. Via compositional binning, a nearly complete genome of this novel organism has been extracted from the metagenome and its metabolism reconstructed, allowing for genome-wide comparisons with a related species. This so far is the most dramatic example of assembling a genome of a species that is a minor member of a community. Thus the method has been dubbed 'high-resolution metagenomics'. In addition, as part of this project, complete genomes of novel bacteriophages have been assembled from the same metagenome and their association with *M. mobilis* has been proposed, suggesting a mechanism for a dynamic control of the *Methylotenera* populations.

METATRANSCRIPTOMICS

Metatranscriptomics, analysis of community transcripts isolated directly from the environment or from microcosms in which the community has been disturbed or manipulated in a certain way, represent the next logical step in the meta- (-omics) approach. This method should enable reaching beyond the community's genomic potential (metagenomic blueprint), and connect more directly the taxonomic make up of the community to its in situ activity (function), via profiling of (most abundant) transcripts and correlating them with specific environmental conditions. For large-scale metatranscriptomics experiments, the next generation sequencing technologies are especially attractive as assembly is not a prerequisite for transcript analysis. The few metatranscriptomic studies published so far (Urich et al., 2008; Frias-Lopez et al., 2008; Gilbert et al., 2008; Poretsky et al., 2009) have employed the 454 sequencing technology, as this technology produces reads of sufficient length to allow for functional predictions based on a single read. These reads were then processed in a gene-centric way. Obviously, all the pitfalls discussed above relating to the analysis and annotation of short metagenomic reads apply to the short metatranscriptomic reads. Thus, with few exceptions (for example when a genome of a cultivated species is well represented in the environment in question and could be used as a scaffold), only general functional predictions can be made, and in most cases no phylogenetic assignments can be made for specific functional genes. In addition, biases and limitations specific to the analysis of RNA molecules apply: often times only very small amounts of the RNA can be isolated, so an amplification step is necessary (Frias-Lopez et al., 2008; Gilbert et al., 2008). The natural abundance of non-messenger RNA can be a blessing (if a careful phylogenetic profiling is desired; Urich et al; 2008) or a curse (if mRNA is the primary target) as efficient separation of mRNA from more abundant ribosomal and transport RNA remains a problem. Of the potential mRNA transcripts, typically only one third can be matched to known genes or functional gene categories while the rest cannot be classified for the lack of any matches in the databases (orphan proteins). Thus, the resolution provided by direct analysis of short reads remains very low.

As a proof of a concept, we tested an approach in which environmental transcripts were matched to a scaffold previously generated for a community from the same study site. A metagenomic scaffold representing the methylotorph community of Lake Washington sediment was employed (Kalyuzhnaya et al., 2008), and transcript sequences from a community sampled from the same study site were generated using the Illumina technology (resulting in ultra-short reads of approximately 40 bp; unpublished data). The metagenomic scaffold we used is mostly represented by contigs of low sequence coverage, reflecting the insufficient sampling that is typical of metagenomic studies (low-resolution metagenome). However, a small part of the metagenome, representing a composite genome of *Methylotenera* species is made up of contigs of much higher sequence coverage (high-resolution metagenome; Kalyuzhnaya et al., 2008). We matched the transcripts separately to the low-resolution metagenome and to the high-resolution metagenome. Only 8% of the almost 25 million Ilumina reads found targets in the metagenome, reflecting that both the metagenome and the metatranscriptome must have been significantly under-sampled. When matching the transcripts against the low-resolution scaffold, we determined that approximately 35% of the metagenome overlapped with the metatranscriptome. However, when matching was done with the much better sampled *Methylotenera* scaffold, we found that over 96% of the composite genome had matches in the metatranscriptome. This result highlights the necessity of well-covered and more complete genomic scaffolds from the environments that are interrogated via metatranscriotpmics, to enable high-resolution analysis.

METAPROTEOMICS

Metaproteomics, analysis of protein profiles of microbial communities, presents an even better opportunity to address the function directly, as proteins are the molecules that ultimately perform the function. However, metaproteomics, even more so than metatranscriptomics, rely on quality metagenomic data. The large-scale MS/MS-based metaproteomics approach has been pioneered (Ram et al., 2005) and further perfected by collaborative efforts between the Banfield and the Hettich groups, establishing the current state-of-the-art of the field (VerBerkmoes et al., 2009a). The power of this approach was first demonstrated on a community of low complexity (Ram et al., 2005), an AMD community that was not identical but similar to the community for which the high quality metagenomic sequence has been previously generated (Tyson et al., 2004). Despite differences between the sequences of predicted proteins in the dataset and those in the actual sample, it was possible to match shotgun MS/MS spectra to peptides, resulting in positive identification of over 2000 proteins that belonged to the five most abundant members of the community. For the most abundant organism, *Leptospirillum* group II, expression of 50% of the proteins was detected by proteomics. Important insights into the respective metabolic contributions of the archaeal and the bacterial members have been revealed, including identification of a novel cytochrome that appears to be central to iron oxidation and AMD formation. The group has now produced over 30 datasets from the AMD system and these have been employed to establish the relationship between the species abundance in communities and the efficiency of protein identification, concluding that, while organisms constituting 30-40% of the community can be sampled to saturation by the metaproteomics approach, the most abundant proteins from members constituting as little as 1% of the population can also be detected. These results are important for planning future proteomics projects, including considerations for specific enrichments if members of low-abundant taxa need to be targeted.

Proteomics has been successfully applied to a community of a higher complexity, the enhanced biological phosphorous removal (EBPR) community, dominated by a single species, *Candidatus* A. phosphatis (Wilmes et al., 2008) similar to the ones described by García Martín et al. (2006), so the reference genomes generated in the latter work have been used as scaffolds for peptide matching, resulting in identification of approximately 2300 proteins, of which approximately 700 have been assigned to *Candidatus* A. phosphatis, enabling extensive analysis of the metabolic pathways central to EBPR.

The most complex metaproteome analyzed to date is the metaproteome of human feces (VerBerkmoes et al., 2009b). Even with an unmatched metagenomic database used for protein detection, on the order of 1000 proteins of bacterial origin have been detected per sample (about 30% of the detected proteins were human proteins), and their relative abundances have been estimated.

The future of metagenomics

Metagenomics is coming of age but still gaining momentum. This coincides with rapid improvement of sequencing technologies and with the realization that the next bottleneck in metagenomics will not be the sequence data production but computation and data storage. From the experience of the past five years, the power of metagenomics is obvious, while its potential is still waiting to be fully realized. It is now quite clear that even communities of limited complexity pose major challenges in terms of genomic exploration, highlighting the necessity of much deeper sampling, and the need for special assembly and analysis tools. Such improvements are possible and imminent, given the fast progress in these areas. However, the cost of metagenomic sequencing to high coverage, even when employing the next generation technologies, remains a major challenge. The price of sequencing is typically advertised as a per-base cost. However, sequences generated by different technologies require different depths of coverage. While the Sanger technology has been brought to state of the art by years of perfecting, producing reads of up to 1 kb with very low error rate, the 454 technology that is predicted to produce reads of similar length in the near future inherently has a much higher error rate. Thus a higher coverage is required to obtain data of similar quality. With the yet more per-base cost effective technologies, such as Illumina or SOLiD, the depth of sequencing needs to be yet much higher (probably 40 to 50 fold) to assure sequence quality. Moving to the next level (in sequencing depth i.e. sequence quality) is necessary to truly understand how complex microbial communities operate, how they evolve and how they respond to the changing environment. It is also essential for gene and pathway discovery via metagenomics.

The stage is now set for Gb-scale metagenomic projects, such as sequencing complex communities to (nearly) saturation. Carrying out such projects will not only test the performance of the newly emerging computational tools for sequence analysis, but will ultimately demonstrate whether we can apply the same or similar 'gold standards' to metagenomic sequences as to single genome sequences. It will also ultimately test the predictions of how much sequencing is necessary to enable delineating (via binning and assembly) single-species genomes that are parts of a community gene pool. Without such 'saturation'-level metagenomic sequencing experiments, comparative analyses of communities over time or space will remain of little value, as only (small) parts of under-sampled communities will be compared to one another. While in the future carrying out such experiments may become routine, at this time it will likely require a concerted community effort. A document published by the U.S. National Academies' National Research Council (NRC) in 2007, entitled "The new science of metagenomics: revealing the secrets of our microbial planet" calls for a new Global Initiative to drive advances in the field of Metagenomics, in a way that the Human Genome Project advanced the mapping of our genetic code (see again Woese, 1998). Such an initiative would help move the field of metagenomics to the new level and toward a brighter future.

Acknowledgements

The author acknowledges support from the National Science Foundation (MCB00604269). Many thanks to Dr. N. Ivanova for critical reading of the manuscript.

References

Brulc, J.M., Antonopoulos, D.A., Miller, M.E., Wilson, M.K., Yannarell, A.C., Dinsdale, E.A., Edwards, R.E., Frank, E.D., Emerson, J.B., Wacklin, P., Coutinho, P.M., Henrissat, B., Nelson, K.E., White BA. (2009) Gene-centric metagenomics of the fiber-adherent bovine rumen microbiome reveals forage specific glycoside hydrolases. *Proceedings of National Academy of Sciences USA* **106**, 1948-1953.

Dinsdale, E.A., Edwards, R.A., Hall, D., Angly, F., Breitbart, M., Brulc, J.M., Furlan, M., Desnues, C., Haynes, M., Li, L., McDaniel, L., Moran, M.A., Nelson, K.E., Nilsson, C., Olson, R., Paul, J., Brito, B.R., Ruan, Y., Swan, B.K., Stevens, R., Valentine, D.L., Thurber, R.V., Wegley, L., White, B.A., Rohwer, F. (2008) Functional metagenomic profiling of nine biomes. *Nature* **452**, 629-632.

Edward, R.A., Rodriguez-Brito, B., Wegley, L., Haynes, M., Breitbart, M., Peterson, D.M., Saar, M.O., Alexander, S., Alexander, E.C. Jr., Rohwer, F. (2006) Using pyrosequencing to shed light on deep mine microbial ecology. *BMC Genomics* **7**, 57.

Frias-Lopez, J., Shi, Y., Tyson, G.W., Coleman, M.L., Schuster, S.C., Chisholm, S.W., Delong, E.F. (2008) Microbial community gene expression in ocean surface waters. *Proceedings of National Academy of Sciences USA* **105**, 3805-3810.

Friedrich, M.W. (2006) Stable-isotope probing of DNA: insights into the function of uncultivated microorganisms from isotopically labeled metagenomes. *Current Opinion in Biotechnology* **17**, 59-66.

García Martín, H., Ivanova, N., Kunin, V., Warnecke, F., Barry, K.W., McHardy, A.C., Yeates, C., He, S., Salamov, A.A., Szeto, E., Dalin, E., Putnam, N.H., Shapiro, H.J., Pangilinan, J.L., Rigoutsos, I., Kyrpides, N.C., Blackall, L.L., McMahon, K.D., Hugenholtz, P. (2006) Metagenomic analysis of two enhanced biological phosphorus removal (EBPR) sludge communities. *Nature Biotechnology* **24**, 1263-1269.

Gilbert, J.A., Field, D., Huang, Y., Edwards, R., Li, W., Gilna, P., Joint, I. (2008) Detection of large numbers of novel sequences in the metatranscriptomes of complex marine microbial communities. *PLoS ONE* **3**, e3042.

Hallam, S.J., Konstantinidis, K.T., Putnam, N., Schleper, C., Watanabe, Y., Sugahara, J., Preston, C., de la Torre, J., Richardson, P.M., DeLong, E.F. (2006) Genomic analysis of the uncultivated marine crenarchaeote *Cenarchaeum symbiosum*. *Proceedings of National Academy of Sciences USA* **103**, 18296-18301.

HANDELSMAN, J. (2004) Metagenomics: application of genomics to uncultured microorganisms. *Microbiology and Molecular Biology Reviews* **68**, 669-685.

HANDELSMAN, J., RONDON, M.R., BRADY, S.F., CLARDY, J., GOODMAN, R.M. (1998) Molecular biological access to the chemistry of unknown soil microbes: a new frontier for natural products. *Chemistry and Biology* **5**, R245-249.

KALYUZHNAYA, M.G., LAPIDUS, A., IVANOVA, N., COPELAND, A.C., McHARDY, A.C., SZETO, E., SALAMOV, A., GRIGORIEV, I.V., SUCIU, D., LEVINE, S.R., MARKOWITZ, V.M., RIGOUTSOS, I., TRINGE, S.G., BRUCE, D.C., RICHARDSON, P.M., LIDSTROM, M.E., CHISTOSERDOVA, L. (2008) High-resolution metagenomics targets specific functional types in complex microbial communities. *Nature Biotechnology* **26**, 1029-1034.

KUNIN, V., COPELAND, A., LAPIDUS, A., MAVROMATIS, K., HUGENHOLTZ, P. (2008) A bioinformatician's guide to metagenomics. *Microbiology and Molecular Biology Reviews* **72**, 557-578.

LAPIDUS, A. (2009) Genome sequence databases (overview): sequencing and assembly. In *The Encyclopedia of Microbiology*. M. Schaechter, pp196-210. New York: Elsevier.

LI, L.L., McCORKLE, S.R., MONCHY, S., TAGHAVI, S., VAN DER LELIE, D. (2009) Bioprospecting metagenomes: glycosyl hydrolases for converting biomass. *Biotechnology for Biofuels* **2**, 10.

LIOLIOS, K., MAVROMATIS, K., TAVERNARAKIS, N., KYRPIDES, N.C. (2008) The Genomes On Line Database (GOLD) in 2007: status of genomic and metagenomic projects and their associated metadata. *Nucleic Acids Research* **36**, D475-479.

MARKOWITZ, V.M., IVANOVA, N.N., SZETO, E., PALANIAPPAN, K., CHU, K., DALEVI, D., CHEN, I.M., GRECHKIN, Y., DUBCHAK, I., ANDERSON, I., LYKIDIS, A., MAVROMATIS, K., HUGENHOLTZ, P., KYRPIDES, N.C. (2008) IMG/M: a data management and analysis system for metagenomes. *Nucleic Acids Research* **36**, D534-538.

MOU, X., SUN, S., EDWARDS, R.A., HODSON, R.E., MORAN, M.A. (2008) Bacterial carbon processing by generalist species in the coastal ocean. *Nature* **451**, 708-711.

PELLETIER, E., KREIMEYER, A., BOCS, S., ROUY, Z., GYAPAY, G., CHOUARI, R., RIVIÈRE, D., GANESAN, A., DAEGELEN, P., SGHIR, A., COHEN, G.N., MÉDIGUE, C., WEISSENBACH, J., LE PASLIER D. (2008) "*Candidatus* Cloacamonas acidaminovorans": genome sequence reconstruction provides a first glimpse of a new bacterial division. *Journal of Bacteriology* **190**, 2572-2579.

PORETSKY, R.S., HEWSON, I., SUN, S., ALLEN, A.E., ZEHR, J.P., MORAN, M.A. (2009) Comparative day/night metatranscriptomic analysis of microbial communities in the North Pacific subtropical gyre. *Environmental Microbiology* **11**, 1358-1375.

PREIDIS, G.A. AND VERSALOVIC, J. (2009) Targeting the human microbiome with antibiotics, probiotics, and prebiotics: gastroenterology enters the metagenomics era. *Gastroenterology* **136**, 2015-2031.

RAM R.J., VERBERKMOES, N.C., THELEN, M.P., TYSON, G.W., BAKER, B.J., BLAKE, R.C. 2ND, SHAH, M., HETTICH, R.L., BANFIELD, J.F. (2005) Community proteomics of a natural microbial biofilm. *Science* **308**, 1915-1920.

SCHLOSS PD AND HANDELSMAN, J. (2005) Metagenomics for studying unculturable microorganisms: cutting the Gordian knot. *Genome Biology* **6**, 229.

SCHLÜTER, A., BEKEL, T., DIAZ, N.N., DONDRUP, M., EICHENLAUB, R., GARTEMANN,

K.H., Krahn, I., Krause, L., Krömeke, H., Kruse, O., Mussgnug, J.H., Neuweger, H., Niehaus, K., Pühler, A., Runte, K.J., Szczepanowski, R., Tauch, A., Tilker, A., Viehöver, P., Goesmann, A. (2008) The metagenome of a biogas-producing microbial community of a production-scale biogas plant fermenter analysed by the 454-pyrosequencing technology. *Journal of Biotechnology* **136**, 77-90.

Schmidt, T.M., DeLong, E.F., Pace, N.R. (1991) Analysis of a marine picoplankton community by 16S rRNA gene cloning and sequencing. *Journal of Bacteriology* **173**, 4371-4378.

Snyder, L.A., Loman, N., Pallen, M.J., Penn, C.W. (2008) Next-generation sequencing-the promise and perils of charting the great microbial unknown. *Microbial Ecology* **57**:1-3.

Staley, J.T. and Konopka, A. (1985) Measurement of in situ activities of nonphotosynthetic microorganisms in aquatic and terrestrial habitats. *Annual Review of Microbiology* **39**, 321-346.

Strous, M., Pelletier, E., Mangenot, S., Rattei, T., Lehner, A., Taylor, M.W., Horn, M., Daims, H., Bartol-Mavel, D., Wincker, P., Barbe, V., Fonknechten, N., Vallenet, D., Segurens, B., Schenowitz-Truong, C., Médigue, C., Collingro, A., Snel, B., Dutilh, B.E., Op den Camp, H.J., van der Drift, C., Cirpus, I., van de Pas-Schoonen, K.T., Harhangi, H.R., van Niftrik, L., Schmid, M., Keltjens, J., van de Vossenberg, J., Kartal, B., Meier, H., Frishman, D., Huynen, M.A., Mewes, H.W., Weissenbach, J., Jetten, M.S., Wagner, M., Le Paslier, D. (2006) Deciphering the evolution and metabolism of an anammox bacterium from a community genome. *Nature* **440**, 790-794.

Thurber, R.V., Willner-Hall, D., Rodriguez-Mueller, B., Desnues, C., Edwards, R.A., Angly, F., Dinsdale, E., Kelly, L., Rohwer, F. (2009) Metagenomic analysis of stressed coral holobionts. *Environmental Microbiology* **11**, 2148-2163.

Tringe, S.G., von Mering, C., Kobayashi, A., Salamov, A.A., Chen, K., Chang, H.W., Podar, M., Short, J.M., Mathur, E.J., Detter, J.C., Bork, P., Hugenholtz, P., Rubin, E.M. (2005) Comparative metagenomics of microbial communities. *Science* **308**, 554-557.

Tringe, S.G., and Hugenholtz, P. (2008) A renaissance for the pioneering 16S rRNA gene. *Current Opinion in Microbiology* **11**, 442-446.

Tyson, G.W., Chapman, J., Hugenholtz, P., Allen, E.E., Ram, R.J., Richardson, P.M., Solovyev, V.V., Rubin, E.M., Rokhsar, D.S., Banfield, J.F. (2004) Community structure and metabolism through reconstruction of microbial genomes from the environment. *Nature* **428**, 37-43.

Urich, T., Lanzén, A., Qi, J., Huson, D.H., Schleper, C., Schuster, S.C. (2008) Simultaneous assessment of soil microbial community structure and function through analysis of the meta-transcriptome. *PLoS One* **3**, e2527.

Venter, J.C., Remington, K., Heidelberg, J.F., Halpern, A.L., Rusch, D., Eisen, J.A., Wu, D., Paulsen, I., Nelson, K.E., Nelson, W., Fouts, D.E., Levy, S., Knap, A.H., Lomas, M.W., Nealson, K., White, O., Peterson, J., Hoffman, J., Parsons, R., Baden-Tillson, H., Pfannkoch, C., Rogers, Y.H., Smith, H.O. (2004) Environmental genome shotgun sequencing of the Sargasso Sea. *Science* **304**, 66-74.

VerBerkmoes, N.C., Denef, V.J., Hettich, R.L., Banfield, J.F. (2009) Systems biology: Functional analysis of natural microbial consortia using community proteomics. *Nature Reviews Microbiology* **7**:196-205.

VerBerkmoes, N.C., Russell, A.L., Shah, M., Godzik, A., Rosenquist, M., Halfvarson, J., Lefsrud, M.G., Apajalahti, J., Tysk, C., Hettich, R.L., Jansson, J.K. (2009) Shotgun metaproteomics of the human distal gut microbiota. *ISME Journal* **3**, 179-189.

Warnecke, F. and Hess, M. (2009) A perspective: metatranscriptomics as a tool for the discovery of novel biocatalysts. *Journal of Biotechnology* **142**, 91-95.

Warnecke, F., Luginbühl, P., Ivanova, N., Ghassemian, M., Richardson, T.H., Stege, J.T., Cayouette, M., McHardy, A.C., Djordjevic, G., Aboushadi, N., Sorek, R., Tringe, S.G., Podar, M., Martin, H.G., Kunin, V., Dalevi, D., Madejska, J., Kirton, E., Platt, D., Szeto, E., Salamov, A., Barry, K., Mikhailova, N., Kyrpides, N.C., Matson, E.G., Ottesen, E.A., Zhang, X., Hernández, M., Murillo, C., Acosta, L.G., Rigoutsos, I., Tamayo, G., Green, B.D., Chang, C., Rubin, E.M., Mathur, E.J., Robertson, D.E., Hugenholtz, P., Leadbetter, J.R. (2007) Metagenomic and functional analysis of hindgut microbiota of a wood-feeding higher termite. *Nature* **450**, 560-565.

Wilmes, P., Andersson, A.F., Lefsrud, M.G., Wexler, M., Shah, M., Zhang, B., Hettich, R.L., Bond, .P.L, VerBerkmoes, N.C., Banfield, J.F. (2008) Community proteogenomics highlights microbial strain-variant protein expression within activated sludge performing enhanced biological phosphorus removal. *ISME Journal* **2**, 853-864.

Woese, C.R. (1987) Bacterial evolution. *Microbiological Reviews* **51**, 221-271.

Woese, C.R. (1998) A manifesto for microbial genomics. *Current Biology* **8**, R781-R783.

Wommack, K.E., Bhavsar, J., Ravel, J. (2008) Metagenomics: read length matters. *Applied and Environmental Microbiology* **74**, 1453-1463.

Woyke, T., Teeling, H., Ivanova, N.N., Huntemann, M., Richter, M., Gloeckner, F.O., Boffelli, D., Anderson, I.J., Barry, K.W., Shapiro, H.J., Szeto, E., Kyrpides, N.C., Mussmann, M., Amann, R., Bergin, C., Ruehland, C., Rubin, E.M., Dubilier, N. (2006) Symbiosis insights through metagenomic analysis of a microbial consortium. *Nature* **443**, 950-955.

Biotechnology and Genetic Engineering Reviews - Vol. 26, 353-370 (2009)

The impact of Structural Proteomics on Biotechnology

BABU A. MANJASETTY*, ANDREW P. TURNBULL[1], SANTOSH PANJIKAR[2]

Proteomics & Bioinformatics Research Group, Research and Industry Incubation Center, Dayananda Sagar Institutions, Bangalore 560 078 India, [1]*Cancer Research Technology Ltd, Birkbeck College, University of London, London, WC1E 7HX, UK and* [2]*EMBL Hamburg Outstation, c/o DESY, Notkestrasse 85, D-22603 Hamburg, Germany*

Abstract

Structural proteomics (SP) projects are capable of producing thousands of protein structures per year by employing semi-automated technologies. It is too early to assess and evaluate the scientific impact of these protein structures, although SP initiatives have substantially changed the traditional way of protein characterization. Many of the methodologies and technologies developed by SP have been adapted by structural biology laboratories and pharmaceutical companies to lower the costs, increase the speed and productivity of structure determination pipelines and to enhance drug discovery programs. The advent of genomic and proteomic technologies have facilitated rapid advances in our understanding of the molecular details of cellular function. The purpose of this review is to consider the impact of these technologies on protein structure analysis and to illustrate how it's directing the focus of research relevant to biotechnology.

* To whom correspondence may be addressed (babu.manjasetty@gmail.com)

Abbreviations: SP, Structural Proteomics; MAD, Multi-wavelength Anomalous Dispersion; SAD, Single-wavelength Anomalous Dispersion; MR, Molecular Replacement; PCR, Polymerase Chain Reaction; MX, Macromolecular Crystallography; PDB, Protein Data Bank.

Introduction

The universe of possible combinations of the 20 basic amino acids permits virtually unlimited diversity in proteins. This has been revealed through the large scale human genome sequencing projects being completed at the end of the last century, representing the first exquisite genetic blueprint of life (Collins *et al.*, 2003). The massive increase in genomic information in public databases arising from these genomic projects has formed the starting point for proteomics research (Cho, 2007). The term *proteomics* was coined by Wilkins and Williams as early as 1994 to describe the large-scale characterization of all the proteins of a cell or tissue. Two-dimensional electrophoresis, mass spectrometry structural proteomics, functional genomics and bioinformatics are the most important technologies underpinning proteomics (Alterovitz *et al.*, 2006). SP is a branch of proteomics science -a large scale scientific project involving cross-disciplinary teams seeking to increase the number of protein structures through the development and implementation of high throughput automated technologies and novel techniques (Banci *et al.*, 2007; Manjasetty *et al.*, 2008). An integrated approach has been applied to protein structure determination at the SP centers that have been established worldwide (Heinemann *et al.*, 2000). The efforts during the first decade of the 21st century will allow high-throughput protein structure determination to become a reality (Blow, 2008; Manjasetty *et al.*, 2007). Protein structural information not only provides the functional inference of the living systems from which they are derived, but also developments in this area of research have had a major impact on biotechnology (Yakunin *et al.*, 2004). Recently, a knowledge base has been created in order to take the SP products / process to industry / academia (Berman *et al.*, 2009). Here, we present an overview of recent developments in high-throughput procedures that have been adapted to genomic-scale structural analyses via an integrated approach. All evidence suggests that the pace of these advances will continue to accelerate in the future.

Protein structure analysis

Protein structure determination relies on very labor-intensive methods and utilizes expensive equipment for gene cloning, protein expression and purification, crystallization, X-ray diffraction data collection and structure determination, refinement and validation. Automation of the entire process is critical to achieve the goals and desired output of structural proteomics.

PROTEIN PRODUCTION

The development of high-throughput cloning, expression and purification systems has played a central role in SP. Recombinant protein yield and solubility are highly dependent on the specific protein sequence and the type of vector, host cell and culture conditions used. High-throughput cloning requires implementing procedures based on the polymerase chain reaction (PCR). A preliminary step involves the design of specific gene PCR amplification primers followed by the screening of potential PCR-amplified clones to verify that the insert is properly orientated. Sequence analysis of positive clones must be performed to confirm that the correct reading frame has been obtained and that no PCR errors have been introduced. Automated systems are available for picking, gridding, micro-arraying and sequencing.

To generate expression vector clones, generic cloning systems can be used based on ligation-independent cloning. This technique has been adapted by several SP centers to be compatible with *E.coli*, yeast and Baculovirus-insect cell host systems (Gileadi *et al.*, 2008; Shrestha *et al.*, 2008). Recently, the ligation-independent cloning *In-Fusion™* system from Clontech Laboratories Inc has been developed, which streamlines the expression cloning process by alleviating the need for restriction digestion of the PCR fragment, ligation or blunt-end polishing, making it suitable for high-throughput expression screening applications (Berrow *et al.*, 2007). For SP, high-throughput re-combinant protein expression systems are commonly based on *E.coli* (Stevens, 2000). Bacterial expression systems are capable of producing high yields of recombinant protein so that the protein of interest is produced in an enriched form. Additionally, *E.coli* protein expression is cheaper and faster than eukaryotic systems. However, *E.coli* lacks the cellular machinery for post-translational modifications necessary for the correct folding and activity of some eukaryotic proteins. Furthermore, the expression of eukaryotic proteins in *E.coli* can result in aggregation, formation of insoluble inclusion bodies and degradation. Therefore, *E.coli* based systems are not always viable for the expression of eukaryotic proteins including some human proteins. In such cases, eukaryotic hosts such as yeast and baculovirus-insect cells are employed. Yeast expression systems (*S. cerevisiae* and *P. pastoris*), combine several advantages of prokaryotic and eukaryotic expression systems: High yields of recombinant protein expression can be achieved and eukaryotic post-translational modifications are introduced (Prinz *et al.*, 2004). Similarly, the recombinant baculovirus-insect cell expression system can accomplish most post-translational modifications including phosphorylation, N- and O-linked glycosylation, acylation, disulphide cross-linking, oligomeric assembly and subcellular targeting, which may be crucial to ensure the biological integrity of some human proteins (Albala *et al.*, 2000).

At SP centers, gene constructs are expressed in a synchronous fashion and affinity tags are employed to facilitate the use of generic purification protocols (Gileadi, *et al.*, 2008). Tags range from large tags such as Nus A (54 kDa), maltose-binding protein (MBP; 40 kDa) and glutathione-S-transferase (GST; 26 kDa) to smaller tags such as the 6xHis tag- the most commonly used tag which allows a generic single step purification using nickel-NTA affinity column chromatography (Bruel *et al.*, 2000; Stevens, 2000). More recently, Bio-Rad Laboratories Inc. have launched the *Profinity eXact* fusion-tag system which provides a novel alternative to existing affinity tag and tag removal techniques, utilizing a modified form of subtilisin protease immobilized onto a chromatographic support to generate tag-free target protein in a single step. An advantage of *in vitro* expression systems is the facilitation of selenomethionine (SeMet) incorporation which can be used for phase determination.

CRYSTALLIZATION

Every protein presents its own crystallization challenge and the growth of protein crystals is a complex and poorly understood process. Crystal growth occurs when molecules are brought into a supersaturated state, which is achieved through the slow removal of solvent (McPherson, 2004). The growth of the best quality crystals requires optimization of the experimental conditions. The protein crystallization process is still largely empirical and initially involves screening a broad range of conditions by adjusting the experimental parameters to identify optimal crystallization conditions to convert soluble proteins into crystals (Wooh *et al.*, 2003). These can include varying temperature, pH, precipitants, additives, protein concentration, and expression and purification conditions (Asherie, 2004). In addition, several strategies such as systematic screening of multiple constructs of the same protein, differently ligated or complex forms of the target of interest, modification of surface residues, removal of disordered regions and varying the orthologs, can help to facilitate crystallization (Page, 2008). Each of these variations represents an individual crystallization experiment, frequently performed in 24- or 96- well plate format. Furthermore, the use of multi-well plates is suitable for the three most commonly used crystallization techniques: sitting drop, hanging drop and the microbatch method.

Over the past years, the crystallization process has become heavily automated through the use of liquid handlers, pipetting instrumentation, drop dispensing and crystallization robotics (Miyatake *et al.*, 2005). Automation has several key advantages including the ability to use smaller amounts of protein sample along with the ability to set up each experiment within minutes. The latest nanolitre robotic liquid dispensing systems are capable of dispensing very small drops, which reduces the amount of protein required for screening conditions and hence the overhead on protein production (Walter *et al.*, 2005).

The traditional seeding techniques such as macroseeding, microseeding and streak seeding are used in protein crystallization to grow better quality crystals (Bergfors, 2003). A microseed-matrix procedure has been recently robotized (D'Arcy *et al.*, 2007; Walter *et al.*, 2008) to promote high-throughput crystallization experiments. An alternative mechanism to achieve nucleation is heterogeneous nucleation. The regularity of the surface of heterogeneous nucleants such as silica or hair can induce crystal growth. Additionally, the addition of insoluble heterogeneous nucleating agents provides a simple method to increase the likelihood of crystal formation when using sparse matrix crystallization screens (Thakur *et al.*, 2007).

A recent development in protein crystallization has been the use of high density chip-based, microfluidic systems, including Emerald Biosystems Microcapillary Protein Crystallization System (MPCS) (Gerdts *et al.*, 2008) and Fluidigm Corporations TOPAZ® system (Segelke, 2005), for crystallizing proteins using the free-interface diffusion method at nano-litre scale. MPCS CrystalCards require less than 4 μl of protein sample to prepare and store approximately 800 individual nanovolume crystallization experiments, whereas TOPAZ screening chips require only approximately 1.4 μl of protein solution per 96 reagents.

The other steps of the crystallization pipeline including plate storage at constant temperature, imaging and data capture of the crystallization drops and data management have also been automated and computerized to increase throughput (Hiraki *et al.*, 2006; Walter, et al., 2005). Despite the strides made in increasing the number of physical crystallization trials throughput, the act of finding just a few crystals among potentially thousands of crystallization experiments still remains a task requiring human input. A number of attempts are being made to automate the crystal detection from the imaged drop, and varying degrees of success have been reported (Liu *et al.*, 2008). Automated crystal recognition has the potential to reduce the time consuming human effort for screening crystallization drop images. Protein crystallization has always been largely trial-and-error, which is why it is sometimes regarded as more of an art than a science. Some proteins crystallize readily whereas a large proportion of proteins stubbornly refuse to produce diffraction quality crystals (Chayen *et al.*, 2008).

X-RAY DIFFRACTION DATA COLLECTION

The developments at synchrotron facilities for X-ray diffraction data collection have made it possible to rapidly collect data with minimum human intervention. The improvements include unprecedented brightness in the soft and hard X-ray spectral regions at the new third generation light sources. The tunable canted undulator radiation sources in multiple sections of the storage ring for macromolecular crystallography work independently. The experimental end-station includes instrumentation to handle micro-crystals of only a few microns in dimension, fast-framing active-pixel detectors for diffraction image recording, such as the novel *PILATUS* hybrid pixel array detector at the Swiss Light Source (**pixel** appar**atu**s for the **S**LS) (Broennimann *et al.*, 2006), automated sample changers, crystal auto-centering and cryogenic apparatus. Computer protocols have also developed to ensure that optimal strategies are used for data collection and to extract the maximum information from the collected data (Bourenkov and Popov, 2006; Popov and Bourenkov, 2003). These bright sources and advanced instrumentation will extend Macromolecular Crystallography (MX) beamlines into unexplored realms of sample size and experimental design. Research in all of these areas has been developed and implemented for high-throughput operations at almost all synchrotron sources worldwide.

The use of automated sample changers at synchrotron sources, combined with sample barcode readers, has facilitated the rapid screening of crystals and data collection. For example, the CATS sample changer at the BESSY-MX beamline BL14.1 (Berlin, Germany) is capable of handling up to 90 frozen samples. Automated sample exchange at EMBL/ESRF beamlines is coupled with *DNA* analysis, an expert system which takes test images of the crystal of interest, autoindexes the images and makes a strategy prediction based on the Laue class and uses the program *BEST* (Bourenkov and Popov, 2006) to suggest the exposure time to be used. It can then collect optimum data and integrate the resulting data set using *MOSFLM* (Leslie, 1992). The *DNA* project aims to completely automate the collection and processing of X-Ray protein crystallographic data. It is now possible for users to control data collection remotely through advancements in end-station control software (Soltis *et al.*, 2008). For example, *MxCube* installed on ESRF end stations provides a unified beamline control module for the MX station and ties together all the operations involved in collecting data for MX experiments in a single user interface. Several instances of the *MxCube* interface can be run simultaneously so that data collection can be controlled remotely from the home lab. Remote access to synchrotron sources is becoming more common place with, for example, 50% of users accessing beamlines at the Stanford Synchrotron Radiation Source (SSRL), USA in this way.

STRUCTURE DETERMINATION METHODS

The central problem in X-ray crystallography is the determination of the phase problem. The phrase, "the phase problem" is used to highlight the difficulties in determining reflection phase information from measured intensities. The intensities of diffracted X-rays can be derived experimentally but their corresponding phase angles have to be determined indirectly. Experimental phasing of crystal structures relies on the accurate measurement of two or more sets of reflections from isomorphous crystals, where the scattering power of a small number of atoms is different for each set. The techniques of isomorphous replacement, anomalous scattering, molecular replacement (MR) and anomalous dispersion (SAD and MAD) are commonly used to solve the phase problem.

Isomorphous replacement requires the introduction of atoms of a high atomic number (heavy atoms such as mercury, platinum, uranium) into the macromolecule under study without disrupting its structure or packing in the crystal. Thus, a perfect isomorphous derivative is one in which the only change between it and the native molecule is the incorporation of one or more heavy atoms. This is commonly done by soaking crystals of native molecules in a solution containing the desired heavy atom. The binding of these atoms to functional groups in the macromolecules is facilitated by the presence of large solvent channels in protein and nucleic acid crystals into which these functional groups protrude. The addition of one or two more heavy atoms to a macromolecule introduces differences in the diffraction pattern of the derivative relative to that of the native. If this addition is truly isomorphous, these differences will represent the contribution from the heavy atoms only; thus, the problem of determining atomic positions is initially reduced to locating the position of a few atoms. Once the positions of these atoms have been accurately determined, they are used to calculate a set of phases for data measured from the native crystals. Although, theoretically, one needs only two isomorphous derivatives to determine the three-dimensional structure of biological macromolecules, in practice more than two are usually necessary owing to errors in data measurement and scaling, and lack of isomorphism.

If the protein contains anomalously scattering atoms, the difference in intensity between the Bijvoet pairs can be exploited for protein phase angle determination. In the multiple-wavelength anomalous dispersion method (MAD) the wavelength dependence of the anomalous scattering is used. The principle of this method is rather old, but it was the introduction of the tunable synchrotron radiation sources that made it a technically feasible method for protein structure determination. Hendrickson and colleagues (Hendrickson *et al.*, 1988; Murthy *et al.*, 1988) were the first to take advantage of this method for protein structure determination. Of course, the protein must contain an element that gives a sufficiently strong anomalous signal. Therefore, the elements in the upper rows of the periodic table are not suitable. Hendrickson showed that the presence of one Selenium atom (Se; atomic number 34) in a protein of not more than 150 amino acid residues is sufficient for a successful application of MAD (Hendrickson *et al.*, 1990) providing the data are of sufficient quality. With more Se atoms, the size of the protein can, of course, be larger. One way to introduce Se into protein is by growing a methionine-auxotroph microorganism in the presence of Se-methionine rather than methionine.

The main advantage of MAD versus MIR is that for MAD, all data sets are collected from the same crystal. Thus, there are no problems with non-isomorphism. A disadvantage is that a very stable and energy-tunable source of radiation is required, typically at a synchrotron site. Another disadvantage is that a number of diffraction data sets have to be collected from one crystal, which runs the risk that the onset of radiation damage may introduce problems with non-isomorphism.

In recent years, great progress has been made in improving the interpretation of poor quality electron density maps. In addition to improvements in refinement procedures for the parameters of anomalous scatterers, solvent flattening and histogram matching as well as statistical density modification algorithms have become more powerful. A few years ago, Rice *et al.*, analyzed 18 MAD datasets and concluded that data collection at one wavelength would have been sufficient in most cases to resolve the phase ambiguity when combined with density modification (Rice *et al.*, 2000). The method called SAD (Single-wavelength Anomalous Diffraction) was originally demonstrated by Wang (1985). Data for SAD is usually collected so that it is highly redundant in order to reduce random variation due to the usual errors associated with data collection; as the anomalous signal is usually of the order of a few percent, accurate measurements are prerequisite to success. However, radiation damage at third generation synchrotron sources is a major factor in the lifetime of crystals and it has been noted that this effect may become dominant after rather short periods of exposure (Ravelli and Garman, 2006). The usable signal deteriorates rapidly after high exposures and this effect becomes dominant over increased redundancy in phasing trails in the experiment. The significant isomorphic variation of the diffracted intensities is induced by X-ray irradiation. These intensity changes also allow the crystal structure to be solved by the radiation-damage-induced phasing (RIP) technique. Furthermore, the use of modern approaches to phasing such as fast halide soaks allow structures to be determined more rapidly than traditional techniques. Recently, it has been demonstrated that ultraviolet radiation can be used to induce intensity differences in protein crystals that can be used to obtain phase information, a technique known as UV-RIP (ultraviolet radiation damage-induced phasing). The feasibility of this method for phasing of macromolecules and their complexes has been successfully demonstrated by Raimond Ravelli and coworkers (Nanao and Ravelli, 2006). The structural changes induced under UV irradiation are more specific compared to those from X-rays and allowed an elegant phasing scheme to be worked out leading to precise experimental phase information (Rudino-Pinera *et al.*, 2007; Schonfeld *et al.*, 2008).

The sulphur SAD phasing method in macromolecular crystallography (Dauter, 2002) uses anomalous scattering from substructures containing heavier elements such as S, P or Cl present in the crystals. This method has an advantage over other phasing methods such as MAD and isomorphous replacement in that it can be used when neither a Se-Met nor a heavy atom is present in the crystal. Generally, the anomalous signal of sulfur atoms in a protein crystal is so weak that it requires accurate, high quality and highly redundant data collection at longer wavelength for phasing to be successful (Mueller-Dieckmann *et al.*, 2007).

The molecular replacement (MR) method (Long *et al.*, 2008) makes use of a known three-dimensional structure as a suitable starting model to provide phase angles for the observed structure factor amplitudes from the unknown structure. When a family of homologous structures is available, it is often preferable to calculate an "average structure" for use as a starting model. Quite often, if the relative positions of the copies are not known, the presence of more than one copy of the unknown structure in the asymmetric unit can complicate the search. In such cases, the phase determination can be facilitated through the use of an oligomeric search model. Placement of the molecule in the target unit cell requires six parameters: 3 rotational and 3 translational parameters. More recently, software aimed at automatically assembling the set of "best" models for MR has been developed. Examples are MrBUMP (Keegen & Winn, 2007) and BALBES (Long *et al.*, 2008).

AUTOMATED COMPUTER PROGRAMS

Crystal structure determination by isomorphous replacement and anomalous scattering techniques is a multi-step process where each step from substructure determination to model building, refinement and validation, requires certain decisions to be made. These decisions dictate the choice of the crystallographic computer programs that are most suitable to perform the specific tasks and the optimal input parameters for each of these programs. The important parameters include the space group of the crystal, the number of molecules in the asymmetric unit, the type of the heavy atom derivative, the extent of derivatisation, the diffraction limit of both the native and the derivatised crystal and the quality of the collected diffraction data. After the collection of the X-ray data (of native and/or derivative crystals), existing and well established crystallographic computer programs for X-ray data processing and scaling, for solving the substructure, for the refinement of heavy atoms, phase calculation, density modification, phase extension, non-crystallographic symmetry (NCS) averaging (if more than one molecule is present in the asymmetric unit) are normally relied upon in order to progress to an interpretable electron density map. The interpretability of the map depends to a large extent on the success of the preceding steps and is generally limited by the resolution of the data and the quality of the phase information. Traditionally, each of the steps described above are carried out by an experienced crystallographer capable of finding a successful structure determination pathway.

High-throughput X-ray structure determination requires software that is automated and designed for minimum user intervention. For example, *Auto-Rickshaw*, an automated crystal structure determination pipeline (Panjikar *et al.*, 2005), provides a number of phasing protocols ranging from experimental phasing, molecular replacement and a combination of both techniques. The pipeline invokes a variety of macromolecular crystallographic program packages during the structure determination process using its machine learning and automatic crystallographic decision making process for the analysis. The pipeline produces partial or complete models in a short time, once a minimal set of input parameters and scaled intensity data are provided by the user. The invoked crystallographic packages are the CCP4 suite of crystallographic programs, *SHARP, SHELX, OASIS, BP3, SOLVE/RESOLVE, CNS, ARP/wARP* and the Uppsala software factory. The pipeline is largely being used for the quick validation of X-ray diffraction experiments at synchrotron sources including the EMBL-Hamburg beamline (Brzezinski *et al.*, 2007; Yu *et al.*, 2006). The system is available as a web server (http://www.embl-hamburg.de/Auto-Rickshaw) and, since April 2008, it has been available to the worldwide crystallographic community to aid efforts in structure determination, quick X-ray data evaluation and optimization of data collection.

There has also been considerable development in programs by various authors with different goals and degrees of built-in automation, such as *SOLVE/RESOLVE* (Terwilliger, 2000; Terwilliger and Berendzen, 1999), *AUTOSHARP/SHARP* (Vonrhein *et al.*, 2007), *BnP* (Weeks *et al.*, 2005) and *HKL2MAP* (Pape and Schneider, 2004) that automatically locate heavy atom sites/anomalous scatterers and generate a phase set, speeding up the whole process of structure determination.

More recently, automated systems for structure determination have been developed that combine different crystallographic computer programs to build a crystal structure determination pipeline. Examples include *ACrS* (Brunzelle *et al.*, 2003), *ELVES* (Holton and Alber, 2004), *CRANK* (Ness *et al.*, 2004) and *PHENIX* (Python-based hierarchical environment for integrated Xtallography) (Adams *et al.*, 2004). Most of these systems require sufficiently high resolution X-ray data and reasonable phase information, mainly because the model building step is based on either the program package *ARP/wARP* (Langer *et al.*, 2008; Morris *et al.*, 2004; Perrakis *et al.*, 1999), *RESOLVE* (Terwilliger, 2000) or *BUCANEER* (Cowtan, 2006). The Bucaneer software for automated model building has new algorithms for the interpretation of electron density maps and automated model building, which enables protein models to be rapidly constructed without the need for user intervention, further simplifying and streamlining the whole process. At present, the success rates for these programs are limited by the resolution of the data set (typically, the resolution must exceed 2.5 Å to build an entire model successfully). The program *COOT* (Emsley and Cowtan, 2004) can be used for manual model building, model completion and validation when automated procedures fail to produce a complete model. *COOT* displays maps and models and allows model manipulations such as idealization, real space refinement, manual rotation/translation, rigid-body fitting, ligand searching, solvation, mutations and rotamer fitting.

STRUCTURE TO FUNCTION

Determining the biological function of the protein from the protein structure is the ultimate goal of SP projects. The protein function may be described at several different levels by combining structural, biochemical and biophysical data. For instance, the cellular function may depend on its temporal and spatial expression in the cell; the molecular function of an enzyme depends entirely on its specificity and the reaction it catalyses and the physiological function on the organ in which the expressing cells are found (van Helden *et al.*, 2000). The structure-based identification of functional elements through bioinformatics tools with the help of powerful computer systems is the first step towards the prediction of function. Hypothetical proteins identified as targets for structure determination bear little or no sequence similarity to known three-dimensional protein structures and sequences, and therefore have a high likelihood of adopting a novel fold.

If a solved target has no sequence similarity and no structural resemblance to any of the 3D structures available in the protein data bank (PDB), the bioinformatics approach to evaluate protein function will fail. In such cases, site-directed mutagenesis or chemical modification of specific residues can be performed to identify key residues critical for binding or catalysis.

If the structure of such a target bares a strong similarity to an already known protein fold, then the alignment of the structures can be used to identify evolutionary conserved residues important for enzymatic function. A number of software tools exist for protein structure and sequence analysis such as the ProFunc(Laskowski *et al.*, 2005) and ProTarget (Sasson and Linial, 2005) servers, which perform automated protein functional annotation using the solved structures. More exciting novel methodologies for assigning protein functions computationally have recently been developed by SP centers (Watson *et al*, 2007; Song *et al*, 2007; Hermann *et al*, 2007). A wiki, named, The Open Protein Structure Annotation Network (TOPSAN) is designed to collect, share and distribute information about protein three-dimensional structures and to provide knowledge about the functions and roles of these proteins in their respective organisms (http://proteins.burnham.org/).

Biotechnological implications

With the advent of the post-genomic age, life science research has evolved from genomics to proteomics. In general, the start up of structural proteomics projects has had a profound impact on structural biology research programs (Kambach, 2007). The high-throughput methods and technologies developed by the SP projects are focused on *what proteins look like*. In contrast, structural biologists who utilize these technologies are focused on *how proteins work*. Solving novel protein structures is regarded as an important analytical tool to complement general biochemical research. Appreciation of the value and power of the techniques available today at SP centers is enormous. The availability of SP technologies developed at the specialized centers and the utilization of the SP results promotes cost effective approaches to cover protein fold space. The ultimate goal of structural proteomics is to determine the structure of all proteins in a cell or organism. SP approaches have led to several thousands of protein structures being determined and deposited into the protein data bank (PDB).

SP projects have solved some 6000 novel protein structures to date with the present rate being approximately 1000 structures per year. SP centers per structure contribution to novel leverage was over 4-fold higher than that for non-SP structural biology efforts during the past 8 years (Nair *et al.*, 2009). These structures can be useful for many purposes although it may still be too early to assess the full importance and impact of these structures. In many cases, SP structures have already proven their value in the broad scientific community.

The protein structures from SP centers can be used to determine the structure of novel proteins using molecular replacement. Most SP structures are from bacterial proteins which are relatively easy to purify and crystallize. Eukaryotic protein structures on the other hand, particularly those from human, can be much more difficult to solve. Their bacterial counterparts can sometimes be used to phase the human homolog. For example, the X-ray structure of *homo sapiens* protein FLJ36880 (PDB ID 1SAW) was solved by molecular replacement using a homology model generated by PDB codes 1NR9 and 1GTT (SP structures). A search of the PDB reveals that several hundreds of SP structures have already been used as homologous probe structures by international investigators to solve novel structures.

SP structures can be used to generate homology models of entire protein families using bioinformatics computational tools and these structural models can be used to understand the functions of these proteins without the need for experimental 3D structures.

SP structures are important for the functional analysis of proteins annotated as hypothetical proteins or with vague functions. For example, the structure-function relationship of the SP protein structure, *E. coli* arabinose isomerase, ECAI (solved at NYSGRC; PDB codes 2AJT and 2HXG) revealed its application in tagatose (a new sweetener) production. The crystal structure of ECAI forms a basis for identifying molecular determinants responsible for isomerization of arabinose to ribulose *in vivo* and galactose to tagatose *in vitro*. Another SP structure, ybeY, (solved at NYSGRC), through its sequence similarity to a number of predicted metal-dependent hydrolases, provides a structural and functional description for this protein family. The function of many of these novel SP structures remains unclear and provides the potential for further functional investigations through bioinformatics and proteomics approaches.

The knowledge of accurate three-dimensional protein structures is a pre-requisite for rational drug design. SP has had a substantial impact on the drug discovery pipeline and has expanded our understanding of the molecular processes governing human diseases at the atomic level (Lundstrom, 2007; Tari *et al.*, 2005). Recent advances in SP technologies have increased the wealth of protein structures available. Further functional studies will aid in the development of effective therapeutics and small molecule inhibitors against life-threatening illnesses and diseases. High-throughput screening has increased the sophistication to prepare protein–drug complexes. Automated robotic crystallization, the remote access to synchrotron facilities from the home institution, and advances in computational facilities have helped to increase the speed in lead identification and ultimately, provide more detailed information on their interactions to focus medicinal chemistry efforts. In recent years, the use of 3D protein structural information in drug discovery research has matured and has been applied at all levels, ranging from target identification by utilizing combinatorial chemistry to the design of suitable drug candidates.

SP projects have significantly increased the number of high-resolution human protein structures and these structures have been used to understand human diseases and to develop selective inhibitors for drug discovery. For example, the SP structure isocitrate lyase, a persistence factor of *Mycobacterium tuberculosis* (Sharma *et al.*, 2000) has been extensively studied by biomedical scientists and has facilitated the development of glyoxylate cycle inhibitors as new drugs for the treatment of tuberculosis (Munoz-Elias and McKinney, 2005; Purohit *et al.*, 2007). Recent findings are likely to stimulate a number of follow-up studies and facilitate the identification of macrophage receptors and signal transduction pathways targeted for immunomodulation by pathogenic mycobacteria (Ehrt and Schnappinger, 2007). The joint Canadian-British-Swedish Structural Genomics Consortium (SGC) operates three SP centers focused on human proteins of medical importance. The structure of phytanoyl-CoA hydoxylase provides the basis for further understanding the major molecular cause of Refsum disease, a peroxisomal disorder with severe neurological symptoms (McDonough *et al.*, 2005). With regard to the protein kinase domain, SP has contributed >50% of all novel kinases structures during the past three years and determined more than 30 novel catalytic domain structures(Marsden and Knapp, 2008). The crystal structures of farnesyl diphosphate synthase (FDPS) complexed with nitrogen-containing bisphosphonates currently used for osteoporosis therapy allowed a molecular mechanism of action to be postulated for drug discovery (Kavanagh *et al.*, 2006). Structural proteomics data of protein families combined with virtual ligand screening and discovery, pocket identification and compound optimization holds particular promise for advancing early stage discovery programs(Nicola and Abagyan, 2009; Weigelt *et al.*, 2008). The availability of technologies for genomics, proteomics, structural genomics, transcriptomics, and high-throughput screening for identification of targeted drugs which were almost unimaginable only a few years ago (Kornek and Selzer, 2009).

For over a century, chemists have developed the ability to control the arrangement of small numbers of atoms inside molecules, leading to revolutions in drug design, plastics, and numerous other areas. Nanotechnology encompasses research and development in the scale of 0.1 to 100 nanometers to create unique structures, devices and systems. Many existing technologies depend critically on processes that take place on the nanometer scale. In a cell, molecules are often organized into functional aggregates, normally with nanometer dimensions. Visualizing and studying these structures-especially as they change dynamically during cycles of function-is one of the key challenges posed to nanobioscience. Information generated by structural proteomics (X-ray crystallography) coupled with molecular simulations has begun to clarify the dynamics of these nanostructures, aiding the design of novel nanobiomachines for medical applications (Moghimi *et al*, 2005; Baumgaertner 2008). Combining the static snapshot structures of the ribosome (Ramkrishnan, 2002; Tajkhorshid *et al*, 2002) with the ribosome simulations has helped to elucidate a crucial molecular mechanism for gene expression, but has also opened the door for simulations of other large molecular machines important for gene expression and drug design (Sanbonmatsu *et al*, 2005).

Many crucially important processes in biology involve the translocation of a nanobiosystem through nanometer-scale pores, such as the translocation of DNA and RNA across nuclear pores and protein transport through membrane channels (Baumgaertner 2008). The study of interactions between nanomaterials such as Carbon Nanotubes (CNTs) and cellular components, such as membranes and biomolecules, is fundamental for the rational design of nanodevices interfacing with biological systems (Yeh and Hummer, 2004; Zimmerli and Koumoutsakos, 2008). For instance, some pathogenic bacteria have a complex injection device (*syringe-like nano-organelle*) comprising many proteins. This molecular syringe has to be attached across two membranes so that proteins can be transferred from the bacterial cells into human cells. In particular, these syringes can be used to deliver toxins into infected cells. Numerous pathogens cause disease, for example, *Salmonella*; *Yersinia* and Intestinal *E.coli*, are both equipped with this nano-organelle. This bacterial syringe represents an excellent potential target for drugs to combat these diseases. In order to understand the function of the syringe, structural knowledge of the many proteins comprising the syringe is required. SP technologies provide the basis to bring about a revolution in the design of drug delivery systems that are small and smart, by understanding structure function relationships of bacterial syringes.

In living organisms, proteins function when they bind to their partners in a specific way in a short physiological timescale. The experimental structure determination of protein complexes is more difficult than that of individual proteins. For example, very little is known about the process of DNA replication in eukaryote cells – cells that have their genetic information contained in a nucleus – which is performed by the complex molecular machine called the replisome. To understand this process at the molecular level requires the protein structure of the entire replisome to be determined, which comprises some 30-40 proteins. Advances in SP technologies allow the structures of the individual components of macromolecular nanomachines like the replisome to be determined (Warren *et al*, 2008) which enable models describing the macromolecular systems to be built. Biologists have realized the impracticality of trying to successfully predict complex molecular mechanisms using intuition.

Accordingly, biologists are now seeking computational methodologies and tools as research kits for structural modeling of complexes of their choice (Vakser and Kundrotas, 2008). Protein complex models obtained by such approaches can be further refined using low-resolution structural data obtained from electron microscopy (EM), small angle X-ray scattering (SAXS) and hydrodynamic studies, which provide further evidence of how proteins within the complexes associate to perform their biological functions.

Conclusion

The collection of complete large-scale sequence mapping of many cellular organisms has resulted in biology suddenly being flooded with genome-based data. Researchers are attempting to develop new technologies to investigate new kinds of questions about the complex nature of living cells. Structural proteomics follows an integrated approach to mass protein structure analysis and represents the most efficient and powerful method for acquiring atomic-level structural information for individual proteins as well as multi-protein complexes. More importantly, SP technologies are trying to decipher biological function of novel proteins using sequence-structural similarity to known proteins- however, this technology is limited to predicting the cellular function of a given protein. In general, the biological view of where SP is impacting biology, particularly in the areas of biotechnology, nanosciences and drug discovery, is marvelous. Furthermore, the advances in SP technologies give a glimpse of the dramatic impact that SP is likely to have on all fields of biological sciences in the future.

References

ADAMS, P.D., GOPAL, K., GROSSE-KUNSTLEVE, R.W., HUNG, L.W., IOERGER, T.R., McCOY, A.J., MORIARTY, N.W., PAI, R.K., READ, R.J., ROMO, T.D., SACCHETTINI, J.C., SAUTER, N.K., STORONI, L.C. AND TERWILLIGER, T.C. (2004). Recent developments in the PHENIX software for automated crystallographic structure determination. *J Synchrotron Radiat* **11**, 53-55.

ALBALA, J.S., FRANKE, K., McCONNELL, I.R., PAK, K.L., FOLTA, P.A., RUBINFELD, B., DAVIES, A.H., LENNON, G.G. AND CLARK, R. (2000). From genes to proteins: high-throughput expression and purification of the human proteome. *J Cell Biochem* **80**, 187-191.

ALTEROVITZ, G., LIU, J., CHOW, J. AND RAMONI, M.F. (2006). Automation, parallelism, and robotics for proteomics. *Proteomics* **6**, 4016-4022.

ASHERIE, N. (2004). Protein crystallization and phase diagrams. *Methods* **34**, 266-272.

BANCI, L., BAUMEISTER, W., ENFEDAQUE, J., HEINEMANN, U., SCHNEIDER, G., SILMAN, I. AND SUSSMAN, J.L. (2007). Structural proteomics: from the molecule to the system. *Nat Struct Mol Biol* **14**, 3-4.

BAUMGAERTNER, A. (2008) Concepts in bionanomachines: Translocators. *J Computational & Theoretical Nanoscience*, **5**, 1-39.

BERGFORS, T. (2003). Seeds to crystals. *J Struct Biol* **142**, 66-76.

BERMAN, H.M., WESTBROOK, J.D., GABANYI, M.J., TAO, W., SHAH, R., KOURANOV, A., SCHWEDE, T., ARNOLD, K., KIEFER, F., BORDOLI, L., KOPP, J., PODVINEC, M., ADAMS, P.D., CARTER, L.G., MINOR, W., NAIR, R. AND LA BAER, J. (2009). The protein structure initiative structural genomics knowledgebase. *Nucleic Acids Res* **37**, D365-368.

BERROW, N.S., ALDERTON, D., SAINSBURY, S., NETTLESHIP, J., ASSENBERG, R., RAHMAN, N., STUART, D.I. AND OWENS, R.J. (2007). A versatile ligation-independent cloning method suitable for high-throughput expression screening applications. *Nucleic Acids Res* **35**, e45.

BLOW, N. (2008). Structural genomics: inside a protein structure initiative center. *Nat Methods* **5**, 203-207.

BOURENKOV, G.P. AND POPOV, A.N. (2006). A quantitative approach to data-collection strategies. *Acta Crystallogr D Biol Crystallogr* **62**, 58-64.

Broennimann, C., Eikenberry, E.F., Henrich, B., Horisberger, R., Huelsen, G., Pohl, E., Schmitt, B., Schulze-Briese, C., Suzuki, M., Tomizaki, T., Toyokawa, H. and Wagner, A. (2006). The PILATUS 1M detector. *J Synchrotron Radiat* **13**, 120-130.

Bruel, C., Cha, K., Reeves, P.J., Getmanova, E. and Khorana, H.G. (2000). Rhodopsin kinase: expression in mammalian cells and a two-step purification. *Proc Natl Acad Sci U S A* **97**, 3004-3009.

Brunzelle, J.S., Shafaee, P., Yang, X., Weigand, S., Ren, Z. and Anderson, W.F. (2003). Automated crystallographic system for high-throughput protein structure determination. *Acta Crystallogr D Biol Crystallogr* **59**, 1138-1144.

Brzezinski, K., Stepkowski, T., Panjikar, S., Bujacz, G. and Jaskolski, M. (2007). High-resolution structure of NodZ fucosyltransferase involved in the biosynthesis of the nodulation factor. *Acta Biochim Pol* **54**, 537-549.

Chayen, N.E. and Saridakis, E. (2008). Protein crystallization: from purified protein to diffraction-quality crystal. *Nature Methods*, **5**, 147-153.

Cho, W.C. (2007). Proteomics technologies and challenges. Genomics Proteomics *Bioinformatics* **5**, 77-85.

Collins, F.S., Morgan, M. and Patrinos, A. (2003). The Human Genome Project: lessons from large-scale biology. *Science* **300**, 286-290.

Cowtan, K. (2006). The Buccaneer software for automated model building. 1. Tracing protein chains. *Acta Crystallogr D Biol Crystallogr* **62**, 1002-1011.

D'Arcy, A., Villard, F. and Marsh, M. (2007). An automated microseed matrix-screening method for protein crystallization. *Acta Crystallogr D Biol Crystallogr* **63**, 550-554.

Dauter, Z. (2002). New approaches to high-throughput phasing. *Curr Opin Struct Biol* **12**, 674-678.

Ehrt, S. and Schnappinger, D. (2007). Mycobacterium tuberculosis virulence: lipids inside and out. *Nat Med* **13**, 284-285.

Emsley, P. and Cowtan, K. (2004). Coot: model-building tools for molecular graphics. *Acta Crystallogr D Biol Crystallogr* **60**, 2126-2132.

Gerdts, C.J., Elliott, M., Lovell, S., Mixon, M.B., Napuli, A.J., Staker, B.L., Nollert, P. and Stewart, L. (2008). The plug-based nanovolume Microcapillary Protein Crystallization System (MPCS). *Acta Crystallogr D Biol Crystallogr* **64**, 1116-1122.

Gileadi, O., Burgess-Brown, N.A., Colebrook, S.M., Berridge, G., Savitsky, P., Smee, C.E., Loppnau, P., Johansson, C., Salah, E. and Pantic, N.H. (2008). High throughput production of recombinant human proteins for crystallography. *Methods Mol Biol* **426**, 221-246.

Heinemann, U., Frevert, J., Hofmann, K., Illing, G., Maurer, C., Oschkinat, H. and Saenger, W. (2000). An integrated approach to structural genomics. *Prog Biophys Mol Biol* **73**, 347-362.

Hendrickson, W.A., Horton, J.R. and LeMaster, D.M. (1990). Selenomethionyl proteins produced for analysis by multiwavelength anomalous diffraction (MAD): a vehicle for direct determination of three-dimensional structure. *EMBO J* **9**, 1665-1672.

Hendrickson, W.A., Smith, J.L., Phizackerley, R.P. and Merritt, E.A. (1988). Crystallographic structure analysis of lamprey hemoglobin from anomalous dispersion of synchrotron radiation. *Proteins* **4**, 77-88.

HERMANN, J.C., MARTI-ARBONA, R., FEDOROV, A.A., FEDOROV, E., ALMO, S.C., SHOICHET, B.K. AND RAUSHEL, F.M. (2007) Structure-based activity prediction for an enzyme of unknown function. *Nature* 448, 762-763.

HIRAKI, M., KATO, R., NAGAI, M., SATOH, T., HIRANO, S., IHARA, K., KUDO, N., NAGAE, M., KOBAYASHI, M., INOUE, M., UEJIMA, T., ODA, S., CHAVAS, L.M., AKUTSU, M., YAMADA, Y., KAWASAKI, M., MATSUGAKI, N., IGARASHI, N., SUZUKI, M. AND WAKATSUKI, S. (2006). Development of an automated large-scale protein-crystallization and monitoring system for high-throughput protein-structure analyses. *Acta Crystallogr D Biol Crystallogr* 62, 1058-1065.

HOLTON, J. AND ALBER, T. (2004). Automated protein crystal structure determination using ELVES. *Proc Natl Acad Sci U S A* 101, 1537-1542.

KAMBACH, C. (2007). Pipelines, robots, crystals and biology: what use high throughput solving structures of challenging targets? *Curr Protein Pept Sci* 8, 205-217.

KAVANAGH, K.L., DUNFORD, J.E., BUNKOCZI, G., RUSSELL, R.G. AND OPPERMANN, U. (2006). The crystal structure of human geranylgeranyl pyrophosphate synthase reveals a novel hexameric arrangement and inhibitory product binding. *J Biol Chem* 281, 22004-22012.

KEEGAN, R.M. AND WINN, M.D. (2007) Automated search-model discovery and preparation for structure solution by molecular replacement. *Acta Cryst* D63, 447-457.

KORNEK, G. AND SELZER, E. (2009). Targeted therapies in solid tumours: pinpointing the tumour's Achilles heel. *Curr Pharm Des* 15, 207-242.

LANGER, G., COHEN, S.X., LAMZIN, V.S. AND PERRAKIS, A. (2008). Automated macromolecular model building for X-ray crystallography using ARP/wARP version 7. *Nat Protoc* 3, 1171-1179.

LASKOWSKI, R.A., WATSON, J.D. AND THORNTON, J.M. (2005). ProFunc: a server for predicting protein function from 3D structure. *Nucleic Acids Res* 33, W89-93.

LESLIE, A.G.W. (1992). Recent changes to the *MOSFLM* package for processing film and image plate data. *Jnt CCP4/ESF±EACMB Newslett. Protein Crystallogr.* 26.

LIU, R., FREUND, Y. AND SPRAGGON, G. (2008). Image-based crystal detection: a machine-learning approach. *Acta Crystallogr D Biol Crystallogr* 64, 1187-1195.

LONG, F., VAGIN, A.A., YOUNG, P. AND MURSHUDOV, G.N. (2008). BALBES: a molecular-replacement pipeline. *Acta Crystallogr D Biol Crystallogr* 64, 125-132.

LUNDSTROM, K. (2007). Structural genomics and drug discovery. *J Cell Mol Med* 11, 224-238.

MANJASETTY, B.A., SHI, W., ZHAN, C., FISER, A. AND CHANCE, M.R. (2007). A high-throughput approach to protein structure analysis. *Genet Eng (N Y)* 28, 105-128.

MANJASETTY, B.A., TURNBULL, A.P., PANJIKAR, S., BUSSOW, K. AND CHANCE, M.R. (2008). Automated technologies and novel techniques to accelerate protein crystallography for structural genomics. *Proteomics* 8, 612-625.

MARSDEN, B.D. AND KNAPP, S. (2008). Doing more than just the structure-structural genomics in kinase drug discovery. *Curr Opin Chem Biol* 12, 40-45.

MCDONOUGH, M.A., KAVANAGH, K.L., BUTLER, D., SEARLS, T., OPPERMANN, U. AND SCHOFIELD, C.J. (2005). Structure of human phytanoyl-CoA 2-hydroxylase identifies molecular mechanisms of Refsum disease. *J Biol Chem* 280, 41101-41110.

MCPHERSON, A. (2004). Introduction to protein crystallization. *Methods* 34, 254-265.

MIYATAKE, H., KIM, S.H., MOTEGI, I., MATSUZAKI, H., KITAHARA, H., HIGUCHI, A. AND

Miki, K. (2005). Development of a fully automated macromolecular crystallization/observation robotic system, HTS-80. *Acta Crystallogr D Biol Crystallogr* **61**, 658-663.

Moghimi, S.M., Hunter A.C., Murray, J.C. (2005) Nanomedicine: current status and future prospects. *The FASEB Journal* **19**, 311-330.

Morris, R.J., Zwart, P.H., Cohen, S., Fernandez, F.J., Kakaris, M., Kirillova, O., Vonrhein, C., Perrakis, A. and Lamzin, V.S. (2004). Breaking good resolutions with ARP/wARP. *J Synchrotron Radiat* **11**, 56-59.

Mueller-Dieckmann, C., Panjikar, S., Schmidt, A., Mueller, S., Kuper, J., Geerlof, A., Wilmanns, M., Singh, R.K., Tucker, P.A. and Weiss, M.S. (2007). On the routine use of soft X-rays in macromolecular crystallography. Part IV. Efficient determination of anomalous substructures in biomacromolecules using longer X-ray wavelengths. *Acta Crystallogr D Biol Crystallogr* **63**, 366-380.

Munoz-Elias, E.J. and McKinney, J.D. (2005). Mycobacterium tuberculosis isocitrate lyases 1 and 2 are jointly required for in vivo growth and virulence. *Nat Med* **11**, 638-644.

Murthy, H.M., Hendrickson, W.A., Orme-Johnson, W.H., Merritt, E.A. and Phizackerley, R.P. (1988). Crystal structure of Clostridium acidi-urici ferredoxin at 5-A resolution based on measurements of anomalous X-ray scattering at multiple wavelengths. *J Biol Chem* **263**, 18430-18436.

Nair, R., Liu, J., Soong, T.T., Acton, T.B., Everett, J.K., Kouranov, A., Fiser, A., Godzik, A., Jaroszewski, L., Orengo, C., Montelione, G.T. and Rost, B. (2009). Structural genomics is the largest contributor of novel structural leverage. *J Struct Funct Genomics.* **10(2)**, 181-191.

Nanao, M.H. and Ravelli, R.B. (2006). Phasing macromolecular structures with UV-induced structural changes. *Structure* **14**, 791-800.

Ness, S.R., de Graaff, R.A., Abrahams, J.P. and Pannu, N.S. (2004). CRANK: new methods for automated macromolecular crystal structure solution. *Structure* **12**, 1753-1761.

Nicola, G. and Abagyan, R. (2009). Structure-based approaches to antibiotic drug discovery. *Curr Protoc Microbiol* Chapter 17, Unit17 12.

Page, R. (2008). Strategies for improving crystallization success rates. *Methods Mol Biol* **426**, 345-362.

Panjikar, S., Parthasarathy, V., Lamzin, V.S., Weiss, M.S. and Tucker, P.A. (2005). Auto-Rickshaw: an automated crystal structure determination platform as an efficient tool for the validation of an X-ray diffraction experiment. *Acta Crystallogr D Biol Crystallogr* **61**, 449-457.

Pape, T. and Schneider, T.R. (2004). HKL2MAP: a graphical user interface for phasing with SHELX programs. *J. Appl. Cryst* **37**, 843-844. .

Perrakis, A., Morris, R. and Lamzin, V.S. (1999). Automated protein model building combined with iterative structure refinement. *Nat Struct Biol* **6**, 458-463.

Popov, A.N. and Bourenkov, G.P. (2003). Choice of data-collection parameters based on statistic modelling. *Acta Crystallogr D Biol Crystallogr* **59**, 1145-1153.

Prinz, B., Schultchen, J., Rydzewski, R., Holz, C., Boettner, M., Stahl, U. and Lang, C. (2004). Establishing a versatile fermentation and purification procedure for human proteins expressed in the yeasts Saccharomyces cerevisiae and Pichia pastoris for structural genomics. *J Struct Funct Genomics* **5**, 29-44.

PUROHIT, H.J., CHEEMA, S., LAL, S., RAUT, C.P. AND KALIA, V.C. (2007). In Search of Drug Targets for Mycobacterium tuberculosis. *Infect Disord Drug Targets* **7**, 245-250.

RAMAKRISHNAN, V.(2002). Ribosome structure and the mechanism of translation. *Cell* 108, 557-72.

RAVELLI, R.B. AND GARMAN, E.F. (2006). Radiation damage in macromolecular cryocrystallography. *Curr Opin Struct Biol* **16**, 624-629.

RICE, L.M., EARNEST, T.N. AND BRUNGER, A.T. (2000). Single-wavelength anomalous diffraction phasing revisited. *Acta Crystallogr D Biol Crystallogr* **56**, 1413-1420.

RUDINO-PINERA, E., RAVELLI, R.B., SHELDRICK, G.M., NANAO, M.H., KOROSTELEV, V.V., WERNER, J.M., SCHWARZ-LINEK, U., POTTS, J.R. AND GARMAN, E.F. (2007). The solution and crystal structures of a module pair from the Staphylococcus aureus-binding site of human fibronectin--a tale with a twist. *J Mol Biol* **368**, 833-844.

SANBONMATSU KY, JOSEPH S AND TUNG CS (2005) Simulating movement of tRNA into the ribosome during decoding *PNAS*, **44**:15854-15859.

SASSON, O. AND LINIAL, M. (2005). ProTarget: automatic prediction of protein structure novelty. *Nucleic Acids Res* **33**, W81-84.

SCHONFELD, D.L., RAVELLI, R.B., MUELLER, U. AND SKERRA, A. (2008). The 1.8-A crystal structure of alpha1-acid glycoprotein (Orosomucoid) solved by UV RIP reveals the broad drug-binding activity of this human plasma lipocalin. *J Mol Biol* **384**, 393-405.

SEGELKE, B. (2005). Macromolecular crystallization with microfluidic free-interface diffusion. *Expert Rev Proteomics* **2**, 165-172.

SHARMA, V., SHARMA, S., HOENER ZU BENTRUP, K., MCKINNEY, J.D., RUSSELL, D.G., JACOBS, W.R., JR. AND SACCHETTINI, J.C. (2000). Structure of isocitrate lyase, a persistence factor of Mycobacterium tuberculosis. *Nat Struct Biol* **7**, 663-668.

SHRESTHA, B., SMEE, C. AND GILEADI, O. (2008). Baculovirus expression vector system: an emerging host for high-throughput eukaryotic protein expression. *Methods Mol Biol* **439**, 269-289.

SOLTIS, S.M., COHEN, A.E., DEACON, A., ERIKSSON, T., GONZALEZ, A., MCPHILLIPS, S., CHUI, H., DUNTEN, P., HOLLENBECK, M., MATHEWS, I., MILLER, M., MOORHEAD, P., PHIZACKERLEY, R.P., SMITH, C., SONG, J., VAN DEM BEDEM, H., ELLIS, P., KUHN, P., MCPHILLIPS, T., SAUTER, N., SHARP, K., TSYBA, I. AND WOLF, G. (2008). New paradigm for macromolecular crystallography experiments at SSRL: automated crystal screening and remote data collection. *Acta Crystallogr D Biol Crystallogr* **64**, 1210-1221.

SONG, L., KALYANARAMAN, C., FEDOROV, A.A., FEDOROV, E.V., GLASNER, M.E., BROWN, S., IMKER, H.J., BABBITT, P.C., ALMO, S.C., JACOBSON, M.P., GERLT, J.A. (2007) Prediction and assignment of function for a divergent N-succinyl amino acid racemase. *Nat. Chem. Biol* 3, 486-491.

STEVENS, R.C. (2000). Design of high-throughput methods of protein production for structural biology. *Structure* **8**, R177-185.

SANBONMATSU, K.Y., JOSEPH, S. AND TUNG, C.S. (2005). Simulating movement of tRNA into the ribosome during decoding. *PNAS*, **44**, 15854-15859.

TARI, L.W., ROSENBERG, M. AND SCHRYVERS, A.B. (2005). Structural proteomics in drug discovery. *Expert Rev Proteomics* **2**, 511-519.

TERWILLIGER, T.C. (2000). Maximum-likelihood density modification. *Acta Crystallogr D Biol Crystallogr* **56**, 965-972.

Terwilliger, T.C. and Berendzen, J. (1999). Automated MAD and MIR structure solution. *Acta Crystallogr D Biol Crystallogr* **55**, 849-861.

Thakur, A.S., Robin, G., Guncar, G., Saunders, N.F., Newman, J., Martin, J.L. and Kobe, B. (2007). Improved success of sparse matrix protein crystallization screening with heterogeneous nucleating agents. *PLoS ONE* **2**, e1091.

van Helden, J., Naim, A., Mancuso, R., Eldridge, M., Wernisch, L., Gilbert, D. and Wodak, S.J. (2000). Representing and analysing molecular and cellular function using the computer. *Biol Chem* **381**, 921-935.

Vakser, I.A. and Kundrotas, P. (2008). Predicting 3D structures of protein-protein complexes. *Curr Pharm Biotechnol* **9**, 57-66.

Vonrhein, C., Blanc, E., Roversi, P. and Bricogne, G. (2007). Automated structure solution with autoSHARP. *Methods Mol Biol* **364**, 215-230.

Walter, T.S., Diprose, J.M., Mayo, C.J., Siebold, C., Pickford, M.G., Carter, L., Sutton, G.C., Berrow, N.S., Brown, J., Berry, I.M., Stewart-Jones, G.B., Grimes, J.M., Stammers, D.K., Esnouf, R.M., Jones, E.Y., Owens, R.J., Stuart, D.I. and Harlos, K. (2005). A procedure for setting up high-throughput nanolitre crystallization experiments. Crystallization workflow for initial screening, automated storage, imaging and optimization. *Acta Crystallogr D Biol Crystallogr* **61**, 651-657.

Walter, T.S., Mancini, E.J., Kadlec, J., Graham, S.C., Assenberg, R., Ren, J., Sainsbury, S., Owens, R.J., Stuart, D.I., Grimes, J.M. and Harlos, K. (2008). Semi-automated microseeding of nanolitre crystallization experiments. *Acta Crystallogr Sect F Struct Biol Cryst Commun* **64**, 14-18.

Warren, E.M., Vaithiyalingam, S., Haworth, J., Greer, B., Bielinsky, A.K., Chazin, W.J. and Eichman B.F. (2008) Structural basis for DNA binding by replication initiator Mcm10. *Structure* **16**, 1892-1901.

Watson, J.D., Sanderson, S., Ezersky, A., Savchenko, A., Edwards, A., Orengo, C., Joachimiak, A., Laskowski, R.A., Thornton, J.M. (2007) Towards fully automated structure-based function prediction in structural genomics: a case study. *J Mol Biol* **367**, 1511-1522.

Weeks, C.M., Shah, N., Green, R., Miller, R. and Furey, W. (2005). Automated web- and grid-based protein phasing with BnP. *Acta Crystallogr A* **A61**, 152-154.

Weigelt, J., McBroom-Cerajewski, L.D., Schapira, M., Zhao, Y. and Arrowmsmith, C.H. (2008). Structural genomics and drug discovery: all in the family. *Curr Opin Chem Biol* **12**, 32-39.

Wooh, J.W., Kidd, R.D., Martin, J.L. and Kobe, B. (2003). Comparison of three commercial sparse-matrix crystallization screens. *Acta Crystallogr D Biol Crystallogr* **59**, 769-772.

Yakunin, A.F., Yee, A.A., Savchenko, A., Edwards, A.M. and Arrowsmith, C.H. (2004). Structural proteomics: a tool for genome annotation. *Curr Opin Chem Biol* **8**, 42-48.

Yeh, I.C. and Hummer, G. (2004) Nucleic acid transport through carbon nanotube membranes, *PNAS,* **101,** 12177-12182.

Yu, Y., Liang, Y.H., Brostromer, E., Quan, J.M., Panjikar, S., Dong, Y.H. and Su, X.D. (2006). A catalytic mechanism revealed by the crystal structures of the imidazolonepropionase from Bacillus subtilis. *J Biol Chem* **281**, 36929-36936.

Zimmerli, U. and Koumoutsakos, P. (2008). Simulations of Electrophoretic RNA Transport Through Transmembrane Carbon Nanotubes *Biophysical Journal,* **94,** 2546-2557.

Biotechnology and Genetic Engineering Reviews - Vol. 26, 371-388 (2009)

Biotechnology as the engine for the Knowledge-Based Bio-Economy

ALFREDO AGUILAR[1*], LAURENT BOCHEREAU[2] AND LINE MATTHIESSEN[3]

[1]Head of Unit Biotechnologies, Directorate Biotechnologies, Agriculture and Food, Research Directorate General, European Commission, 1049 Brussels, Belgium; [2]Head of Section, Science, Technology and Education, Delegation of the European Commission to the United States, 2300 M Street NW, 20037-1434, Washington DC, USA, and [3]Head of Unit Horizontal Aspects and Coordination, Directorate Biotechnologies, Agriculture and Food, Research Directorate General, European Commission, B-1049 Brussels, Belgium

Abstract

The European Commission has defined the Knowledge-Based Bio-Economy (KBBE) as the process of transforming life science knowledge into new, sustainable, eco-efficient and competitive products. The term "Bio-Economy" encompasses all industries and economic sectors that produce, manage and otherwise exploit biological resources and related services. Over the last decades biotechnologies have led to innovations in many agricultural, industrial, medical sectors and societal activities. Biotechnology will continue to be a major contributor to the Bio-Economy, playing an essential role in support of economic growth, employment, energy supply and a new generation

Disclaimer: This publication expresses the views of the authors and should not be regarded as a statement of the official position of the European Commission or of its Directorate General for Research.

* To whom correspondence may be addressed (alfredo.aguilar-romanillos@ec.europa.eu)

Abbreviations: AG: Advisory Group, CWG: Collaborative Working Groups, EC: European Community, EC-US TF: European Community – United States Task Force on Biotechnology Research; EP: European Parliament, ERA-NET: European Research Area – Network, ETP: European Technology Platform, EU: European Union, FP: Framework Programme; KBBE: Knowledge-Based Bio-Economy, LMI: Lead market Initiative, R&D: Research and Development, SCAR: Standing Committee for Agriculture Research, WP: Workprogramme.

of bio-products, and to maintain the standard of living. The paper reviews some of the main biotechnology-related research activities at European level. Beyond the 7th Framework Program for Research and Technological Development (FP7), several initiatives have been launched to better integrate FP7 with European national research activities, promote public-private partnerships and create better market and regulatory environments for stimulating innovation.

Introduction: FP7 - the EU Instrument for Research

FP7 is the short name for the Seventh Framework Programme for Research, Technological Development and Demonstration Activities (European Union 2006). This is the EU's main instrument for funding research in Europe and it will run from 2007 to 2013. Based on a co-decision procedure between the Council of European Research Ministers and the European Parliament, FP7 has a total budget of EUR50.5 billion, representing an average 40% increase from FP6.

FP7 is made up of 4 main blocks of activities forming 4 specific programmes, namely: Cooperation - Collaborative Research; Ideas – European Research Council; People – Human Potential, Marie Curie actions; and Capacities – Research capacities. In this article we will focus, almost exclusively on the Cooperation – Collaborative programme. The reader is invited to consult the CORDIS web site where updated and more specific information on any of the activities related to FP7 can be found (See References).

The "Cooperation" programme, with a total budget of EUR32.4 billion, consists of 10 thematic areas, corresponding to major fields in science and research. Under the Cooperation programme, research activities are supported that address European social, economic, environmental, public health and industrial challenges, serve the public good and support developing countries.

THE FP7 FOOD, AGRICULTURE AND FISHERIES AND BIOTECHNOLOGY THEME

Complementary to several of the 10 FP7 cooperation themes (e.g. health, nanotechnologies) that provide significant support to biotechnology, the theme Food, Agriculture and Fisheries, and Biotechnology has the clear objective of building a European knowledge-based bio-economy by bringing together science, industry and other relevant stakeholders. The term "Bio-economy" includes all industries and economic sectors that produce, manage and otherwise exploit biological resources (such as agriculture, forestry, fisheries and other bio-resource industries) and related services (supply and consumer industries).

With a budget of EUR1.9 billion (2007-2013), it aims at mobilizing new and emerging research opportunities that address social, environmental and economic challenges: the growing demand for safer, healthier, higher quality food and for sustainable use and production of renewable bio-resources; the increasing risk of epizootic and zoonotic diseases and food related disorders; threats to the sustainability and security of agricultural, aquaculture and fisheries production; and the increasing demand for high quality food, taking into account animal welfare and rural and coastal contexts and response to specific dietary needs of consumers.

The advancement of knowledge in the sustainable management, production and use of biological resources (microbial, plant and animal) provides the basis for safer, eco-efficient and competitive products and services for agriculture, fisheries, feed, food, health, forest-based and related industries. Important contributions to the implementation of existing and prospective policies and regulations in the area of public, animal and plant health and consumer protection are anticipated. New renewable energy sources are supported under the concept of a European knowledge-based bio-economy. Rural and coastal development is addressed by boosting local economies whilst preserving our heritage and variety of cultures. Research is also carried out on the safety of food and feed chains, diet-related diseases, consumer food choices and the impact of food and nutrition on health.

The approach is to maximise what is known as European Added Value by bringing together all relevant actors (appropriate research disciplines and industrial sectors, farmers, forest owners, consumers, etc.) and by supporting: (i) research which goes beyond national or bi-national interest such as the optimal management of open sea fisheries, animal and plant diseases, and prevention of diet-related disease; (ii) global issues such as climate change and agriculture, food and health (e.g. obesity); (iii) European policies e.g. Common Agriculture Policy, Health and Consumer Protection and Renewed Sustainable Development Strategy; (iv) a critical mass achieved through multilateral collaboration; (v) variety and diversity in Europe such as the understanding of food habits and attitudes; (vi) common EU research policy in issues such as IPR, communication, and public-private cooperation, and SME participation.

The creation of a European Bio-Economy is expected to open the way for innovations and competitiveness by developing new, sustainable, safer, affordable, eco-efficient products. These objectives underpins the Lisbon strategy and are fully in line with the European strategy on life sciences and biotechnology (European Commission, 2002, 2007a). It is specifically expected to promote the competitiveness of European agriculture and biotechnology, seed , forestry , fisheries and food companies and in particular high-tech SMEs, while improving social welfare and well-being, reducing environmental footprints , and supporting EU policies. The Programme Food, Agriculture and Biotechnology is being implemented through periodical annual call for proposals, in which, consortia comprising research institutions from different European and other countries (universities, research institutes, SME, industries, etc.) are invited to submit proposals to be evaluated following independent peer-review assessments. Table 1 summarizes the call for proposals so far launched in FP7. An interim catalogue of the 27 projects selected in the Biotechnology areas (Novel Sources of Biomass and Bioproducts; Marine and Fresh-water Biotechnology; Industrial Biotechnology; Biorefinery; Environmental Biotechnology and Novel Trends in Biotechnology) in the 2007-2008 calls is available from the authors and at the CORDIS website. This interim catalogue lists the approximately 335 institutions participating in the projects and it will be regularly updated incorporating new projects from successive calls.

International scientific cooperation

Science and technological development have always been an international endeavour, but increasing global challenges such as intensified economic globalisation, the rise

Table 1. Outline of the Calls for Proposals, FP7 – Food, Agriculture and Biotechnology.

	FP7-KBBE-2007-1	FP7-KBBE-2008-2A	FP7-KBBE-2008-2B	FP7-KBBE-2009-3	FP7-KBBE-Biorefinery[3]	FP7-KBBE-2010-4	Total
Date of publication	22/12/2006	15/06/2007	30/11/2007	03/09/2008	03/09/2008	30/07/2009	
Deadline	02/05/2007	11/09/2007[1] 19/02/2008[2]	26/02/2008	15/01/2009	02/12/2008[1] 05/05/2009[2]	14/01/2010	
Indicative budget (in MioEuro)	192.09	110	98.85	188.85	57	190.01	836.80
Indicative budget for Biotechnologies (in MioEuro)	45.50	30	21.56	45.68	10	56.02	208.76

1 Stage 1
2 Stage 2
3 Joint call from Themes: Agriculture and Fisheries, and Biotechnology (10M€). Nanosciences, Nanotechnologies, materials and new production technologies (7M€), Energy (30M€) and Environment (10M€)

of new global players and the provision of global public goods (food security, health threats, climate change, and energy security) reinforce the case for a new approach to international cooperation in science and technology from a European perspective.

The need for critical mass and large-scale infrastructure for advancing research in many areas increasingly call for strong international partnerships. European research institutes seek to learn and benefit from good practice in research and innovation links elsewhere in the world. Researchers and students, both in Europe and the rest of the World are looking beyond training opportunities in European countries and the USA, seeking world-class centres of learning and research.

FP7 places new emphasis on international research cooperation which is increasingly seen as being at the centre of Community policies. There are significant opportunities for the EU to put its scientific and technological expertise to the forefront in meeting its political, social, economic and humanitarian commitments in sustainable development fields ranging from global climate change, food security and biodiversity to fulfilling the Millennium Development Goals. S&T may also play a role in the implementation of international agreements where the EU is a party, such as on biodiversity and climate change.

In December 2008, the European Council has called the Member States and the Commission to form a European Partnership in the field of international scientific and technological cooperation with the view to developing better coherence and synergies between the various international scientific and technological cooperation activities carried out in Europe by Member States and the European Community. A Strategic Forum for International Cooperation has been since put in place. (European Commission, 2008a; Council of the European Union, 2009)

INTERNATIONAL COOPERATION IN FP7

The new approach to international cooperation in FP7 aims to rise to these challenges by way of innovative mechanisms for promoting international research collaboration. It aims to address three interdependent objectives: (i) supporting European scientific and economic development through strategic partnerships with third countries in selected fields of science and by engaging the best third country scientists to work in and with Europe; (ii) facilitating contacts with partners in third countries with the aim of providing better access to research carried out elsewhere in the world and (iii) addressing specific problems that third countries face or that have a global character (e.g. by contributing towards Millennium Development Goals, addressing global climate change, food security, combating biodiversity loss, water and energy scarcity).

BASIC PRINCIPLES

Three basic principles have been adopted in order to expand the international collaboration:

(i)   Programming: unlike previous RTD framework programmes, FP7 includes both a broad opening of international research collaboration across the whole Framework Programme and a programming of specific priorities for third countries and regions in different calls for proposals across the thematic work programmes;

(ii)   Targeting: by defining specific actions for collaboration with third countries and regions in each of the thematic programmes, FP7 ensures that budgets for international cooperation are built in at the level of each of the relevant calls for proposals; and,

(iii) Partnership and Dialogue: the principle of partnership will be a particular focus of the international cooperation actions for third countries and regions under FP7.

EXAMPLES

Examples of this new approach are: (a) Co-ordinated Calls – implying the co-ordination of parts of Food, Agriculture and Fisheries, and Biotechnology Calls with simultaneous Calls from Third countries, on agreed topics and with co-financing of the projects. Two co-ordinated Calls have been implemented by this Theme, respectively with Russia in 2008 (2 projects in the area of biotechnology, on industrial enzymes and plant-produced vaccines) and with India in 2009 (2 projects in the area of food, functional foods and food by-products); (b) Twinning of projects between sets of on-going FP7 projects and similar sets of projects supported by a third country programme. This has been implemented with Canada, in the areas of bioproducts and food since 2008, and with Argentina and MERCOSUR in the area of soils, plant and food research in 2009; (c) Cross thematic calls. An example is the call involving several Themes aimed at addressing specific priorities for third countries and regions, such as the EU-Africa call for proposals which will mobilize 64 millions of Euros from the Themes Health, Food, Agriculture and Fisheries, and Biotechnology and Environment towards research priorities in the areas of health, food and water security which have been identified jointly between Europe and the African Union.

The strong FP7 international opening also allows for the European Commission to launch research initiatives on global issues such as animal health, or plant abiotic resistance. These global initiatives aim to co-ordinate actions among multiple S&T programme managers world-wide (thus involving national programmes from the EU and from Third countries as well as from international programmes) in order to jointly tackle global challenges via multilateral international alliances.

EC-US TASK FORCE ON BIOTECHNOLOGY RESEARCH

Since 1990, the EC-US Task Force on Biotechnology Research has been coordinating transatlantic efforts to promote research on biotechnology and its applications for the benefit of society. Established in June 1990 by the European Commission and the White House Office of Science and Technology, the Task Force has acted as an effective forum for discussion, for coordination and for developing new ideas on the future of biotechnology while challenging the scientific communities on both sides of the Atlantic to expand their thinking beyond specific scientific disciplines.

Through sponsoring workshops, and other activities, the Task Force brings together scientific leaders and early career researchers from Europe and the United States to forecast research challenges and opportunities and to promote better links between researchers. Over the years, by keeping a focus on the future of science, the Task Force has played a key role in establishing a diverse range of emerging scientific fields, including bioinformatics, neuroinformatics, nanobiotechnology, neonatal immunology, synthetic biology, systems biology and applications of biotechnology to sustainable energy (Aguilar et al, 2008).

In 2009 and 2010, the Task-Force will continue to sponsor several new workshops on a wide range of trans-disciplinary themes such as virtual tissues, early life programming of obesity, bioinformatics and high throughput technologies which will help identify research priorities that ought to be addressed through efforts at international level. Societal issues which have an impact on biotechnology, such as bioethics, the role of women in science, and public perception of biotechnology have always been an integrated part of the discussions. In this context the Task Force is organising in 2009 in San Francisco a workshop on "A Global Look at Women's Leadership in Biotechnology Research". Additional information on the EC-US Task Force on Biotechnology Research can be found at: http://ec.europa.eu/research/biotechnology/ec-us/index_en.html.

Coordination with European National Policies and Programmes

In 2000, the EU decided to create the European Research Area (ERA). Such an ERA should inspire the best talents to enter research careers in Europe, incite industry to invest more in European research, contributing to the EU objective to devote 3% of GDP for research, and to the creation of sustainable growth and jobs.

Nine years on, the creation of ERA has become a central pillar of the EU research policy and a number of specific initiatives as been taken to promote the establishment of ERA in Food, Agriculture and Fisheries and Biotechnology.

THE ERA-NET SCHEME

More than 90% of the EU's public-funded research in Food, Agriculture and Fisheries, and Biotechnology is conducted at national or regional level. To overcome the resultant fragmentation and duplication of efforts across the EU, the Commission introduced within FP6 the ERA-NET scheme, with the aim of promoting the networking of national or regional research programmes and encouraging the mutual opening of these programmes. The scheme also enabled national systems to exchange good practice in programme management and take on tasks collectively that they would not have been able to take on independently.

Because the intention was to empower the Member States, Associated States and regions themselves, instead of imposing top-down central control, the ERA-NET scheme addressed programme owners and managers in national ministries, regional authorities and funding agencies. Actions undertaken by the ERA-NETs related to KBBE (out of a total of 71 actions) included mapping of national research activities and funding, education and training activities, foresight activities, best practices regarding IPR etc. Table 2 outlines the ERA-NETs in the area of the KBBE.

Although the launching of joint calls for transnational research activities was not specifically foreseen at the start, of the ERA-NET scheme, this approach has been adopted with increasing enthusiasm. All the ERA-NETs in the KBBE succeeded in raising their efforts up to the joint call level. So far, a total budget of EUR309million of national funds have been committed in these joint calls (ranging from EUR1.7-38 million per call).

In contrast to the FP6 approach, the ERA-NET scheme is no longer seen as a "bottom-up" action in FP7. Instead, it has become primarily an implementation tool for strategic areas identified through dialogue with Member States and associated countries. So far, 4 new ERA-NETs have been launched in FP7 in the area of KBBE (Table 2).

Table 2. ERA-NETs in the KBBE area (Food, Agriculture and Fisheries and Biotechnology).

Acronymes	Title	Web page Links
ACENET	Applied Catalysis	www.sysmo.net www.acenet.net
ARIMNET-MED(FP7)		www.arimnet.net/
BIODIVERSA	An ERA-NET in Biodiversity Research (BiodivERsA)	www.eurobiodiversa.org/
BIOENERGY	Bioenergy from renewable bio-resources	www.eranetbioenergy.net
CORE Organic (Fund. Res)	Coordination of European Transnational Research in Organic Food and Farming	www.coreorganic.org
EMIDA (FP7)	Animal Health	www.emida-era.net
ERA-ARD	Agricultural Research for Development	www.era-ard.org/
ERA-IB	Industrial Biotechnology	www.era-ib.net
ERA-PG	Plant Genomics	www.erapg.org
ERASysBio	Towards a European Research Area for Systems Biology - A Transnational Funding Initiative to Support the Convergence of Life Sciences with Information Technology & Systems Sciences	www.erasysbio.net/
ERATRANS-BIO (Soc. Sc)	EUROpean network of TRANS-national collaborative RTD for SME's projects in the field of BIOtechnology	www.eurotransbio.net/
EUPHRESCO	Coordination of European Phytosanitary (Statutory Plant Health) Research	www.euphresco.org/
ICT-AGRI (FP7)		http://ictagri.eu/
MARIFISH		www.marifish.net/
NANOSCI-ERA	Nanosciences	www.nanoscience-europe.org
RURAGRI(FP7)		
SAFEFOODERA	Food Safety - Forming a European platform for protecting consumers against health risks	www.safefoodera.net/
WOODWISDOMNET	Wood material science	www.woodwisdom.net

THE STANDING COMMITTEE ON AGRICULTURE RESEARCH (SCAR)

Complementary to the ERA-NET scheme, other coordination initiatives have emerged through dialogue with Member States and associated countries in the context of the Standing Committee on Agriculture Research (SCAR) (European Commission, 2008b). Established in 1974 to provide scientific support to the Community Agriculture policy, SCAR was reactivated in 2005 with a renewed mandate from the EU's Agriculture Council of Ministers to play a major role in the coordination of agricultural research efforts in Europe.

Among its main initiatives, SCAR adopted a structured approach for prioritisation of research topics for further collaboration, through the establishment of a number of Member/Associated State Collaborative Working Groups (CWGs). The establishment of CWGs is an alternative, more flexible, mechanism to the ERA-NET scheme with the objective to increase research collaboration between funders and programme managers on key-research areas. Since 2005, fourteen CWGs have been set up by the Member/Associated States engaging voluntarily and on a variable geometry basis in the definition, development and implementation of common research agendas based on a common vision on how to address major challenges in the field of agricultural research. CWGs are working in a similar way to ERA-NETs, following the same step-by-step approach – focussing on information exchange during the early stages, the identification of gaps in research and priority areas for collaboration and, where applicable, they should be aiming for joint activities and/or common research calls.

However, the dynamism of several CWGs in terms of commitment paved the way for opening ERA-NET opportunities in FP7 so that five CWGs, declared their intention to submit ERA-NET proposals in the first relevant calls for proposals. ERA-NETs are also able to re-apply for Commission support to extend and/or reinforce their integration e.g. by broadening their partnership or increasing the types of collaboration (Table 2).

NETWORK ON THE KNOWLEDGE BASED BIO-ECONOMY WITH THE REPRESENTATIVES EU MEMBER STATES, ACCEDING AND CANDIDATE COUNTRIES (KBBE-NET)

Given that not only our traditional competitors, the US and Japan, but also competitors in the Asia-Pacific region and South America are increasingly investing in life sciences and biotechnology research, the European Commission recognises the need to increase our collaboration at EU level. Therefore, the European Commission proposed in its third Progress Report (European Commission, 2005) on the implementation of the Life Sciences and Biotechnology Strategy (European Commission 2002, 2007a) the establishment of a network with EU Member States to help coordinate the development and implementation of a European Research Policy for a Knowledge-Based Bio-Economy (KBBE) in co-ordination with the Standing Committee on Agricultural Research (SCAR).

In Spring 2006, a European wide KBBE-NET with high-level officials from Member States, acceding and candidate countries, was established on the initiative of Commissioner for Science and Research Mr. J. Potočnic. Each meeting is organised in cooperation with the EU presidency.

The main role of the KBBE-NET is to support the Commission and the Member States to achieve a coordinated effort in the development and implementation of a European research policy for a Knowledge Based Bio-Economy. This involves: (i) Strategic discussion and recommendations for establishing a European Research Agenda in the long term (FP7, and beyond) which should allow the construction of a European Knowledge Based Bio-Economy. This work should also contribute to the midterm review of the EU Life Sciences and Biotechnology Strategy and its implementation; (ii) Enhancing the exchange of information between Member States regarding national research policies and mapping of activities including international cooperation; (iii) Enhancing cooperation between Member States (joint research programmes, common infrastructures, training programmes, etc.). In the same manner as SCAR, the KBBE-NET has established Collaborative Working Groups (CWGs) in order to strengthen coordination among member states in emerging areas of biotechnology. There are currently two CWGs: on Marine Biotechnology and on Synthetic Biotechnology.

An Expert Group on Food and Health was established in 2008 to seek independent expert advice on the development of a long-term strategic approach in the shaping of national multidisciplinary programmes in the food and health area at the European level. The Expert group will identify key action lines where Members States can encompass and enhance cross-border themes and disciplines to foster public research programmes on food and health.

TOWARDS JOINT PROGRAMMING

The recent concept of Joint Programming (European Commission, 2008c) goes a step further than the ERA-NET scheme and elicits direct cooperation of Member State public programmes defining common visions, strategic research agendas, and the pooling of resources to address major societal challenges. All participating public authorities orient their programmes and funding to contribute in a coherent manner to the implementation of a joint research agenda. The full tool box of public research instruments (National and regional research programmes, Intergovernmental research organisations and collaborative schemes, Research infrastructures, Mobility schemes, etc.), should be explored and used to implement the individual Joint Programming Initiatives.

Some specific major societal problems in Europe linked for example to climate change, the energy crisis or food supplies would benefit from a critical mass of public research efforts. This trend was emphasised at the informal Competitiveness Council meeting on 17-18 July 2008 in Versailles, where food and agriculture were identified as one of four main challenges facing society today. On this basis, the Council is currently reflecting on this new approach and is in the process of identifying research sectors suitable for future Joint Programming Initiatives.

Cooperation with Industry

THE EUROPEAN TECHNOLOGY PLATFORMS

The European Technology Platforms (ETP) in the area of Food, Agriculture and Fisheries, and Biotechnology embrace most, if not all, of the industries that produce, manage or otherwise make use of biological resources, including wastes. In Europe alone, they represent a market size of over EUR 1.5 trillion, employing more than 22 million people. The KBBE European Technology Platforms work together in an innovative multidisciplinary and transdisciplinary approach to scientific research. The integration among the many disciplines represents a key challenge for the future and it will contribute to sustainable development in a new ecological and holistic perspective.

The European Technology Platform (ETP) initiative brings together all interested stakeholders to develop a long-term vision to boost Europe's growth, competitiveness and sustainability. The ETPs address a specific challenge, create a coherent, dynamic strategy to achieve that vision and steer the implementation of an action plan to deliver agreed programmes of activities and optimise the benefits for all parties. In fostering effective public-private partnerships, ETPs also contribute to the development of the Lisbon strategy and the European Research Area (ERA) of knowledge for growth.

STRATEGIC RESEARCH AGENDAS

Each ETP develops a Strategic Research Agenda (SRA). The SRA details the common vision of stakeholders active in the specific sector for the next decade, and includes recommendations needed to fulfil certain goals. An ETP should, in a medium to long term perspective, generate sustainable competitiveness and world leadership for the EU in the field concerned, by stimulating increased and more effective investment in R&D, accelerating innovation and eliminating the barriers to the deployment and growth of new technologies.

SRAs have provided valuable input in a number of cases, including the design of the Food, Agriculture and Fisheries, and Biotechnology theme of the Seventh Framework Programme (FP7), and they have contributed to the annual Work Programmes in order to better meet the needs of industry. The ETPs under the area of the Knowledge Based Bio-Economy chart the future strategic R&D path for key European industries. Their objective is to cover the entire life cycle of bio-economy products and services, and enhance European excellence in the KBBE fields.

OPEN AND TRANSPARENT PROCESS

Each ETP is structured according to a set of rules outlining the responsibilities of proper practices and procedures for new and existing members. Broad-ranging participation is encouraged, including the involvement of small and medium-sized enterprises (SMEs), public authorities, civil society organisations, universities, public research institutions, users and consumers. In support of greater access to information, communication and transparency, each ETP has a website which contains strategic documents, forecast activities, past events, and rules for participation.

ETPs can raise overall RTD investment and ensure the consistency of European efforts in the fields of Food, Agricultural and Fisheries, and Biotechnologies' research by sharing a common vision and a consistent strategic framework at EU level for both RTD funding and deployment initiatives. Table 3 summarises the main features of the ETPs in the area of the Knowledge-Based Bio-Economy. Further information on the ETPs is also available on the following websites:

http://cordis.europa.eu/fp7/kbbe/home_en.html
http://cordis.europa.eu/technology-platforms/individual_en.html
http://ec.europa.eu/research/agriculture/index_en.html

Table 3. European Technology Platforms (ETPs) in the area of the Knowledge-Based Bio-Economy (KBBE).

Name of the ETP	Number of stakeholders[a]	ETPs website
Aquaculture	30	www.eatip.eu
Biofuels	44	www.biofuelstp.eu
Food for Life	69	www.etp.ciaa.eu
Forest-Based Sector	4	www.forestplatform.org www.ftpdatabase.org
Global Animal Health	63	www.ifaheurope.org/euplatform/platform.htm
Manufuture (Agricultural Engineering and Technologies)	50	www.manufuture.org/collective_initiatives.html
Plants for the future	15	www.plantetp.org
Sustainable Chemistry (Industrial Biotechnology)	17	www.suschem.org www.bio-economy.net
Sustainable Farm Animal Breeding and Reproduction	16	www.fabretp.org

[a]The ETPs are open to all stakeholders, both organisations and individuals. A full list of members is available on the ETP website (http://cordis.europa.eu/technology-platforms)

OPPORTUNITIES FOR RESEARCH INTENSIVE SME'S

Small and medium–sized enterprises (SMEs) with a research focus are the economic powerhouse behind scientific and technological developments in the food, agriculture and biotechnology sectors. They play a pivotal role in the success of a European Knowledge-Based Bio-Economy. FP7 aims to strengthen the innovative capacity of SMEs and their contribution to the development of new technology-based products and markets. Around 15 % of 1.9 billion EUR allocated to theme Food, Agriculture and Fisheries and Biotechnology has been targeted to facilitate research conducted by SMEs. Therefore the programme endeavours to identify topics of particular interest to SMEs on the principle that inclusion of SMEs should add to the science and/or technological excellence, and particularly, to increasing the chances for successful exploitation of results.

Furthermore, in order to create a favourable environment for SMEs, simplified financial and administrative procedures have been introduced under FP7. The new SME rules allow funding of up to 75% of total costs for R&D, and for management and training activities SMEs might receive up to 100% financial support. A guarantee fund has been set up to cover the financial risks of defaulting project participants, and strong intellectual property rules provide effective protection with particular attention to the special needs of SMEs.

To improve access to debt financing, the Risk Sharing Finance Facility (RSFF) extends the ability of the European Investment Bank to provide loans or guarantees for research to companies deemed to be too risky under normal banking practice. In addition to the activities offered in the Cooperation Programme for research-intensive SMEs, the Capacities Programme of FP7 has an activity on "Research for the benefit of SMEs". The objective is to strengthen the innovative capacity of European SMEs and their contribution to the development of new technology-based products and markets. It also aims at bridging the gap between research and innovation by helping SMEs outsource research, increase their research efforts, extend their networks, better exploit research results and acquire technological know-how. The programme will help them outsource research, increase their research efforts, extend their networks, better exploit research results and acquire technological know-how, bridging the gap between research and innovation. This activity on less research-intensive SMEs nicely complements the one on research-intensive ones developed in the theme Food, Agriculture and Fisheries and Biotechnology. It is worth recalling that SMEs represent 99% of all enterprises in Europe, they contribute more than two thirds of European GDP and provide 75 million jobs in the private sector. They are therefore key to the implementation of the renewed Lisbon strategy for economic growth and employment.

More information on SMEs is available at:

http://cordis.europa.eu/fp7/kbbe/home_en.html
http://cordis.europa.eu/fp7/capacities/research-sme_en.html

Innovation and commercialization

THE LEAD MARKET INITIATIVE FOR EUROPE

While Europe plays a leading role in science and in fostering science, it seems less successful in converting science-based findings into commercially and societal valuable innovations. At the same time, markets are increasingly recognised as important drivers of innovation. More innovation-friendly market framework conditions are necessary in Europe to reduce the time-to-market of new goods and services and to enable emerging sectors to grow faster. This is called demand-side policy. As a result, companies will see a quicker return on their R&D and innovation investment and public investment in R&D and innovation programmes should attain greater outputs as measured by, for example, jobs, new-to-market products and patents.

The Lead Market Initiative for Europe will foster the emergence of lead markets of high economic and societal value. The LMI rests on two main pillars: the six lead market areas themselves and the implementation of their action plans (policy coordination). A lead market is understood as the market of a product or service in a given geographical area, where the diffusion process of an internationally successful innovation (technological or non-technological) first took off and is sustained and expanded through a wide range of different services (European Commission, 2007b).

Six markets have been identified – eHealth, protective textiles, sustainable construction, recycling, bio-based products and renewable energies. These markets are highly innovative, respond to customers' needs, have a strong technological and industrial base in Europe and depend more than other markets on the creation of favourable framework conditions through public policy measures. For each market a plan of actions for the next 3-5 years has been formulated.

The LMI consists of coordinated priority actions in each market area, which should lower barriers to bring new products or services onto the market. These measures are described in the action plans. The added-value of the initiative is about developing a prospective, concerted and tailored approach of regulatory and other policy instruments: legislation, public procurement, standardisation, labelling, certification and complementary instruments. The LMI is not primarily about funding programmes, but the EU's 7th Framework Programme for research and technological development (FP7), the Competitiveness and Innovation Framework Programme and national and structural funds may fund activities in support of the LMI.

Public authorities can promote the quick take-up of innovations by implementing a number of "instruments" or policy initiatives. For example, public organisations could put measures in place, within the existing legal framework, to procure more innovative goods and services. Demand-side innovation policy complements supply-side policy which mainly uses public investment through grants to stimulate innovation in the EU, in Member States, in regions or cities. The LMI uses a number of demand-side policy instruments that work in synergy. These instruments are tailored, following extensive consultations with stakeholders, to bring down barriers for innovative goods and services in six market areas. The result will be to give industry the opportunity to turn these innovations into world-wide leading products or services in new high-growth markets ("lead markets") powered by the EU's dynamic industry and innovation systems.

LEAD MARKET FOR BIO-BASED PRODUCTS: AN INNOVATIVE USE OF RENEWABLE RAW MATERIALS

Bio-based products are made from renewable, biological raw materials such as plants and trees. The long term growth potential for bio-based products will depend on their capacity to substitute fossil-based products and to satisfy various end-used requirements at a competitive cost. Europe is well placed in the markets for innovative bio-based products, building on a leading technological and industrial position. Perceived uncertainty about product properties and weak market transparency, however, hinder the swift take-up of products.

The Commission's action plan for this Lead market integrates all necessary actions in a synchronised way in order to favour the innovation of the new products and services (European Commission, 2007c, 2007d). The actions range from improving the implementation of the present targets for bio-based products through standardisation, labelling and certification for ensuring quality and consumer information on the new products, to harnessing purchases by public authorities to demonstrate the way to the future. Table 4 outlines the Action Plan for this Lead Market Initiative.

European citizens will greatly benefit from reduced dependency on fossil products and of reduced emission of pollutants, through the wider use of these bio-based products. In the medium term, additional capacity could also help to reduce the average prices of goods.

Conclusions

Recognising the great potential for biotechnologies, Europe has, over the past decades, adopted supporting research programmes and promoted engaged dialogue with all

Table 4. Action plan for the lead market initiatives on bio-based products.

Policy Instruments	Objectives	Actions	Actors
Legislation	Ensure the coherent, comprehensive and coordinated development of policies and regulations that impact the development of bio-based product markets.	Establish a high-level advisory group, to assist the thematic inter-service task force on bio-based products in the follow-up of the present action plan.	EC Stakeholders
Public procurement	Encourage Green Public Procurement for bio-based products.	Establish a network between public purchasers of bio based products to apply the Commission guide on public procurement for innovation, to identify good practices in the field of bio based products and promote their application across the EU.	EC Member States Industry
		Member States to consider developing milestones and roadmaps for increasing the use of bio-based products within National Action Plans on Green Public Procurement.	Member States
Standardisation, labelling, certification	Aggregate demand for bio-based products through a coordinated approach for standard setting and labelling.	Establish standards/labels for specific bio-based products involving all relevant actors.	EC CEN Industry Other stakeholders
Complementary actions	Communication of policies regarding bio-based products as well as the benefits of bio-based products.	Conduct an information campaign via different media with focus on SMEs.	EC
	Support access to finance for R&D&I.	Promote the establishment of strategically important bio-refinery pilot plants and demonstrators involving all actors and investments at EU, national and regional level.	EC Member States Stakeholders

For more extensive information consult: COM(2007)860 final; SEC(2007)1729 and http ://ec.europa.eu/enterprise/policies/innovation/policy/lead-markets-initiative/index-eu.htm

interested parties. More recently, the launch of the European Technology Platforms has contributed to create shared strategic visions between the main public and private actors and to mobilize efforts towards agreed goals. Increased coordination of public research efforts at national and European levels and improved market and regulatory environments should foster research investment and accelerate innovation.

FP7 provides the research community with funding certainty over the next few years. One of the FP7 thematic priorities is dedicated to the strengthening the European knowledge-based bio-economy bringing together science, industry and relevant stakeholders from Europe and the rest of the world. The conditions are, therefore, favourable towards the sustainable development and deployment of biotechnologies as an engine for the knowledge-based bio-economy.

References

Aguilar, A., Bochereau L., Matthiessen L. (2008). Biotechnology and sustainability: the role of transatlantic cooperation in research and in innovation. Trends in Biotechnology. **26**, 163-165.

Cordis. Website of the European Union Framework Programmes for R&D. http://cordis.europa.eu/home_en.html.

Council of the European Union (2009). Conclusions of the Council concerning a European partnership for international scientific and technological cooperation. Official Journal of the European Union, **C18**, 24.1.2009, 11-13.

European Commission (2002). Life Sciences and Biotechnology: A strategy for Europe (COM(2002)27. Communication from the Commission to the European Parliament, the Council, the Economic and Social Committee and the Committee of the Regions. Luxembourg: Office for Official Publications of the European Communities.

European Commission (2005). Life Sciences and Biotechnology – A Strategy of Europe, Third Progress Report and Future Orientations. COM(2005)286 final. Report from the Commission to the European Parliament, the Council, the Committee of the Regions and the European Economic and Social Committee. Brussels: European Commission.

European Commission (2007a). Mid-term review of the Strategy on Life Sciences and Biotechnology. COM 175 final. Communication from the Commission to the Council, the European Parliament, the European Economic and Social Committee and the Committee of the Regions. Brussels: European Commission.

European Commission (2007b). A Lead Market Initiative for Europe. COM(2007)860 final. Communication from the Commission to the Council, the European Parliament, the European Economic and Social Committee and the Committee of the Regions. Brussels: European Commission.

European Commission (2007c). A Lead Market Initiative for Europe. SEC(2007)1729. Annex I to the Communication from the Commission to the Council, the European Parliament, the European Economic and Social Committee and the Committee of the Regions. Brussels: European Commission.

European Commission (2007d). A Lead Market Initiative for Europe, Explanatory Paper on the European Lead Market Approach: Methodology and Rationale. SEC(2007)1730. Annex II to the Communication from the Commission to the Council, the European Parliament, the European Economic and Social Committee and the Committee of the Regions. Brussels: European Commission.

European Commission (2008a). A strategic European Framework for International Science and Technology Cooperation. COM(2008)588 final. Communication from the Commission to the Council and European Parliament. Brussels: European Commission.

EUROPEAN COMMISSION (2008b). Towards a coherent strategy for a European Agricultural Research Agenda. COM(2008)862 final. Communication from the Commission to the Council, the European Parliament, the European Economic and Social Committee and the Committee of the Regions. Brussels: European Commission.

EUROPEAN COMMISSION (2008c). Towards Joint Programming in Research: Working together to tackle common challenges more effectively. COM(2008)468 final. Communication from the Commission to the European Parliament, the Council, the European Economic and Social Committee and the Committee of the Regions. Brussels: European Commission.

EUROPEAN UNION (2006). Decision n° 1982/2006/EC of the European Parliament and of the Council of 18 December 2006 concerning the Seventh Framework Programme of the European Community for Research, Technological Development and Demonstration Activities (2007-2013). Official Journal of the European Union, **L412,** 30.12.2006, p1-41.

Biotechnology and Genetic Engineering Reviews - Vol. 26, 389-406 (2009)

Patent Reform in the United States

*ANN E. MILLS AND PATTI M. TERESKERZ

Center for Biomedical Ethics and Humanities, Program in Ethics and Policy in Healthcare Systems, University of Virginia, Charlottesville, VA 22901, USA

Abstract

The recent financial meltdown has muted the patent reform debate in the United States. But given that President Obama, as well as many members of Congress, support patent reform, we expect the debate to resurface. In this essay, we look carefully at reports from three prestigious organizations which have been enormously influential in the debate. We examine the empirical basis contained in these reports upon which proposed legislative changes are based. We conclude that the empirical data being used to justify the need for reform either has serious methodological limitations or is non-existent. Moreover, we review recent court decisions which have already altered the patent environment calling into further question whether the limited data that exists is still applicable. The effect of these recent decisions has not been adequately evaluated or assessed. Thus, we recommend other empirical studies are needed to inform public policy as to whether patent reform is necessary.

Introduction

During the last decade, issues associated with patent reform have ignited the imagination of the public, academics, and Congress. The public is regularly told that the patent system is dead or broken (McDermontt, 2007), is in the hands of "trolls" (Bulkeley, 2005), supports the proliferation of patents - which reduces innovation (Jeffe and Lerner, 2004), or results in high costs, which are passed on to the eventual consumer (Heller and Eisenberg, 1998), etc. There is a huge body of academic literature on the subject of patent reform, and members of Congress have proposed changes that would have a dramatic effect on the patent environment.

* To whom correspondence may be addressed (AMH2R@hscmail.mcc.virginia.edu)

Abbreviations: USPTO, United States Patent and Trademark Office; CAFC, United States Court of Appeals for the Federal Circuit; NIH, National Institutes of Health; TSM, teaching-suggestion-motivation

The *Patent Reform Act of 2007* was introduced in April 2007 (Patent Reform Act of 2007). Among its controversial provisions are: limitation of damages to the economic value of the improvement associated with the patent; reform of the doctrine of "inequitable conduct;" the imposition of new requirements on patent applicants; and the initiation of post-grant opposition proceedings. (Other provisions include: first-to-file rights; provisions to facilitate filing a patent application without inventor cooperation; venue limitations; authority to the USPTO to create additional regulations.)

The financial meltdown of the past year has distracted members of Congress, and the bill has been taken off the docket, meaning that it will not be considered any time soon. But the reform debate continues. Both houses in Congress have recently introduced patent reform legislation which in many ways mirrors the *Patent Reform Act of 2007* (the new legislation does not address inequitable conduct). If enacted, the *Patent Reform Act of 2009* will have a significant impact on the U.S. patent system and the various industries that rely on it to supply incentives to innovate (Patent Reform Act of 2009).

Moreover, President Obama has expressed interest in patent reform. For instance, in both his campaign strategy website and his transition website President Obama promises to "update and reform our copyright and patent systems to promote civic discourse, innovation, and investment while ensuring that intellectual property owners are fairly treated" (Organizing for America) (Change.gov).

The patent system supplies many industries the incentives needed to innovate. So before new measures are adopted, it is important that there is compelling empirical evidence justifying the need for reform. In this essay, we examine the empirical basis upon which the proposed legislative changes are based. For reasons described below, we conclude that the empirical data being used to justify the need for reform either has serious methodological limitations or is non-existent. Ironically, the limited studies which do exist, at least in the biotechnology industry, have not found that the patent system imposes serious impediments to innovation and successful commercialization. As it stands, it cannot be said, with any degree of certainty, that there are potential impediments to innovation and successful commercialization.

We begin by providing background material on the three major reports upon which reform proponents have relied. (Much of what follows is reprinted with permission from a report which we prepared for the Biotechnology Industry Organization.) (Mills, Tereskerz, 2008). We then examine the empirical studies relied upon by these reports to make their recommendations and comment upon the validity of this data. Finally, we look at recent judicial decisions which may have completely changed the landscape and bring into question whether the limited data, which do exist, are still applicable. Moreover, the effect of these recent decisions has not been adequately evaluated or assessed. Thus, we recommend other empirical studies are needed to inform public policy as to whether patent reform is necessary.

Background

Supporters of patent reform generally cite three reports from prestigious institutions to justify their position. They are the Federal Trade Commission's report, *To Promote Innovation: The Proper Balance of Competition and Patent Law and Policy,* (hereafter called "FTC") (Federal Trade Commission, 2003); the National Academy of Science's Committee on Intellectual Property Rights in the Knowledge-Based Economy, *A Patent System for the 21st Century*, (hereafter called "NAS") (Committee on Intellectual Property Rights in the Knowledge-Based Economy, 2004;) and the National Research Council Committee on Intellectual Property Rights in Genomic and Protein Research and Innovation. *Reaping the Benefits of Genomic and Protemic Research: Intellectual Property Rights, Innovation, and Public Health*, (hereafter called "NRC") (National Research Council, 2006).

The three reports have different goals but, common to the three reports, are concerns associated with an anti-commons or patent thicket and the associated issues and potential problems associated with poor patent quality or questionable patents. An anti-commons (an expression coined by Heller and Eisenberg) raises the concern as to whether or not the multitude of patents that have been granted actually inhibits, rather than facilitates, the transfer of technology (Heller and Eisenberg, 1998). An anti-commons occurs when multiple owners hold the right to exclude each other from a scarce resource, so that no one holds an effective right of entry, and under-use of the resource results. An example may be the problem of royalty stacking, where an inventor must obtain multiple licenses to commercialize a product. Carl Shapiro has discussed similar concerns using the term "patent thickets." He contends that technologies that depend on the agreement of multiple parties can be held up by any one of them, making commercialization difficult (Shapiro, 2001).

Questionable patents or patents of poor quality are those patents which have been granted that might be deemed invalid if challenged, either by litigation or reexamination because they fail to meet the statutory requirements of novelty, non-obviousness, or utility, or because they contain claims that are unclear, not enabled across their full scope, or suffer from an insufficient technical description (Holman, 2006). For instance, a claim might be worded so as to make its boundaries indefinite, perhaps allowing the patentee to stretch the terminology to cover a large number of technologies, including even later-arising technologies that the inventor could not have envisioned at the time of disclosure (Sullivan and Loretto, 2004). In other words, a questionable patent is a patent that may have been improperly granted by the USPTO and that, more likely than not, would be declared invalid upon legal review by a court.

The issue of questionable patents imposes costs on firms that can be seen as unnecessary or illegitimate. For instance, even though a questionable patent would likely be found invalid if challenged, it may be that a firm prefers to pay a licensing fee that is lower than the risk-adjusted cost of litigation (Barker, 2005). It may be that the firm chooses to design around a patent, or a firm may decide that it is in its interest to cease the allegedly infringing activity altogether (Apple, 2005). Thus, the issuance of questionable patents may result in higher costs in the process of innovation and successful commercialization.

The focus of the three reports on issues associated with an anti-commons or patent thicket and poor quality patents means that some of the recommendations made for patent reform are similar. *But more importantly, without compelling evidence, each report concluded that there are serious and systemic problems with the patent system.* Thus, this conclusion, coming from prestigious organizations, is often used by those who wish to reform the patent system. Below we discuss the reports.

Three Reports

TO PROMOTE INNOVATION: THE PROPER BALANCE OF COMPETITION AND PATENT LAW
AND POLICY

The Federal Trade Commission and the Department of Justice, over the course of 2002, interviewed more than 300 representatives and academics who are referred to as "panelists" from the biotechnology industry, the pharmaceutical industry, the computer hardware industry, (including semiconductors) and the software and internet industries, in an effort to understand whether or not the patent system, and the framework of laws and regulation governing competition, promote innovation (Federal Trade Commission, 2003, p. 9). These panelists included business representatives from large and small firms and the independent inventor community; leading patent and antitrust organizations; foremost antitrust and patent practitioners; and scholars in economics and antitrust and patent law (Federal Trade Commission, 2003, p.3-4). Given the differences between the industries, for instance, in the way innovation occurs; the costs associated with innovation; entry costs; and the length of time needed to bring a product to market, it is not surprising that panelists disagreed as to whether or not the patent system was accommodating their industries by promoting innovation and successful commercialization. For instance, representatives from both the pharmaceutical and biotechnology industries agreed that the patent system was essential to create incentives for innovation. Representatives from the computer hardware industry and the software and internet industries were less certain that the patent system promoted or provided incentives for innovation. One panelist in the software and internet industries said, "Compared to the effect of competition in this industry, the current patent system has relatively little effect on the motivation to innovate" (Federal Trade Commission, 2003, Chapter 3, p. 49).

Panelists also disagreed with the nature and degree to which the patent system might be hindering innovation specifically in regard to the issues of an anti-commons or patent thicket and the issue of questionable or poor quality patents. For instance, panelists from the pharmaceutical industry did not seem concerned about the possibility of a patent thicket or anti-commons occurring in the industry. One panelist noted that patent thickets in the pharmaceutical industry are generally not problematic because pharmaceutical products are based on a small number of patents (Federal Trade Commission, 2003, p.5). In the biotechnology industry, some, but not all, panelists believed that the biotechnology industry can be threatened by a patent thicket or anti-commons – especially in regard to research tools. "A research tool is a technology that is used by pharmaceutical and biotechnology companies to find, refine, or otherwise design and identify a potential product or properties of a potential drug product" (Federal Trade Commission, 2003, Chapter 3, p. 18). While in the computer hardware industry, "… none of the panelists disputed the existence of densely overlapping patent rights in the computer hardware industries" (Federal Trade Commission, 2003, Chapter 3, p. 34). And in the software industry, a number of panelists agreed about the existence of a patent thicket in the software industry (Federal Trade Commission, 2003, Chapter 3, p. 52). Additionally, panelists were split about the issue of poor quality patents, depending on the industry they represented.

The FTC report also contains the findings of Walsh et al. (2003) which other panelists echoed (hereafter called the "first Walsh study"). The first Walsh study found that, in biotechnology, industry participants use a number of "mechanisms" to avoid the risk of an anti-commons or patent thickets (Federal Trade Commission, 2003, Chapter 3, p. 24-25). Mechanisms include relying on a research exemption, obtaining a license, or inventing around patents. Moreover, the authors suggest that new USPTO guidelines governing utility (in January 2001, the USPTO adopted new utility guidelines to clarify patentability standards for emerging technologies, such as gene-related technologies, where uses for new materials that have not been fully characterized are not readily apparent), (United States Patent and Trademark Office Utility Examination Guidelines, 2001; Barton, 2000) active intervention from the NIH (the NIH has a number of initiatives aimed at promoting greater access to research tools) (National Institutes of Health, 1999) and overall shifts in the courts' attitudes toward research tool patents also have lessened the risk of a patent thicket or anti-commons: Walsh *et al* report that their respondents are particularly concerned about the *University of California v. Eli Lilly & Co.* in which California tried to argue that its patent on insulin which was based on it work on rats meant that Lilly was infringing because it covered Lilly's human-based bioengineered insulin production process. The CAFC ruled that the claim was not valid because California did not possess the claimed invention at the time of filing. Cockburn et al, (2003) find that the CAFC went from upholding the plaintiff in about 60 percent of cases to finding for the plaintiff in only about 40% of cases in recent years. Moreover, as cited in the FTC's report, the first Walsh study also observes the "…very high technological opportunities in the biotechnology industry, which enables firms to redirect their research efforts to areas less encumbered by patent claims to avoid possible infringement issues" (Federal Trade Commission, 2003, Chapter 3, p. 25).

The first Walsh study warned, however, that the Federal Circuit case, *Madey v. Duke University* (*Madey v. Duke University*, 2003) which emphasized the narrow scope of the so-called "research exemption" available to universities, might make these findings inapplicable. (The case is discussed below.)

A PATENT SYSTEM FOR THE 21ST CENTURY

In 2004, the NAS released its report on the patent system, "*A Patent System for the 21st Century*" which undertook to critique the patent system against seven desirable criteria, two of which are directly concerned with issues associated with poor quality patents and an anti-commons or patent thicket. They are:

- The patent system should reward only those inventions that meet the statutory tests of novelty and utility, that would not at the time they were made be obvious to people skilled in the respective technologies, and that are adequately described; and

- Access to patented technologies is important in research and in the development of cumulative technologies where one advance builds upon one or several previous advances.

The report recognized that, at least theoretically, scholars agree how important it is for the patent system to appropriately reward inventors, and it pointed to other scholars who suggest that questionable patents may cause investment in research to decline or be abandoned (Committee on Intellectual Property Rights in the Knowledge-Based Economy, 2004, p. 46). The NAS report offers examples of some poor quality patents such as a patent for cutting or styling hair using scissors or combs in both hands (Committee on Intellectual Property Rights in the Knowledge-Based Economy, 2004, p. 47). These concerns are similar to the concerns raised in the FTC report.

In regard to the USPTO which has come under sharp attack in recent years for issuing poor quality patents, the NAS notes "…the claim that quality has deteriorated in a broad and systematic way has not been empirically tested" (Committee on Intellectual Property Rights in the Knowledge-Based Economy, 2004, p. 48). Nevertheless, the report claims, "…a nontrivial number of errors in judgment are inevitable in a system whose output by 3,000 individual examiners is 167,000 patents annually" (Committee on Intellectual Property Rights in the Knowledge-Based Economy, 2004, p. 48).

The report points to three seemingly direct measures of quality. They are: 1) the ratio of invalid to valid patent determinations in infringement lawsuits; 2) the error rate in USPTO quality assurance reviews of allowed patent applications; and 3) the rate of claim cancellation or amendment or outright patent revocation in re-examination proceedings in the USPTO (Committee on Intellectual Property Rights in the Knowledge-Based Economy, 2004, p. 48). Studies using these measures give mixed results with some studies suggesting that validity is being upheld more than invalidity and other studies indicating the reverse, and the error rate of the USPTO is currently on a downward trend at around 4 percent (Committee on Intellectual Property Rights in the Knowledge-Based Economy, 2004, p. 48). But, as pointed out by the NAS, there are serious deficiencies with each measure. There are selection bias effects associated with litigation. For instance, parties will rationally avoid futile litigation over patents that are very strong, patently invalid, or commercially worthless. Generally, only patents in the middle, those whose validity can be rationally disputed with a reasonable expectation of success, will be found in litigation. More importantly, however, the numbers of patents associated with any of these procedures is very small. At the time of the report, the litigation rate was just over 1 percent; only 2-3 percent of a year's patents are reviewed by the USPTO; and re-examined patents represent about 0.3 percent of the total number of patents (Committee on Intellectual Property Rights in the Knowledge-Based Economy, 2004, p. 49).

But, despite the lack of evidence showing a systematic decrease in the quality of patents issued – and some evidence showing that the quality of patents has not declined, but is actually improving – the NAS believes that the USPTO is issuing more poor quality patents. The NAS cites four reasons to justify this belief. First, the NAS points to studies which show that that the number of examiners per 1,000 applications is down about 20 percent over the past few years, and the NAS notes the complexity of patents is increasing (Committee on Intellectual Property Rights in the Knowledge-Based Economy, 2004, p. 51). In regard to the productivity of examiners, the NAS reports that, at the time of writing, even though examiners have access to scientific and technical databases capable of automated searches for prior art, automated filing and processing of applications is only now being implemented. Second, the NAS believes that the patent approval rate is significantly higher in the United States, relative to other industrialized countries, (Committee on Intellectual Property Rights in the Knowledge-Based Economy, 2004, p. 52) even though the

NAS acknowledges available studies show differing rates of patent approval. For instance, Quillen and Webster estimate that the approval rate is between 85 percent and 97 percent, while Clark's analysis puts the approval rate at closer to 75 percent (Committee on Intellectual Property Rights in the Knowledge-Based Economy, 2004, p. 53). The USPTO reported that in 2006 its approval rate was 54% (United States Patent and Trademark Office, 2006). More recent studies put the approval rate at about 67%. (Lemley and Sampet, 2007). The NAS admits that the inability to follow individual applications and application families from original filing to final disposition of all members may mean that arriving at a precise approval rate may be elusive (Committee on Intellectual Property Rights in the Knowledge-Based Economy, 2004, p. 54). Third, although the USPTO has implemented changes in policy regarding the treatment of genomic (United States Patent and Trademark Office Utility Examination Guidelines, 2001), and business method patents (Committee on Intellectual Property Rights in the Knowledge-Based Economy, 2004, p. 56), partly in response to criticisms associated with the quality of patents it issues, the NAS claims to be unable to assess the impact of these changes (Committee on Intellectual Property Rights in the Knowledge-Based Economy, 2004, p. 56). Fourth, the NAS believes that there may have been some dilution in the application of the non-obvious standard as a result of court decisions and their incorporation in the examination guidance compiled in the USPTO's Manual of Patent Examining Procedure (Committee on Intellectual Property Rights in the Knowledge-Based Economy, 2004, p. 59).

Specifically, the NAS believes that the guidance for determining obviousness, as set forth in the three cases known as the "Graham trilogy," (*Graham v. John Deere Co.* 1966; *Calmar, Inc. v. Cook Chemical Co.*, 1966; *United States v. Adams,* 1966) can lead to conceptual confusion, depending on how it is applied. (The CAFC has supplemented this guidance with the TSM test. This test was designed to eliminate hindsight bias in the determination of obviousness. Critics argue that applying the TSM test effectively lowers the bar that Congress set for patentability when it codified the pre-existing law of non-obviousness because the USPTO and litigants essentially have to prove that an invention is not non-obvious, as opposed to proving that it is obvious.) But, the recent *KSR International Co. v. Teleflex Inc. et al.* decision as discussed below, is expected to lower significantly the burden for patent examiners and litigants to establish the obviousness of a claimed invention (*KSR International Co. v. Teleflex Inc. et al.,* 2007).

The NAS specifically states, "Neither USPTO resources in relation to its workload, nor patent approval rates, nor changes in the treatment of genomic and business method inventions and the non-obviousness standard are, separately, conclusive evidence that patent quality is too low or declining. However, taken together they lead the committee to conclude that there are reasons to be concerned about both the courts' interpretations of the substantive patent standards, particularly non-obviousness, and the USPTO's application of the standards in examination" (Committee on Intellectual Property Rights in the Knowledge-Based Economy, 2004, p. 62).

With regard to questions of access (an anti-commons or patent thicket), the NAS noted that there was only one area – biotechnology research and development, primarily when applied to human health – where it was repeatedly suggested that there might be a significant problem of access to patented technology (Committee on Intellectual Property Rights in the Knowledge-Based Economy, 2004, p. 71). In order to understand whether or not access to research tools was being prevented by an excess of patents, the NAS commissioned Walsh et. al. to undertake the survey discussed above. (This is the first Walsh study, which was used in the FTC report.) And, as we discussed above, the first Walsh study found that there did not appear to be an anti-commons or patent thicket, but rather industry actors were finding working solutions to any potential anti-commons or patent thicket.

REAPING THE BENEFITS OF GENOMIC AND PROTEOMIC RESEARCH

The report published by the NRC was more limited in scope than the two early reports discussed in that it only focused on the trends in the patenting of genomic and protein related inventions (National Research Council, 2006). It did, however, express similar concerns as those discussed above as to the effects of patenting on innovation and successful commercialization. To address these issues to the widest extent possible within the constraint of limited resources, the NRC consulted the existing research literature and received testimony from scholars in the field. In addition, it engaged in three original research efforts:

1. a search for issued patents and published patent applications in selected biotechnology categories;

2. a small survey of university licensing of selected categories of patents;

3. a survey of biomedical research scientists to ascertain their experience with intellectual property and its effects on research (National Research Council, 2006, p.100).

We discuss the NRC's findings below.

A SEARCH FOR ISSUED PATENTS AND PUBLISHED PATENT APPLICATION IN SELECTED BIOTECHNOLOGY CATEGORIES

NRC staff consulted with USPTO supervising examiners to develop search algorithms for each category identified. The identified categories are: gene and gene regulation, haplotype/SNPs, gene expression profiling, protein structure, protein-protein interactions, modified animals, software, algorithms, databases, EGF pathway, CTLA[4] pathway, NF-kB pathway. These searches were run on the patent claims field to obtain the number of U.S. patents and assignees, assignee countries, inventor countries, applications years, and ultimate assignees over the period from January 1, 1995 to February 1, 2005. Similar searches were run on the extensive genomic and genetic database maintained by Georgetown University Staff with support from the National Institutes of Health (National Research Council, 2006, p.107). The results corresponded very closely but not exactly.

Researchers found that the numbers of issued patents declined in most categories beginning in 2000-2001, with the exception of protein structures, where the numbers were low to begin with. The NRC asked whether or not a decline in patenting was expected to continue. They noted that public funding was not declining, nor was research productivity and that it was unlikely that the economic environment played a role, at least not initially, because patents that issued after 2000 derived from applications filed two or more years earlier (National Research Council, 2006, p.100). The NRC acknowledged that, while hard to assess, greater conservatism on the part of USPTO was almost certainly a factor in the decline (National Research Council, 2006, p.100). So it may be that in spite of the claimed inability of the NAS to assess the impact of the USPTO's utility guidelines, that their implementation in which the claimed utility of an invention must be "specific, substantial and credible" (United States Patent and Trademark Office Utility Examination Guidelines, 2001) might be having an effect on the numbers of patents issued and the quality of issued patents.

A SMALL SURVEY OF UNIVERSITY LICENSING OF SELECTED CATEGORIES OF PATENTS

Obtaining information on licensing is not easy as firms and universities consider licenses and licensing terms proprietary information that they voluntarily disclose very selectively and only when it is to their advantage. Nevertheless, the NRC references a survey done at the beginning of 2003 on 30 US academic institutions owning 75 or more of the DNA-based patents (Pressman et al, 2006). Nineteen institutions provided data on licensing frequency for about 2,700 patents. In sum, the survey found that "interview respondents reported practices broadly consistent with the NIH guidelines issued in 1998 and with the Guidelines for Licensing of Genomic Inventions, which were in draft form and published for comment at the time the survey was conducted and the results analyzed. Further, university technology transfer offices reported considering the NIH guidelines de facto regulations binding on grantee institutions" (Pressman et al, 2006)

The NRC also conducted its own small survey. It obtained the cooperation of the five universities assignees with the most patents (in all but one case) related to the three molecular pathways (identified below) to supply data on the licensing of these inventions. The NRC reports results were similar to the survey discussed above – that is, practices were broadly consistent with the NIH guidelines (National Research Council, 2006, p.119).

A SURVEY OF BIOMEDICAL RESEARCH SCIENTISTS TO ASCERTAIN THEIR EXPERIENCE WITH INTELLECTUAL PROPERTY AND ITS EFFECTS ON RESEARCH

To collect more extensive (but still preliminary data) on the whether or not patents contribute to the problems associated with an anti-commons the NRC arranged with Walsh and colleagues to undertake a more extensive survey than the one described above (National Research Council, 2006, p.119). This study is referred to as the second Walsh study. A sample of 1,124 persons included, among others, investigators in universities, government laboratories, and other non-profit institutions; industry scientists; and researchers working on one of the signaling proteins. Industry researchers were over-sampled to ensure they made up one third of the total. And to ensure that the survey respondents contained sufficient numbers of individuals who work in the fields of biomedical sciences of high commercial interest (because of their association with normal and disease-associated cellular processes) a specially selected sample of approximately 100 researchers working on each of the three molecular pathways (EGF, CTLA4 and Nf-kB) was also included (National Research Council, 2006, p.120). The unadjusted response rate was 33% (National Research Council, 2006, p.120).

The findings of the second Walsh study are similar to the first Walsh study. For instance, drawing from the responses, those whose research goals were "drug discovery" "basic research" or "other" (to differentiate them from survey respondents working on the three molecular pathways) yielded a sample of 274 persons. In this sample, Walsh found that the top five reasons for project abandonment, were, in order of frequency, "lack of funding" (62%), "conflict with other priorities" (60%), "a judgment that the project was not feasible" (46%), "not scientifically important (40%)," and "not that interesting" (35%) (National Research Council, 2006, p.123-124).

Technology access issues (issues associated with a patent thicket or poor quality patents) were cited much less frequently as a reason for project abandonment across all types of scientists and researchers. For instance, in the sample above, "unreasonable terms for obtaining research inputs" was cited by 10% of survey respondents and "too many patents covering needed research inputs" was cited by only 3% of survey respondents as reasons for project abandonment (National Research Council, 2006, p.123-124).

Overall, that is across all types of respondents, the number of projects abandoned or delayed as a result of technological access difficulties is extremely small, as is the number of occasions in which investigators revise their protocols to avoid intellectual property issues or pay high costs to obtain one (National Research Council, 2006, p.123). For instance, no academic respondent abandoned work as a consequence of either delay or inability to receive permission from a patent owner (National Research Council, 2006, p.123). Of those seeking permission from a patent owner (24 out of 32 academic respondents) only one encountered a demand for licensing fees, in the range of $1 - $100 (National Research Council, 2006, p.123).

But, even though this report produced no evidence of a patent thicket or that researchers were being unreasonably challenged by questionable patents, the NRC echoes the concern voiced by both the FTC report and the NAS report in regard to the *Madey* case (*Madey v. Duke University,* 2006). The concern being that *Madey* might cause academic researchers and industry participants to become more proactive in asserting their intellectual property rights, thus allowing a patent thicket or anti-commons to emerge.

The recommendations of the reports

The three reports we have discussed above were written from different perspectives and with differing goals. But, common themes emerge. All three reports were concerned with issues associated with the possibility of an anti-commons emerging and ensuring patent quality and so avoiding uncertainty, potential litigation, and other unnecessary costs associated with poor quality patents.

Despite lack of conclusive evidence, the FTC report concludes that poor quality patents are being issued and urges a number of reforms to address the potential problems. These reforms that are aimed directly at improving the issuance of quality patents include:

1. Enact legislation to create a new administrative procedure to allow post-grant review and opposition to patents;

2. Enact legislation to specify that challenges to the validity of a patent are to be determined based on a "preponderance of evidence" rather than presuming an issued patent is valid;

3. Tighten certain legal standards to evaluate whether a patent is obvious;

4. Provide adequate funding for the USPTO;

5. Modify certain USPTO rules and implement portions of the USPTO's 21st Century Strategic Plan (Federal Trade Commission, 2003 p.1-16).

Similarly, despite lack of evidence showing the existence of an anti-commons, the NAS proposes addressing the issues of an anti-commons and patent quality through:

1. Reinvigorating the non-obvious standard

2. Creating a procedure for third parties to challenge patents for a limited time after their issuance

3. Providing more resources for the USPTO (Committee on Intellectual Property Rights in the Knowledge-Based Economy, 2004, p.81-83)

The NRC report, although concerned only with genomics and proteomics, contains a recommendation that non-obviousness should not be based on the absence of structurally similar molecules and instead should be evaluated:

> "by considering whether the prior art indicates that a scientist of ordinary skill would have been motivated to make the invention with a reasonable expectation of success at the time the invention was made" (National Research Council, 2003, p. 142).

In addition, the NRC also encouraged universities to be:

> "…familiar with the USPTO utility guidelines and should avoid seeking patents on hypothetical proteins, random single nucleotide polymorphisms and haplotypes, and proteins that have only research, as opposed to therapeutic, diagnostic, or preventive, functions"
> (National Research Council, 2003, p. 144).

Although this analysis focuses on issues associated with an anti-commons or patent thicket and poor quality patents, it should be noted that some of the recommendations of the reports have found their way into past proposed and currently proposed legislation. For instance, both the FTC and the NAS reports recommend creating a post-grant opposition proceeding, and both the FTC and NAS reports favor limitations on when damages may be trebled for willfulness.

Patent reform: lack of empirical justification

A problem that has not been adequately addressed is that the call to reform the patent system to date is based on conjecture, anecdote, and individual publicized cases. This may be particularly true when it comes to patents in the biotechnology industry. Caufield and colleagues point out that in biotechnology areas such as gene patenting, where ethical, legal, and economic concerns proliferate, the timing of major policy documents illustrates that policy has largely been driven by social unease, preliminary data, and literature on adverse practical ramifications, as well as several highly publicized patent protection controversies (Caulfield et al, 2006).

The primary empirical studies offered by the FTC, NRC, and NAS reports are the first and second Walsh studies discussed above which investigated the biotechnology industry, which has been perceived to be the industry most likely to suffer from an anti-commons. Although both studies report subjective findings based on interviews and surveys, both studies concluded that an anti-commons is being avoided in the industry. We note that qualitative studies such as this are often used in order to better understand a perceived problem or to gain a new perspective (Strauss and Corbin, 1990). Thus, qualitative methods are appropriate in situations where one might need to first identify the variables that could later be quantitatively studied. None of the reports relied upon to justify overhaul are based on objective, empirical data, other than the NRC's commissioned study limited to DNA patents, which found issued patents declining in this area. *And, it should be noted that the entire FTC's report was, in fact, a qualitative study.*

As discussed above, the first Walsh study involved interviews of a small sample of intellectual property attorneys, business managers and scientists, with 24 from the pharmaceutical industry, representing 10 firms, 18 from biotech, representing 15 firms, 13 from six universities and 15 others. While the small number of individuals interviewed means that it is impossible to determine if the results are representative of the national experience, the study found no evidence that the proliferation of patents was causing an anti-commons in the biotechnology industry. In addition, methods describing the study period, questions posed, the sampling method to select respondents, or rate of those agreeing to be interviewed are not provided in the publication describing the results which makes it difficult to assess the validity of the findings.

While the second Walsh study suffers from a modest response rate, again putting into question whether the existing research truly represents the national experience, the second study supports the findings of the first study, namely that there is no evidence that the proliferation of patents was causing an anti-commons in the biotechnology industry.

These studies, while limited in that the data are not combined with representative objective data, do not indicate that there is an anti-commons in the biotechnology industry, and therefore, they do not justify support for the proposed patent system reform. However, given the methodological issues associated with the studies and the lack of any representative empirical data, it is impossible to categorically state what is or is not occurring with absolute certainty. Further, there is another problem. Court cases recently decided have now changed the landscape significantly. The impact of such decisions has not been studied or was studied too soon after the decisions to draw certain conclusions.

Recent court cases that may alter patent environment

MADEY V. DUKE UNIVERSITY

John Madey was employed by Duke University to direct a lab using equipment developed from his two patents related to free electron laser technology. Following internal disagreements, Duke removed him as a director but continued to use his patents. Madey sued claiming patent infringement. The lower court dismissed his claim based on the common law experimental use doctrine. (The experimental use exemption was articulated in 1813 by Judge Story. In *Whittemore v. Cutter*, Judge Story used the term "philosophical" instead of "scientific" to describe the experimental use exemption from patent infringement. The essential component of the court's reasoning was that those skilled in such "useful arts" are free to use the knowledge imparted by a patent disclosure for amusement, to satisfy idle curiosity or for strictly philosophical inquiry) (*Whittemore v. Cutter*, 1813). On appeal, the CAFC reversed and remanded. The U.S. Supreme Court in 2002 denied Duke's petition seeking review of the Federal Circuit's decision in the University's case with John Madey (*Madey v. Duke University*, 2002).

The CAFC held that the experimental use exemption does not apply to research that furthers universities' "business objectives," including research, educating and enlightening students and faculty, holding that "so long as the act is in furtherance of the alleged infringer's legitimate business and is not solely for amusement, to satisfy idle curiosity or for strictly philosophical inquiry, the act does not qualify for the very narrow and strictly limited experimental use defense" (*Madey v. Duke University*, 2002, 1362). Moreover, the profit or non-profit status of the user is not relevant (*Madey v. Duke University*, 2002, 1362).

The *Madey* court did not provide any examples of the types of uses that would qualify for the experimental use defense (those acts performed solely for amusement, to satisfy idle curiosity or for strictly philosophical inquiry). It did, however, delineate the boundaries of the defense by examining what uses would not qualify for the experimental use defense. The court found that the experimental use exception should not insulate commercial research from claims of patent infringement. This applied to Duke who as Judge Gajarsa noted, was "…not shy in pursuing an aggressive patent licensing program from which it derives a not insubstantial revenue stream" *Madey v. Duke University*, 2002, 1367).

Moreover, the Court reasoned that the experimental use exemption might unfairly advance Duke's business interests. The court used a broad definition of business interest:

> [M]ajor research universities, such as Duke, often sanction and fund research projects with arguably no commercial application whatsoever. However, these projects unmistakably further the institution's legitimate business objectives including educating and enlightening students and faculty participating in these projects….(which) serve to increase the status of the institution and lure lucrative research grants, students and faculty (*Madey v. Duke University*, 2002, 1362).

The FTC, NAS and NCR reports, as well as the Walsh reports, question whether the absence of an observed anti-commons is associated with researchers' lack of knowledge about the *Madey* decision. The speculation is that as more researchers and their institutions become more knowledgeable about the decision, they, as well as private industry, may become more proactive in protecting their intellectual property rights, allowing an anti-commons to emerge. The NAS states that:

> An informal poll of research institutions reported to a September 30, 2002, meeting organized by the Association of American Universities, American Association of Medical Colleges, Council on Government Relations, and National Association of State Universities and Land Grant Colleges, revealed that a number of institutions were receiving more notification letters with respect to patent infringement in the aftermath of the decision. The organizations have arranged with the American Association for the Advancement of Science to continue to monitor universities' experience in this regard (Committee on Intellectual Property Rights in the Knowledge-Based Economy, 2004, p 76).

We are, however, unaware of any follow-up on the part of these organizations. But it should be noted that just as before the *Madey* decision when people posited that the possibility of anti-commons was occurring with no empirical evidence, in the aftermath of the *Madey* decision people are once again positing that an anti-commons may occur with no empirical evidence.

EBAY INC. V MERCEXCHANGE, LLC

In *EBay Inc. v MercExchange, LLC* a patent owner alleged patent infringement for a method of conducting internet sales (*Ebay Inc., et al., Petitioners v. MerEchange,* 2006). The lower court found the patent had been infringed but denied permanent injunctive relief. The CAFC reversed, finding the District Court abused its discretion by denying the permanent injunction, applying its general rule of granting injunctions against patent infringement except in exceptional circumstances. On appeal to the U.S. Supreme Court, the Court found that the District Court was incorrect in its categorical denial of injunctive relief and likewise the CAFC erred in categorically granting such relief. Accordingly, the Court remanded the case ordering that the traditional four-factor framework that governs the award of injunctive relief be applied. This four-pronged test requires a plaintiff to demonstrate that: 1) it has suffered irreparable injury; 2) remedies available at law are not adequate to compensate for the damage incurred; 3) upon balancing hardships between plaintiff and defendant, a remedy in equity such as injunctive relief is warranted, and 4) that the public interest would not be disserved by granting injunctive relief.

This ruling is most problematic for patentees who do not themselves practice their inventions and who have in the past relied on the high likelihood of an injunction following a finding of infringement.

KSR INTERNATIONAL CO. V. TELEFLEX INC. ET AL.

In 2002, Teleflex filed suit against KSR International for patent infringement of its electronic floor pedals. KSR argued that the pedals were nothing more than a combination of existing technologies and were therefore obvious and unenforceable (*KSR International Co. v. Teleflex Inc. et al.*, 2007). The District Court found for KSR, holding that the patent was obvious. The Court of Appeals reversed, holding that if the TSM test for obviousness were appropriately applied, the patent would have been found to be non-obvious. Applying this test means a patent is obvious only if prior art, the problem's nature, or knowledge of a person with ordinary skill in the art indicate motivation or suggestion to combine the prior art teachings. The Supreme Court reversed, holding that the patent was invalid because it was obvious to a person of ordinary skill in the field to combine the existing technologies, finding that KSR provided convincing evidence that combining the existing technologies was well within the grasp of a person of ordinary skill in the field and the benefit of doing so was obvious (*KSR International Co. v. Teleflex Inc. et al.*, 2007, p. 1727). While acknowledging that the TSM test provides helpful insight in that a patent existing of combined elements is not obvious merely by proving that each element was independently known as prior art, the Court, counseled against confining the obviousness analysis to a formal conception of "teaching, suggesting, and motivation." Thus, the Supreme Court effectively raised the bar for obviousness and new guidance to the USPTO was issued in 2007. (United States Patent and Trademark Office, 2007)

Conclusion

We are left in a situation in which there is a lack of compelling evidence that justifies reforming the patent system in a way that could potentially disrupt the incentives of industries that rely on patents to innovate. Indeed, the Walsh studies suggest that there is no basis to believe that the proliferation of patents is hindering research. Moreover, the courts, particularly the Supreme Court, have taken the lead in initiating reforms to the patent system and the industries that are affected by it. The *Madey* decision blurred the boundaries between universities and their missions and industry. But, despite the worry voiced by the reports discussed above, the only evidence we have that the consequences of *Madey* might contribute to an anti-commons through increased litigation activities is an informal poll showing that some universities have received more notification letters warning of infringement (Committee on Intellectual Property Rights in the Knowledge-Based Economy, 2004, p 76).

In both *EBay* and *KSR*, the Supreme Court has increased uncertainty in the business environment. As to *EBay*, there is now less certainty as to whether injunctive relief will be granted when patents are infringed. And while this might weaken patent rights, it should effectively have eased the concerns of some of the FTC panelists, particularly those in the computer hardware and semi-conductor industries, who are concerned that findings of infringement could result in automatic injunctive relief. And, the *KSR* decision has, in essence, agreed with one of the major recommendations for patent reform made by the FTC, the NAS, and the NRC -- that the obvious standard should be strengthened.

We don't know how these judicially-enacted patent reforms will play out. Theoretical arguments can be made on both sides of the anti-commons and patent quality debate. Most observers of the patent system welcome and acknowledge the need for patent reform in some areas, such as ensuring the USPTO has the resources it needs to ensure patent quality. But, reforming the patent system in a way that would weaken patent rights is controversial and should be carefully assessed and balanced with the potential untoward impact such changes may have on various industry sectors, particularly those whose primary assets are patents, not manufactured products. In these sectors, significant changes that introduce uncertainty and weaken patent rights may reduce investment in these industries, which in turn may adversely impact innovation and successful commercialization.

There is no doubt that the explosion of innovation which has resulted in large numbers of patents being issued has strained the patent system. But, the patent system supports and protects innovation by providing the incentives for innovation, and legal mechanisms protect intellectual property rights. We have seen that as innovation cycles have occurred, the system has shown itself to be resilient and self-correcting. For instance, as mentioned by the reports, discussed above, the USPTO has revised its guidelines in response to criticisms and will has implemented new standards to accommodate the *KSR* ruling. And, as mentioned by the reports, industry participants are finding solutions to problems associated with possible patent thickets and poor quality patents. Implementing reform measures aimed at weakening patent rights and enforcement mechanisms is dangerous because innovation often depends on strong patent rights and enforcement mechanisms. The danger to innovation increases when reform is implemented without methodologically sound empirical data that also adequately takes into account the effect of confounding variables such as recent court decisions.

References

Apple, E. (2005) The Coming US Patent Opposition, *Nat Biotechnol* **23**., 245-245.

Barker, D.G. (2005). Troll or no Troll? Policing Patent Usage with an Open Post-Grant Review, *Duke L. & Tech. Rev.* 9-10.

Barton, J. (2000) Intellectual Property Rights. Reforming the Patent System. *Science* **287**, 1933-1934.

Bulkeley, W.M. (2005) Aggressive Patent Litigants Pose Growing Threat to Big Business, *Wall St. J.*, Sept. 14, at A1.

Calmar, Inc. v. Cook Chemical Co., 383 U.S. 1(1966). Decided together with *Graham v. John Deere Co.*

Caulfield, T., Cook-Deegan, R.M., Kieff, F.S. and Walsh, J.P. (2006) Evidence and Anecdotes: Analysis of Human Gene Patenting Controversies. *Nature Biotechnology*, **24**, 1091-1094.

Change.Gov Available http://change.gov/agenda/technology_agenda/

Cockburn, I.M., Kortum, S. and Stern, S. (2003) Are All Patent Examiners Equal? Examiners, Patent Characteristics and Litigation Outcomes in in (Cohen W.M. and Merrill, S eds) *Patents in the Knowledge-Based Economy* 285. -Available at http://books.nap.edu/book/0309086361/html/285.html#pagetop in which Cockburn et. al. find that the CAFC went from upholding the plaintiff in about 60 percent of cases to finding for the plaintiff in only about 40% of cases in recent years.

Committee on Intellectual Property Rights in the Knowledge-Based Economy. (2004) *A Patent System for the 21ˢᵗ Century*. Washington DC: National Academy of Science Available http://lab.nap.edu/nap-cgi/discover.cgi?term=a%20patent%20 system&restric=NAP

EBAY INC., ET AL., PETITIONERS V. MERECHANGE, L.L.C. 547 U.S. (2006).

FEDERAL TRADE COMMISSION. (2003). *To Promote Innovation: The Proper Balance of Competition and Patent Law and Policy.* Available http://www.ftc.gov/os/2003/10/ innovationrpt.pdf

GRAHAM V. JOHN DEERE CO., 383 U.S. 1 (1966)

HELLER, M. AND EISENBERG, R.S. (1998) Can Patents Deter Innovation? The Anticommons in Biomedical Research. *Science* **280**, 698-701.

HOLMAN, C.M. (2006). Biotechnology's Prescription for Patent Reform, 5 *J. Marshall Rev Intell Prop L* (Spring). 317-319.

JEFFE, A.B. AND LERNER, J. (2004) *Innovation And Its Discontents: How Our Broken Patent System Is Endangering Innovation and Progress, and What To Do About It.* Princeton University Press, Princeton N.J at 2.

KSR INTERNATIONAL CO. V. TELEFLEX INC. ET AL 127 S. Ct. 1727, 2007

LEMLEY, M. AND SAMPET, B.N. (2007). Is the Patent Office a Rubber Stamp? *Standford Public Law and Legal Theory Working Paper Series.* July, available at http://papers. ssrn.com/sol3/papers.cfm?abstract_id=999098#PaperDownload

MADEY V. DUKE UNIVERSITY, 307 F. 3d 1351, 1362(Fed Cir. 2002), *cert denied*, 123 S. Ct. 2639 (2003)

MCDERMONTT, W. (2007) The Patent System is Broken, *post.gazette.com*, July 27, Available http://www.post-gazette.com/pg/07208/804719-109.stm

MILLS, A. TERESKERZ, P. (2008). Proposed Patent Reform Legislation: Limitations of Empirical Data to Inform the Public Policy Debate. Available http://bio.org/ip/ domestic/UVA_Limitations_of_Empirical_Data.pdf

ORGANIZING FOR AMERICA, AVAILABLE http://www.barackobama.com/issues/ technology/

NATIONAL INSTITUTES OF HEALTH. (1999) Policy on Research Tools. Available http://ott. od.nih.gov/policy/policies_and_guidelines.html

NATIONAL RESEARCH COUNCIL. (2006) Committee on Intellectual Property Rights in Genomic and Protein Research and Innovation. *Reaping the Benefits of Genomic and Protemic Research: Intellectual Property Rights, Innovation, and Public Health.* Washington D.C.: National Academies Press Available http://fermat.nap. edu/catalog/11487.html?onpi_newsdoc11172005

PATENT REFORM ACT OF 2007, Available H.R. 1908 [110th]: Patent Reform Act of 2007 (GovTrack.us)

PATENT REFORM ACT OF 2009, Available S. 515: Patent Reform Act of 2009 (GovTrack. us)

PRESSMAN ET.AL. (2006) Patenting and Licensing Practices for DNA Based Patents at US Academic Institutions. *Nature Biotechnology* **24**, 31-39.

SHAPIRO, C. (2001) Navigating the Patent Thicket: Cross Licenses, Patent Pools, and Standard Setting. In A. Jeffe, J. Lerner and S. Stern, eds. *Innovation Policy and the Economy.* Available at SSRN: http://ssrn.com/abstract=273550

STRAUSS, A. AND CORBIN, J. (1990) *Basics of Qualitative Research: Grounded Theory Procedures and Techniques.* Newbury Park, CA: Sage Publications, Inc.

SULLIVAN, J.D. AND LORETTO, D. (2004) *Symbol Technologies v. Lemelson: Prosecution Laches, and the Unmet Challenge of Junking "Junk Patents,"* 86 J Pat & Trademark

Off Soc'y (September), 748-757.

UNITED STATES V. ADAMS, 383 U.S. 39 (1966)

UNITED STATES PATENT AND TRADEMARK OFFICE. (2001) Utility Guidelines, 66 Fed. Reg. 1092.

UNITED STATES PATENT AND TRADEMARK OFFICE. (2006). Press Release. Fiscal Year 2006: A Record-Breaking Year for the USPTO: Patent and trademark quality best on record in over 20 years Available http://www.uspto.gov/web/offices/com/speeches/06-73.htm

UNITED STATES PATENT AND TRADEMARK OFFICE. (2007) Examination Guidelines for Determining Obviousness Under 35 U.S.C. 103 in View of the Supreme Court Decision in *KSR International Co.* v. *Teleflex Inc.* 72 Fed. Reg. 57526.

WALSH, J.P., ARORA, A. AND COHEN, W. (2003). Effects of Research Tool Patents and Licensing on Biomedical Innovation, in (Cohen W.M. and Merrill, S eds) *Patents in the Knowledge-Based Economy* 285. -Available at http://books.nap.edu/book/0309086361/html/285.htm1#pagetop

WHITTEMORE V. CUTTER, 29 F. Cas. 1120 1121 (C.C.D. Mass 1813)

Index